CANADA

EXPLORING NEW DIRECTIONS

CANADA
EXPLORING NEW DIRECTIONS

AUTHORS

Leonard A. Swatridge
&
Ian A. Wright

William Hildebrand

Clifford A. Oliver

Gary D. Pyzer

Eugene D'Orazio

Gregory Pearson

Fitzhenry & Whiteside

Canada: Exploring New Directions
Fourth revised edition

Fitzhenry & Whiteside Limited
195 Allstate Parkway
Markham, Ontario L3R 4T8   905-477-9700
e-mail: godwit@fitzhenry.ca
website: www.fitzhenry.ca

Illustrators: Patricia Code, Steven Corrigan, Irma Ikonen

Art Direction: Darrell McCalla

The publisher wishes to thank J.E. Reid Barter of
 Upper Canada College for his review of this text.

Printed and bound in Canada by Maracle Press Limited

Second frontispiece: reprinted with permission from *Voices from the Bay*
McDonald, Miriam, Lucaasie Arragutainaq and Zack Novalinga, 1997.
*Voices from the Bay: Traditional Ecological Knowledge of Inuit and Cree
in the Hudson Bay Bioregion.* Ottawa: Canadian Arctic Resources Committee
and Environmental Committee of Municipality of Sanikiluaq.

Canadian Cataloguing in Publication Data

Main entry under title:
   Canada: exploring new directions   Rev. ed.
Previous ed. by Leonard A. Swatridge and Ian Wright.
Includes index.

ISBN 1-55041-377-5

1. Canada–Geography.   2. Physical geography–Canada.   3. Human
   geography–Canada.   I Swatridge, Leonard A., 1931-

FC57.S92 1998   917.1   C98-931319-0
F1008.2.S92 1998

The most important thing about Spaceship Earth (an instruction book didn't come with it). **Buckminster Fuller**

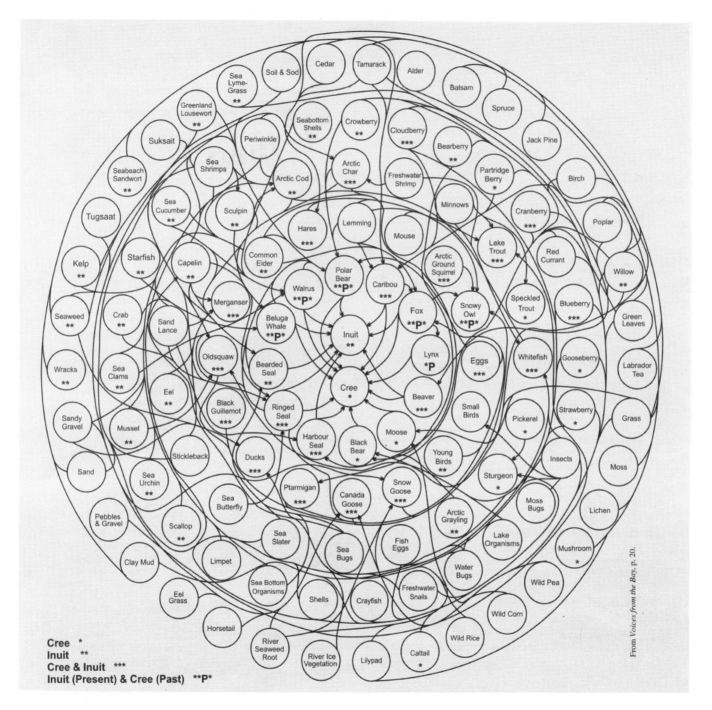

Cree `*`
Inuit `**`
Cree & Inuit `***`
Inuit (Present) & Cree (Past) `**P*`

From *Voices from the Bay*, p. 20.

Hudson Bay Food Web

Inuit and Cree living in Hudson Bay, James Bay and Hudson Strait are part of the natural order of relationships connecting the largest animals to the smallest organisms. This graphic of the Hudson Bay Food Web illustrates the detailed knowledge Inuit and Cree have of the environment in which they live, and the animal and plants upon which they depend.

"Everything in [our] environment has a place and use by people and wildlife. So, damage to any of these causes problems."

Lucassie Arragutainaq, ***Voices from the Bay***, *p. 5*

CANADA
EXPLORING NEW DIRECTIONS

The Physical Pattern

*"If some countries have too much history,
then we have too much geography."*

William Lyon Mackenzie King, Prime Minister of Canada, 1936

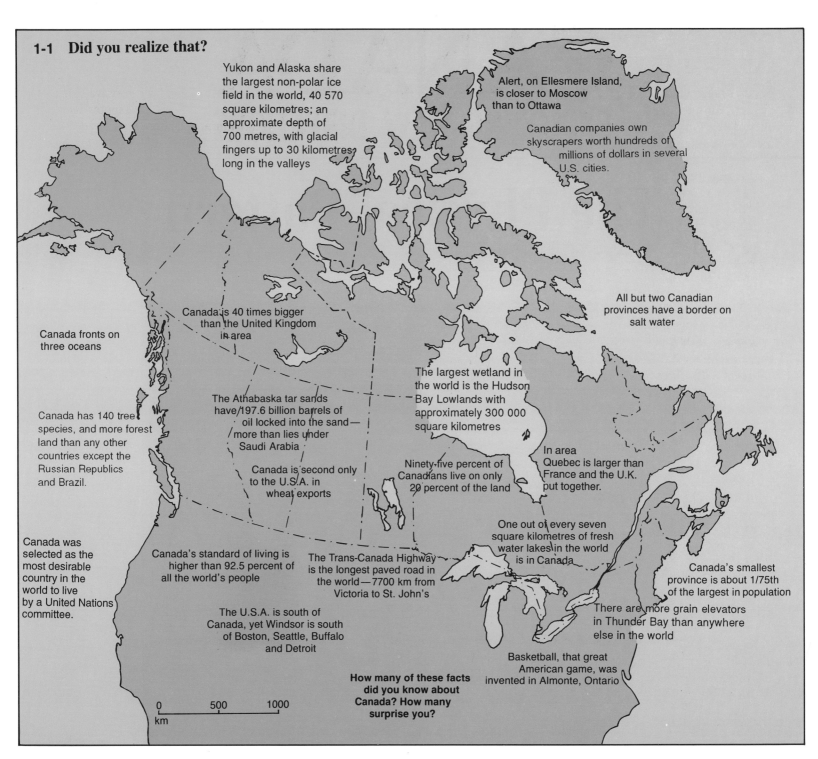

1-1 Did you realize that?

Yukon and Alaska share the largest non-polar ice field in the world, 40 570 square kilometres; an approximate depth of 700 metres, with glacial fingers up to 30 kilometres long in the valleys

Alert, on Ellesmere Island, is closer to Moscow than to Ottawa

Canadian companies own skyscrapers worth hundreds of millions of dollars in several U.S. cities.

Canada is 40 times bigger than the United Kingdom in area

All but two Canadian provinces have a border on salt water

Canada fronts on three oceans

Canada has 140 tree species, and more forest land than any other countries except the Russian Republics and Brazil.

The Athabaska tar sands have 197.6 billion barrels of oil locked into the sand— more than lies under Saudi Arabia

The largest wetland in the world is the Hudson Bay Lowlands with approximately 300 000 square kilometres

Canada is second only to the U.S.A. in wheat exports

Ninety-five percent of Canadians live on only 20 percent of the land

In area Quebec is larger than France and the U.K. put together.

Canada was selected as the most desirable country in the world to live by a United Nations committee.

Canada's standard of living is higher than 92.5 percent of all the world's people

The Trans-Canada Highway is the longest paved road in the world—7700 km from Victoria to St. John's

One out of every seven square kilometres of fresh water lakes in the world is in Canada

Canada's smallest province is about 1/75th of the largest in population

There are more grain elevators in Thunder Bay than anywhere else in the world

The U.S.A. is south of Canada, yet Windsor is south of Boston, Seattle, Buffalo and Detroit

How many of these facts did you know about Canada? How many surprise you?

Basketball, that great American game, was invented in Almonte, Ontario

0   500   1000
km

Canada as a Whole

Geographers enjoy geography for many reasons. The main reason for many of them is its variety. Geography includes the study of many different and interesting topics. Too many people believe that it only involves mountains, rivers, climate and a great many place names. The truth is that there is also a geography of people, of trade, of sports, of wealth, of resources, of transportation — and of much more.

The map of Canada, 1-1, provides a glimpse of some of the many topics that are involved in the geography of Canada. Do any of them surprise you by being included? Perhaps you thought some were not "geographical"? This book will begin with a study of Canada as a whole — its location, size, political pattern and wealth. Other chapters will look more fully at the various topics suggested on map 1-1. All of them are part of the geography of Canada.

Geographers generally ask one question before any other. That question is "where?" This study of Canada's geography, then, will begin with a look at its location.

A Northern Land

Relative Location   There are two basic ways to locate a feature on the earth's surface. One is by giving its **relative location**, that is, its location in relation to something else. Almost anyone who has given directions using **landmarks** has actually given a relative location. If you say to someone, "That house is past the park on this side of the store," you have given a relative location. The house is now located in relation to the park and the store. Both house and store are landmarks.

Canada's relative location can be readily seen on the **circumpolar map**, 1-3, and the world map, 1-5. Canada, which forms the northern portion of North America, is bordered by three oceans, the continental U.S.A. and Alaska. Canadians can easily forget how fortunate they are to have peaceful "undefended" borders, 8900 km long, with their one friendly neighbour. Several of the countries in chart 1-2 have had serious border disputes with one or more neighbours. Some of these disputes have led to costly wars. India, for example, has had disputes with both Pakistan and China. For decades the infamous Iron Curtain, with its guards, barbed wire and minefields, formed West Germany's eastern border. In the past, border disputes have been a major cause of conflict and war. Even today such disputes hit the front pages of our newspapers.

1-2   Canada Compared To Selected Countries – 1996

COUNTRY (Region)	AREA 1,000 km sqd.	POPULATION (1,000)	G.N.P. ($ per capita)	LIFE SPAN Male / Female		URBAN POP POP.	LARGEST CITY (1000 pop)
CANADA (North Am.)	9,976	29,963	26,420	75	82	77%	TORONTO (4,338)
U.S.A. (North Am.)	9,363	265,455	37,145	72	79	76%	NEW YORK CITY (7,333)
JAPAN (East Asia)	372	125,612	51,867	77	83	78%	TOKYO (11,772)
U.K. (Europe)	245	58,784	24,854	74	80	90%	LONDON (6,968)
GERMANY (Europe)	357	81,891	34,533	73	80	85%	BERLIN (3,478)
RUSSIA (Europe/Asia)	17,075	148,070	3,578	58	72	73%	MOSCOW (8,400)
KENYA (East Africa)	583	29,137	351	57	57	20%	NAIROBI (1,505)
KUWAIT (South West Asia)	18	2,070	25,704	74	79	97%	AL-JAHRA (139)
INDIA (South Asia)	3,288	952,969	419	59	60	27%	BOMBAY (9,926)
HONG KONG (East Asia)	1	6,304	31,158	76	81	100%	HONG KONG (6,304)
BRAZIL (South America)	8,512	157,872	4,550	57	67	78%	SAO PAULO (16,417)
JAMAICA (Caribbean)	11	2,505	1,917	71	76	50%	KINGSTON (104)
AUSTRALIA (Australasia)	7,687	18,287	24,273	75	81	85%	SYDNEY (3,773)

* G.D.P. for 1994

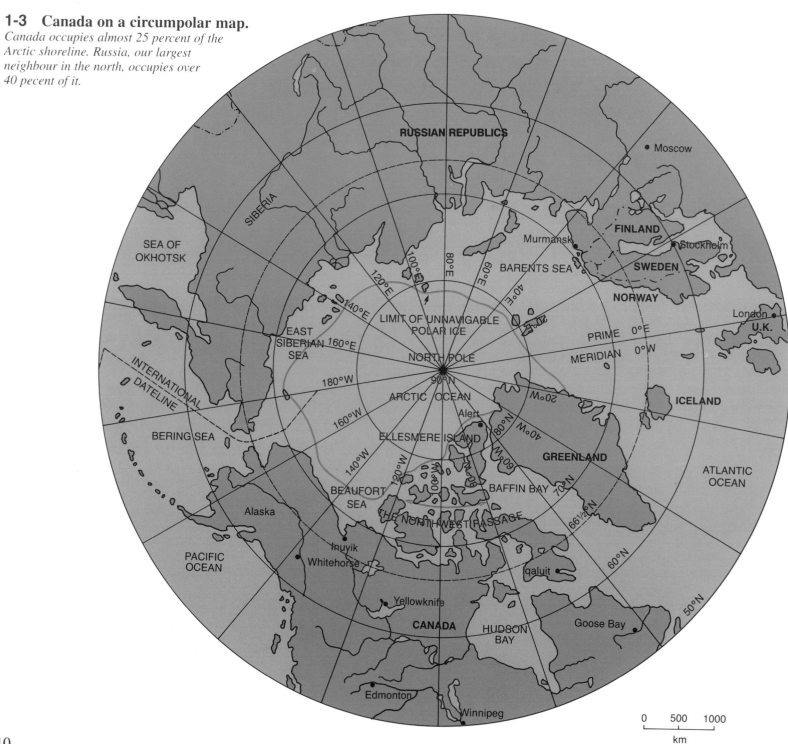

1-3 Canada on a circumpolar map.
Canada occupies almost 25 percent of the Arctic shoreline. Russia, our largest neighbour in the north, occupies over 40 pecent of it.

The circumpolar map shows that Canada has long coastlines on three oceans. Though not considered a great sea power, it has a coastline of 244 000 km — one of the world's longest. The defence of Canada's airspace, and of the islands and surrounding waters of the **Arctic archipelago**, is an important concern to Canadians and to their national government.

Canada is also located between Western Europe and East Asia. Its people have come from both of these regions. It is believed that the native peoples originally settled North America by crossing a ''land bridge'' from East Asia. Thousands of years later, in the seventeenth century, Europeans began to settle in Canada. Since then, more immigrants have come to Canada from Europe, Asia and the other continents. The world's greatest economic powers today are all in the northern **hemisphere**. Canada has many advantages because it is located right in the middle of all of them.

1-4 A typical landscape in Canada's High Arctic. *This picture of Victoria Island, NWT, was taken in early summer.*

Absolute Location Canada can also be located according to its **absolute location**. This refers to its location on the earth's **latitude** and **longitude lines**. Latitude lines, called **parallels** because they are parallel on the globe, are the horizontal lines running east to west. They begin at the **equator** — the centre or zero line — and measure, in degrees, locations to the north and south. The vertical lines, which run from pole to pole on the globe, are the longitude lines, or **meridians**. They measure east and west, in degrees, from a central line called the **prime meridian**. This line is also known as the **Greenwich meridian**, since it passes through Greenwich, a suburb of London, England. The grid is very simple as it appears on **Mercator maps**, such as map 1-5. A globe, however, is the true representation of the earth. On a globe, all these apparently straight lines are drawn on a curved surface. The half-globe 1-6 shows, on the western hemisphere, the major latitude and longitude lines. Map 1-3, which is centred on the North Pole, will also help you understand the earth grid as it appears on the globe.

To provide an absolute location using the earth's grid, it is necessary to give the latitude (the distance in degrees north or south of the equator) and longitude (the distance in degrees east or west of the prime meridian). On map 1-3, Moscow's location can be seen to be 56°N, 38°E, and Edmonton's as 54°N, 114°W.

The diagram 1-7 illustrates how the

11

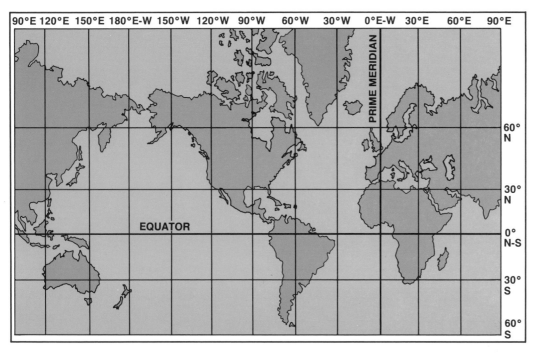

90°E 120°E 150°E 180°E-W 150°W 120°W 90°W 60°W 30°W 0°E-W 30°E 60°E 90°E

PRIME MERIDIAN

60° N

30° N

EQUATOR 0° N-S

30° S

60° S

1-5 A Mercator map. *The curved latitude and longitude lines on a globe appear as a simple grid of straight lines on a Mercator map. Note that on the Mercator projection the parallels of latitude appear farther apart the farther they are from the equator.*

1-6 The earth's grid as it appears on the globe.

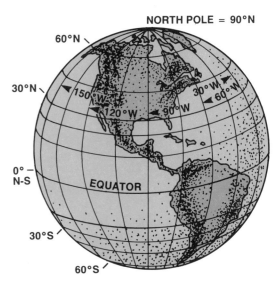

NORTH POLE = 90°N

60°N

30°N

150°W 120°W 90°W 30°W 60°W

0° N-S

EQUATOR

30°S

60°S

The globe is shown tipped slightly, so the longtitude lines can be seen converging on the North Pole.

grid lines are measured in degrees from the centre of the earth, as angles.

1. **Using an atlas:**
   (a) **Find the latitude of the southernmost point of Canada's land mass.**
   (b) **Compare answer (a) to the latitude of the northernmost point of California.**
   (c) **Find the longitude of the most western and eastern points of Canada's land mass.**
   (d) **The 100°W meridian divides Canada roughly in half. Using a straight edge, extend it to Russia on map 1-3. What east meridian does it meet at the pole? Does this suggest a**

simple mathematical relationship between east and west meridians?
2. **Determine the relative location of your school by its distance and direction from at least two landmarks.**
3. **Determine the absolute location of your school and of Canada's capital to the closest degree.**
4. **Using a dictionary, find the derivation of the term** *meridian*, **and explain why it is used to identify the lines of longitude.**
5. **Which countries in chart 1-2 are located along these lines of the earth grid?**
   (a) **60°W**
   (b) **20°S**
   (c) **the prime meridian**
   (d) **the equator**
   (e) **47°E**

6. (a) **Name the two parallel lines that serve as borders separating Canada from the U.S.A.**
   (b) **Name the border that parallels and is very close to 140°W.**

New Words to Know

Relative location
Landmarks
Circumpolar map
Iron Curtain
Arctic archipelago
Hemisphere
Absolute location
Meridians of longitude

Latitude lines
Longitude lines
Parallels of latitude
Equator
Prime meridian
Greenwich meridian
Mercator map

1-7 How the earth's grid lines are measured.

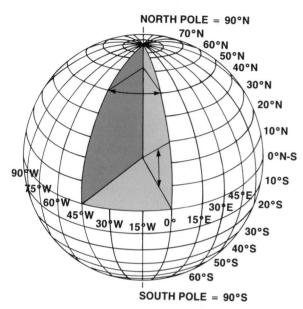

NORTH POLE = 90°N
70°N
60°N
50°N
40°N
30°N
20°N
10°N
0°N-S
10°S
20°S
30°S
40°S
50°S
60°S
SOUTH POLE = 90°S

90°W
75°W
60°W
45°W
30°W
15°W
0°
15°E
30°E
45°E

In this cutaway the angles used to name the various grid lines can be seen. Which parallel and which meridian are being measured?

A Huge Country

One of the best ways to begin understanding any country is to compare it with others. Chart 1-2 provides a summary of mainly statistical data on a group of countries. Every continent except Antarctica is represented. These countries are also located on the political map of the world, 1-5. Both the chart and the map should be referred to when reading the following sections.

Canada spreads over 9 970 000 km^2 of land and fresh water. This makes it 40 times the size of the United Kingdom (U.K.). Canada is the second-largest country in the world after Russia, which is more then double Canada's size. To most people, the sheer size of Canada is overwhelming. Jamaica, Kuwait and two Hong Kongs would all fit easily into Great Bear Lake in Canada's far north. Even some provinces are larger than many countries. Quebec and Ontario are each larger than France, West Germany and the U.K. put together. Europeans are used to measuring distances in their countries in hundreds of kilometres. They are amazed to learn that Canada's greatest east-west distance is 5514 km, and that its greatest north-south extent is 4634 km.

Canada's immense size has created problems for all of its people, but particularly for its business people, politicians and students.

Business people have difficulty trying to distribute products over a widely spread national market. Transportation is costly. Though Canada has modern methods of transportation, the higher costs are still added to the prices consumers must pay for goods. The way in which Canadians are distributed across the country makes these costs even higher. There are four major clusters of population in Canada:

1. southern Ontario and Quebec, where many of the nation's large cities are located;
2. the southern part of the Prairie provinces;
3. the southern coastlands of British Columbia; and,
4. the Atlantic provinces, especially their southern coastlines.

Even within these areas, however, the population is scattered through many cities and towns. Outside these four areas there are far fewer people, and they are even more widely scattered. This wide dispersion of people throughout millions of square kilometres of territory has made Canada more than normally difficult to develop.

Politicians find that Canada's size and small, scattered population create a number of quite different problems. The huge size makes it costly to defend. The country is very difficult to govern because its people are concentrated in four widely separated parts of the country. This has led to the development of four distinct **regions** within the country. The economy and lifestyle in Lunenburg, Nova Scotia, for example, are quite different from those of Calgary, Alberta, or St-Jean, Quebec. On top of that, the people themselves are varied — in their language, culture, ethnic background and income. To govern a country with all this diversity widely spread "from sea to sea" is no easy matter. The development of a truly distinctive Canadian identity has taken longer because of the geographical distances that separate Canadians. But there is also no doubt that a strong spirit of Canadianism has developed in spite of, rather than because of, the country's huge size. In more recent decades this growing spirit has been aided enormously by the incredible advances in communications technology. Many ties, from national television systems to airlines travelling regularly from coast to coast, have fostered this growth.

Geography students have a problem grasping Canada's size. Since no one can learn and understand everything about Canada, choosing and under-

1-8 **The provinces and regions of Canada.**

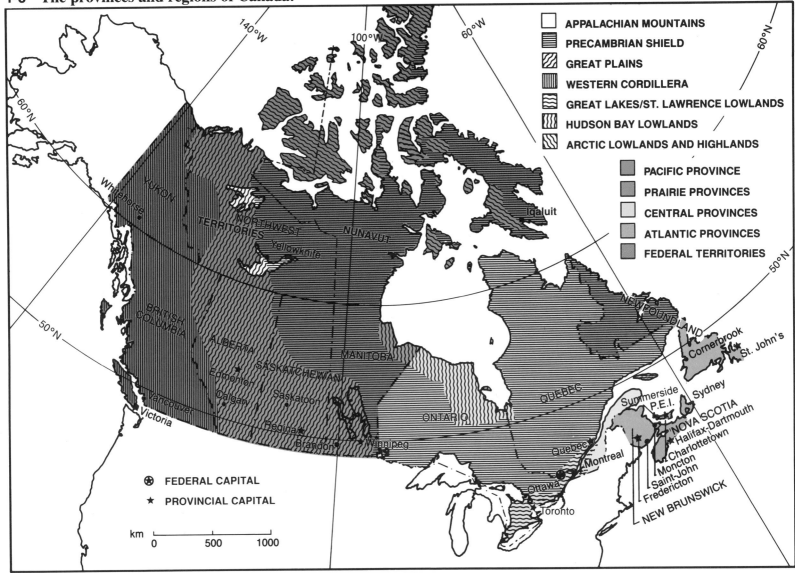

Unlike some countries, which are much the same throughout, Canada is very much a country of regions. The map suggests at least three different ways in which Canada is divided into distinct regions.

1. **Four provinces are located entirely between two parallels of latitude.**
   (a) **What provinces are they?**
   (b) **Name the parallels.**
   (c) **Name the three meridians between the provinces.**
2. **Name five provinces in which the Appalachians are found.**

3. **Name the major physical regions found in Ontario, in order of their area.**
4. **Look carefully at the "prairie" provinces and explain why that term might give a person from another country the wrong impression.**
5. **Using an atlas, determine which of the provinces borders on:**
   (a) **Vermont** (b) **North Dakota** (c) **Washington** (d) **Minnesota.**
6. **What is the latitude of the most northerly provincial city on the map?**
7. **Prove that Halifax-Dartmouth is not due east of Toronto.**
8. **What Ontario city is closest to being due west of Halifax-Dartmouth?**

14

1-9 Harvesting wheat near Regina.

1-10 The city of Moncton, *New Brunswick, population in excess of 100,000, has become a major Canadian communications hub. Royal Bank, Purolater, Federal Express, CP Hotels, and United Parcel Service (among others) use Moncton as their customer call service base.*

standing the essentials becomes even more important when studying this country. Perhaps Mackenzie King was correct in his statement (quoted at the beginning of this chapter) about Canada's excess of geography.

1. **(a) How many times larger is Canada in area than each of Japan, Jamaica and India? (Use the data in Chart 1-2.)**
    (b) How many times smaller is Canada in population than each of Japan, India, the U.S.A. and the U.K.? (Use chart 1-2.)

2. **Among all countries, Russia and Canada have the largest land areas in the world. Using an atlas, name the next four largest countries.**

New Words to Know
Region

The Political Pattern

One of the most important aspects of Canada is its political system. The provinces are very important in Canada. According to the constitution, the national government in Ottawa must

share a great deal of economic and social power with the provincial governments. This means that Canada has a **federal political system,** as has the U.S.A.

Chart 1-11 provides basic statistics and information about Canada's provinces and territories; the political map 1-8 shows their location. Both the map and the chart group the provinces into **regions**. These groupings are commonly used by Canadians across the country. They are as follows:

15

1-11 Canada's Provinces and Territories

REGION	PROVINCE	AREA (km²)	POPULATION (1000s) 1986 census	POPULATION (1000s) 1996 census	LARGEST URBAN CNTRS (Capital City)
Atlantic	Newfoundland	405 720	578.1	570.7	St John's Cornerbrook
	Prince Edward Island	5 660	128.8	137.3	Charlottetown Summerside
	Nova Scotia	55 490	892.1	942.8	Halifax-Dartmouth Sydney
	New Brunswick	73 440	727.7	762.5	Saint John Moncton (Fredericton)
Central	Quebec	1 540 680	6 733.8	7 389.1	Montreal Quebec
	Ontario	1 068 580	9 477.2	11 252.4	Toronto Ottawa
Prairie	Manitoba	649 950	1 094.0	1 143.5	Winnipeg Brandon
	Saskatchewan	652 330	1 032.9	1 022.5	Saskatoon Regina
	Alberta	661 190	2 438.7	3 885.1	Edmonton Calgary
Pacific	British Columbia	947 800	3 020.4	3 855.1	Vancouver Victoria
Federal Territories	Yukon	483 450		31	Whitehorse
	Northwest Territories	1 305 372		42	Yellowknife
	Nunavut	2 120 948		25	Iqaluit
Canada		9 970 610	26 203.8	29 963.6	Toronto (Ottawa)

* Though Fredericton is the capital of New Brunswick and Ottawa is the capital of the nation, neither is one of the two largest urban centres in its particular region.

The population figures are actual, based on the count taken during the 1986 census and the 1996 census. Simple statistical projections can be used to obtain estimates of population changes in the years since 1996.

1. **Name in order the three largest, and the three smallest provinces (a) in area, and (b) in population.**
2. **How much greater is the population of Ontario than the population of Prince Edward Island?**
3. **How much greater than the area of Ontario is the area of Nunavut?**
4. **(a) Determine the total population of each of the five regional groups.**
   (b) Rank the groups in order of population size, and construct a small bar graph.
5. **By what percentage has each province's population increased? Rank the provinces accordingly.**
6. **Identify the province with the largest increase. What factor may account for this increase?**

The Pacific province British Columbia is the only province on the Pacific Ocean. Sometimes it is included with the Prairie provinces under the term "Western provinces."

The Prairie provinces These include Alberta, Saskatchewan and Manitoba, all of which have large areas of flat, treeless land called **prairie**.

The Central provinces Ontario and Quebec make up this region.

The Atlantic provinces This region includes Newfoundland and the three Maritime provinces of New Brunswick, Prince Edward Island and Nova Scotia.

The outline of the country's major physical regions has been drawn onto the political map 1-8. This provides considerable help when it comes to understanding the provinces: physical features have a great influence on the resources available, the economic development possible, the type of settlement pattern, and the size of the population that can be supported.

The vast majority of Canadians lives in the south of the country. The location of the major urban centres confirms this. The environment is much better there — the land is flatter, the climate warmer and the soils deeper. All of these factors make for favourable conditions for agriculture and settlement. In the south the industries are much closer to the U.S.A., which is of particular importance to the central provinces. The U.S.A. is the market for 70 to 80 percent of Canada's exports. About 90 percent of Canada's people "hug" the American border, living within 200 km of it. This will be discussed later in much more detail.

Canada's Ecozones

Figure 1-8 introduces three different ways in which Canada is divided into regions. A region can be defined as an area which shares one or more characteristics. Regions can be determined by single factors like language, vegetation or population: the Canadian map in Figure 1-91, for example, shows six major soil regions in this country. Sometimes regions are also defined by a combination of different characteristics working together — rock structure and climate affect soil type, which in turn has an influence on agriculture, which then affects the population pattern, and so on. The integration of elements within a region is an important area of study since everything living depends on everything else, and can be much more fruitful than isolated research on specific system components.

The term **ecosphere** was created because scientists wanted to study our planet as an interdependent series of environments. We know that earth's physical geography (geology, climate, soils, etc) influences and is affected by earth's vegetation, which in turn affects and is affected by the animals (including human beings) who live here. The earth ecosphere is constantly changing.

When Canadian ecologists and geographers sought to understand and manage the environmental systems within Canada, they realized the ecosphere concept would have to be broken down to various smaller, more manageable levels. Thus, while ecosphere is a global concept, **ecozone** refers to an area where organisms and their physical environment endure as a system — on a continental or even a national level.

Scientists study ecozones by integrating environmental information with social and economic data — climate, human activity, vegetation, soils, and geological and physical features of the landscape. Ecoprovinces, ecoregions, and ecodistricts are subdivisions of an ecozone, and an eco-element is a location-specific component of an eco-district.

Canada is one of the largest nations on earth. We have great variety of lands and waters, with huge areas of forest and treeless Arctic barrens. We have massive lake and river systems, and complex coastal areas. We have vast plains, impressive mountain ranges, rainforests, deserts, wilderness, and large urban centres. Our landscape is amazing in its biodiversity.

Canada has fifteen land-based and five marine ecozones, which in turn are subdivided into 47 ecoprovinces, 177 ecoregions, and 5 395 ecodistricts.

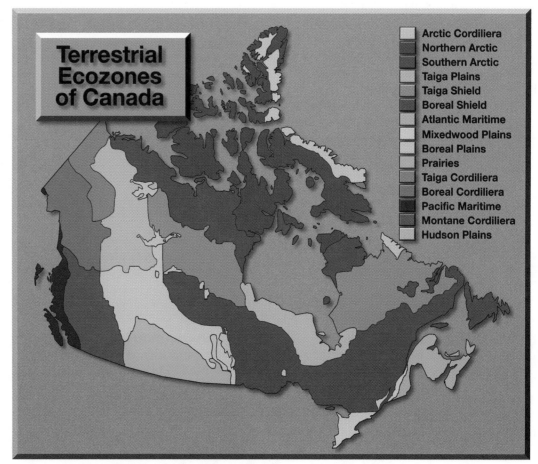

1-12 Terrestrial Ecozones of Canada.

Canadian ecozones are usually large, defined by similar environments and life forms, and, for the most part, they are very complex, including, many differing environmental elements. Rock, plant, or animal components (to name but a few) have interacted with one another, and with other ecoelements, for billions of years to form interdependent and unique environmental webs.

Some Canadian ecozones are greater than 200 000 km2, and the criteria used to define them must be broad-based — thus we speak of a mountain range rather than a specific peak. Sizes of our ecozones vary: the Atlantic Maritime zone is much smaller than the Boreal Shield. Boundaries on ecozone maps are transition areas where one characteristic ecology (climate region, plants and animals, or human activity, for example) is gradually replaced by another. Canada also shares ecozones with other countries. The Canadian Prairies ecozone (450 000 km2) is part of the Global Temperate Grasslands. Canada includes the northern tip of that region which extends into the United States. The Prairies ecozone is less diverse than the Pacific Maritime ecozone which has temperate rain forest over much of its lower slopes, and alpine meadows and mountain tundra at higher elevations. The Pacific Maritime ecozone has, in addition, a very diverse marine ecosystem along its coast and extending out and around the thousands of offshore islands.

Ecozones change over time. Forest fires and insect infestations cause significant damage in the Boreal Shield. Old growth trees disappear to be replaced by grass, bushes, and fast-growing species like poplar. Native animals depart, and even

1-13 Canada's Terrestrial Ecozones.

| | Total Area | | | |
	% of Canada	Land km2	Fresh Water km2	Dominant Cover
Arctic				
Arctic Cordillera	2.5%	230 973	19 717	perennial snow/ice
Northern Arctic	15.2%	1 361 433	148 477	barren lands
Southern Arctic	8.3%	773 041	59 349	arctic/alpine tundra
Taiga				
Taiga Plains	6.5%	580 139	66 861	coniferous forest
Taiga Shield	13.7%	1 253 887	112 513	transitional forest
Hudson Plains	3.6%	353 364	8 996	transitional forest
Boreal				
Boreal shield	19.5%	1 782 252	164 118	coniferous forest
Atlantic				
Atlantic Maritime	2.0%	193 978	19 772	mixed forest
Great Lakes / St. Lawrence				
Mixedwood Plains	2.0%	138 421	56 009	agricultural cropland
Central Plains				
Boreal Plains	7.4%	679 969	57 831	coniferous forest
Prairies	4.8%	469 681	8 429	agricultural cropland
Pacific/Western Mountains				
Taiga Cordillera	2.7%	264 480	380	coniferous forest
Boreal Cordillera	4.7%	459 680	4 920	coniferous forest
Pacific Maritime	2.2%	205 175	13 805	coniferous forest
Montane Cordillera	4.9%	479 057	13 053	coniferous forest

soil layers may alter for a while. Nevertheless, with the passage of time, the boreal forest will grow back — strengthened in some cases — because certain trees actually depend upon forest fires for their survival.

Each ecozone provides different opportunities for, or restrictions on, human activities and uses — like mining, farming, transportation, and recreation. Mining, for example, although found all across the country, is predominantly associated with the fifteen Shield ecozones, which are sparsely populated.

The majority of this country's population can be found in four of the smallest ecozones that are located in the south: Pacific Maritime; Prairies; Mixed Wood Plains; and Atlantic Maritime. These regions changed dramatically as a result of human settlement and resource exploitation activities. They will never revert to their original state.

The five marine ecozones (Pacific, Arctic Archipelago, Arctic Basin, Northwest Atlantic, and Atlantic) cover parts of the Arctic, Atlantic, and Pacific Oceans. The Atlantic and Pacific ecozones are generally ice-free except for pockets of seasonal landfast ice. The Northwest Atlantic, and

Arctic Archipelago have seasonal ice cover, while the Arctic Basin holds permanent pack ice.

Here are a few ecozone facts:

• The coastal temperate rain forest of the Pacific Maritime ecozone covers more than 10 million hectares; some of its ecosystems have the highest **biomass** (total volume of living matter) per hectare found on earth.

Some 2.2 million hectares of this rainforest have been logged over the last 120 years. In early 1995, 52% of the major forest cover was comprised of trees over 140 years old, and 8.4% (900 000 ha) of the forest had protected status.

• Most salmon stocks in the Pacific Marine ecozone are in relatively good condition; Sockeye, Pink, and Chum salmon have increased coastwide since the 1960s. Numbers of these species in the Fraser River (despite published reports of 'missing sockeye') are close to historic highs. Chinook and Coho salmon stocks, however, are depressed, owing to overfishing, habitat damage, and natural factors. Although many Chinook stocks are responding well to rebuilding efforts, declines in Coho stocks persist and have been the focus of a major rebuilding program.

• Canada has 25% of the world's major wetlands: the Boreal Plains, 8.7% in the Atlantic Maritime, and 1.7% of the Montane Cordillera. The Montane Cordillera has the most diverse conditions of all Canada's ecozones. Its weather ranges from the driest to the hottest, and the wettest to the coldest in the country.

• 34% of all Canadian threatened and endangered species can be found in the Mixedwood Plains ecozone, 15% in the Boreal Shield, 12% in the Atlantic Maritime, 8% on the Prairies, 7% in the Montane Cordillera, and 24 % in the ten remaining ecozones.

• The Arctic Archipelago Marine Ecozone does not consist of open seascapes, like most marine zones, but is, instead, a vast network of islands, island chains, straits, sounds, bays, fjords, channels and gulfs; the Hudson Plains ecozone contains the "largest extensive area of wetlands in the world...with extensive peatlands, and shallow open waters less than 2m deep" (Environment Canada).

• The Boreal Shield's forested land component was 85.6% compared to 80.4% for the Taiga Plains, 78.1% for the Atlantic Maritimes, 76.2% for the Boreal Plains, and 73.7% for the Montane Cordillera.

• The Prairies ecozone contains 16 major dam sites, and future water developments must be planned to sympathize with eco zone requirements as the quality of its current water is "already near the extremes of acceptability for human use," according to the Prairie Provinces Water Board, which has developed water quality objectives on interprovincial rivers.

• Almost half of Canada's population lives in the Mixedwoods Plains ecozone, which also "contains the most productive soils in Canada" (Environment Canada).

1. **What is an ecosystem and where is the study of ecospheres an important area of analysis?**
2. **How do scientists study ecozones?**
3. **Name the four ecozones that contain the majority of Canada's population.**
4. **How does the fragility of the Prairies water supply effect organic life? Cite three examples.**
5. **Why are the majority of Canada's endangered species located in the Mixed Wood Plains?**

New Words to Know

ecosphere   ecozone

biomass

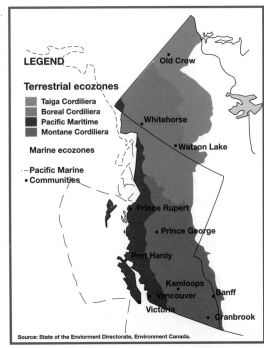

1-14 Canada's West Coast Ecozones

Building Canada's Landforms

The major landform regions of Canada are outlined on map 1-8. Before discussing them in detail, however, it is necessary to explain something about how the surface of the earth was fashioned. It has taken the forces of nature hundreds of millions of years of **geological** history to "make" Canada. Throughout that time there have been at work **tectonic forces** (which build the landscape) and forces of **erosion** and **deposition** (which grade the landscape). When the nature of the forces acting upon the earth is better understood, and the time frame is grasped, the descriptions of the landform regions become more meaningful.

Geological Time and Landform Development

People often refer to "this old earth," but they might be surprised to learn just how old it is. Most scientists believe that the earth became a planet 4 600 000 000 years ago. This means that virtually everything on the earth is that old too, since the materials that make up its plants and animals (including people) have all been recycled from planet earth. This fact has given rise to another commonly heard term, "Mother Earth."

For the hundreds of millions of years of its existence, the earth has been under constant change. The statement

"Nothing is constant except change," definitely applies to the surface, or **crust**, of the earth. Even though the earth is old, some areas seem new, since the earth keeps changing and re-forming itself. There are rocks in northern Manitoba and Ontario that geologists believe haven't changed over three billion years, but there are also other rocks being formed right now in countless rivers, streams and lakes throughout the world. Right now, in a nearby river or stream, rocks are being broken up, and their particles are being deposited elsewhere as new, unconsolidated "rock," often simply called sand or silt.

1-16   Chart of geological time.

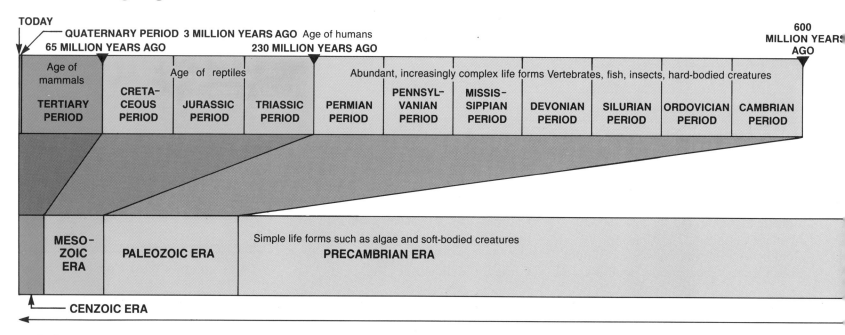

At the beginning of this century, the earth was thought to be about 100 million years old. However, many new discoveries have led scientists to believe that the earth is actually 4.6 billion years old. It is also believed that human beings did not appear on earth until the

Because the span of geological time is so vast, geologists have organized it into segments. The large segments shown on chart 1-16 are called **eras**. They in turn are subdivided into **periods**. The transition from one period or era to another, in most instances, is marked by stretches of mountain-building activities. One long-lasting, extensive mountain-building period ushered in the present era, the Cenozoic. During that time, about 65 million years ago, the high Rocky Mountains were formed along the eastern edge of the Western Cordillera.

In addition, the **fossil** records may change between one period or one era and another. Most scientists believe that life on earth began during the Precambrian time (all the geological time that came before the Cambrian period). The very limited record of life from that era shows that there were simple marine plants, and **invertebrate** animals in small numbers. However, from the Cambrian period onward, there is ample evidence of more complex and abundant life forms.

It was emphasized earlier that the earth's surface has been undergoing constant change. Mountains exist today where sea floors may have been

1-17 Pangaea, the supercontinent.

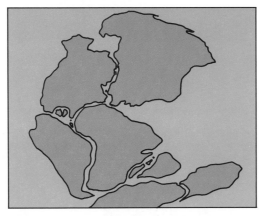

This is what scientists believe the earth looked like 200 million years ago, when forces in the mantle brought all the continents together in one great land mass. The collision of two continental plates caused the formation of the Appalachian Mountains at this time. Can you identify the two continents involved?

deposited. Plains or low hills may be found now where mountains once stood. One change on the earth's surface that scientists now agree must have occurred — though it's only in the past few decades they have done so — relates to the movement of the continents. People are still surprised when they look at a map like 1-17 and see the world the way scientists now believe it looked

1-18 A slice of the earth's interior.

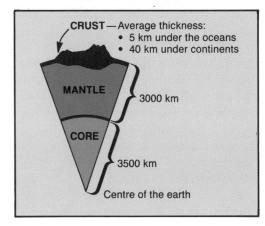

200 000 000 years ago. Back then the continents were all grouped together as one huge land mass. This master continent has been named **Pangaea**. The continents have not always been fixed in their present positions, according to the theory of **plate tectonics**. This theory holds that the continents have moved about on the earth's surface, and are still moving today.

The **continental plates** have shifted considerably from their positions of many millions of years ago. They have been pushed about on the surface of the earth

4600 MILLION YEARS AGO

3700 MILLION YEARS AGO

PRECAMBRIAN ERA

Date of the oldest rocks found in the Earth's crust

PRECAMBRIAN ERA

Formation of the Earth as a planet

ological eras are drawn to scale; periods within the eras are not to scale.

beginning of the most recent period, the Quaternary, about 3 million years ago. On the scale of this chart, all of human history is too small   *to show as a separate area.*

by forces arising in the **mantle**, the zone of hot, semi-liquid rock just below the crust. The process of "drifting" has been likened to the way the crust on top of milk or stew shifts in a saucepan on a stove: it is pushed about by **convection currents** in the liquid being heated from underneath. When one realizes the huge size of the mantle and the core, it is easier to understand how forces generated there could move the massive continental blocks.

On map 1-19 the **mid-ocean ridges** are indicated. It is the welling up along these ridges of molten **magma** from the mantle that enabled scientists to determine just how the process of moving continents worked. North America has shifted westward thousands of kilometres because of the mantle's action; it continues to move slowly to the northwest. This motion has been the major cause of the extensive mountain-building and earthquake activity along the western and leading edge of the North American plate. The coast of British Columbia is near the edge of the huge **Ring of Fire** — the volcano and earthquake zone that encircles the Pacific Ocean.

The first part of North America to be formed into solid rock on the once completely molten surface of the planet was the Precambrian, or Canadian, Shield. The rest of the continent was added to, or plastered onto, the edge of this ancient land mass, which is the core of the present continent. The plains that formed around its edges were built up in part from sediments carried off the shield and deposited in the shallow seas that once surrounded it. Just beyond those plains, movements of continental

1-19 Map of the continental plates in their present locations.

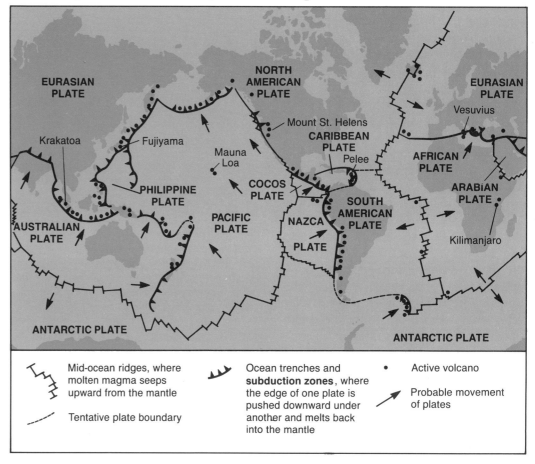

plates, and upheavals in the crust caused by volcanic activity, created the Appalachians and the Western Cordillera. Since the Appalachians were formed about 200 million years ago, they are sometimes referred to as old **fold mountains**. The Rockies of the Western Cordillera are sometimes called *young* fold mountains, since, to a geologist, 65 million years ago is quite recent! The forces of erosion added further deposits from these mountains to the seas surrounding the shield. During this long period of time, the shield, which once included mountain ranges as high as the Himalayas are today, was worn down continually by the forces of wind, water and ice. Today it has been reduced to mere "mountain roots" — little more than a rough, tough remnant of what it used to be.

The final stage in the sculpting of Canada's landforms has taken place over the past million years. The sculptor was the **continental icecap**, which advanced and then retreated across

1-20 The origin of Canada's landforms.

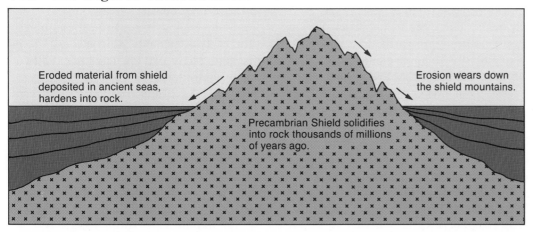

Eroded material from shield deposited in ancient seas, hardens into rock.

Erosion wears down the shield mountains.

Precambrian Shield solidifies into rock thousands of millions of years ago.

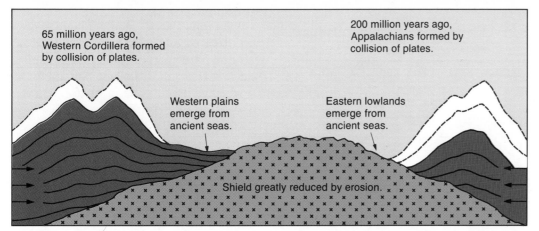

65 million years ago, Western Cordillera formed by collision of plates.

200 million years ago, Appalachians formed by collision of plates.

Western plains emerge from ancient seas.

Eastern lowlands emerge from ancient seas.

Shield greatly reduced by erosion.

Canada many times. During these movements the ice spread south, then melted back to the north, always under the control of changing climate conditions. This was the **Pleistocene ice age**. The ice didn't make its final retreat back to the far north until about 6000 years ago. It is hard to imagine 1500 metres of ice perched on top of Canada's landscape, but geologists are convinced that at several different times during the ice age, Canada looked much like Greenland does today.

The Pleistocene ice age had a great influence on the present-day surface of much of Canada. In the north, it filed and scoured the surface rocks and structures that remained after millions of years of erosion. In the south, it deposited vast amounts of materials and shaped much of the landscape that we see today in southern Canada. Scientists do not think that the ice age is over, but since icecaps take a long time to develop — thousands of years — no one should expect major changes in their lifetime.

1. Almost everything on the earth is 4 600 000 000 years old. What name is given to the materials that landed on earth as recently as this year?
2. (a) List the geological eras from oldest to youngest.
   (b) In which era would the year 1999 be found?
3. (a) Why is "master continent" a good name for Pangaea?
   (b) Throughout the world there are examples of both old and young fold mountains. Where is there an example of each in North America? How much difference is there between their ages?
   (c) Explain why the leading edge of a continental plate tends to resemble the front edge of a car after a head-on collision.
4. Which is Canada's oldest and most extensive landform?
5. Scientists say that the continental ice sheets could advance again. Explain what that means, in your own words.

New Words to Know
Geology
Tectonic forces
Erosion
Deposition
Era
Period
Fossil
Invertebrate
Pangaea
Plate tectonics
Crust
Continental plate
Mantle
Convection current
Mid-ocean ridge
Magma
Ring of Fire
Fold mountain
Continental icecap
Pleistocene ice age

Building the Landscape

It was mentioned previously that to understand Canada's landforms, it is necessary to have some knowledge of geological time, and of the forces that build and grade the earth. In the following sections, the building materials and rocks that make up the crust, and the forces that act upon them, will be briefly explained.

The Building Materials of the Crust

Rocks are the inorganic matter that makes up by far the greatest proportion of the earth's crust. Sometimes this material is solid — like the most common rock in the crust, **granite**. Sometimes it is unconsolidated — like the **sand** on a beach.

A careful look at sand will help provide an understanding of the characteristics of rocks. Anyone who has ever been on a beach has noticed that sand contains different kinds of particles. Some will be of a clear or milky-coloured crystal, some will likely be brown or tan in colour. Both types of particles are really tiny pieces of **minerals**: the clearer crystals are **quartz**, the brown ones are **feldspar**. Both could easily have been broken off a granite rock. Feldspar and quartz are the main constituents of granite. Granite generally contains other minerals as well.

From this simple example comes the generalization that rocks are formed from a mixture of minerals, which are often consolidated — bonded together by heat or pressure. In some rocks the different minerals are clearly visible.

Feldspar and quartz are, in that order, the most common minerals in the earth's crust. And like rocks, they too can be broken down, into the true building blocks of everything on earth, the **elements**. There are 92 elements that occur naturally on earth. Everyone is acquainted with some elements, such as **oxygen**, gold, copper, **hydrogen**, and carbon, which is called **diamond** in its crystal form. Many people don't realize that these substances are elements. Though minerals may often be found separately in the earth's crust, elements rarely are; they are generally combined with other elements, particularly oxygen. Elements combine chemically in nature to form minerals, though sometimes, as with the diamond noted above, a mineral may consist of only one element. Quartz is a chemical combination of two elements, **silicon** and oxygen. These two elements are the two most common in the earth's crust; by weight, oxygen makes up almost 50 percent of the crust. Unlike minerals, elements cannot be broken down into separate substances by ordinary chemical means. They can only be broken down by smashing their atomic structures. That is no easy task and is dangerous besides! The energy released by an atomic bomb is the result of atom-smashing.

The terms "rock" and "mineral" are too often used carelessly. People tend to call any hard, naturally occurring substance a rock; most do not pay attention to whether it is a mineral or not. They may even say, about a diamond in an engagement ring, "That is some rock." Calling a diamond an elegant element or a marvellous mineral is much more accurate.

Rocks form the bulk of the crust, and there is a huge variety of them. For example, there are scores of different kinds of granite alone. In addition, any granite can be re-formed in many different ways. The rock cycle shown on 1-24, which uses granite as an example, gives a better idea of how rocks change, how they "come and go," and how they are classified.

Magma is the molten material of which the earth is formed. **Igneous rocks** are simply magma that has cooled and become solid. They are sometimes nick-named "fire rocks." Igneous rocks were formed first; all other rocks, in all their variety, are formed from them. Some-

1-21 Elements form minerals form rocks.

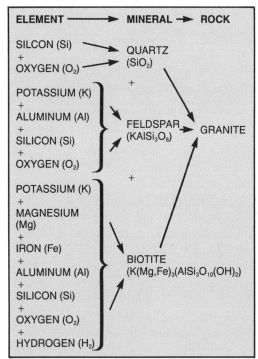

ELEMENT ⟶	MINERAL ⟶ ROCK
SILICON (Si) + OXYGEN (O_2)	QUARTZ (SiO_2)
POTASSIUM (K) + ALUMINUM (Al) + SILICON (Si) + OXYGEN (O_2)	FELDSPAR ($KAlSi_3O_8$)
POTASSIUM (K) + MAGNESIUM (Mg) + IRON (Fe) + ALUMINUM (Al) + SILICON (Si) + OXYGEN (O_2) + HYDROGEN (H_2)	BIOTITE ($K(Mg,Fe)_3(AlSi_3O_{10}(OH)_2)$)

GRANITE

The letters in brackets are the chemical symbols for the elements involved. Some minerals can be very complex.

1-22 A typical view of the Precambrian Shield by Lake Superior. *These rocks are the oldest in Canada, and among the oldest in the world.*

1-23 The Albion Hills, near Bolton, Ontario. *These hills, just north of Toronto, were deposited by glaciers very recently, in geological terms.*

times igneous rocks formed under the surface of the earth (granite is formed this way); sometimes they cooled on the surface in a flow of lava (**basalt** is an example of this).

Igneous rocks on the surface would have been subjected to the forces of erosion. A river might have eroded particles of the rock and deposited them in a lake, thus making **sedimentary rock** or "layered rock."

If heat or pressure acted on the igneous or the sedimentary rock, then **metamorphic rock** might have been formed. Metamorphic rocks, such as asbestos or **granite gneiss** (pronounced "nice"), are also known as "changed rocks." They often have their minerals lined up in a parallel way. This gives them a "banded" appearance. As diagram 1-24 shows, igneous granite may be broken down, and then, by erosion and pressure, be made into a sedimentary **sandstone**; this could then be re-formed into a metamorphic granite gneiss. These changes have been taking place on the earth over billions of years of geological time, so some rocks now in the crust have been through many changes.

1. (a) **Name the three rock types and give two or more examples of each.**
   (b) **Name four or more elements.**
2. **Explain the following terms:**
   (a) **Element**
   (b) **Mineral**
   (c) **Rock**
   (d) **Metamorphic rock**
3. **Why might a geologist say that rocks are like a stew, but minerals more like a cream soup?**

New Words to Know

Rocks	Diamond
Granite	Silicon
Sand	Igneous rocks
Mineral	Basalt
Quartz	Sedimentary rocks
Feldspar	Metamorphic rocks
Element	Granite gneiss
Oxygen	Sandstone
Hydrogen	

25

1-24   The rock cycle.

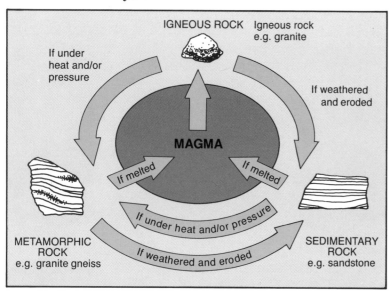

A particle of granite, an igneous rock, could easily become a part of a sedimentary rock, then a metamorphic rock, then a sedimentary rock, and so on. However, why is the rock "cycle" not a true cycle?

1-25   A granite gneiss. *This is a metamorphic granite. Heat and pressure melted the original rock, and the minerals ran together in bands, the black being biotite, the red feldspar, and the white quartz.*

Tectonic Forces That Build Landscapes   There are two basic kinds of **tectonic forces** that build or fashion the landscape. The first is **vulcanism**, which had much to do with forming the crust in the first place. The second includes those forces that cause the crust, once formed, to be twisted and warped in various ways.

The forces of vulcanism arise in the mantle, where pressure may cause molten rock to rise to the surface of the crust. Such **extrusions**, as they are called, may create a violent volcano, or spread more gently over the countryside as a **lava flow**. Large areas in the Western Cordillera were formed by such lava flows. **Volcanoes** have been very active throughout the history of the **cordillera**, the shield and the Appalachians;

today, however, only the cordillera still has **dormant volcanoes**. One of them, Mount St. Helens in the State of Washington, erupted in 1982. The dust cloud rose 13.5 km into the atmosphere, causing widespread damage.

Sometimes the lava does not reach the surface; instead, it is forced between the layers of rock, which are called **strata**. When it cools, the rocks it forms are known as **intrusions**. This process will sometimes cause a rise in the surface rocks. The Western Cordillera has a number of these intrusive features. The core of the Coastal Ranges, for example, is an extensive exposed area that has been pushed upward. This core, once a large intrusion, has been revealed on the surface by erosion. It is 1760 km long and 145 km wide.

The second group of tectonic forces do not add to the crust, but do shape it by exerting powerful stresses and strains in the crust itself. The crust is constantly in motion — very slow motion — as it "floats" on the heavier, semi-liquid, molten magma below it. Forces and movements in the mantle shift the continental plates.

Sometimes the tension pushes the layers of rock together and makes them bulge upward at the centre like a dome. If the magma beneath an area of rock flows away, the crustal rocks may lose their support and drop, forming a depression. Depending on the forces, distances, and types of rock involved, the rock layers may slowly begin to fold in order to relieve the pressure. When this happens, ranges of folded hills or

1-26 Sedimentary rocks, Hog's Back, Ottawa.

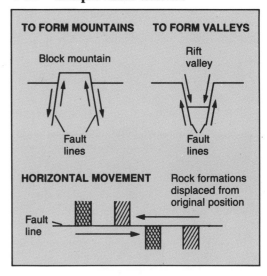

TO FORM MOUNTAINS   TO FORM VALLEYS

Block mountain   Rift valley

Fault lines   Fault lines

HORIZONTAL MOVEMENT   Rock formations displaced from original position

Fault line

mountains are created, sometimes stretching for thousands of kilometres. The Appalachians, which run from Newfoundland to Georgia, are a good example of this.

The rock strata may be pulled in a variety of directions. This can cause cracks or fractures, called **joints**. These are by far the most common result of tensions in the crustal rock. Often there is movement, called **displacement**, along a joint or set of joints. This is known as **faulting**, and the joint is then known as a **fault**. The movement may be horizontal — pushing along, against or away from the fault line. But often it is vertical — up or down. The Canadian Shield is well known for the variety and number of its faults — there are millions of them. As a result, the shield has examples of all the fault-line features: depressions where a block has fallen between two parallel faults (causing a long depression called a **rift valley**); the oppo-

site — land raised between two faults into hills or mountains; and **fault scarps**, which are lands along one side of a fault that have risen or fallen, forming cliffs. If the displacement is horizontal, rocks may be displaced for many kilometres.

Fortunately for Canadians, fault action in the shield is now largely complete — the ancient shield has stabilized. But movement still continues along active faults in the Western Cordillera. The most famous is probably the San Andreas fault in California. Sudden crustal movement is the main cause of the devastating **earthquakes** that can occur when there is action along the fault lines. These movements mean, of course, that the crust in those areas is still under tension and is restless and unsettled. This is to be expected along the extensive faults which separate the continental plates. Along these edges there is often considerable earthquake and volcanic activity.

1. **Vulcanism builds the landscape by causing intrusions and extrusions.**
   (a) **How is each formed?**
   (b) **Name the features that may result from them.**
2. (a) **Lay several sheets of paper on your desk. Describe how you would make a "fold" without lifting the paper off the desk.**
   (b) **What do the several sheets of paper represent, and what would a geologist call them?**
3. **Name three possible results of vertical pressure on a fault, and briefly sketch each one.**

New Words to Know

Tectonic forces	Intrusion
Vulcanism	Joint
Extrusion	Displacement
Lava flow	Faulting
Volcano	Fault
Cordillera	Rift valley
Dormant volcano	Fault scarp
Strata	Earthquake

1-28 Mountains before and after erosion.

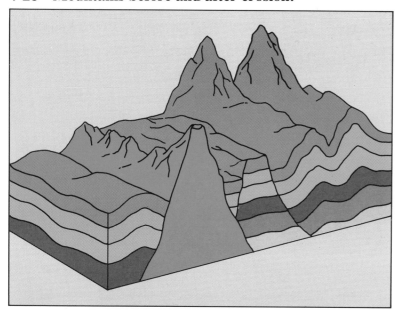 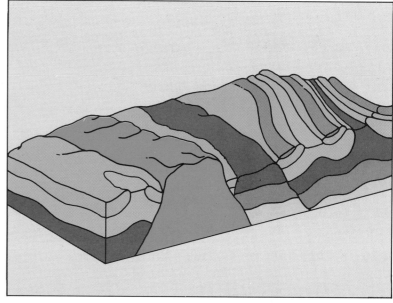

Can you pick out examples of volcanic, block and fold mountains?
How would you describe the changes in the appearance of the landscape after erosion?

Forces That Grade the Landscape

The forces that grade or level out the landscape are illustrated in 1-29. There are three steps to the **cycle of erosion** shown: the breaking up or **weathering** of the rock, the transporting of the rock materials, and the depositing of the materials. This is a true cycle, for the process may repeat itself again and again. Deposits may remain stationary for thousands of years, or may be eroded away within minutes to form another rock somewhere else.

The net result of these processes is the smoothing out of the surface of the earth. The higher and more prominent parts of the earth's surface are particularly vulnerable to the work of water, wind and ice. These are the main **agents of erosion**. They chew away at the earth's prominent features, reducing their height and moving the loosened materials elsewhere. With the depressions and lower regions of the surface, the opposite happens: down into them flow the water, wind and ice, depositing materials they are carrying. The overall effect is that higher areas are literally broken up and moved to raise lower areas, thus smoothing the earth's surface. This is exactly the purpose of a grader along a road or at a construction site; the agents of erosion act like graders.

Even while the volcanic mountains of the Canadian Shield and the fold mountains of the Appalachians were being formed, the agents of erosion were working to reduce their height. The shield's mountains, which were formed two billion years ago, are no more. The Appalachians remain, but they are "only" 200 000 000 years old, and nature has already reduced their bulk considerably. Nature has all the time in the world.

A statement made by the famous British geologist, Charles Lyell, still holds true after 150 years: "The present is the key to the past." The forces of erosion are at work today.

The first stage in the cycle of erosion is the weathering of the rock. This is often done **chemically**, by rainwater and streams in particular; they simply dissolve the rock the way tea dissolves a sugar cube. There is also **mechanical** breaking down of rock. The many joints in the surface of the crust open it up to weathering. Water may get into joints, or into the tiniest cracks in a rock; when it freezes it also expands, causing pieces to break off. Pieces may be broken off the side of a mountain or hill by growing plant roots, or wandering animals, and fall down the slope. As they fall, they can easily chip other rocks, or dislodge them. The latter then fall and repeat the process.

Running water not only transports rock, but also continues to break it up mechanically and chemically. Rocks frozen in a glacier increase the glacier's power to scour and file more rock from the surface. The wind too, as it carries small particles of earth or sand, has considerable abrasive power.

Once a piece of rock has been chipped, filed or dissolved from a larger one, the second stage of erosion — transportation — becomes easier. During the ice age and earlier, huge quantities of rock were carried south by the continental ice sheets. Though the ice sheets have melted back for now, smaller **alpine glaciers**, particularly in the cordillera, have continued to advance. They slowly flow down from the mountains, breaking up, scouring and transporting the walls and floors of their valleys. These smaller glaciers, pulled by gravity, have played a major role in reducing the size of Canada's mountains and highlands.

When it comes to transporting eroded materials, running water is far more important than glacial action. Running water is at work throughout the country, moving earth materials that measure in the thousands of tonnes, for thousands of kilometres. Rivers and streams of all sizes will carry rock in **suspension** and in **solution**; they will even roll and bounce the larger stones and boulders along the bottom of the stream bed. In flood time anyone can see the tremendous carrying power of a river. However, every minute of every hour of every day, streams are quietly and slowly eroding and moving the landscape from the high places for deposition in the lowlands.

1-29 The erosion cycle.

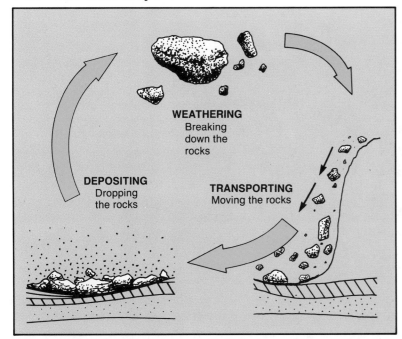

There are three parts to this cycle; two of them tear down or erode the landscape, while one builds it up again. Why can these processes together be considered a true cycle?

1-30 Niagara Falls. *Tough Lockport dolomite caps the Niagara Escarpment. The falls erode the softer rock underneath the dolomite, undercutting it until it breaks off and falls into the gorge.*

29

The final stage in the erosion cycle is deposition. Though part of the cycle of erosion, deposition alone is not considered an eroding process; unlike the erosive forces, deposition builds up the land rather than tearing it down.

When the wind drops, the water slows down, the glacier begins to melt and these agents of erosion deposit the materials they have been carrying. The result may take many different forms — a **sand dune**, a **delta** at the mouth of a river, or a **moraine** deposited where a glacier melted.

Canada has excellent examples of almost every kind of deposition. Sediments from the high mountains of the shield and cordillera were deposited in ancient seas, which later became the Great Plains. When the land rose and the sea drained away, the Great Plains remained, with hundreds of metres of sedimentary rock beneath them. The St. Lawrence Lowlands, in a similar way, were first built up by sea sediments, becoming land when the sea drained eastward. The Great Lakes Lowlands too are underlain by marine sediments; covering their surface are the additional deposits left by the continental ice sheets.

The dry lands of Alberta, Saskatchewan and the interior of British Columbia all have extensive areas of dunes and wind-blown soils. Dust storms remain a problem in such areas — a problem sometimes worsened by poor farming practices. Smaller, wind-deposited sand hills may be found in Ontario, for example in Prince Edward County.

1. **Why might a geologist say that the Appalachians are "only" 200 000 000 years old?**
2. **Name the three actions that form the cycle of erosion, and briefly describe how water carries out each one.**
3. **Explain why it can be said that the cycle of erosion acts like a grader.**

New Words to Know

Cycle of erosion	Rock in suspension
Weathering	Rock in solution
Agents of erosion	Sand dune
Alpine glacier	Delta
Chemical weathering	Moraine
Mechanical weathering	

1-31 How a river carries its load.

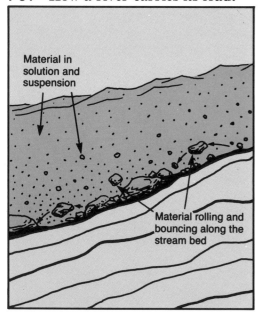

Material in solution and suspension

Material rolling and bouncing along the stream bed

Though this diagram is meant to illustrate how a river transports material, what other part of the erosion cycle is also taking place?

1-32 Sandbanks Provincial Park, Prince Edward County, Ontario.
These wind-deposited sand dunes may be over 20 m high.

1-33 Canada's major landform regions.

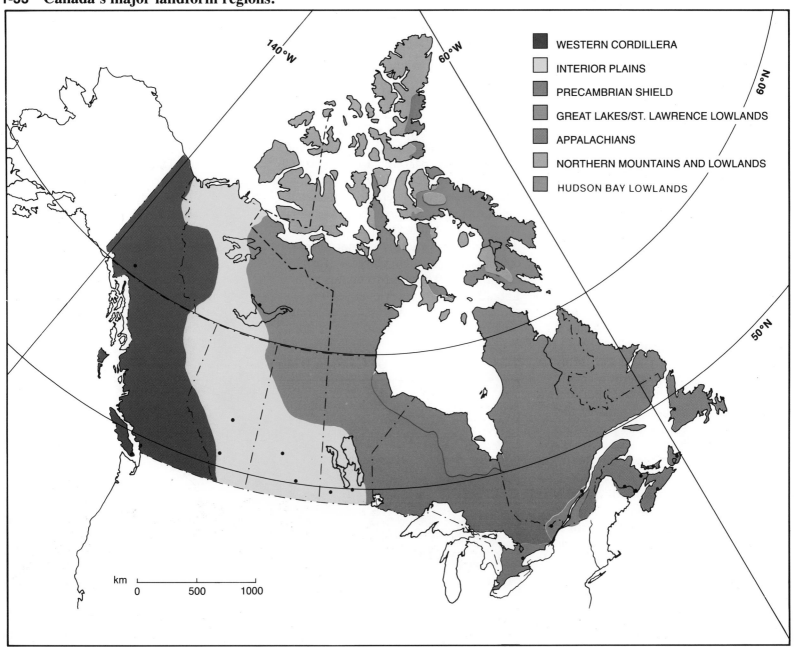

WESTERN CORDILLERA
INTERIOR PLAINS
PRECAMBRIAN SHIELD
GREAT LAKES/ST. LAWRENCE LOWLANDS
APPALACHIANS
NORTHERN MOUNTAINS AND LOWLANDS
HUDSON BAY LOWLANDS

It is clear from this map how much the Precambrian Shield dominates Canada's landforms. It was also very important in the formation of the other regions. Only one of Canada's landform regions does not extend into the U.S.A. Of the regions shared by the two countries, which is the only one that is bigger in Canada?

31

Canada's Landform Regions

The map and diagram, 1-33 and 1-34, provide two different perspectives on Canada's landform regions. The map shows the country as a fringe of lowlands and mountains surrounding a large, dominating shield: the fringe on the west is considerably larger than the one on the east. The cross-section from east to west shows the country as a huge lowland between two mountainous areas. Since the **Precambrian Shield** is by far the most obvious Canadian landform region, this section will start by describing it.

The Precambrian Shield

This shield covers almost half of Canada — about 4 800 000 km². That means it is almost equal in area to all the rest of Canada put together. The shield forms the core of the North American continental plate. It actually underlies much of the rest of Canada and parts of the U.S.A., so it is really much larger than even the landform map shows. The map shows only the area exposed on the earth's surface.

The name *Pre*cambrian indicates that the rocks of the shield were in place *before* the Cambrian period of earth's geological history. ("Cambria" was the ancient Latin name for Wales, today a part of the United Kingdom. The age of certain rocks found in Wales is now used to date rocks worldwide.) All of geological time which came before the Cambrian period, 600 million years ago, is considered "Precambrian."

The shield is very, very ancient — it contains Canada's oldest rocks — and has seen a great number of cycles of mountain-building and erosion. Only the roots of mountains remain. The region is extremely complex, for it has undergone repeated faulting, folding and vulcanism. While geologists agree they know a great deal about the shield, the records in the rock are so poor that they admit there is much more they do not know. The agents of erosion have combined to reduce the area to a **peneplain**, which means "almost a plain." Today the shield's elevation is seldom over 600 m. Only in a few places — for example, the Torngat Mountains of Labrador, which reach over 1600 m — can the term "mountains" be accurately used to describe the shield.

The photographs of the shield country show that though the landscape is very rugged, it is also quite low in relief. The term landscape may be a misnomer, since the region is more like a **lakescape**. Lakes are numerous due to the complex folded and faulted nature of the rock formations. The Pleistocene ice ages which affected almost all of Canada added to the distortions in the drainage patterns. Because the shield is like a saucer in shape and form, the river and lake systems tend to drain toward Hudson Bay, in its centre.

The Precambrian Shield is also known by other names — the Canadian Shield, Canada's Vacation Land, Canada's Mineral Storehouse. It definitely is a

1-34 Cross-section of Canada from west to east.

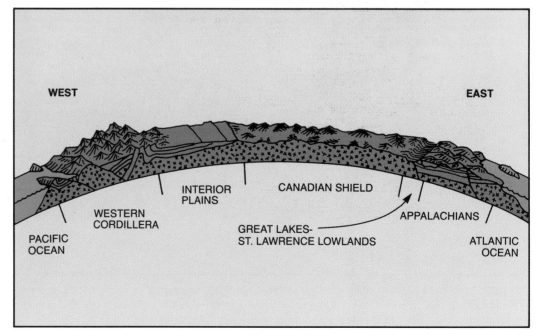

WEST EAST

PACIFIC OCEAN
WESTERN CORDILLERA
INTERIOR PLAINS
CANADIAN SHIELD
GREAT LAKES- ST. LAWRENCE LOWLANDS
APPALACHIANS
ATLANTIC OCEAN

1-35 *Islands, Canoe Lake*, **by Tom Thomson, 1916.**
Thomson was a member of the "Group of Seven" artists, who became famous in part for their interpretation of Canada's north country, particularly areas of the shield.

1-36   Typical landscape in the shield, near Sudbury, Ontario.

playground for all seasons and a great source of wealth. The varied geological activity that created metamorphic rocks also created many minerals. In illustration 1-37, an intrusion forcing its way up to the surface, through a fault in the crust, creates a potentially rich mining area. The molten magma melts the rock along the sides of the intrusion, forming minerals and often concentrating their metallic content. Geologists seek out these mineralized areas, often finding ores of gold, copper, nickel, lead, zinc, iron and other metals, as well as other valuable minerals. It is interesting to note, however, that for all its vast size

1-37   Mineralization along an intrusion.

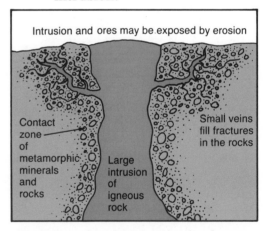

Intrusion and ores may be exposed by erosion

Contact zone of metamorphic minerals and rocks

Large intrusion of igneous rock

Small veins fill fractures in the rocks

This diagram shows an intrusion after the magma has cooled and erosion has exposed the ores at the surface. Such intrusions, with their high temperatures, have created many of the ores of iron, lead, gold, copper, zinc, and other metals found in the shield. They can change the rocks for a kilometre, or for a mere fraction of a centimetre.

and resources, the shield has the lowest population of Canada's four major southern landform regions. That small population is widely scattered across the shield. Most of the people live in or near the major mining and forestry centres and along the transportation lines which link them.

The shield is not exclusively Canadian. Small portions project south into Minnesota, Wisconsin, Michigan and New York. Still, because of its huge extent and great importance to Canada, when foreigners think of this country, they generally picture the land and lakescape of the Precambrian Shield.

1. (a) Using an atlas and map 1-33, name the landform regions found in Canada which are also found in the U.S.A.
   (b) Which of these are larger in Canada than in the U.S.A.?
2. (a) Name the Canadian provinces and territories which are at least partly on the shield. Name those that are partly in the Appalachians.
   (b) Estimate the percentage of each province found on the shield, and the percentage found on the Appalachians.
3. Explain why the shield is called "Precambrian."
4. How did the shield become a peneplain?

New Words to Know

Precambrian Shield   Lakescape
Peneplain

The Great Plains

The Great Plains extend over a vast portion of central Canada. They stretch 1500 km from near the Ontario-Manitoba border to the Rockies, and over 1800 km from the 49th parallel to the Mackenzie Delta. On these plains live the vast majority of the four-and-one-half million people of the Prairie provinces as well as most of the tiny population of the northern territories. These same provinces also take up a significant portion of the Precambrian Shield, and a small slice of the Western Cordillera. Most of their population, however, lives on the Great Plains, and

1-38 Ordovician seas in North America.

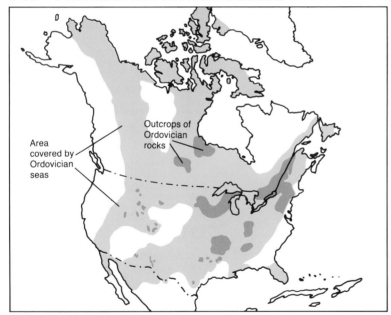

During the Ordovician period, approximately 400 to 500 million years ago, seas covered much of what is now central North America. Virtually the entire Great Lakes/St. Lawrence Lowlands have Ordovician **outcrops.** *According to the map, which other major landform regions would also have them? How might geologists determine that the areas shown were covered by the inland seas?*

1-39 Maximum extent of glaciation in North America.

Only one small portion of Canada wasn't affected by glaciation. Can you find it on the map? Of course, all the areas which were glaciated were not affected in the same way. For example, which areas received most of the deposition?

34

most of their economic activity takes place there.

Though often compared to the immense steppes of Russia, Canada's Great Plains are only a small part of the much greater Central Plains of North America. The shield is almost entirely in Canadian territory; most of the Central Plains are in the U.S.A.

It makes sense to discuss the Great Plains just after discussing the Precambrian Shield. This is because the plains were formed after the shield; in fact, thousands of metres of sedimentary rock which were on or remain on the plains, were deposits from that shield. Shallow seas invaded North America during the early eras. They teemed with hard-shelled sea life. These shells formed part of thick strata of **limestone** and chalk. More material from the shield — and during the Cenozoic Era, from the cordillera — later eroded, adding to the rock layers of the plains.

The term Great Plains can be confusing, for they are by no means completely flat. They actually have a great variety of features, several of which show considerable relief. There are three different levels on the plains. Two escarpments separate the levels. The Manitoba Escarpment, which averages about 300 m in height, separates the First and Second **Prairie Levels**. It also divides the region of Paleozoic rocks from that of Mesozoic ones. The harder rocks of the second level withstood the forces of erosion better than those farther east. Similarly, the Third Prairie Level was protected by tougher **cap rock** than the second. The third level also includes much younger sediment from the Rockies. The Missouri Coteau,

1-40 Aerial view of a typical Saskatchewan wheat farm. *In what season was the photo taken? What clues give the answer to this question? What feature crosses the picture in the lower left corner?*

1-41 Aerial panorama of the plains.

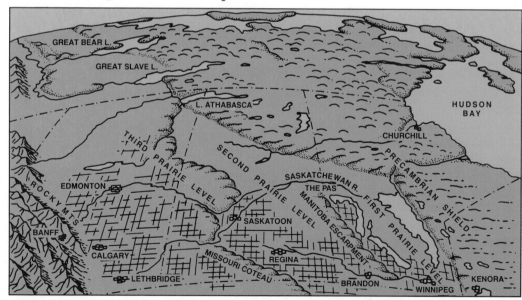

This is not a photo, but a sketch map. What region would be directly under the viewer?

another escarpment, is less distinct than the Manitoba Escarpment; it appears more like a low row of hills, most of them under 200 m high.

The Great Plains grow higher as they approach the cordillera. They start at about 330 m above sea level in the east, rise to about 650 m at the Second Prairie Level, and finally reach 1300 m near the foothills of the Rockies. Other **relief features** include some very deep valleys cut by rivers. Since the area was covered by glaciers in the last ice age, there are scattered morainic features. In southeastern Alberta, the arid **badlands** have been severely eroded by winds. The cordilleran edge of the plains is gently folded into foothills. Other hills, such as the Cypress Hills, which cut across the Alberta-Saskatchewan border, are erosional remnants.

Because it is underlaid by sedimentary rocks, the Great Plains is not a source of metallic minerals. But there *is* natural gas and oil. These have collected in the region's porous rocks (such as limestone). Both have brought great wealth to Alberta and Saskatchewan. The **tar sands** and **heavy oil** deposits in these two provinces have only recently been exploited to any degree. As well, buried deep under the surface of Saskatchewan are huge deposits of **potash**, the remains of an ancient seabed. It is now sold throughout the world as fertilizer.

The greatest physical resource of the Prairie provinces has always been, of course, the rich soil. It has made the region Canada's breadbasket. Some of these soils have developed from glacial deposits, others from the flat beds of huge glacial lakes. Where these lakes have drained away, rich, flat agricultural land has been left behind.

Outside of the several large prairie cities, the settlement pattern has generally been determined by the agriculture. Generally, the population densities are low, with the amount of rainfall being the main factor determining just how low. Certain river valleys and ancient lake floors, which are particularly favourable for agriculture, may have quite high population densities.

1. **Explain five major differences between the shield and the Great Plains.**
2. **Name the two escarpments which separate the Prairie Levels, and explain how they differ.**
3. **Though the plains stretch well beyond the 60th parallel, why are most of the people in the southern portions?**
4. **Explain the connection between ''shallow seas'' and the various strata beneath the Great Plains.**

New Words to Know
Limestone
Prairie Levels
Cap rock
Relief features
Outcrop
Badlands
Tar sands
Heavy oil
Potash

The Western Cordillera

The Western Cordillera forms the western boundary of the Great Plains, just as the shield forms the eastern boundary. In a way, the Great Plains region is a ''child'' of the other two regions; materials were eroded from both to form the plains.

The Western Cordillera is by far the highest, most rugged and therefore most

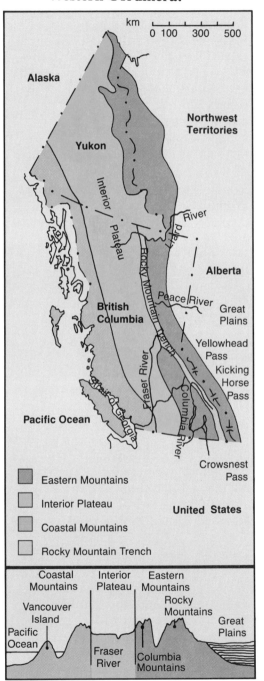

1-42 Map and cross-section of the Western Cordillera.

36

formidable barrier to trade and movement anywhere in Canada. It was such an imposing barrier that before British Columbia would agree to become a Canadian province it demanded a railway to link it with the rest of Canada. Otherwise, the Pacific coast would have been almost completely isolated from eastern Canada by the huge distances and the rugged cordillera. Today the cordillera remains a barrier, separating the bulk of B.C.'s people from the rest of the country in many ways. Even though the transportation and communication links have been greatly improved since the nineteenth century, the barriers of time and higher cost remain.

Like Canada, British Columbia consists of a large interior **plateau** between two mountain ranges. Adding to the complications, the Eastern Mountains are separated into two major ranges by a deep trough, as are the Coastal Mountains.

The Eastern Mountains consist of two ranges running northwest to southeast, separated by the deep Rocky Mountain Trench. This down-faulted trench has mountains several thousand metres high on both sides; the trench itself is only 900 m — sometimes less — above **mean sea level**. The Columbia Mountains on the west side of this huge gash are almost as high as the Rockies along the east side. The latter range has extensive areas over 3000 m, with peaks rising well over that. The Columbia Mountains are older, more faulted and more folded; they include more intrusives than the Rockies. Both ranges are formed mostly of folded sedimentary rock, though there was much more metamorphic action in the Columbia Mountains. In both, however, fossils may be found several thousand metres above sea

1-43 Peaks in the Rocky Mountains. *Once an ancient sea floor, these rocks have been raised thousands of metres into the air. The snow helps outline the sedimentary layers, and the folds created by geological forces.*

1-44 The Okanagan Valley. *Although the interior plateau is relatively flat compared to B.C.'s mountain ranges, valleys such as the Okanagan cut deeply into it, helping make the plateau a rugged landscape in its own right.*

level. Both mountain areas were once sea floors.

The Rockies were folded because of the movement of the North American plate toward the Pacific one. Sedimentary rocks were folded over a period of time about 65 million years ago, during the worldwide young fold mountain-building period. Since then, the Rockies have been eroded by water, wind and ice. Yet they still have peaks over 3500 m. Mount Robson, at 3954 m, is the highest peak in the Rockies chain.

Because the North American plate rides over the Pacific plate just off the B.C. coast, there is a great deal of volcanic activity nearby. The Coastal Mountains are split into two. On the west are the offshore islands, on the east, the mainland mountains along the coast. Between them is a trough filled by the Pacific Ocean. Massive intrusions in the earth's mantle have raised the crust, and are sometimes exposed on the surface; these are the Coastal Mountains. To the north, in the Yukon, in the very high St. Elias Mountains, is Mount Logan, the highest peak in Canada at 5951 m.

The Interior Plateaus are a highland area that has been uplifted to elevations from 1300 m to over 2000 m. Though they are not as rugged as the mountains on either side, the landscape of the plateaus does resemble an area of high hills or low mountains. Rivers have cut deeply into the land here. The Okanagan Valley, for example, is well over 1000 m below the surface of its surrounding plateau. The plateaus are very rugged. Some favoured areas, however, on glacial or river deposits, provide good farmland.

The mountains of the cordillera are sparsely populated. The people are concentrated in the valleys and troughs. Transportation lines remain costly to build in the interior, so many follow river valleys linking scattered mining and farming towns. Several of the largest cities are along the coast, built on narrow ledges or, as is the case with Vancouver, on delta lands.

1. **Name four mountain ranges in British Columbia.**
2. **Using an atlas, find the names of three bodies of water along the trough of the Pacific that splits the Coastal Mountains in two.**
3. **In your own words, briefly describe the location and appearance of the Interior Plateaus.**
4. **(a) Name three major passes through the Rockies. Why are these important?
(b) Why are the Rocky Mountains better known than the St. Elias Range, even though the latter has many peaks higher than any in the Rockies?**

New Words to Know
Plateau
Mean sea level

The Great Lakes/St. Lawrence Lowlands

The smallest of the four southern landform regions, the Great Lakes/St. Lawrence Lowlands, has by far the largest population. This small region, sometimes referred to as the nation's **heartland**, includes the densely populated portions of Ontario and Quebec. In this region live close to three-fifths of Canada's people. The Great Lakes Lowlands and the St. Lawrence Lowlands together form one region.

They are, however, separated from each other by a southerly projection of the shield into eastern Ontario.

The lowlands are more heavily populated than the rest of Canada because geographically they are the most favoured landform region in Canada. The landscape of gently rolling hills and flat plains provides an excellent physical base for agriculture and settlement. The lowlands reach farther south than any other region of Canada, have a more pleasant climate, and are more suited to agriculture than most of the rest of the country. In terms of location, this region is close to the huge markets of Canada's main trading partner, the U.S.A.

The Great Lakes/St. Lawrence Lowlands were formed in much the same way as the Great Plains. The region was submerged under inland seas. Sediments were deposited on it from the shield and the Appalachians. When the seas drained away, they left sometimes hundreds of metres of material. The deeper sediments had been deposited early in the Paleozoic Era, and had hardened into varied strata of limestones, sandstones and **shales**. Toronto stands atop sedimentary Ordovician rock, which was formed over 425 million years ago on top of the ancient Precambrian igneous rock. Of course, the present strata beneath the surface of Ontario were not the only ones laid down; erosion has removed much of what was once there.

One clear reminder of the erosive action that has changed the landscape through time is the Niagara Escarpment. This is the most obvious and extensive landform feature in southern Ontario.

1-45 The rocks beneath Toronto.

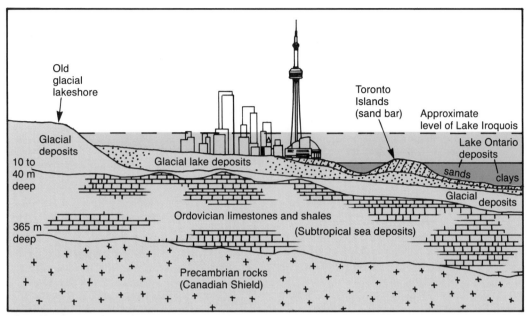

Old glacial lakeshore

Toronto Islands (sand bar)

Approximate level of Lake Iroquois

Glacial deposits

10 to 40 m deep

Glacial lake deposits

Lake Ontario deposits

sands clays

Glacial deposits

365 m deep

Ordovician limestones and shales (Subtropical sea deposits)

Precambrian rocks (Canadian Shield)

This sketch provides a generalized view of both the consolidated and the unconsolidated rocks beneath the downtown area. Because the solid rock may be deep, what problem is created for the builders of skyscrapers?

1-46 A kame in the Baden Hills, near Kitchener, Ontario. *Kames are formed where a river flows off the edge of a glacial ice cap, depositing debris in the conical shape seen in the photo.*

Erosion removed the softer strata east and north of the escarpment; the tougher **Lockport dolomite**, which caps the escarpment, resisted those forces. Thus the escarpment remains above the plain below it. The Niagara River flows over this cap rock, producing the world-famous falls.

As diagram 1-45 illustrates, the rock in place above the old sea floors is mostly an unconsolidated or unsorted glacial moraine. This material was deposited quite recently, "just a moment ago" in geological terms. The entire lowland area was once buried beneath the ice of the Pleistocene period; the final ice sheet did not melt and retreat from the area until a mere 9000 years ago.

The continental ice sheets also finished the scouring and deepening of the present Great Lakes. These lakes were even greater just after the ice melted back, for they were filled to overflowing by the massive meltwaters, and the ice partially blocked their drainage. This means that the shorelines of the lakes used to be far above where they are now. As a result, old shorelines can be seen along hundreds of kilometres of today's Great Lakes. Lake Ontario was once the much larger Lake Iroquois; into that **glacial lake** materials were deposited as the rivers drained the melting ice sheet. Downtown Toronto was once covered by the water of Lake Iroquois. Portions of the previous lake bed can be seen in diagram 1-45.

In most areas of the Great Lakes Lowlands, the surface cover of the landscape consists of ice-deposited materials. When the ice sheets melted back, much of the debris they were carrying simply fell to the ground, covering the area like a rough blanket. This post-glacial blanket is known as **ground moraine**. It is unconsolidated, unsorted (which means that throughout it are particles of various sizes), and may be many metres thick. In other areas such as the Oak Ridges moraine, the moraine was built up into a small hill, or row of hills, like a low ridge. This type of feature is the result of two sections of an ice sheet depositing material between them over a period of time. A great variety of post-glacial features can be found throughout southern Ontario and to a lesser extent in southern Quebec. They can also be found in certain other parts of the world, wherever ice sheets passed and melted back.

39

Mont Saint-Hilaire Map Study

Mont Saint-Hilaire is one of the Monteregian Hills. It is east of Montreal, near the Richelieu River in southern Quebec. This shows how a typical Monteregian Hill appears on a topographic map.

Symbols The conventional symbols used on Canadian topographic maps are reproduced on page 405. Students should familiarize themselves with the most common symbols when starting to use the map, and should refer to the key as needed.

1. **Sketch the following symbols, and find one of each on the map:**
   - (a) embankment
   - (b) campsite
   - (c) school
   - (d) dual highway
   - (e) bench mark
   - (f) quarry

Locating Features Canadian and other topographic maps use a square grid (the numbered blue lines) to locate features. Every place on the map can be located using a six-figure **map reference**. For example, the D in Domaine-Rouville, just south of the hill, is at 442423. The 442 is the **easting**; the 44 is the vertical blue line of the grid, and the 2 is two-tenths of the distance east from that line to the next one, 45. The 423 is the **northing**,

because it measures the distance to the north. The 42 is the horizontal blue line, and the 3 is three-tenths of the distance north to the next line. The easting is always given first.

2. **What is located at (a) 448507 (b) 447449 (c) 450466 (d) 411434 (e) 407476 (f) 456416?**
3. **What would the grid references be for:**
   - (a) the top of Le Pain de Sucre
   - (b) the school in northern Beloeil
   - (c) the Hôtel de Ville in Beloeil
   - (d) the curling club in Otterburn Park
   - (e) the school just north of the hill
   - (f) the church in Otterburn Heights?

Elevation The **elevation**, or height of the land above sea level, is shown in two ways. Sometimes a black number with an arrow is printed on the map. This indicates that the exact height of that spot has been determined by surveying, and a metal plate called a **bench mark** has been placed there to record that information. There is one on the bridge at 411504, where the elevation is exactly 47 feet.

The other way of showing elevation is by **contour lines**—curving, light-brown lines that follow the shape of the land. Each point on a particular contour line is at the same elevation above sea level, with a regular vertical

distance, called the **contour interval**, between it and the lines above and below. Contour heights and the contour interval are given in feet or metres. Many contour lines are needed to show the changes in height on the hill; very few contour lines appear on the plains around it, for they are flat or gently sloping.

4. **Heights are given in feet on this map. What is the elevation at 455423? Find other contour lines that are labelled, and work out the contour interval of the map.**
5. **What is the highest contour line actually labelled on the hill?**
6. **What is the elevation of the top of the hill Le Pain de Sucre?**
7. **According to the bench marks on the bridges, which way is the Richelieu River flowing?**
8. **Explain in your own words, or by using a sketch as well, why the contour lines would be close together when the land is steep, and far apart when it is flat or gradually sloping.**
9. **What happens to the contour lines when they cross the streams on the hill? Draw a sketch map to illustrate your answer.**
10. **Estimate the heights of land on which these features stand: (a) the building at 420454 (b) the school at 437475 (c) the arena at 418476.**

Scale This topographic map has a scale of 1:50 000. This means that one centimetre on the map represents 50 000 centimetres on the ground. Because the 1:50 000 tells you the fraction the map represents of the actual distances involved, it is called the **representative fraction**, or simply, the **scale**. This map also has a linear scale, in which distances on the map are marked in yards and metres along a line.

11. **Using the linear scale on page 405, measure in kilometres:**
    - (a) The distance between the two highway bridges across the Richelieu River.
    - (b) The width of the hill east to west, using the 300 foot contour as its base.
    - (c) The length of the Trans-Canada Highway (Highway 20) which appears on the map.

1-47 Mont St-Hilaire, due east of Montreal, is typical of the Monteregian Hills. *The photo was taken due west of the hill. Compare it to the map, noting how the two areas of higher elevation appear. Where would the Richelieu River be? Why is the landscape around the hill so flat?*

1-48   Part of Beloeil topographic map 31 H/11.
Contour interval is 25 feet. For scale and key, see page 405.

Culture This map is of an area in southern Quebec, and the area's culture is most obvious in the almost complete use of French names.

12. English-speaking students should not have much difficulty in finding the French words for: **(a)** arena **(b)** equestrian centre **(c)** discharge **(d)** mountain **(e)** cemetery **(f)** City Hall.

13. How can the Quebec long-lot system of landholding, as described in chapter two, be detected on this map?

New Words to Know

Map reference	Contour line
Easting	Contour interval
Northing	Representative fraction
Elevation	Scale
Bench mark	

1-49   Formation of the Monteregian Hills.

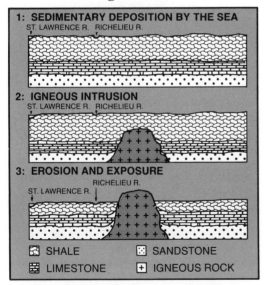

Hot magma intruded into the sedimentary deposits of the sea floor and cooled. Much later, this intrusion was exposed by the erosion of the softer sedimentary rock.

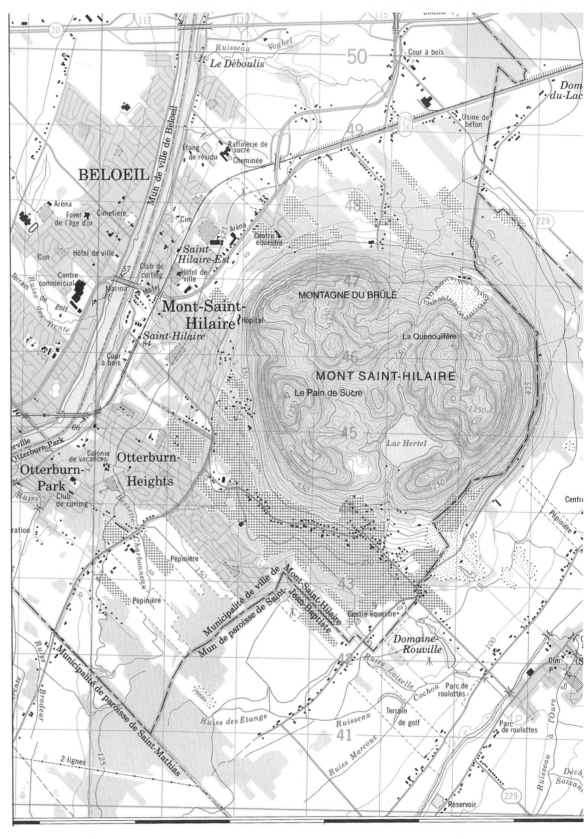

1-50 Geological map of southern Ontario.

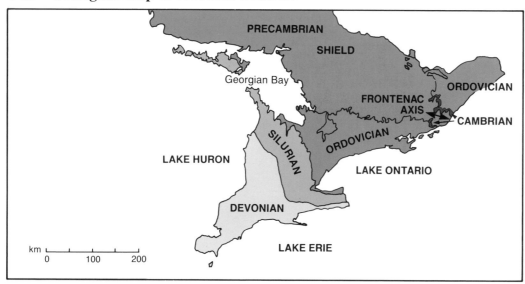

The Precambrian Shield is found on or near the surface in most of Ontario. In the south, the seas of the Paleozoic era deposited layers of limestone and sandstone over it in many areas. In much more recent times, wind, water and particularly ice have deposited further layers of loose earth and debris on top of these layers. The Frontenac Axis is an extension of the shield into the U.S.A. It creates the Thousand Islands where it crosses the St. Lawrence, and forms the Adirondack Mountains in northern New York State.

1-51   The Niagara Escarpment, near Milton, Ontario. *In places the escarpment, which stretches hundreds of kilometres across southern Ontario, is steep, as at the left; in others, as at the right, it has been weathered into a slope.*

The geological development of both the Great Lakes Lowlands and the St. Lawrence Lowlands has just been explained. But the latter has one additional feature that the former does not. Above its glacial deposits, the St. Lawrence Lowlands have a top layer of sediment that was laid down, not by an ice sheet, but by the Champlain Sea. Because the weight of the ice had depressed the land, the sea was able to flood inland, covering the entire St. Lawrence Lowlands area. The land later rebounded, forcing the water off. This was fortunate; the marine deposits in the Champlain Sea made the lowlands extremely flat and provided a base from which excellent soils have developed.

One feature in the St. Lawrence Lowlands deserves special mention. The Monteregian Hills are scattered across them, mostly to the east and southeast of Montreal. The city itself is named after the most famous of these hills, Mount Royal, which overlooks the city. These hills, such as Mont Bruno and Mont St-Hilaire, are exposed igneous intrusions, and stand sometimes as much as 470 m above the flat plains that surround them. (See pages 40-41.)

1. (a) Name the parts of the crust below downtown Toronto.
   (b) Which of the answers to (a) are not found beneath the crust of most areas of the Great Lakes Lowlands?
2. How was the formation of the Great Plains similar to the formation of the Great Lakes Lowlands?
3. (a) What is the approximate age of the surface rock just west and south of the Niagara Escarpment?
   (b) What is the approximate age of the bedrock below Cornwall, London and Parry Sound?

42

4. Explain the formation of the Monteregian Hills.

New Words to Know

Heartland	Lockport dolomite
Shale	Glacial lake
Kame	Ground moraine

The Appalachians

One of the most famous faults in Canada is known as **Logan's Line**. It separates the flat, low-lying St. Lawrence Lowlands from the Appalachian Mountains, along a line from Lake Champlain to Quebec City. Canada's Appalachians are only a small part of the whole formation. The part of this complex lying in Canada is perhaps best referred to as "upland," for rarely does it reach the height of true mountains. Seldom do they rise above 500 m. Their highest point, Mont Jacques Cartier on Quebec's Gaspé Peninsula, is just over 1260 m.

The geological development of the Appalachian region in Canada was similar to that of several other regions already described. Sediments were laid down in the Appalachian area, much as they were in the shallow seas that later became the plains, the lowlands and the cordilleran regions. Fold mountains were formed due to the collision of the North American and European plates. This was over 200 000 000 years ago, during the old fold mountain-building period, early in the Mesozoic era. Geologists believe that Atlantic Canada was once joined to northern Europe; there are rocks and rock formations in both Wales and Scotland that are similar to some found in Nova Scotia (which literally means "New Scotland") and Newfoundland.

The Appalachians do continue out onto the **continental shelf**, but there they are, of course, submerged.

As in the cordillera, vulcanism occurred as the continental plates collided. Volcanoes erupted, and magma intruded into the existing sedimentary layers. There was also considerable faulting, the remains of which can be seen today in several of the region's uplands, such as the Cobequid Hills and the Cape Breton Highlands.

Because the major mountain-building period in the Appalachians occurred 200 000 000 years ago, the forces of erosion have had time to greatly change the original formations. Erosion has reduced much of the Canadian portion of the Appalachians to a peneplain; this is very similar to what happened on the shield. Today the region shows the results of the final sculpting by the ice sheet. As in the lowlands, glacial moraines are common in the landscape; as in the shield, some areas are deeply scoured and there are numerous glacial valleys.

The terrain is generally rugged, though ice and water deposits have levelled out some areas and favoured them with deep soil. The population is concentrated in these favoured areas; strings of settlement often follow the river valleys and seacoasts. The larger cities are located on deep harbours, many carved out by glacial action.

Because of the great variety of geological activity, and the great variety of rocks found here, the mineral resources are also varied. Base metals are mined from the igneous rocks, asbestos from the metamorphic rocks, and coal and salt from the sedimentary ones.

1-52   The Cabot Trail winds along the coast of Cape Breton Island.
The highlands of Cape Breton Island are part of the Appalachian Mountain system.

The Northern Mountains and Lowlands

This is the least-known, the least-settled and, economically, the least-used of all of Canada's major landform regions.

The mountains, formed by several periods of mountain-building, are mostly the result of folding and faulting. They are found on the northernmost islands of the Arctic archipelago. Some are quite high. The highest, Barbeau Peak, reaches 2616 m.

The lowlands are in two quite separate areas. One is the Hudson Bay Lowlands, on the mainland and quite far south compared to the main body of the region. While larger than the Great Lakes/St. Lawrence Lowlands, it has a very low population. The other lowlands are mostly on the more southern islands of the Arctic archipelago, and support even fewer people.

1. What and where is Logan's Line?
2. Why is it not very accurate to say that part of Canada is in the Appalachian Mountains?
3. (a) Name two relief features formed by faulting in Atlantic Canada.
   (b) In which provinces are they located?
4. (a) Briefly explain the settlement pattern in Atlantic Canada.
   (b) For what reasons have the larger towns and cities developed in Appalachian Canada? (An atlas might provide some ideas.)
5. Give two major reasons why the northern mountains and lowlands are not more fully utilized.

New Words to Know
Logan's Line
Continental shelf

1-53   Fjord in Gros Morne National Park, Newfoundland. *The steep valley sides and rounded bottom are typical of valleys carved by glaciers.*

1-54   The Alberta Badlands, in the Red Deer River Valley, *were created over 18 000 years ago when the retreat of Ice Age glaciers exposed rock formations from the Cretaceous Period (76 000 to 64 000 million years ago) or the time of the dinosaurs. The wealth of dinosaur fossils in the area led to the designation of Dinosaur Provincial Park as a World Heritage site in 1980.*

Weather and Climate

"Canada has a climate ten months' winter and two of poor sledding." (Nineteenth-century saying in Eastern Canada.)

People have always been concerned about the weather and climate. Canadians sometimes feel that not only do they have too much geography, they have too much weather and climate. What they are really saying is that climatic conditions interfere with their daily lives, and make them more difficult. The scores of TV programs and radio announcements, and the many weather maps and descriptions in the newspapers, show that everyone is interested in the weather. But concern about the weather and climate is not new. As far back as 100 B.C., there was a weather vane on top of a temple roof at the Acropolis in Athens, Greece!

Weather and **climate** do affect peoples' lives in countless ways. What people wear may depend on whether it is raining or snowing, hot or cold. Where people go and what they do when they get there will often depend on whether it is too hot, too cold or just right. Which team will win the football game may again depend on the weather — some teams are good "mudders" and some are not. It is often said that Canadians don't do anything about the weather — they just complain about it. This is far from the truth, of course: building a domed stadium or putting on a raincoat is doing something. Still, it is not yet possible to *control* the weather; Canadians construct buildings to at least keep bad weather at bay.

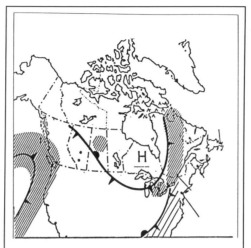

This is a simplification of the maps actually used by meteorologists, and is similar to maps published daily in some newspapers. It shows polar air pushing southward, even in May, causing rain in a large area in eastern Canada. Notice how the lows are associated with precipitation, and the highs with dry weather.

1-55a   Simplified weather map of North America.

On a national level, weather and climate make living in Canada more expensive. The cost of heating a home can easily exceed $1500 a year even in Canada's most southerly areas. The transportation systems must constantly be on the alert for the dangers of our weather, and the expenses involved in dealing with it. The latter can include the purchase of more snowploughs, or the extra costs when materials arrive late

1-55b   Satellite photo of a cold front advancing over eastern Canada.

or weather-damaged. A few companies suffer if it is cold, others if it is hot. The natural-gas companies make little or no profit during a mild winter; a hot summer can see pool installers and chemical companies thriving. But it is the farmers who are perhaps most concerned about the weather, in that it affects their daily work and their livelihood so directly. A farmer's crops can be destroyed by one hailstorm, or made more profitable if the rain comes at a suitable time.

In recent years, a very serious concern has arisen about the weather and climate. It is no longer possible to speak of the "purifying" winds and "cleansing" rains; like so much of our environment, they too have become polluted. One of the key problems is **acid rain**, which results from increased industrial and automobile emissions. Another is increased levels of **radiation**. The Chernobyl nuclear disaster in 1986 made people realize that pollution is an international problem. These concerns will be dealt with more fully in Chapter Four.

Very young children soon learn to say with a frown, "It's cold out," or with a smile, "It's snowing." But they do not always understand. What is *cold*, or what is *snowing*. The satellite photo 1-55 makes it clear that "it" is the **atmosphere**. The envelope of air that surrounds the earth is the "it" that is cold, or drops rain or snow upon us. Weather and climate are formed and occur only in the lower part of the atmosphere, the part affected by the earth's surface.

This section will explore the general topic of weather and climate, and then review Canada's major climate regions.

What Is Weather and Climate?

Weather is generally defined as the condition of the atmosphere considered over a short period of time — usually one day. Though people often say that a day has "good" or "bad" weather, many changes can take place in the weather conditions hour by hour, sometimes even minute by minute, as the day proceeds.

1-56   Selected climate statistics.

		J	F	M	A	M	J	J	A	S	O	N	D	Year
VANCOUVER, B.C.	TEMP.	3	4	6	9	13	16	18	18	14	10	6	4	10°C
49°N 123°W	PRECIP.	140	120	96	58	49	47	26	35	54	117	138	164	1044 mm
											Annual Temperature Range			15°C
WINNIPEG, MANITOBA	TEMP.	−18	−16	−8	3	11	17	20	19	13	6	−5	−13	3°C
50°N 97°W	PRECIP.	26	21	27	30	50	81	69	70	55	37	29	22	517 mm
											Annual Temperature Range			38°C
ST. JOHN'S, NFLD.	TEMP.	−4	−5	−3	1	6	10	15	15	12	7	3	−2	5°C
48°N 53°W	PRECIP.	153	163	135	121	99	94	89	101	120	138	163	174	1550 mm
											Annual Temperature Range			20°C
COPPERMINE, NWT	TEMP.	−29	−30	−26	−17	−6	3	9	8	3	−7	−20	−26	−11°C
68°N 115°W	PRECIP.	12	8	13	10	12	20	34	44	28	26	15	11	234 mm
											Annual Temperature Range			39°C
MANAUS, BRAZIL	TEMP.	28	28	28	27	28	28	28	28	29	29	29	28	28°C
3°S 76°W	PRECIP.	249	231	262	221	170	84	58	38	46	107	142	203	1811 mm
											Annual Temperature Range			2°C

Climate is defined as the condition of the atmosphere considered over a long period of time. Climate statistics generally include annual figures averaged over many years. Climate is the sum of all those daily weather conditions experienced throughout the four seasons. Climate, like the weather, can change. It changes, however, much more slowly and far less dramatically. Climatic changes generally take place over many years — even many decades — rather than hour by hour, as with the weather.

There are two major conditions of the atmosphere that interest **meteorologists**, who study the weather, and **climatologists**, who study climate. These two are its water content and its temperature. These scientists are concerned with basically the same things as the average citizen, though they want to know the full details. By studying present conditions, they seek ways to predict or forecast future conditions.

When scientists study weather and climate conditions, they often begin by studying statistical charts similar to the ones shown in 1-56. In these charts, the temperature and **precipitation** for each month and for the entire year are listed. The monthly temperature given is a monthly **mean**, or average. The monthly precipitation figure is the total amount of all the different forms of water that fell from the atmosphere that month, whether it came as rain, **snow, sleet, hail** or mixtures of these. Here is a brief explanation of the various statistics:
Mean Monthly Temperature (M.M.T.) This is determined by averaging the **mean daily temperatures** that occurred during the month in question. The daily mean is the average of the highest and lowest temperatures recorded in a day.

Mean Annual Temperature (M.A.T.)
This figure is the average of the 12 monthly means.

Annual Temperature Range (A.T.R.)
This is the difference between the highest and lowest monthly means in a year. This usually, but not always, is the difference between the January and July M.M.T.s.

Total Monthly Precipitation (T.M.P.)
This is the total precipitation that falls in the month.

Total Annual Precipitation (T.A.P.)
This is the total precipitation that falls in the entire year.

1. **Determine the following statistics given in chart 1-56:**
   (a) **The monthly temperature mean in July for Winnipeg.**
   (b) **The total annual precipitation for Vancouver.**
   (c) **The January and July monthly means for St. John's.**
   (d) **The lowest monthly mean for Coppermine.**
   (e) **The lowest monthly precipitation for Coppermine.**
2. **Complete the chart below using the statistics for the five weather stations in chart 1-56.**

3. **Which station has the largest temperature range? Why?**
4. **What station has the highest annual mean temperature? Why is this so?**

	Highest	Lowest
Latitude		
Mean Monthly Temperature		
Mean Annual Temperature		
Total Annual Precipitation		
Annual Temperature Range		

New Words to Know

Weather	Climatologist
Climate	Precipitation
Acid rain	Mean
Radiation	Snow
Atmosphere	Sleet
Meteorologist	Hail

Mean Monthly Temperature
Mean Daily Temperature
Mean Annual Temperature
Annual Temperature Range
Total Monthly Precipitation
Total Annual Precipitation

1-57 Climate graphs.

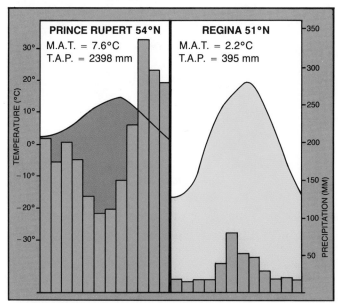

*Also called **climographs**, these graphs show the monthly conditions of a weather station for a year. Temperature is shown with a line graph, precipitation with a bar graph. Snowfall is included in total water precipitation. Ten mm of snow equal approximately 1 mm of rain.*

Using the climate statistics, work out what climate regions these stations are in. Choose a station from each of the other climate regions and prepare climate graphs for them.

Three Ways of Causing Precipitation

Precipitation occurs when moist air is forced to rise. When it rises, the air pressure decreases, and the air cools as a result. If the air is moist enough, or cooled enough, the water vapour **condenses** into water droplets and forms clouds. If the droplets are further cooled, or are knocked together by movement, they may become heavy enough to fall as rain. When these processes take place below the freezing point, then ice crystals form and snow may fall. Hail forms when water

1-58 Relief precipitation.

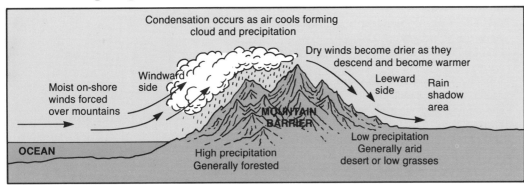

This is very common along mountainous coasts, such as that of British Columbia.

1-59 Frontal or cyclonic precipitation.

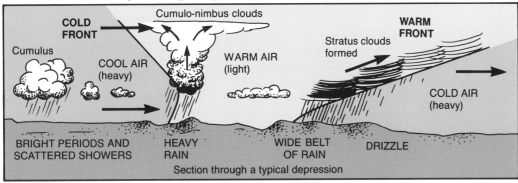

This is caused by either a cold or warm front, often associated with a low-pressure area. As the low passes through, the cold front "noses" the lighter warm air upwards, often violently enough to cause thunderstorms. The warm air rises over the heavier cool air causing a longer but gentler precipitation period.

1-60 Convectional precipitation.

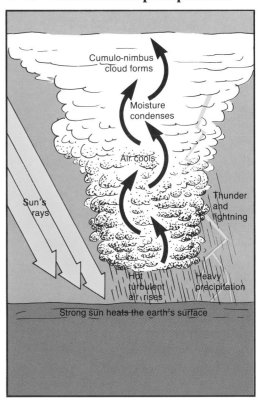

This is common in the summer on the prairies, and in southern Ontario and Quebec. Why would this be the most common type of precipitation in the equatorial regions?

droplets are forced upward and frozen, forming small balls of ice. Sleet is generally a mixture of ice and water formed in the atmosphere.

All precipitation forms in the same basic way. But there are three different reasons why the air may be forced to rise. This means there are three *types* of precipitation, which differ according to what first caused the air to rise.

Relief Precipitation (diagram 1-58) This type is caused by winds carrying air up over a mountain barrier. It is the high mountains of the Western Cordillera, for example, that cause the large amount of precipitation along Canada's west coast.

Convectional Precipitation (diagram 1-60) When the sun is powerful enough to heat the earth's surface to a high temperature, convectional precipitation may result. The earth's hot surface heats the air above it, forcing it to rise and causing precipitation. This is the com-

mon cause of the very large amount of precipitation in the tropics. It can also cause summer rains in many parts of Canada.

Frontal Precipitation (diagram 1-59) This type of precipitation is caused by moving masses of warm and cold air. The leading edge of the moving air mass is called a **front**. Whether **cold** or **warm**, such fronts may cause air to rise, and thus cause precipitation. The diagram shows how each front has a different profile or cross-section. Frontal precipitation occurs from time to time all year long, in most parts of Canada (weather map 1-70).

Clouds and Humidity

Precipitation is not the only form of water considered part of the weather and climate. Two other forms are often mentioned in weather reports, namely **clouds** and **humidity**. Clouds, as has already been noted, are made of water droplets that aren't heavy enough to fall. There are two basic cloud types: **stratus**, or layered clouds, and **cumulus**, or puffy clouds. (The latter resemble cotton batting.) Probably the most famous of all clouds are the **cumulonimbus**, or "thunderheads," which are

formed when air rises rapidly and violently. Such clouds often form when there are strong convection currents forcing the air up, or when a cold front is rapidly advancing.

Humidity is the amount of water vapour in the air, sometimes described as the "dampness" of the air. It is most commonly expressed as **relative humidity**. This is the amount of water vapour in the air relative to the amount it is able to hold at that temperature.

To determine this figure expressed as a percentage, the following calculation is made:

$$\frac{\text{The amount of water vapour in the air}}{\text{The amount of water vapour the air could hold (capacity)}} \times 100$$

The higher the relative humidity, the damper the air. Canada has some very damp climates — the Atlantic provinces are a good example — and some dry ones too, like the Arctic in winter.

1. **Explain five ways in which the weather has affected your life in the past month, for better or for worse.**
2. **Name three ways in which precipitation can form, and explain what is similar about them. (You can consult diagrams 1-58, 1-59, 1-60.)**
3. **With which precipitation types would you expect to find the following cloud form?**
   (a) Cumulonimbus
   (b) Stratus
4. **If there were 40 grams of water in the air in the form of water vapour, and that air at that temperature could hold 50 grams, what would the relative humidity be?**
5. **If the relative humidity of a mass of air was at 100 percent:**
   (a) what would begin to form if the temperature dropped by 10°C?
   (b) what is this process in (a) called?

New Words to Know
Condense
Relief precipitation
Convectional precipitation
Frontal precipitation
Front
Cold front
Warm front
Clouds
Humidity
Stratus clouds
Cumulus clouds
Cumulonimbus clouds
Relative humidity

1-61 Annual average precipitation in Canada.

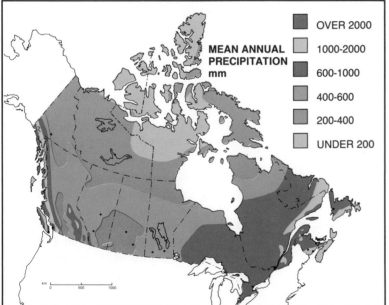

MEAN ANNUAL PRECIPITATION mm

OVER 2000
1000-2000
600-1000
400-600
200-400
UNDER 200

Canada is wet on the west and east coasts, and particularly dry on the north coast and in the southern interior. The Arctic region is Canada's driest, the Pacific coast the wettest. Can you explain why?

Lines which join points receiving the same amount of precipitation are called **isohyets**. *What would be the precipitation along the line between the two green areas on the map?*

Factors Affecting Weather and Climate

There are many factors which affect weather and climate. These include latitude, elevation (or altitude), winds and air masses, nearness to large bodies of water, the presence of physical features, and ocean currents.

Latitude: An Influence That Is Everywhere The relationship between latitude and the weather and climate is very straightforward: as latitude increases, whether north or south of the equator, temperature decreases. For centuries, geographers have broken the world into three temperature belts because of this fact. These are illustrated in map 1-62. Two of the zones are extreme — the **Tropical** and the **Polar**. Between them the conditions mix to create the **Temperate Zone**.

There are three causes for these zones. First, the earth is like a ball, not like a flat piece of plywood. Second, the **earth's axis** is tilted at an angle of 23.5 degrees to the direct rays of the sun. Third, the earth travels around the sun in an **orbit**.

Because the earth is round, or *spherical*, the sun's rays are more **direct** near the equator than they are near the poles. The result is that less heat is received in the polar areas and much more in the equatorial areas. Diagram 1-63 shows two sun rays of the same size striking the earth. Clearly, the ray in the polar area must spread its heat over a larger area. The beam of a flashlight or heat lamp can be used to show this effect (illustration 1-64). Furthermore, the diagram of the earth shows that the **indirect ray** must pass through more of the atmosphere. This means it loses more heat than the direct ray before reaching the earth's surface.

The second and third factors work together, and so must be considered

1-63 Effect of the sun's direct and indirect rays.

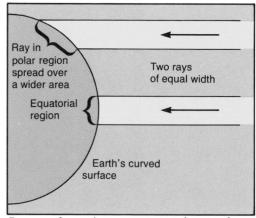

Because the sun's rays are spread over a larger area in the polar region, each square kilometre there receives less heat and light than a similar-sized area near the Equator.

together. As the earth revolves in its orbit around the sun, the axis tilt of 23.5 degrees to the **plane of the orbit** remains the same. As a result, the most direct rays of the sun, which together are called the **heat equator**, shift between the two tropic lines.

On every June 21 they are on the **Tropic of Cancer**, and on every December 22, on the **Tropic of Capricorn**. The equator, then, really is under the heat equator only twice a year. These are the **equinox** dates, March 21 and September 23. The tilt of the axis is also the cause of the phenomenon called **Land of the Midnight Sun**. In summer, in the northern polar areas, the sun does not set, but merely dips toward the horizon line. The same thing happens at the south pole six months later. Another effect of this shifting heat equator is to give the world its seasons.

Chart 1-56 shows many of the differ-

1-62 Major temperature zones of the earth.

In this view the earth is tipped slightly south to enable the North Pole to be seen. How are the border lines between the zones determined?

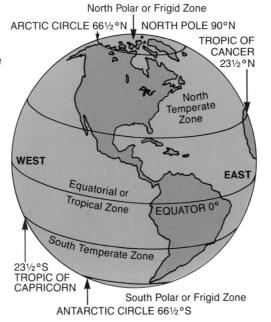

North Polar or Frigid Zone
ARCTIC CIRCLE 66½°N — NORTH POLE 90°N
TROPIC OF CANCER 23½°N
North Temperate Zone
WEST
EAST
Equatorial or Tropical Zone
EQUATOR 0°
South Temperate Zone
23½°S TROPIC OF CAPRICORN
South Polar or Frigid Zone
ANTARCTIC CIRCLE 66½°S

1-64 Direct and indirect light from a flashlight.

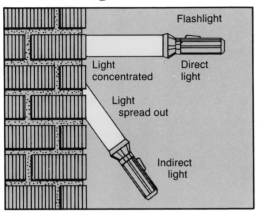

Flashlight

Light concentrated — Direct light

Light spread out

Indirect light

ences between Coppermine in Canada's far north, and Manaus, in equatorial Brazil. The other three stations on the chart are temperate, or "in between." It's very clear that Winnipeg, Vancouver and St. John's — the other three cities on the chart — while very close in latitude, vary greatly in their January temperatures. Obviously, all places with similar latitudes do not have the same mean temperatures. The reasons for this are explained by the next five factors, each of which also influences temperature.

Elevation: the Higher, the Colder

Temperatures drop when the elevation rises because air expands as it rises from the earth's surface. This decrease in air pressure results in cooling. The rate of cooling is about 1°C for every 100 m in height. (As described earlier, this temperature drop also causes precipitation.)

1-65 Canada's climate regions — a summary

Temperate West Coast
Similar Climate:
British Isles, New Zealand
Major Factors Affecting:
Along Pacific coast onshore Westerlies increase maritime effect
High mountain barrier parallels coast
Warm Alaska current offshore
Conditions
Very moderate temperatures; winters above freezing
Warm summers, cool winters
Precipitation received all year - high in winter
Examples
Vancouver, Prince Rupert, Victoria

Semiarid Prairie
Similar Climate:
Ukraine, eastern interior of Australia
Major Factors Affecting:
Great distance from maritime influences
Cordillera a barrier to maritime air from Pacific
Shield and plains open area to influence of dry, cold Arctic air mass
Conditions
Cold, relatively long winters
Warm, relatively short summers
Semiarid, generally less than 500mm annually
Some precipitation all year, but particularly summer
Examples
Winnipeg, Regina, Calgary

Subarctic
Similar Climate:
Siberia, northern Sweden
Major Factors Affecting:
Winter influenced by Polar North Easterlies
Southern portions influenced by South Westerlies in summer
Shield, plains and lowlands enable polar winds to reach far south
Influenced greatly by the cold, dry Arctic air mass
Little maritime influence — frozen Arctic Ocean
Conditions
Winters lasting six months or more
Cool, short summers
Precipitation low to moderate, mainly in summer
Examples
Schefferville, Kapuskasing, Yellowknife

Mountain
Similar Climate:
Andes, European Alps
Major Factors Affecting:
Complex barriers and throughways
Westerlies from the Pacific
Great relief in many areas
Great elevation in many areas
Conditions
Great variation throughout area
Major changes in conditions over short distances
Extreme conditions - areas very cold, dry, or windy
Western slopes wetter than eastern ones
Examples
Kamloops, Whitehorse

Arctic
Similar Climate:
Siberia, Antarctica
Major Factors Affecting:
High latitude
Polar High Pressure Zone
Polar North Easterlies
Little maritime influence
Conditions
Extremely cold winters, with -20 to -30°C common
Long 8-10-month winters; cold, very short summers
Low precipitation - a "cold" dessert
Examples
Coppermine, Resolute, Inuvik

Temperate East Coast
Similar Climate:
Northeast China, northeast U.S.A.
Major Factors Affecting:
Polar North Easterlies dominate in winter, South Westerlies prevail in summer
Cold, dry continental air mass dominates in winter; warm, moist Gulf air prevails in summer
Open to influence of winds and air masses, from all directions
Western portion closer to continental influences, eastern to maritime influences
Conditions
Warm summers, particularly in the more inland areas
Cold winters, particularly in the more inland areas
Precipitation moderate and year-round, but higher in eastern area
Examples
St. John's, Halifax, Montreal, Toronto

1-66 Birth of an Avalanche.

Slab avalanches are the most common type of slide. These pictures show the chronology of many slides which occur on steep slopes in the backcountry, away from avalanche control programs.

Early December
The rain soaked surface of snow feezes into a hard crust of ice

Mid-December
Hoar frost forms on top of the crust, creating a loosely packed, unstable layer.

Early January
New snowfalls occur, but the layer of hoar frost and ice crust prevent a strong bond with the underlying snow

Mid-January
Rain soaks the snow, increasing its weight. A slab of heavy snow breaks loose, sliding on the slick layer of hoar frost and ice, and the avalanche begins

It should be noted that in chart 1-56, the only weather station of the five that is over 100 m in elevation is Winnipeg, which is 250 m above mean sea level. By how much would its average temperatures rise if the city were at sea level?

1. **What would the world's temperatures be like if the earth was like a piece of plywood, and presented a flat surface to the sun?**
2. **Explain in your own words why direct rays heat the earth's surface more than indirect rays.**
3. **(a) Assume that Manaus is on the equator, and then make note of its latitude and mean annual temperature. Make note of the same two figures for Winnipeg. What is the special statistical relationship between these figures for these two cities? (Use chart 1-56.)**
   (b) How well does this relationship apply for the other weather stations given?
   (c) Should it apply for the other three stations in relation to Manaus?
4. **(a) Mount Logan is Canada's highest peak, at 5951 m. If the temperature at sea level near the mountain was 14°C on a summer day, what would the temperature be at the summit?**
   (b) On that same day, at what elevation on Mount Logan would the temperature be at the freezing point?

New Words to Know

Tropical Zone	Plane of the orbit
Polar Zone	Heat equator
Temperate Zone	Tropic of Cancer
Earth's axis	Tropic of Capricorn
Orbit	Equinox
Sun's direct rays	Arctic Circle
Sun's indirect rays	Antarctic Circle
Land of the Midnight Sun	
Elevation	

Winds and Air Masses: Moving the Air and Weather Around **Wind** is air moving horizontally across the earth's surface. Winds are the great "movers of the atmosphere." They can move cold air to moderate a hot area, or moist air to bring rain to a dry one. Winds follow one basic principle: they move dense, heavy, falling air from **high pressure areas** toward areas of **low pressure**, where the air is less dense, lighter and rising. Since cold air is heavier and

1-67 The wind circulation as it should be.

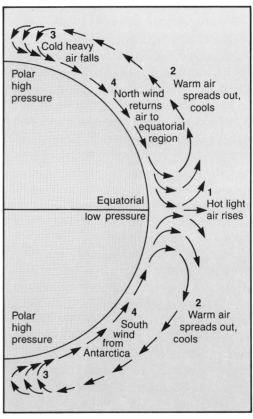

Since the earth has cold polar regions, and hot equatorial ones, this is the way one might expect the wind and pressure zones to develop.

denser than warm air, the air in the polar areas has a high pressure. In the hot equatorial areas, the opposite is true. Therefore, the basic air movement should be across the earth from the high pressure to the low, from the poles to the equator. Diagram 1-67 shows this simple and expected wind circulation pattern. The next figure, however, shows that the expected pattern is not the actual one.

Diagram 1-68 shows the prevailing wind and pressure belts of the world. The movement is always toward the low-pressure areas, and away from the high-pressure ones. Canada is located in the zones of the Polar North Easterlies, the Subpolar Low and the South Westerlies.

Two other factors besides air pressure must be noted, for they also affect the direction of winds throughout the world. One is the movement of the heat equator between the tropic lines. As this heat equator shifts, the tropical low-pressure area naturally shifts with it; after all, the former causes the latter. This shifting causes all the wind belts to shift northward in the northern summer, and southward in the southern summer. During the Canadian winter, the Polar Easterlies and the Subpolar Low are drawn southward to cover much more of the country with cold air. In summer the reverse happens: the colder air shifts north, and the warmer Westerlies affect much more of the country.

In addition, all winds in the northern hemisphere swing to the right, and all those in the southern hemisphere to the left. This is known as the **Coriolis force**, and is caused by the rotation of the earth. As the earth rotates from east to west, north-flowing winds — the Wester-

1-68 Generalized pattern of prevailing winds.

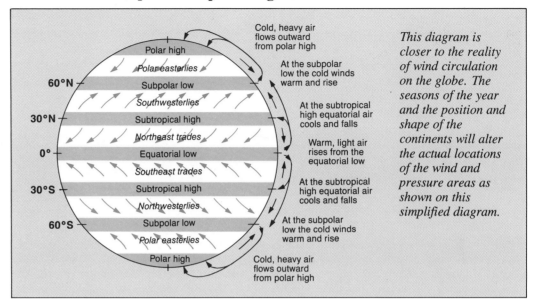

This diagram is closer to the reality of wind circulation on the globe. The seasons of the year and the position and shape of the continents will alter the actual locations of the wind and pressure areas as shown on this simplified diagram.

lies for example — tend to drag behind the rotation. The result is that all winds tend to veer to the right. The so-called Westerlies are more properly called, then, the South Westerlies (though the term Westerlies is much more common).

Diagrams 1-67 and 1-68 show only a very generalized pattern of wind circulation. Other, more local winds can move in the opposite direction to the prevailing ones from time to time. A large lake can create a pressure system on a small scale, and cause its own local winds. (This will be explained later.) On a continental and seasonal scale, land and sea breezes become the massive, powerful monsoons of Asia.

Diagram 1-69 shows that a highrise apartment or office tower can have an important effect locally on the wind direction. Anyone who has ever walked along a street downtown in one of Canada's larger cities has noticed this.

Just turning a corner can make all the difference in the wind's direction, and in its power. (Large cities can have a considerable effect on the weather and climate as a whole, affecting temperatures and precipitation as well as the winds.)

1-69 Tall buildings affect winds on a small scale.

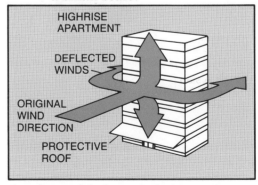

A roof around the base of a building will help to deflect the winds away from pedestrians at ground level.

1-70 *Violent July storms dumped record rainfalls (between 155–280 millimetres) on Quebec's Saguenay-Lac-Saint-Jean region over a 5 day period in 1997. The result: floods which destroyed or damaged 1718 houses, 900 cottages, and forced the evacuation of 16 000 people. Value of the damage was $800 000 000.*

1-71 *In April 1997, Manitoba's Red River flooded over 101 000 hectares of land from Emerson (pictured above), north to the outskirts of Winnipeg. Damage was estimated in excess of $150 000 000 as 28 000 people were evacuated. Winnipeg, which was protected by the Red River Floodway (the 47 kilometre long diversion channel around the east end of the city) escaped unharmed.*

Air masses are huge volumes of air which have certain characteristics. These characteristics are caused by air staying in one place long enough to be affected by the surface it is sitting over. Thus, an air mass, if over an ocean in the tropics, would become warm and moist; if over a polar ice cap, it would become cold and dry. These air masses are forced into movement by the same factors which cause the winds. They tend to move with the winds to areas of low pressure, and away from areas of high pressure.

Map 1-72 shows the pattern of the major air masses of North America. The warm, moist tropical air mass that develops over the Gulf of Mexico is commonly referred to as **Gulf air**. It tends to move north and east over the U.S.A. and then over central Canada. When Gulf air moves northward, its leading edge is a warm front. When cold, dry polar air moves southward, its leading edge is a cold front. The cross-section 1-59 shows how these fronts cause precipitation. They also change temperatures, as might be expected.

Canada is a battleground for the polar and tropical air masses and fronts. Map 1-55 shows that when these fronts move and clash, **depressions**, or low-pressure cells, often develop around them. As these **lows** move toward an area, the forecasts generally warn of rain or snow. The water in that precipitation probably has come from the Gulf of Mexico. As **highs** or high pressure areas approach, the air becomes heavier and falls. That makes them popular, for falling air becomes warmer as it falls, causing dry, sunny conditions. It is rising air that causes precipitation.

1-72 Major air masses and ocean currents affecting Canada.

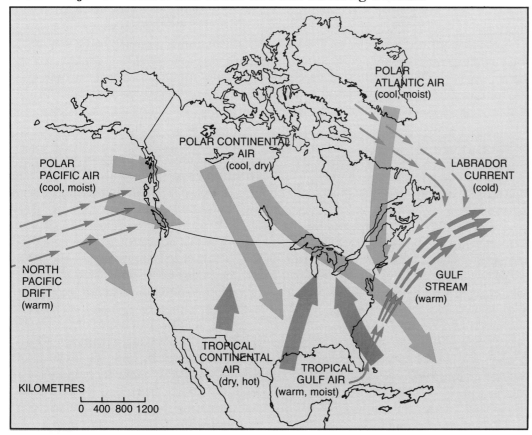

POLAR
ATLANTIC AIR
(cool, moist)

POLAR CONTINENTAL
AIR
(cool, dry)

POLAR
PACIFIC AIR
(cool, moist)

LABRADOR
CURRENT
(cold)

NORTH
PACIFIC
DRIFT
(warm)

GULF
STREAM
(warm)

TROPICAL
CONTINENTAL
AIR
(dry, hot)

TROPICAL
GULF AIR
(warm, moist)

KILOMETRES

0 400 800 1200

This single map can only show the general location and thrust of the air masses and ocean currents. Naturally they move northward in summer and southward in winter, following the shifts of the heat equator.

1. **Describe briefly and in your own words the basic cause of winds.**
2. **Why would rapidly rising air be called an "updraft," and not a wind?**
3. **Name and explain two other factors that can alter the basic direction of the world's winds.**
4. **(a) Why is Canada a meteorological battleground?**
   (b) What does the fighting?
5. **(a) Describe the air in a polar air mass.**
   (b) What two changes would take place if that air mass passed over your area?

New Words to Know
Wind
High pressure area
Low pressure area
Coriolis force
Air mass
Gulf air
Depression
Low
High

1-73 Unstable air masses sometimes cause tornadoes to form.

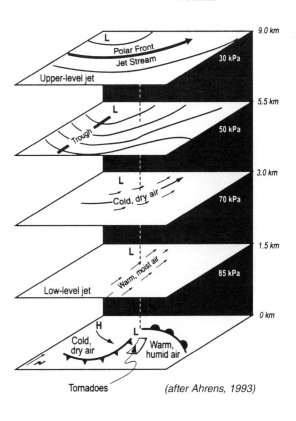

L

Polar Front
Jet Stream

Upper-level jet

9.0 km

30 kPa

L

Trough

5.5 km

50 kPa

L

Cold, dry air

3.0 km

70 kPa

L

1.5 km

Warm, moist air

Low-level jet

85 kPa

0 km

H

Cold,
dry air

L

Warm,
humid air

Tornadoes

(after Ahrens, 1993)

Canada has more tornadoes than any other country in the world, except for the U.S. Southwestern Ontario is Canada's most active tornado zone, but cyclones or twisters can occur anywhere in the country. The Fujita Tornado Intensity Scale rates a tornado's severity from Weak (F-0) with wind speeds of 64 – 116 km/hr to Violent (F-5) with wind speeds of 420 – 512 km/hr. One of Canada's worst tornadoes hit Edmonton in July 1987, killing 27 people, injuring 300, causing $300 million in damage, and leaving hundreds homeless.

55

1-74   Land and sea breezes.

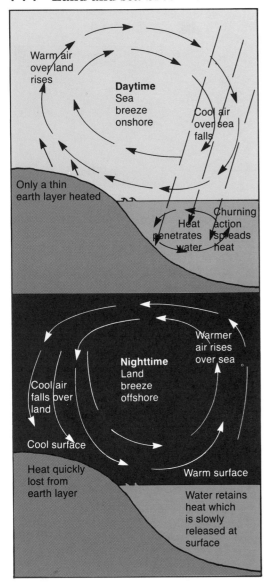

Warm air over land rises

Daytime
Sea breeze onshore

Cool air over sea falls

Only a thin earth layer heated

Heat penetrates water

Churning action spreads heat

Warmer air rises over sea

Nighttime
Land breeze offshore

Cool air falls over land

Cool surface

Heat quickly lost from earth layer

Warm surface

Water retains heat which is slowly released at surface

Near the sea and large bodies of water, land and sea breezes are common. These are actually convection currents in the air, which change direction during a day. The change is caused by the rapid heating and cooling of the land, compared to water bodies, which heat and cool much more slowly.

Nearness to Large Bodies of Water: The Climate Moderators   The oceans, the world's largest bodies of water, have a tremendous influence on the weather and climate. Naturally, they influence the air above them, and then that air may be carried inland. Large lakes such as the Great Lakes can also exert a considerable influence on the air around them.

The great water bodies tend to act in the opposite way to the great land masses. Water heats and cools much more slowly than the land. As a result, the oceans and large lakes tend to be cool when the land masses are warm, and vice versa. When an air mass moves inland from over an ocean, it warms the cold areas in winter, but cools the warm ones in summer. This **maritime** or **marine effect**, felt most strongly along coastal areas, is said to be **moderating**. It keeps the continents from becoming too hot or too cold. The opposite or **continental effect** is found deep in the continental interiors. Because land masses heat and cool quickly, the air over them heats up quickly to a high temperature, and cools quickly to a low one. As a result, continental climates have **extreme temperature** ranges.

As might be expected, areas of maritime influence almost always have more precipitation. Continental areas have a drier weather and climate; for precipitation, they depend mainly on at least some maritime air reaching them.

Canada has both continental and maritime areas. Along the west coast, the Westerlies bring the maritime air inland. Therefore the coast is very wet and moderated. In the prairies, far from the sea and protected by the cordillera, temperatures are much more extreme and precipitation is lower. In eastern Canada, conditions are mixed, as the continental and maritime influences mix. Those areas closer to the Atlantic experience a much stronger maritime effect. The statistics in chart 1-56 for Vancouver, Winnipeg and St. John's illustrate these varied conditions well.

Physical Features: Barriers or Throughways   Differences in relief can have a considerable influence on the weather and climate. The relief precipitation diagram 1-58 illustrates one of the more obvious effects of a mountain barrier. The Coastal Mountains of the Western Cordillera, along with the prevailing Westerlies, give the west coast its high precipitation. These mountains keep the interior plateaus of British Columbia dry, making irrigation necessary in many areas. The Columbia and Rocky Mountains again force moisture from the air on their west or **windward** sides, creating dry conditions on the east or **leeward** sides. The cordillera, then, prevents a strong Pacific maritime influence from reaching the plains.

What is sometimes forgotten is that the *absence* of relief features also has an important influence on atmospheric conditions. In winter, the generally flat and open plains enable the cold, dry air masses from the north to reach deep into North America. At times this cold air can reach south as far as Florida and damage the orange groves. A tongue of Arctic air reaching into southern Canada can make people there think they are in the Arctic. Only distance separates most Canadians from the chilling blasts from the polar areas. Of course, the lack of

1-75 *Northern Lights or Aurora Borealis occur when charged subatomic particles from solar flares enter the earth's upper atmosphere to interact with its oxygen and nitrogen. The resultant shimmering colours in the night sky are an unforgettable sight.*

relief also enables the hot, dry air from the interior to flow to Central and Atlantic Canada, bringing continental warmth to maritime regions.

Ocean Currents.

The effect of the oceans has already been examined. **Ocean currents** can increase or decrease that effect. If the currents are warm, as the Alaska Current or the Gulf Stream are (map 1-72), they will keep both winter and summer conditions on the land somewhat warmer than expected. The cold Labrador Current, of course, has the opposite effect.

The heavy and frequent fogs along coastal areas in Atlantic Canada occur when there is a mixing of air above the cold and warm water currents. The cooling of the warm, moist air above the Gulf Stream leads to condensation and foggy conditions.

In summary, Canada is affected by the same factors that influence weather and climate throughout the world. Weather and climate take place in the atmosphere, and that air envelope is in constant circulation.

New Words to Know

Maritime (marine) effect	Windward
Moderate temperatures	Leeward
Continental effect	Ocean current
Extreme temperatures	

1. Using the statistics in chart 1-56, prove that both Vancouver and St. John's are more maritime in climate than Winnipeg.
2. What if, instead of north and south, the cordillera ran east and west across the 19th parallel. How would climatic patterns change? Where would the most dramatic changes occur?
3. Research the term 'Iceberg Alley.' Where is it? How is it related to ocean currents?

1-76   Climate regions of Canada.

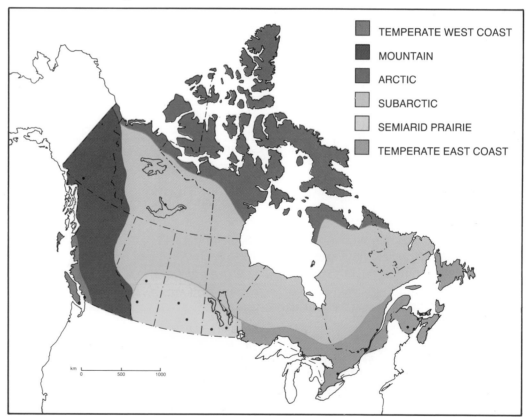

TEMPERATE WEST COAST

MOUNTAIN

ARCTIC

SUBARCTIC

SEMIARID PRAIRIE

TEMPERATE EAST COAST

Unlike some climate maps, this one has the zones extending over the lakes. Why is this a perfectly logical way to show climate? The colours are also chosen in an atttempt to be logical. Can you explain why the prairie is shown as yellow, the temperate zones as green (light and dark), the subarctic light blue, etcetera?

The Climate Regions of Canada

This section has investigated, one by one, the factors that affect weather and climate. The problem is that these factors don't operate separately — they operate as a team. When summer temperatures reach 20°C in Winnipeg, many factors are at work. The latitude and elevation might be lowering the temperature, but the distance from any oceans, and the hot winds sweeping eastward across the open plains, might be raising it. On another day winds and pressure systems might bring a warm, moist air mass to the Winnipeg area, causing frontal rain.

The average climate statistics of a place (chart 1-77) are the result of all the factors working together to produce a set of temperature and precipitation conditions. Since these factors create different conditions in literally millions of different locations, climatologists have developed the idea of climate regions. Throughout a given climate region, conditions are similar in temperature and precipitation, but not exactly the same. In each region, conditions vary from place to place; for example, the annual precipitation might range generally from 200 mm to 500 mm within a region, but it would be unusual for any place within that region to receive only 100 mm or over 600 mm.

By grouping the thousands of places with roughly similar conditions under one region and one name, climate is made much easier to understand.

A summary of Canada's climate regions is on page 51. For each of these regions, an explanation of the major factors affecting it (the

"Why?"), and a description of its conditions (the "What?"), is provided. The table will be more easily understood if reference is made to the statistics in charts 1-56 and 1-77, and to map 1-76.

New Words to Know

Climograph   Isohyet

1-77 Climate statistics for selected Canadian stations.

TEMPERATE WEST COAST		J	F	M	A	M	J	J	A	S	O	N	D	Annual
PRINCE RUPERT, B.C.	T	2	2	4	6	10	12	13	14	12	19	5	3	7.6°C
(54°N)	P	224	177	196	173	130	109	117	149	217	336	293	278	2399 mm
VICTORIA, B.C.	T	4	5	7	10	12	14	16	16	14	11	7	6	10.1°C
(48°N)	P	112	82	57	31	24	27	15	18	32	72	100	128	698 mm
MOUNTAIN		J	F	M	A	M	J	J	A	S	O	N	D	Annual
WHITEHORSE, YUKON	T	−28	−26	−18	−8	4	12	16	14	7	−1	−14	−24	−5.4°C
(61°N)	P	13	12	13	10	17	16	36	35	30	28	23	21	254 mm
KAMLOOPS, B.C.	T	−6	−3	3	10	14	18	21	19	15	8	2	−2	8.3°C
(51°N)	P	36	22	9	6	15	40	21	22	18	16	19	24	247 mm
ARCTIC		J	F	M	A	M	J	J	A	S	O	N	D	Annual
RESOLUTE, NWT	T	−33	−34	−31	−23	−10	0	4	3	−5	−15	−24	−28	−16.4°C
(75°N)	P	3	3	3	6	9	12	26	30	18	15	6	5	136 mm
INUVIK, NWT	T	−29	−29	−24	−15	−1	10	13	10	3	−7	−21	−27	−9.7°C
(68°N)	P	20	10	17	14	18	13	34	46	21	34	15	19	261 mm
SUBARCTIC		J	F	M	A	M	J	J	A	S	O	N	D	Annual
SCHEFFERVILLE, QUEBEC	T	−23	−21	−14	−6	2	9	13	11	6	−1	−9	−18	−4.4°C
(55°N)	P	47	41	41	41	40	90	92	95	80	74	65	40	746 mm
KAPUSKASING, ONTARIO	T	−18	−15	−9	0	8	14	17	16	11	5	−5	−14	1°C
(49°N)	P	55	45	54	53	80	95	85	87	90	72	83	60	859 mm
YELLOWKNIFE, NWT	T	−28	−26	−18	−8	4	12	16	14	7	−1	−14	−24	−5.4°C
(62°N)	P	13	12	13	10	17	16	36	35	30	28	23	21	254 mm
SEMIARID PRAIRIE		J	F	M	A	M	J	J	A	S	O	N	D	Annual
REGINA, SASKATCHEWAN	T	−17	−15	−8	3	11	15	19	18	12	5	−5	−12	2.2°C
(50°N)	P	19	17	21	21	40	83	55	49	34	18	20	17	394 mm
CALGARY, ALBERTA	T	−10	−9	−4	4	10	13	17	15	11	5	−2	−7	3.6°C
(51°N)	P	17	20	26	35	52	88	58	59	35	23	16	16	445 mm
TEMPERATE EAST COAST		J	F	M	A	M	J	J	A	S	O	N	D	Annual
HALIFAX, N.S.	T	−3	−4	0	5	10	14	19	19	16	10	5	−1	7.5°C
(45°N)	P	141	119	113	112	109	94	94	96	117	120	143	126	1384 mm
TORONTO, ONTARIO	T	−4	−4	0	7	13	19	22	21	17	11	4	−2	8.7°C
(44°N)	P	67	59	67	66	70	63	74	61	65	60	63	61	776 mm
MONTREAL, QUEBEC	T	−10	−9	−3	6	13	19	21	20	16	9	2	−7	6.5°C
(46°N)	P	83	81	78	72	72	85	89	77	82	78	85	89	971 mm

1. (a) In your own words, and using the statistics in charts 1-56 and 1-77, describe the climate conditions in the Temperate East Coast region.
   (b) Explain how and why the conditions in the Atlantic portions of the region differ from those in southern Ontario.
2. Explain why neither winds nor air masses bring moderation to the Arctic climate region in winter.
3. Study the climate-station statistics in chart 1-77 to determine:
   (a) in order, the top five stations for precipitation.
   (b) the two with the lowest precipitation.
   (c) Using the same chart, rank the three stations given as examples of the Subarctic region according to annual temperature range, from highest to lowest.
   (d) Give the names and latitudes of the two southernmost stations.
4. (a) Briefly explain why the five cities in the answer to 3(a) have so much precipitation.
   (b) Why do the cities in the answer to 3(b) have so little precipitation?
5. You are packing for a three-day trip to any four of the following places in Canada. State the season of your visit and describe what you would wear and the clothing you would pack in your small suitcase.
   • Victoria, British Columbia.
   • Moose Jaw, Saskatchewan.
   • Whitehorse, Yukon Territory.
   • Banff, Alberta.
   • Corner Brook, Newfoundland.
   • London, Ontario.
   • Pelly Bay, N.W.T.

The Dark Zone

During the week of January 4-10, 1998, up to 85 mm of freezing rain fell over much of south-eastern Canada and the north-eastern United States, causing huge accumulations of ice.

Most of the ice was concentrated in a band up to 200 km wide between Kingston, Ontario and Montreal, Quebec, but there was also substantial accumulation from Georgian Bay to the Atlantic coast. The result caused havoc to the power grid in these regions.

Electricity travels along high tension wires which form a spider-web or power grid designed so that, if one line is broken, others can be used to by-pass the break. This system failed in January 1998 because the devastation was so widespread, that there were not sufficient lines in operation to cover for those damaged. Some 3 million people were without light, heat, phones and water. Water pumping stations don't work when the power is out. Telephones are without power. People depend on refrigeration to keep their food

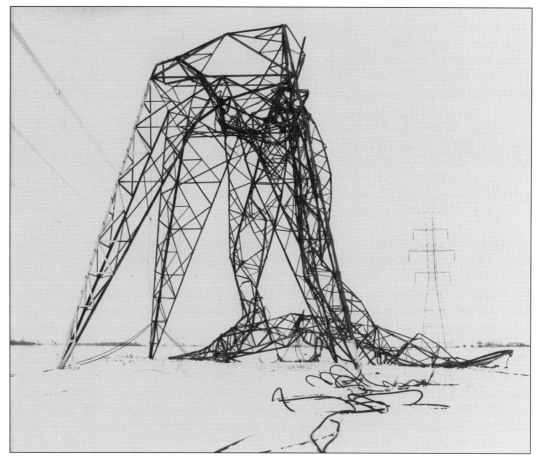

1-80   *Boucherville, Quebec, after the ice storm in 1998.*

fresh and without power millions of pounds of meats and produce were spoilt. Keeping even a few rooms warm was impossible for those homes with forced air furnaces or electric heating. "It was just like living in the 1800s!" said one survivor.

Ice storms are caused when a warm air mass rides up over a colder air mass. Rain falls through the warm and cold air down on to supercooled surfaces. This rain or frontal precipitation immediately freezes into layers of heavy ice. Ice storms are common in Canada but the storm of 1998 was unusually severe. Some ice deposits were over 80 mm thick.

Meteorologists are blaming the storm on El Nino — a climatic event caused by a reversal of pressure systems (high and low air pressure) over the Pacific Ocean which alters basic weather patterns around the globe. The El Nino phenomenon was originally recognized (and named) by fishermen off the coast of South America. They called it "the little boy" or "Christ-child" because its effects tend to appear in that region around Christmas. El Nino takes place when normally cool water off the coasts of Ecuador and Peru becomes warmer. The resultant change in atmosphere moves the jet stream and can cause it to split as it travels across North America. January 1998 was close to the peak of an El Nino year. There have been 23 El Nino and 15 La Nina (the opposite effect — a cold water mass) events between 1911 and 1998. Unusually we have had 3 El Nino's since 1990. El Nino's effects on the winter of 1998 in North America included the warmest December ever in Edmonton, little snow and warm temperatures in Southern Ontario, tornadoes and heavy rainfall in Florida, and severe storms along the Pacific coast.

The 7-day ice storm in 1998 left almost a quarter of a million people in Ontario and some 1.5 million people in the province of Quebec without power. 1000 pylons and 24000 hydro poles in Quebec and 300 pylons and 11000 hydro poles in Ontario were toppled. Four of the five major power lines into Montreal were cut. Some people in southern Montreal were without power for over three weeks. Even when power was restored, sudden over-demand caused overload breaks and additional power losses. Close to 600 000 people were without power in the provinces of New Brunswick and Nova Scotia and the states of Maine, New Hampshire, New York and Vermont during this same period.

Only 25 people died, 4 in Ontario, 21 in Quebec. The victims were crushed by falling ice, electrocuted by downed power lines, slipped in front of vehicles or poisoned by carbon monoxide. Many, many other people were injured, and still others were psychologically scarred after having been evicted or driven out from their homes by cold and hunger.

Short-term economic losses in Canada total over $1.6 billion according to estimates by the Insurance and Conference Boards of Canada. 5500 diary farmers in Ontario and Quebec dumped some $7.8 million in milk and hundreds of cows died of mastitis because there were no electric milking machines. Damage to Christmastree, fruit-tree and maple syrup farms was as high as 30% of production. Many manufacturers also lost significant income because factories were closed and shipping cancelled. Damage in Southern Ontario and Quebec was estimated at over $500 million. Federal emergency funding was arranged by Ottawa to help those in need. The United States suffered as well. U.S. authorities estimate up to $200 million in property claims and repairs will result from the ice storm. A state of emergency was declared in four north-eastern states.

People survived the storm because everyone worked together on the emergency. Neighbours with fireplaces and generators provided shelter to those without. Stores extended credit to customers needing candles, kerosene lamps, fuel and food. The railways drove diesel locomotives into city streets to act as generators for schools, churches and other temporary shelters. People from all over Canada donated tools, foodstuffs and other essentials, while trucking firms provided free transportation. The largest mobilization of Canadian Forces in our history (15 875 troops) was pressed into service to cope with the storm's aftermath. In addition, thousands of volunteers — police, fire fighters, health-care professionals, electrical workers, arborists and people from every walk of life rushed into the afflicted area to help out. General Maurice Baril, during a visit to the area, described the devastation along Montreal's south shore as, "Sarajevo without the bullets".

Canadians are demanding that our hydro-electric suppliers closely examine and rethink their emergency safeguards and procedures to ensure that a power outage of this magnitude does not reoccur.

Canada's Natural Vegetation and Wildlife

Introduction

Map 1-81, which shows Canada's natural vegetation regions, could easily confuse a young student. For example, it might show that the student's school is in a broad-leafed deciduous forest region, even though all the student can see out the window is a field of corn or a shopping plaza. While the reality may not seem to match the map, the map is not wrong, for its title includes the word *natural*. This word implies that the vegetation is what would be there naturally, that is, if human beings had not interfered with nature's work. In nature, the conditions of climate — the amount of heat and precipitation received, and the length of the growing season — largely determine the **natural vegetation**. Of course, people have been altering nature's plans for centuries, by clearing away the natural vegetation to plant crops or build cities.

The natural vegetation regions in Canada follow quite closely the pattern of the climate regions on map 1-76. The very close relationship between precipitation and vegetation can be seen by studying maps 1-61 and 1-81 together. Climate conditions determine whether there is enough temperature and precipitation for this type of tree or that type of grass. Perhaps there isn't sufficient precipitation for any tree, or even for any grasses.

Map 1-81 shows only three types of vegetation, or **flora**. Though there are seven regions, they all consist of forest, grassland or **tundra**. The forest is by far the most important of these. Canada has

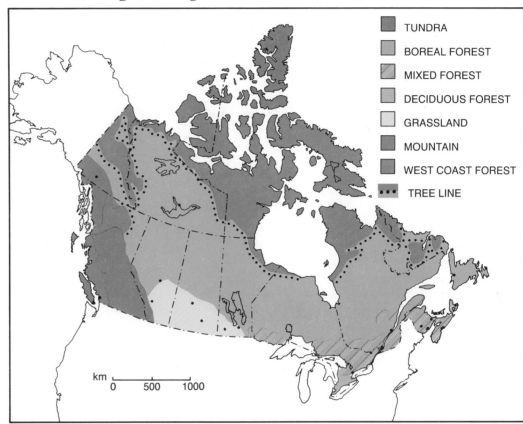

1-81 Natural vegetation regions of Canada.

TUNDRA
BOREAL FOREST
MIXED FOREST
DECIDUOUS FOREST
GRASSLAND
MOUNTAIN
WEST COAST FOREST
••• TREE LINE

What are the only two non-forested regions in the country?

140 species of trees, and its forests cover more land than almost any other country's. (Only Russia and Brazil have more.) The tundra and grassland regions are small compared with the forest lands.

On the map the vegetation types are separated from each other with thin lines. It should be understood, however, that these lines are in the middle of

zones or belts. These areas, known as **transition zones**, are what really separate the regions. The one between the **boreal forest** and the grassland regions, for example, would have a mix of both grassland and forest. The lines drawn on most maps — whether climate, vegetation, soil or urban maps — generally do not represent a sudden, complete change between the features they separate.

Instead they represent zones in which the characteristics of one region gradually give way to the characteristics of the neighbouring one. In a way, the mixed forest region can be considered a transition zone between the **coniferous** and **deciduous forest** regions. It includes a mix of both kinds of trees. Because it covers such a large area, however, it is shown as a separate region.

The wildlife, or **fauna**, found in a given region is directly related to the food and cover provided by the vegetation. For this reason, it is described briefly with the vegetation regions.

The Natural Vegetation Regions of Canada

Tundra   This region has very minimal vegetation cover. That is all that can grow in the extremely dry, cold Arctic climate. The growing season here is very short; in fact, it is measured best in weeks, not months. During that short summer, a thin layer of soil above the **permafrost**, which is the permanently frozen soil, thaws just long enough for a few mosses, **lichens**, low bushes, and Arctic grasses and flowers to grow. The flowers develop, flower and produce seeds in a matter of weeks. Then they lie dormant for the long, 8- to 10-month winter. The **tree line** (diagram 1-84) separates the tundra region from the boreal forest region. North of this line, only in a few very favoured areas — perhaps along a sheltered river valley — can small, stunted trees even exist.

The mosses and lichens of the tundra provide minimal food and cover. This restricts the amount and variety of wildlife that can survive there. Some of the animals, such as the lemming (a kind of rodent) and the arctic fox, are rather small in size. Much larger animals include the caribou and the muskox. The latter are found as far north as the Arctic islands. (Their numbers have dwindled greatly in this century.) The birdlife consists of species that stay in the area year round, such as the great snowy owl, and those that fly in only for the short summer, such as the Canada goose.

1-82   Typical tundra landscape.

1-83   Lichens.

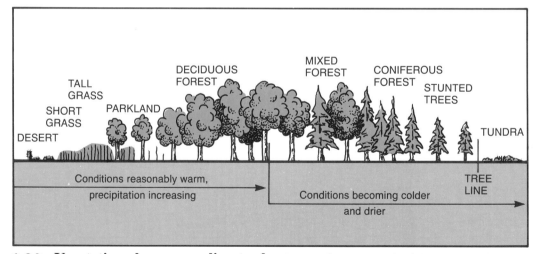

1-84   Vegetation changes as climate changes. *Where are each of these types of vegetation found in Canada?*

Boreal Forest This huge region, the largest vegetation zone in Canada, follows much the same borders as the Subarctic climate region. The Subarctic provides enough heat and moisture, and a long enough growing season, for trees to survive. This region's forests rival those of Siberia in their vastness, and are mainly coniferous. Coniferous trees are cone-bearing trees, and generally have a triangular shape and needle-like leaves. They are often referred to as **softwood** or **evergreen** trees. Except for the larches, they are "ever green." Larches lose their leaves in the fall; all the other coniferous trees lose some of their needles throughout the year, but are never bare. (If you have ever been in a coniferous forest, you will remember having walked on a carpet of brown needles on the forest floor.)

The trees of the boreal forest region are specially adapted to withstand severe climatic conditions and often poor soil conditions. Both are found in this region. These trees have a tough bark, a gummy sap and waxy needles to help them retain moisture. They can produce food for themselves, and grow during times of the year when the less hardy deciduous species are still dormant.

Some of the most common coniferous species found in the boreal forest are the black and white spruce, the jack pine and the larches. Though the coniferous species dominate, there are also some hardy deciduous trees in this region. These include the trembling aspen, the white birch and the balsam poplar.

The coniferous forests provide a much better habitat for wildlife than the sparse tundra. They contain far too many species to name here, but some of the most abundant and wide-ranging are rabbits, squirrels, beavers, black bears, moose, and the white-tailed deer. The birds include a great variety of ducks, smaller songbirds, and predators such as owls and hawks. Songbirds are numerous toward the south of the region.

1-85 *Reifel Bird Sanctuary, one of the many wetlands being preserved in Canada, located near Vancouver, British Columbia. The term "Bird Sanctuary" is a bit of a misnomer, for wetlands (marshes and swamps) also provide protection for thousands of plant and animal species. They preserve entire ecological systems for future generations to enjoy. Wetlands also help maintain water table levels for agriculture and provide beautiful, natural scenic retreats for local visitors and tourists.*

Deciduous Forest This is by far the smallest of Canada's major vegetation regions. It is found well to the south, in southwestern Ontario. It is a tiny, specially favoured part of this country's temperate East Coast climate region. Though it has less precipitation than the Atlantic portion of that region, it has ample to support many deciduous species. This small region has higher summer temperatures and a longer growing season than many other areas of the temperate region. It is actually the northern edge of the large deciduous-forest region of the northeast United States.

The deciduous trees in this region grow larger than the coniferous trees of the boreal forest. Deciduous trees are round in shape and have large, thin, broad leaves, which fall in the "fall." They are called **hardwoods** because many have tough, dense wood. In winter these trees are dormant, with their sap stored in the roots below the soil's frost line.

The dominant species in the deciduous forest region are the maples, oaks and beeches. The famous sugar maple, which is often found in the region, inspired Canada's Maple Leaf flag. This famous Canadian tree ranges all the way to the Maritimes, but does not grow naturally west of the Ontario border. In less favoured areas of this region, where soils are thin or the land swampy, conifers are also found.

So much of this forest has been removed that many animal species once found here in abundance have virtually disappeared due to the lack of cover. Smaller animals and birds are still abundant in the remaining forests. The wildlife found here is very similar to that of the forests to the north, though some animals and birds, such as the wild turkey, are only found in the warmer, deciduous environment.

Mixed Forest As might be expected, this region is found between the coniferous and deciduous forests. It straddles the Subarctic and the temperate East Coast climate regions. It includes both deciduous and coniferous species of trees, and forms a transition zone between the two.

The mixed and deciduous forest regions are part of, or close to, the most densely settled part of the country. As a result, their forests have been fully utilized, and vast areas of the natural vegetation cover have been removed. The wood has been used for many purposes, but today the hardwoods are used mostly for furniture, and the softwoods mostly for pulp and paper.

The fauna in the mixed forest region is similar to that of the boreal forest, but there is a larger variety of species and a greater number of them. For example, there are many more deer and birds.

Grasslands In the southern Prairie provinces, grasslands are the result of the conditions fostered by the semiarid Prairie climate region. It is generally too dry for trees, though some are found along the river valleys and in the moister, cooler areas near the boreal forest. Because trees are scattered on the northern edges of the grasslands, those areas are given the name **parklands**. There is considerable variety in the grassland area. In the central and south-

1-86 Mixed forest.

ern portions of the region — the driest area — the sparse vegetation is like that of a **semidesert**, and includes cactuses and sagebrush. Toward the moister edges of the region, the grass becomes much higher and is called **tall grass**. It is in these moister areas that grain production — especially of wheat — is highest.

Because there are almost no trees to provide cover, few large animals, except

for small deer and antelope, are found in the Grasslands region. The bison, which had few natural enemies did roam the North American plains in the millions before the coming of the settlers. But these herds were slaughtered almost to extinction in a very sad chapter of this continent's history. Through the hard work of a very few people, there are now protected herds of bison that are sufficiently large to guarantee the survival of the species. The most common animals in the grasslands region are "gophers." This term is applied to a range of small species, ranging from prairie dogs and squirrels to "real" gophers.

Mountain Vegetation The varied climatic conditions in the cordillera naturally result in greatly varied vegetation there. Diagram 1-87 shows how the vegetation changes as the elevation increases. Tree species will grow where their particular needs are best found. Thus, the broad-leafed maple, a deciduous tree, might be found along the eastern slopes of the Coastal Mountains, ponderosa pine in the interior plateaus, and along the west side of the Rockies, Western white spruce. Forestry is very important in this region, as will be discussed in Chapter Four.

Some of the most majestic animals in Canada, such as the Rocky Mountain sheep, are found in the cordillera. Because much of this region's natural forest still remains, wildlife is abundant. A variety of deer, bears, goats and sheep are found in different parts of this mountain environment. There are also large cats, such as the cougar and lynx, as well as many smaller tree- and

1-87 Vegetation changes with elevation.

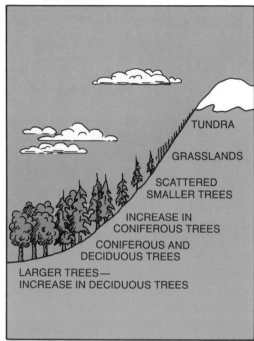

TUNDRA

GRASSLANDS

SCATTERED SMALLER TREES

INCREASE IN CONIFEROUS TREES

CONIFEROUS AND DECIDUOUS TREES

LARGER TREES— INCREASE IN DECIDUOUS TREES

ground-dwelling animals. As might be expected in a forest area, the birdlife is abundant as well.

West Coast Forest The coastal areas of B.C. have the high precipitation and moderate temperatures of the temperate West Coast climate. These conditions make it possible for very large coniferous trees to grow, such as the well-known Douglas fir. The dense forests of huge trees along Canada's west coast are sometimes referred to as **temperate rainforests**. They are Canada's richest source of lumber. In addition to the Douglas fir, species include the sitka spruce and the western red cedar.

The fauna in this area is similar to that of the Mountain region.

1-88 The rain forest of the temperate west coast.

1. Name the vegetation region which is:
   (a) Canada's largest.
   (b) Canada's smallest.
   (c) a result of very wet and moderate conditions.
   (d) a transition zone between the boreal and deciduous forest regions.
   (e) similar to a semidesert in places.
2. Explain why the regions where certain types of wildlife are found are often closely related to the vegetation regions.

3. Why is a coniferous tree able to withstand the severe conditions of the Subarctic climate?
4. Name at least five species of trees found in your local area, or in the forest region closest to it.
5. You are a photographer and want to photograph each of the following animals in their natural environment. Where would you go to photograph:
   (a) a bald eagle and a mountain goat?
   (b) a moose and a loon?
   (c) a snowy owl and a polar bear?
   (d) a prairie dog and a turkey vulture?
   (e) a wild turkey and a white-tailed deer?

New Words to Know

Natural vegetation	Lichens
Flora	Tree line
Tundra	Softwood
Transitional zone	Evergreen tree
Boreal forest	Hardwood
Coniferous trees	Parklands
Deciduous trees	Semidesert
Fauna	Tall grass
Permafrost	Temperate rainforest

1-89 Tall Grass Prairie

Soils of Canada

Early in this chapter it was suggested that holding beach sand in your hand would help you understand rocks and minerals. In a similar way, holding garden **soil** in one hand and beach sand in the other reveals some of the basic facts about soil. The sand, if it is merely particles of rock, is not a soil. Soil does include rock particles, but it obviously includes much more. A careful examination of soil will reveal that it contains the following:

(a) Rock particles
(b) Dead and decaying plant and animal materials
(c) Live animals and plants
(d) Water
(e) Air

To form a soil which will support plant life, these five items must be available in a balanced mixture. Weathering breaks down the bedrock to form rock particles. These have been mixed with dead and decaying plant and animal materials or **humus**. This humus is vital to the soil, for it holds the water needed by all living things. It also provides the materials to support the all-important plants, animals and **bacteria** in the soil. A handful of soil may include earthworms and visible insects; but the bacteria, which are too small to see with the naked eye, perform even more important work than the more visible life forms.

These bacteria, which can only be seen through a microscope, carry out the key soil process: they break down the soil materials into simple nutrients that the plants can absorb through their roots. Water and air are also essential in a soil. They are needed to keep the bacteria alive. If they aren't present the soil literally dies. Unlike the dead sand from the beach, true soil is a living thing.

Processes in the Soil

Diagram 1-90 shows a typical **soil profile**, as well as some of the processes which take place in soils. There are basically three **horizons**, or sections of a soil profile. The A horizon is the topsoil. It receives the humus, which becomes mixed with the rock particles. In the B horizon, the humus and rock particles are broken down and more fully mixed with each other. The C horizon is the **parent material**, which is being broken up into rock particles.

The water in the soil plays a major role in combining the rock and humus materials. When it rains, gravity moves the water down through the soil to the **ground water table**. As this happens, the water dissolves some of the soil material and moves it down. This process is known as **leaching**, and can take valuable materials so deep into the soil that plant roots can't reach them. The reverse process is the upward movement of water by **capillary action**. This can only occur if there are small enough air spaces to enable the water to rise by surface tension. A good soil will actually have 35 to 50 percent of its volume in air spaces.

The presence of water and air makes the third process possible. This third process is **decay**. As noted above, the bacteria reduce the complex materials in the humus and rock to simple, soluble food that the plants can absorb and use. The water, plant roots, earthworms and other animals all assist in breaking up the soil materials. Once broken up or dissolved, the material is more easily attacked by the bacteria. The bacterial action is greatly increased if the soil is warm.

1-90 Simplified soil profile.

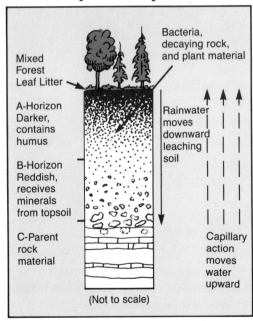

Mixed Forest Leaf Litter

A-Horizon Darker, contains humus

B-Horizon Reddish, receives minerals from topsoil

C-Parent rock material

Bacteria, decaying rock, and plant material

Rainwater moves downward leaching soil

Capillary action moves water upward

(Not to scale)

In this diagram can be seen the three soil horizons, and the three major processes that work on the soil.

Soil Regions of Canada

Tundra Soils The cold conditions in the Arctic result in a permanently frozen subsoil called permafrost (map 1-91). The surface soil will thaw for the brief summer, but will generally be waterlogged, since the evaporation rate is so low. As a result, the soils are very thin and can support only very minimal vegetation. This means that little humus is added to the soil. Agriculture is virtually impossible in such soils.

1-91 Natural soil regions of Canada.

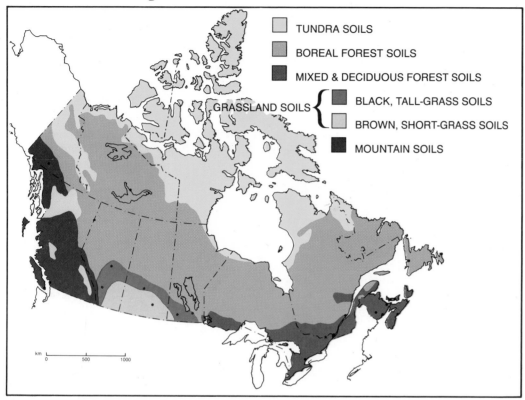

Legend:
- TUNDRA SOILS
- BOREAL FOREST SOILS
- MIXED & DECIDUOUS FOREST SOILS
- GRASSLAND SOILS {
  - BLACK, TALL-GRASS SOILS
  - BROWN, SHORT-GRASS SOILS
- MOUNTAIN SOILS

km 0 500 1000

As expected, these soil regions show a very close relationship to the vegetation regions, except that the grassland has been divided into two.

Grassland Soils   Grasses add a great deal of high-quality humus to soils. Leaching in these soils is not a problem, since these soils develop in semiarid climate regions. The amount of humus in the tall-grass areas is naturally greater than that found in the short-grass regions. Under the tall grasses, the soils are dark-brown to black in colour and very fertile. As the precipitation decreases, so does the grass height, the humus content, the colour and the fertility. The darker soils are excellent for grain crops; the lighter soils are used mainly for cattle-raising. In some of the drier, semidesert areas, ground water moves upward and evaporates from the surface, leaving salt deposits in the soil.

Mountain Soils   Mountain soils reflect localized relief, climates and vegetation. In some areas, slopes can be so steep that soils cannot develop; if these slopes are forested, the forests may remain immature due to erosion. If precipitation is high enough, the valleys may provide excellent forest or grassland soils. At higher elevations some tundra soils may be found.

1. Name and briefly describe four processes that are important to the formation of soils.
2. (a) Compare Canada's coniferous-forest soils with those of the grasslands.
   (b) Why do differences exist?

New Words to Know

Soil	Ground water table
Humus	Leaching
Bacteria	Capillary action
Soil profile	Decay
Soil horizon	Podzols
Parent material	

Boreal Forest Soils   These soils are generally infertile, as coniferous trees add little useful humus to the soil. The needles of these trees, decaying at the top of the A horizon, make the soil very acidic. This strengthens the leaching effect of water and results in the removal of many minerals from the upper soil level. The leached minerals give the B horizon a reddish-brown colour, but the leached A horizon becomes grey. The Russians named these soils **podzols**, meaning "ash grey."

Mixed and Deciduous Forest Soils
Both of these soil groups are called grey-brown podzols. Since deciduous trees add more useful humus to the soil, the soil colour becomes a darker brown as the percentage of deciduous trees increases. Similarly, the soils become more fertile as the humus content goes up. Because it is the humus that holds the water, fewer minerals are leached out of the top layers of these soils than out of grey podzols. As a result, these soils have more colour.

67

Mind Benders and Extenders

1. (a) In regard to former Prime Minister Mackenzie King's statement regarding Canada having "too much geography," what are the advantages and disadvantages of the nation's huge area?
   (b) Arrange for a class debate on the value of Canada's size and diversity.

2. Prepare a brief report (500 to 600 words) on the history, ideas and usefulness of geopolitics — one of the many subdivisions of geography.

3. Where did the term ecozone come from? What are Canada's Marine ecozones? Break into small groups to research each of this country's marine ecozones, and report your results back to the class? What are the particular problems confronting our marine ecozone systems? the world's marine ecozone systems?

4. Describe briefly the relationship that might exist between some of the following characteristics, all of which describe Canada:
   (a) free and democratic
   (b) major trading nation
   (c) well-developed resource base
   (d) temperate climate
   (e) northern hemisphere

5. List in chronological order, from oldest to youngest, the following:
   (a) The Proterozoic era
   (b) The first hardening of the Canadian Shield
   (c) The Ordovician period
   (d) The earth was a molten ball
   (e) The Pleistocene ice age
   (f) 65 000 000 years ago

6. Prepare a research report (200 to 300 words) or seminar presentation on the Theory of Plate Tectonics. Include at least one map and one diagram.

7. Describe the various ways in which the components of the earth's crust can be classified.

8. The first International Fog and Fog Collection Conference was held in Vancouver, B.C. in July, 1998 with over 300 delegates from 70 countries. One goal of the meeting was to define the role of international institutions in using fog as a water supply throughout the next century. What other uses of fog might be important?

9. Explain the similarities of the effect of running water and ice in the erosion cycle.

OR

Describe the action of ice erosion and deposition in your own province (and locality if applicable).

10. Debate this topic: "The Canadian Shield has been and is more of a hindrance to Canada than a help."

11. Explain five of the following six terms as they relate to the Great Lakes/St. Lawrence Lowlands:
    (a) Paleozoic era
    (b) Lockport dolomite
    (c) Ordovician period
    (d) Surface marine deposits
    (e) Intrusions
    (f) Pleistocene ice age

12. Compare the geological development and present appearance of Appalachian Canada with the much larger Appalachian region in the U.S.A.

13. (a) Prepare two climate graphs, one for the Arctic region and one for the station listed which is closest to your own community. Then compare the conditions illustrated.
    (b) If you were to be transported to the Arctic, what would you pack for a summer visit? For a winter visit?

14. The Subarctic climate region is Canada's largest, and is considered by many to be the typically Canadian one. Describe its conditions, using statistics to reinforce your answer, and explain the factors which cause them. Note also differences in conditions of various areas within the region, and explain the reasons for them.

15. (a) Prepare one map of Canada showing the temperatures of the climate stations in 1-77 and 1-56 in January, and a second map for July. The monthly means for both months should be colour-coded, that is, with large dots of red for 20°C to 30°C, orange for 10°C to 19°C, and so on.
    (b) Briefly describe the patterns on the two maps, and the reasons for them.

16. Compare coniferous and deciduous trees using this organizer:

Leaves	Tree Shape	Required Climate Conditions	Uses

17. (a) Describe in detail the natural vegetation in your area. Note the various species of trees and plants and describe them briefly.
    (b) What are the major crops being grown in your local area, or on the farms closest to it?
    (c) Describe statistically the local climate conditions which make the natural and cultivated vegetation possible.

18. "Humus is the most important part of soil." Give the arguments which could be used to support this statement.

19. As a class project, develop a booklet entitled "The Physical Background to this Community." Various teams of students could prepare the text, maps, illustrations, slides, and so on, for various topics. The topics might include landforms, climate, vegetation, wildlife, and soils, and additional subdivisions of these headings. This chapter would provide a beginning to such a booklet, but additional resources would be needed, augmented by field study.

CANADA
EXPLORING NEW DIRECTIONS

The Human Pattern

*"Canada has never been a melting pot;
more like a tossed salad."*
Arnold Edinborough, Canadian Author

Canada's Diverse Population

Compared with the giants of the world, Canada has a small population. Of the 5.7 billion people who inhabited this planet in 1996 , Canada was home for only 29 963 631 — about .52 percent of the world total. China's population in 1995 was 1 203 970 268; the U.S.A. had 265 455 000 people. Even the city-region of Tokyo-Yokohama had more people than all of Canada!

Physically, Canada is the second-largest country in the world, after Russia. Canada has a huge land mass of over 9.9 million square kilometres, yet it contained just over 30 million people in 1998. In this sense, Canada is one of the most underpopulated countries in the world. However, much of its land mass is uninhabitable. The reality is that Canada consists of small, scattered regions of high population density, most of which are concentrated along the southern border. Almost two-thirds of Canada's total population lives in the two southernmost provinces, Ontario and Quebec.

Looking at it another way, over three-quarters of all Canadians live in urban areas — in villages, towns and cities. In fact, about one-half of all Canadians live in Canada's ten biggest cities. This is in contrast with Canada before the 1920s, when only 25 percent of Canadians were urban.

Canada's population is growing, but not because there are more births than deaths. In fact, this country's birth rate is so low that if it did not attract immigrants, its population would be decreasing. Like many parts of Western Europe and North America, and some parts of Asia, Canada has significantly reduced the number of deaths among children and young people in just a few generations. Fewer infants are dying, but offsetting this, fewer Canadians are being born. Unlike many of the countries in Africa or South-East Asia, in recent years Canadian women have been having fewer babies. Canadian families continue to have fewer children.

Canadians, as a group, are also getting older. Not only are seniors living longer, but the proportion of the population under the age of 25 is decreasing. The aging of the population provides some advantages such as increased opportunities for young employees, and also some problems such as an increasing demand for social and recreational facilities for those over 65. This pattern is likely to continue as Canada moves into the twenty-first century.

In the most literal sense, all Canadians are descended from immigrants. Even the Aboriginal Peoples are thought to be descended from migrants who came here from Asia thousands of years ago. Immigration is possibly the most important factor in Canada's population change in the 1980s-1990s. It is immigration that provides the people to keep Canada's population growing. It is immigration that adds to Canada's cultural diversity, and provides the pieces of its cultural mosaic. Canada's immigration officials screen all applicants before they come to Canada, and select only the most appropriate as immigrants. The same officials also send back those who are considered to be inappropriate.

Many of Canada's laws, traditions and cultural patterns reflect the various immigrant groups that have created this country. Particular settlement patterns have developed in different regions; some of these were imposed by the governments of the time, while others have developed in response to local needs and resources. Many of these settlement patterns are unique, and help us to define who we Canadians are.

Canada has a small, diverse, urban and multicultural population scattered over a huge northern land mass. Geographers want to know what pattern trends are significant in Canada's population, and why the trends differ between one part of Canada and another.

The Canadian Census

The study of population, its growth, its distribution, and its future, is called **demography**. One of the best sources for information about a country's people is a **census**. The census is a means used to gather this information. The word census comes from the Latin *censere*, meaning "to assess." A careful study of population statistics can provide important clues to explain who Canadians are and why they are that way.

Every five years Canada takes a census of population. This is a survey of every man, woman and child living in Canada. Canadians living in embassies, and those who are employed overseas in the military, are also included.

The idea of counting and collecting information about people is not new. Over 2000 years ago the Romans were relying on census data to assist them in

2-1   Seniors pose for a picture while attending Elderhostel.
Elderhostel is an international organization offering year-round general education at a minimal cost to persons over age 60.

managing their widespread empire. In Canada, the British North America Act of 1867 called for a census to be taken in 1871 and every ten years from then on. Canadian society changes so rapidly in modern times that the **decennial** (ten-year) census has been found to be inadequate. The five-year (intercensal) census was first taken in 1956, and became mandatory in 1971. Now Canada has a census in every year ending in the number 1 or 6 (1991, 1996, 2001 and so on).

Census information is essential for governments and businesses in planning

Canada's future. Knowing that there will be a higher percentage of elderly people, for example, allows politicians to plan additional medical, housing and recreational services. Similarly, business people can develop new products and programs, and revise their advertising and promotional material for the senior market.

Census information can be analyzed to develop tables and graphs — such as charts of income distribution across Canada — or to generate written commentary on social trends. One problem facing **Statistics Canada** (the government

department in charge of the census) is that it takes time to process the tremendous amounts of data. In some cases statistics from one census are only just being published as the questionnaire for the next census is being distributed. Many of the statistics in this book would be more up-to-date if the census data were made available more quickly. It's a huge job, however, because Canada is a vast land with a widespread and diverse population. The 1991 census cost Canadians about $285 million to enumerate just over 28 million people across Canada.

Each census is organized around geographical areas called **census subdivisions**. In areas of dense population, some of these subdivisions are further divided. In more remote areas, subdivisions may be grouped together. There are 35 different categories of census subdivision alone, and the definition of each type can change from one census to the next. One of the most common data units is the **census tract**. This is the small data-collection area into which large cities are subdivided.

Census data is so important that every effort is made to keep it accurate. Under Canadian law, people who refuse to answer census questions, can be penalized. So can those who give incomplete or misleading information. The data is collected by questionnaires given to each **household**. Until 1971, census representatives knocked on every door and filled in the answers as they were provided. Since then, householders have been provided with questionnaires to be filled out on **enumeration** day. In remote areas, however, the census data is still collected through a home interview.

Most households have to answer just nine short questions about who lives in the household, the language spoken, and so on. Some households, chosen at random, receive a longer version, which includes additional questions on immigration, education, and the like.

During the 1996 census, some First Nation reserves and settlements were not completely enumerated. In some cases the bands would not permit the enumeration; in others, the process was interrupted before it could be completed. The effect was to make some data for 1996 incomplete, and other data (such as changes

2-2   Seniors enjoy a game of lawn bowling in Toronto.

from 1991-1996) less exact.

Some Canadians don't understand the need for census data. Others are concerned about how the raw data might be used. Without accurate census data, communities would not receive the appropriate amounts of "transfer payment" money from the federal government. Businesses, telephone companies, fire and ambulance services, and other social agencies need the census information to make appropriate decisions to better serve their communities. For example, how can a phone company effectively plan new telephone exchanges without knowing how quickly the population is growing?

Information about household heating equipment and fuel is useful for planning energy conservation and housing programs. Mother-tongue data can be used to establish programs to protect and promote the rights and privileges that were established by the Official Languages Act and the Canadian Charter of Rights and Freedoms. The data is essential for all of these and much more.

Census information is confidential. The people who work with the raw data must take an oath of secrecy, and can be fined up to $1000 or jailed for up to six months if they leak census information. Census data published by Statistics Canada has been compiled by computer from many individual census forms to build an **aggregate**. This "hides" everyone's responses, and so maintains the privacy of each individual.

1. **List the advantages and disadvantages of self-enumeration over interviews.**
2. **How could census data help to:**
   (a) find information about new telephone needs?
   (b) decide where to build a new home for seniors?
   (c) locate a new video-game parlour?
   (d) decide where to license a new French-language television station?
3. **What is meant by the word "confidential"?**

New Words to Know

Demography	Household
Census	Enumeration
Decennial	Aggregate
Statistics Canada	
Census subdivision	
Census tract	

Population Growth

Population totals can change not only through immigration, but also through **natural increase** or **natural decrease** as the birth and death rates change. The **crude birth rate** is calculated by dividing the total number of live births in a 12-month period by the population and then multiplying by 1000. The figure is called "crude" because the calculation includes males, children of both sexes, and non-reproductive females, even though (obviously) only women of reproductive age can give birth!

If birth rates increase, the population will grow, as long as the number of deaths does not exceed the number of births. The **crude death rate** is calculated in the same way as the birth rate: divide the total number of deaths in a year by the total population and multiply by 1000. This is also a "crude" rate, in that it includes all of the country's

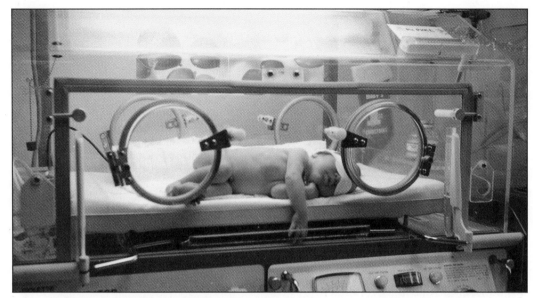

2-3 Canada's infant mortality rate is extremely low compared to less developed countries. *Premature babies, like the infant above, receive sophisticated medical care until they are able to live on their own.*

people, the young and the old, the healthy and the sick.

Canada's birth rate was 28.0 in 1956 during what is often referred to as the post-war baby boom, when there were more babies born than in any other decade of this country's history. The birth rate dropped to 15.7 in 1976, and 12.7 in 1996. Canada's death rate has undergone a similar decline from 9.4 in 1956 to 8.2 in 1976 to 7.2 in 1996.

Rapid advances in medical care, improved diet and nutrition, changes in lifestyle, and greatly improved public sanitation have combined to keep the death rate relatively flat through the 1990s. Heart disease has been the major cause of death for many years; regularly accounting for over 40 percent of Canadian deaths. Cancer accounts for just over 30 percent

(up from 20 percent in the 1970s). AIDS killed slightly under one percent of Canadians during the 1990s, but it was a leading cause of death among young males aged 30-50. Accidents were the leading cause of death for young people aged 15-34.

Canada's birth and death rates are very low. Uganda (in Africa), for example, has a birth rate of 46 and a death rate of 20.7. Jordan (in the Middle East) has a birth rate of 37 with a death rate of 4. Japan's birth rate is 10, and its death rate is 7.7. Haiti's birth rate is 38, and its death rate is 15.9. The United States has a birth rate of 14.8, and a death rate of 8.8

When the birth rate and the death rate are in perfect balance, the natural increase of the population is zero. In order for Canada to maintain zero population growth, every woman of child-bearing years must give birth to an average of 2.1 children. (That extra 0.1 is necessary because not all infants survive to have children of their own.) This figure is the **fertility rate**. In 1996 Canada's fertility rate was only 1.8.

Compared with other countries, the **infant mortality rate** is also extremely low. This tells how many children die within the first year after birth. By the late 1990s infant mortality in Canada had fallen to just over six deaths per thousand live births. The infant mortality rate declined by about 83 percent between 1953 and 1996. In recent years, it has been unusual for Canadian children to die soon after they are born. In comparison, Uganda has an infant mortality rate of 99 deaths per thousand live births! Such a high rate is similar to the one in Canada in the 1920s. In those days it was common for children to catch infectious diseases, like measles or whooping cough, and die in their first year. Canada's infant mortality rate has been reduced through prenatal care, good hygiene, good nutrition and improved medicine, but above all through inoculations against disease.

The **life expectancy at birth** of the average Canadian has increased steadily during this century. This is the number of years the average newborn child is expected to live. It is a good indicator of the general health of a population, because people in poor health generally do not live as long as those in good health. As a general rule babies born in Canada today are expected to live for 77 years. Females tend to live longer than males, as shown in chart 2-4. The

2-4 Selected population statistics (rates per 1000 population)

	1921	1951	1971	1981	1991	1996
Birth Rate	29.3	27.2	16.8	14.9	14.1	12.7
Death Rate	11.5	9.0	7.3	6.8	6.8	7.2
Life Expectancy (males)	58.8	66.4	69.4	71.8	74.6	74.6
Life Expectancy (females)	60.6	70.9	76.5	79.1	81.0	81.0

1. The total populaltion of Canada in 1995 was 29,615,325. That year there were 379,295 births and 215,740 deaths. Calculate the birth and death rates for 1995.
2. Using the statistics in the chart, show that the rapid decline of the birth rate occurred long after the death rate had started to decline.
3. Give three reasons why the birth rate would start to decline only after the death rate has declined significantly. Think in terms of why children are born.
4. Compare the figures for male and female life expectancy; suggest three reasons why there might be such a gap between the two sets of figures.
5. Is the gap between male and female life expectancy getting bigger or smaller? Would you expect this trend to continue? Explain your answer.
6. Choose a country in each of western Europe, southeast Asia, South America, and Africa, and find comparable figures on their birth and death rates and their life expectancies. Compare these data with the Canadian figures above, and identify significant similarities or differences. At least five significant factors should be noted.

figures for life expectancy have not improved as dramatically as those for infant mortality. Between 1951 and 1996, the life expectancy for Canadians increased by thirteen percent. What is more significant, however, is the number of people who are living longer than 65 years.

According to Statistics Canada, the population of this country was 29 963 631 as of July 1, 1996. This is an increase of only 348 306 people over the previous year. Although year after year there are more people in Canada, the population growth rate is continuing its declining trend. Canada's population continues to grow, but at a slower rate than in the past.

According to Statistics Canada, the population growth rate in the 1960s and 1970s slowed down because the widespread use of **contraceptives** resulted in fewer births. The continued slowdown in

population growth in the 1990s is said to be largely a result of the reduction in the number of immigrants being allowed to enter Canada.

Examine graphs 2-5 and 2-6, which show the growth of the Canadian population between 1867 and 1987. The first is a simple bar graph that shows the *total* number of people in the country each year. This kind of graph is called an **arithmetic graph**, because the numbers increase by simple addition. The second graph displays the same information in a different way. It is called a **semi-logarithmic** graph because the numbers on the side increase not by addition (10, 20, 30 . . .), but by multiplication (10, 100, 1000. . .). A semi-logarithmic graph shows the *rate* of growth much more clearly and accurately than a "normal" arithmetic graph. This is because it takes into account not only the number being added but also the size of the number it is being added to. Adding 200 new items to 400 existing items is a much bigger rate of increase than adding 200 new items to 5000.

1. Canada's population was 27 297 000 in 1991, 29 963 631 in 1996, and just over 30 000 000 in 1998. Using the two examples provided below, prepare up-to-date arithmetic and semi-logarithmic graphs of Canada's population growth.

2. The population increased from 3 689 000 in 1871 to 4 325 000 in 1881, and to 21 568 000 in 1971. The arithmetic graph shows different rates of increase than the semi-logarithmic graph for the same periods. Using this formula:

$$\text{Percentage increase} = \frac{\text{population increase} \times 100}{\text{population at start}}$$

calculate the actual percentage increases from 1871 to 1998. Why is a semi-logarithmic graph better for showing population growth?

3. Which of your two graphs supports the statement "Canada's population has actually grown at a fairly steady rate since Confederation?"

4. (a) Use information from both graphs to give evidence that Canada's population growth rate is slowing down.
(b) When did this slowdown in population growth start?
(c) Has a slowdown happened before in Canada's history?

New Words to Know
Natural increase
Natural decrease
Crude birth rate
Crude death rate
Baby boom
Fertility rate
Infant mortality rate
Life expectancy at birth
Contraceptive
Arithmetic graph
Semi-logarithmic graph

2-5 Arithmetic graph of Canada's population growth, 1867-1987.

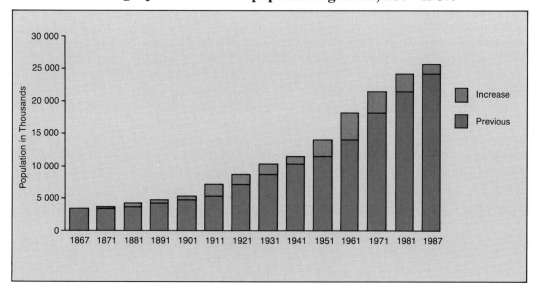

2-6 Semi-logarithmic graph of Canada's population growth, 1867-1991.

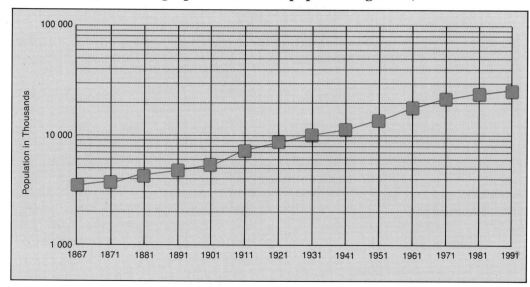

75

Population Distribution and Density

An important characteristic of Canada's population is its uneven **distribution**. Over 90 percent of all Canadians live within 200 km of the southern border; this leaves much of the north sparsely populated. Geographer J. Lewis Robinson of the University of British Columbia has suggested that Canadians live in "islands" of population strung along the southern border. The islands form an archipelago or "chain" of settled areas in a vast sea of rocks and forests. These islands of people are generally separated from each other by considerable distances and by harsh natural environments, and are linked by thin transportation and communication corridors.

Almost 90 percent of Canada is without permanent settlement. In the Atlantic provinces, for example, only Prince Edward Island is completely settled. The other east coast provinces have large areas of their interiors vacant. On the island of Newfoundland, the population rims the coast while the interior is mainly empty of people.

Draw a line on a map of Canada between Quebec City and Sault Ste. Marie. Only 2.2 percent of Canada's total land area, but almost 60 percent of its people, will lie to the south of this line. Even in this southern section of the Great Lakes/St. Lawrence Lowlands, where the population density is the highest in the country, the land is not continuously settled. There are still large tracts of vacant land. But here also are some of the largest **census metropolitan areas (CMAs)** — Quebec City, Montreal, Ottawa, Toronto, Hamilton, St. Catharines, London, Kitchener-Waterloo and Windsor. Together, these nine CMAs account for well over one-third of all of Canada's people; six of Canada's ten largest cities are in this group.

In terms of physical size, the Prairie region is the largest population island in Canada. The population there, however, is less than one-third that of the Great Lakes/St. Lawrence region. The Prairie region has fewer people spread over a larger area, so the population density is lower. This region covers about six percent of Canada's total land area and contains about 15 percent of the total population. For the first 100 years of European occupation, the settlers lived on scattered farmsteads or in widely separated villages and towns. In recent decades major urban centres have developed on the Prairies. Today, by far the largest percentage of Prairie people live in large CMAs such as Winnipeg, Edmonton and Calgary, or in smaller ones like Brandon, Moose Jaw and Medicine Hat. In 1996 over seven percent of all Canadians lived in one of the three largest metropolitan areas of the Prairies.

Most of British Columbia's people

2-7 Ten Largest CMAs, 1996

Toronto	4,263,757
Montréal	3,326,510
Vancouver	1,831,665
Ottawa-Hull	1,010,498
Edmonton	862,597
Calgary	821,628
Québec	671,889
Winnipeg	667,209
Hamilton	624,360
London	398,616

have settled on long strips of land along the river valleys and the coastal plains. Many of these strips are connected by high mountain passes, by narrow routes along river gorges, or even by tunnels. The settled areas of British Columbia account for only about one percent of Canada's total land area, but contain almost ten percent of its population. Half of these people live in a single, densely settled area in the lower Fraser River valley, which is, of course, the Vancouver CMA.

Outside these main "islands" of population are scattered settlements often associated with resource exploitation. People settled in these because there was easy access to water, wood, mineral ore or some other raw material. Some of these isolated resource centres have quite large populations. For example, Thompson, Manitoba, a nickel-mining centre, is home to 15 000 people. Yet around this city are thousands of square kilometres with no permanent habitation.

Half of all Canadians live in Canada's ten largest metropolitan areas. Today 76 percent of Canada's people live in urban areas; in 1921 less than 33 percent were urban-dwellers.

Map 2-8 shows the distribution and **density** of the Canadian population. Density refers to the number of people living in a definite area. It is calculated by dividing the population by the total land area. Canada's population density in 1996 was only 3.0 people per square kilometre. In reality, however, Canadians live in only a small part of their vast country, so the population density

2-8 Canada's population distribution.

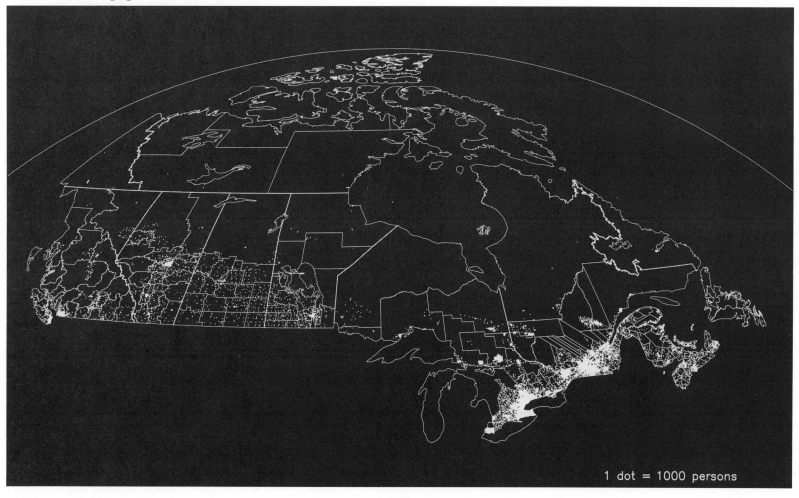

1 dot = 1000 persons

of the "inhabited area" of Canada is much higher. In some metropolitan areas the population density per square kilometre can be over 10 000. In some Arctic regions the population density over hundreds of square kilometres may be zero.

1. (a) Select the province of Canada with the highest population density.
   (b) What is its population density?
   (c) Explain why its population density is the highest of all provinces.

2. (a) Select two provinces of Canada with low population density. Explain why.
   (b) What are their population densities?

3. Select one province of Canada that has many small and isolated communities involved with one industry.

4. Select one province of Canada which includes a transportation corridor between areas of high population density.

5. As the prime minister of Canada it is your job to consider the viewpoints of all Canadians from across the country. Give five reasons why the distribution of Canada's people makes your job difficult.

New Words to Know
Population distribution
Census metropolitan area (CMA)
Density

Population by Province and Interprovincial Migration

Canada's population distribution by province changed dramatically between 1951 and 1996. Graph 2-9 shows that a larger portion of Canadians lived in Ontario, Alberta, British Columbia in 1996, and a smaller proportion in the other provinces.

Each province has its own trends in terms of overall population growth. The patterns of natural increase (birth and death rates) have become more or less uniform across the country; major changes in provincial demography result more often from the **migration** of people from one province to another. Canada's population is becoming less evenly distributed because of internal migration: many people are leaving the less well-off provinces and moving to provinces where there are greater opportunities. Most provinces showed considerable change in their interprovincial **net migration** figures between 1981 and 1996.

The Atlantic provinces, for example, had a negative migration rate. This was largely because many east-coast areas are economically disadvantaged: there are fewer jobs, and the economy is not as diversified as that of other regions. Between 1981 and 1984 more people left Quebec than settled there. In 1985 there was a significant turnaround reducing the flow of migrants from that province, but by 1996 there was a further significant migration from the province.

2-9 Percentage distribution of population by province, 1951-1996

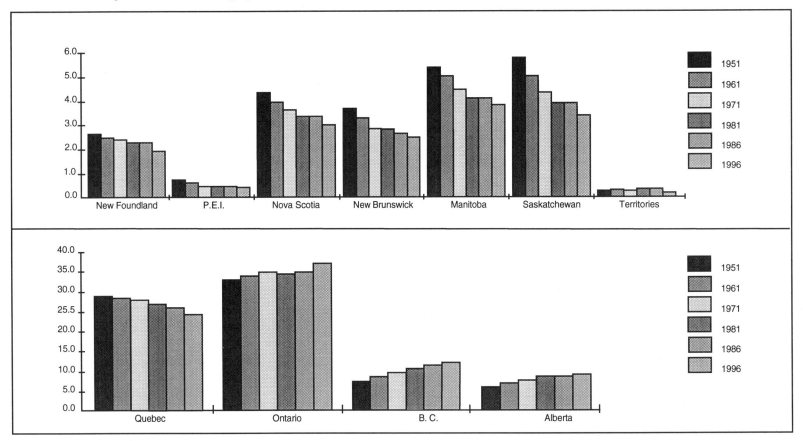

2-10 Net interprovincial migration, 1981-1996

	1981-82	1985-86	1991-92	1995-96
Newfoundland	−5693	−5716	−1,961	−7573
Prince Edward Island	−856	−76	−1,553	880
Nova Scotia	−1936	−1321	987	−893
New Brunswick	−2842	−1891	−2,377	−228
Quebec	−25790	−5349	−12,259	−13217
Ontario	−5665	33562	−6,604	−5579
Manitoba	−2625	−2297	−7,663	−1946
Saskatchewan	−323	−6939	−9,829	−743
Alberta	35562	−3831	7,264	5739
British Columbia	8705	−4501	33,447	23460
Yukon	81	−545	489	801
Northwest Territories	362	−1096	59	−701

1. (a) Which provinces had the highest out-migration? Why?
   (b) Which provinces had the highest in-migration? Why?
2. In small groups, discuss issues faced by provinces where there is significant out-migration. Compare those provinces to those with significant in-migration.

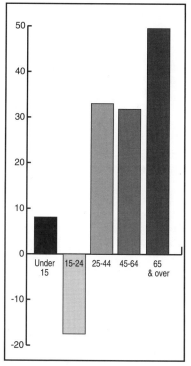

2-11a *Percentage change in the population of selected Canadian age groups between 1981 and 1995*

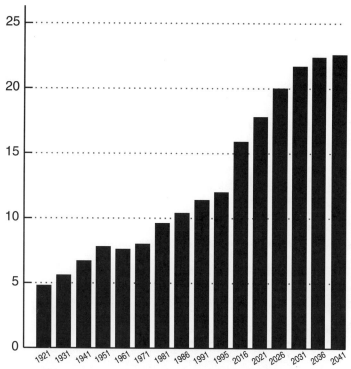

2-11b *Seniors as a percentage of the total Canadian population, 1921–1995, with projections to 2041.*

Population Age and Sex

On average, Canadians are getting older than other peoples of the world. In some countries of Africa and South America during the mid-1980s, almost 50 percent of the population was under 15 and only four percent was over 65. In Canada, 21 percent of the population was under 15 and 11 percent over 65. The people in these two age groups — under 15 and over 65 — together make up the **dependency load**, the proportion of the population that is probably not in the work force. Many other countries have large populations of young people in their dependency load. In Canada, however, a large part of the dependency load is made up of people over 65, and their numbers are growing. The population aged 65 or over was 1.39 million in 1961; this had increased to 3.6 million by 1995. The rate of increase for this age group is more than twice that for Canada's population as a whole.

The distribution of males and of females, and of young and old, can be shown on a **population graph**. This is made from two horizontal bar graphs that are drawn outward from a common vertical axis. It is read horizontally from the central or zero axis. The vertical axis is divided into age groups — usually for every five years; the horizontal axis shows percentages. The left side records the number of males, the right side, the number of females; the proportion of males to females can be seen by comparing the lengths of the bars on each side of the vertical axis. Both bars can be added together to obtain the total number of people in one age group. Population graphs using percentages of the total population show the proportion of

2-12 Population graphs for selected years: (a) 1921 (b) 1966 (c) 1986.

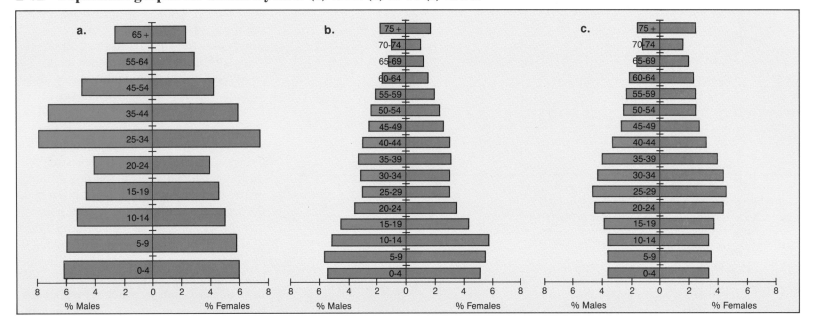

people in each age and sex group. Bar graphs help demographers compare different countries, different time periods and different population distributions.

A population graph gives a "snapshot" of how a population is distributed by age and sex at one particular time. Examining a series of graphs that represent the population distribution at different times can provide clues to how a population is changing.

1. **Look at the total number of people in the 0-to-9 age group in 1921. Are there more or fewer of them in 1986? Why?**

2. **(a) On the 1966 graph, which age groups are the baby-boomers?**
   (b) Where do they appear in 1986?

3. **(a) Draw a population graph to show what you think the pattern will look like in 2010. Provide reasons to support your decision.**
   (b) Where will you be on the graph in the year 2010?

4. **Use the statistics below to create a population graph showing the age-sex distribution of population in Canada in 1996.**

Develop a comparison organizer on the similarities and differences between your graph and the graph of Canada in 1986.

5. **(a) Calculate the dependency load for Canada in 1996.**
   (b) How is this different from the dependency load for Canada in 1986?

New Words to Know

Migration   Dependency load
Net migration   Population graph

2-13 Age-sex population statistics for Canada, by percent, 1996.

Age	0-4	5-9	10-14	15-19	20-24	25-29	30-34	35-39	40-44	45-49	50-54	55-59	60-64	65-69	70-74	75+
Males	6.8	6.9	7.0	6.9	7.0	7.6	9.0	9.1	8.0	7.3	5.6	4.5	4.0	3.6	2.9	3.9
Females	6.3	6.5	6.5	6.5	6.6	7.3	8.6	8.7	7.9	7.1	5.5	4.4	4.1	3.9	3.6	6.3

2-14   Pilot Nancy Kyro flies into Armstrong in northern Ontario. *Many more women now work in non-traditional occupations than did in previous decades.*

The Male-Female Ratio

50.4 percent of Canada's population was female in 1997, and 49.6 percent was male. The numbers are wider apart as the population ages. There are 21 women over the age of 65 to 158 men; Canadian women over the age of 85 outnumber their male counterparts by 263 to 116.

One result of this difference, is that the vast majority of elderly women live alone, and over half are widows.

There has been a tremendous increase in the number of working women in Canada since 1901. That year, woman formed only 16.1 of the **labour force participation rate**. This rate measures the number of people who *are* employed as a percentage of those who *could* be employed. In 1996, the participation rate for females was 57.6 percent (down slightly from the 1991 figure of 59.9 percent) but still a healthy number. There were 6 844 females eligible

for employment in 1996 out of a total female population of 11 887 000. Many women now work in non-traditional roles. They dominate the service-producing industries, financial and insurance industries, health/ social services industries, and education. Women represent only 11 percent of the fishing and trapping industries; 12 percent of mining, quarrying, oil wells; and 24 percent of the utilities industries.

There were 1 223 345 self-employed males in 1996 compared with 579 015 self-employed females. There were 21 185 male unpaid family workers in 1996, compared with 50 540 females in the same situation.

In the 1990s many more women obtained university degrees than ever before. Women received 58 percent of all the undergraduate degrees awarded in 1991, up from 47 percent in 1976. The number of degrees women received in business

law, veterinary sciences, architecture and engineering doubled between 1976 and 1991. In 1991 less than 18 percent of women occupied management positions within the engineering profession, although 1 325 were granted engineering degrees. As we move into the twenty-first century, however, women are becoming more visible managers in all aspects of the business world.

Population Implications

It is always difficult to predict what demographic changes will occur in a population. A war, a depression, or a change in immigration laws, refugee policies or child tax benefits can quickly alter existing patterns and make present-day trends change direction, or even reverse themselves.

In 1996, the Canadian population that was 15 years old or older totalled 22 628 925. Of that, 18 659 825 people were not attending school; 2 801 280 were attending school full time, and 1 167 820 were attending school part time. 3 000 965 had a university degree, and 2 812 015 had less than a Grade 9 education. Enrollment in public and secondary schools was 5 440 334 in 1995/6 (up from 5 141 003 9 in 1990/1). Public school enrollment was 5 095 901. 277 704 students were enrolled in private schools, and 64 268 were enrolled in federal schools. There were 2 461 students in Canadian schools for the deaf and blind in 1995/6.

Examine chart 2-15, which gives statistics on the number of families, the average number of persons in families, and the average number of children in families from 1941-1996. The chart shows there

2-15 Number of families and average number of persons and children per family 1941-1996.

Year	Number of families	Average number of persons	Average number of children
1941	2 509 664	3.9	1.9
1951	3 287 384	3.7	1.7
1961	4 147 444	3.9	1.9
1971	5 070 682	3.7	1.7
1981	6 324 980	3.3	1.3
1991	7 497 400	3.1	1.3
1996	7 879 700	3.0	n/a

1. Study the figures in each column. Do the numbers increase or decrease? Why?
2. Which year had the largest number of families? the largest average number of persons? the largest average number of children?
3. Suggest three reasons why the number of children per family should decrease dramatically between 1971 and 1991.
4. Why do you think the number of families could continue to increase while the number of children decreases?
5. What would you project to be the average size of the Canadian family in the year 2001? Give evidence from the chart to support your answer.

1. In small groups, discuss the changes that are resulting from the decrease in numbers of Canada's youth. What additional changes can you project over the next twenty years. Each group should prepare a two-minute presentation for the class.
2. In small groups, discuss the changes that are resulting from the increase in Canadian seniors. Each group should prepare a two-minute presentation for the class.
3. How has the labour force participation rate for women changed during this century?

New Words to Know
Labour Force Participation rate
Demographic trends

were more families in Canada each year, and that the number of persons per family remained fairly constant until 1971, when it started to decrease. Note how the average number of children per family followed suit.

The relative decline in the number of young people since the 1960s has changed many features of Canadian life. Schools have closed, or been relocated, or even adapted to other uses. The market for children's and teen goods is not growing as rapidly as other markets, and this affects the kinds of goods and services that are available. Businesses watch **demographic trends** carefully so they can adapt.

One obvious trend is the "greying" of Canada. As Canada enters the twenty-first century, there are more and more people moving into the "over 60" age group. Seniors made up 12 percent of the Canadian population in 1995, with an estimated 3.6 million people over the age of 65. Statistics Canada predicts that this number will rise to 23 percent of the population by 2030. Seniors represented only five percent in 1921, and ten percent in 1981. 31 percent of seniors live in urban areas. 58 percent of seniors over the age of 65 were women, and women comprised 70 percent of those over 85.

Six percent of Canada's seniors remained in the workforce in 1995 (in the agricultural and religious sectors.) Most seniors live at home, many of them alone. Eight percent of seniors live with extended families.

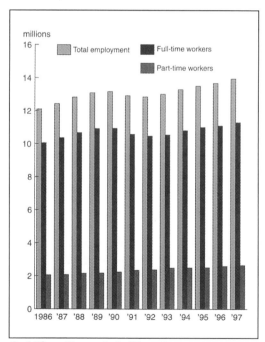

2-15b Canada's Employment Picture 1986-1997.

Special People

Terry Fox, Rick Hansen and Steve Fonyo are household names for Canadians. All three were **disabled**, yet all three travelled across the country to raise money for research and to make other Canadians more aware of the needs of handicapped people. Terry Fox died in 1981 at the age of 22, leaving unfinished his run, which he called the "Marathon of Hope." Despite having lost a leg to cancer, he still ran from Newfoundland to Thunder Bay, Ontario, raising $25 million along the way for cancer research. Steve Fonyo, who had also lost a leg to cancer, raised almost $13 million for cancer research. In his "Journey For Lives" he ran across Canada in just 425 days, finishing in May, 1985. Rick Hansen propelled himself around the world in a wheelchair during his two year "Man in Motion" tour, which ended in May 1987. It raised more than $15 million for spinal cord research and rehabilitation.
The ninth Paralympic Games were held at the same time as the summer Olympics of 1992 in Barcelona. More than 4,000 athletes and staff from 96 countries participated in the world's highest competition for athletes with physical and sensory disabilities. Participants include the blind and partially sighted, amputees, those with cerebral palsy and wheelchair athletes. The Paralympic Games, like the Olympic Games, are held every four years. The first Paralympics held in Rome in 1960 had mostly wheelchair events, but today's participants compete in a variety of sports.

Some of the events are specially designed for those with specific disabilities. In the cycling event, for example, a sighted guide shares a tandem bicycle with the visually impaired athlete. At the 1992 event 137 Canadian athletes with disabilities competed in 11 of the 15 events against the best in the world to bring back a total of 75 medals. Canada stood sixth overall among competing nations, winning a total of 28 gold, 21 silver and 26 bronze medals.

Other Canadians perform in different fields. A Canadian theatre group, the Famous People Players, was founded in 1974 by Diane Dupuy. Its members are mentally handicapped. They have toured the world performing critically acclaimed shows that they arranged and produced themselves. Another group of disabled young actors called Rolling Thunder, has performed in *A Very Special Gala* on CBC television.

In 1991, the Health and Activity Limitation Survey estimated that some 4.2 million Canadians were disabled. That figure represents about 15 percent of all Canadians who were over the age of 15, and who were not in hospitals or other institutions. Almost half of the disabled were over 65 years of age, but seven percent of all Canadians under the age of 15 are disabled in some way. The majority of that group are aged 10-14.

2-16 Terry Fox began his heroic Marathon of Hope in Newfoundland to make people more aware of the handicapped and to raise money for cancer research.

The 1991 survey defined a disabled person as one who had trouble performing some common daily living activities and people with mental handicaps and those limited in the kind or amount of activity they can perform because of a long-term physical condition or health problem. People who had difficulty with various activities solely due to mental illness, as well as those in institutions, were not included in the disabled population.

The 1984 Canadian Health and Disability Survey remains the most comprehensive study of the disabled to date. It estimated rougly 66 percent of the disability problems reported in the survey related to bodily movement or mobility, 14.8 percent related to hearing and 7.7 percent to sight. Many who responded to

the survey said they had difficulty using kitchen and bathroom equipment; 121 000 reported that they could not travel at all. Of the 634 000 Canadians with hearing problems, 168 000 said they could not hear a normal telephone conversation. Some 34 000 people reported they were legally blind. Another 70 000 had difficulty communicating with strangers.

The survey also found that only 42 percent of disabled people of working age were employed, compared to 67 percent of all other working-age Canadians. This is despite the fact that many employers had made provisions for disabled employees. (In some parts of Canada the provision of equal access is required by law.) The employment figures for the disabled worsened dramatically as the severity of the disability

2-17 Canadian disabled persons by nature of disability

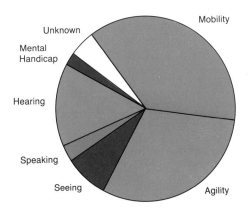

increased. Only four percent of the severely handicapped were able to be employed.

In 1996, a Federal Task Force on Disability Issues' report recommended several ways that Canadians with disabilities could be helped to be fully included as active participants in their communities.

In 1998, the Ontario government prepared legislation for its Ontarians With Disabilities Act, and created its Ontario Disability Support Program to provide income and economic support. Ontario School Boards receive over one billion dollars a year for special education.

Some of the programs helping Canada's disabled include:
- Supported Independent Living with supervision and assistance for adults with developmental disabilities living within the community
- Group Homes
- Sheltered workshops
- Outreach or Community Care

2-18 Canadian disabled according to degree of disability and labour force status

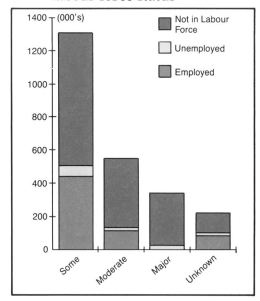

- Assistive Devices programs to pay for selected special equipment or supplies
- Home oxygen programs
- Income and financial assistance
- Child care subsidies
- Lifeskills programs
- Children's Treatment Centres

Canada's disabled are also eligible for many other social security programs such as Unemployment insurance and disability pensions.

In addition to government help, disabled people can also turn to private Canadian organizations like Active Living Alliance for Canadians with a Disability, INDIE (Integrated Network of Disability Information and Education), The Canadian Hearing Society Foundation, The Canadian Paraplegic Association, or Canadian Special Olympics.

The disabled still face many barriers. A simple curb at the edge of a street may make it impossible to visit a friend. Getting a wheelchair into a regular-sized bathroom can be an impossibility. Canada's disabled population is no longer shut away or kept out of sight in institutions, but there remain many attitudes and problems to overcome before all of them achieve equality in our communities.

1. **Why were the efforts of Terry Fox and others so important to Canada's disabled?**
2. **(a) Why is it difficult for the disabled to find permanent employment?**
   (b) How does the information in graph 2-18 make this situation evident?
3. **How might you be of assistance to someone who is disabled?**

New Words to Know
Disabled

The First Canadians

When the first European explorers arrived in Canada in the 1500s, they found a land populated by between 400 000 and 500 000 **Aboriginal peoples**. The **Inuit**, or Arctic peoples lived along the coastal edges and islands of the Arctic. The **First Nations**, with over eighty distinct tribal bodies speaking eleven different language groups, inhabited the rest of this land. The Beothuk Nation lived in what would become known as Newfoundland. The Mi'kmaq Nation lived in the Atlantic region; the Malecite, Montagnais and Nsakapi Nations (Innu) lived in Quebec. The Haudenonsaune (Iroquois) Nations lived in the northeast woodlands along the St. Lawrence and through the Great Lakes region (the Haudenosaume population of Ontario was estimated to be 60 000 at time of contact). The Ojibway lived in the lands stretching from Georgian Bay to the Prairies. The Canadian Plains were home to the Blackfoot, Peigan, Blood, Gros Ventre, Plains Cree, Assinboine, Sioux and Sarcee Nations. There were six Nations established in the western interior plateau region – Interior Salish, Kootenay, Chilcotin, Carrier, Tahltan, and Tagish – and another six Nations – Haida, Tsimshian (Gitksan and Nisga'a), Nootka, Coast Salish, Kwakiutl and Bella Coola – lived and thrived in the Pacific Coastal region. Some estimates place the West Coast Aboriginal population as high as 200 000 at time of contact. Twelve Nations, including the Dogrib, Dene, and Chipewyan, lived in the Mackenzie and Yukon River basins.

Early Settlement

Where did these people come from, and when – no one knows for sure. Aboriginal

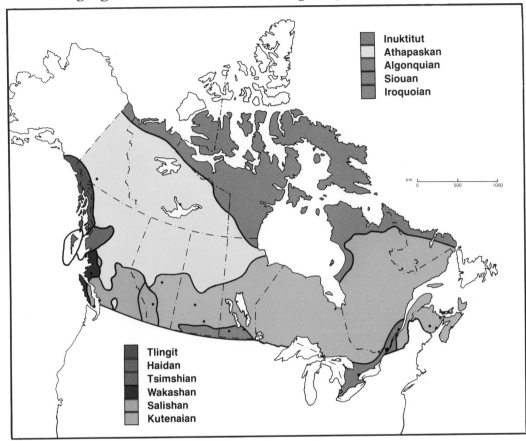

2-19   Language families of Canada's Aboriginal peoples.

Inuktitut
Athapaskan
Algonquian
Siouan
Iroquoian

Tlingit
Haidan
Tsimshian
Wakashan
Salishan
Kutenaian

creation myths suggest that First Nations originated here, with the help of a spirit. **Anthropologists** believe that human life first developed in Africa and Asia, and that all Aboriginal peoples of the Americas migrated to this continent. Many North American Aboriginals may have used the **Bering Land Bridge**, created over 70 000 years ago when glaciers covered this continent and world sea levels dropped more than 90 metres exposing a land connection between Asia and North America (Alaska) through the Bering Strait. It is true that

North American Inuit share a common language base with their Siberian counterparts, and there is some suggestion of Inuit life in the Yukon area of the Arctic as early as 55 000 years ago, although scientists believe most migration waves occurred around 25 000 years ago.

Because glaciers covered most of Canada, initial settlements were largely confined to Alaska and the Arctic. Gradually, however, as the icecap began to diminish, the Alberta – Saskatchewan plains area became available, and later,

2-20 Canada's aboriginal people before European contact.

	Arctic	Subarctic	Northwest Coast	Plateau	Great Plains	Northeast Woodlands
Environment	Arctic coast and sea ice, treeless tundra, arctic climate.	Boreal forest region, long cold winters, short hot summers.	Rugged Pacific coast, mild wet climate, huge forests.	Mountains and valleys of B.C. interior; varied climates from moist to semi-arid.	Rolling grasslands with few trees; hot dry summers, cold winters.	Great Lakes mixed forest region, moderate climate, fertile soils.
Tribal groups	Inuit.	Algonquian and Athapaskan.	Haida, Coast Salish, Bella Coola, Kwakiutl, Nootka, Tsimshian.	Interior Salish, Kootenay and Athapaskan.	Sioux, Algonquian and Athapaskan.	Huron and Iroquoian.
Social organization	Nomadic family groups or small bands, based on mutual cooperation for survival.	Small migratory bands with leaders of demonstrated skill or character.	Stratified society of nobles, commoners and slaves; complex patterns of wealth and inheritance; family hierarchies.	Some tribes similar to coastal groups; others had hereditary leaders and councils of advisors.	Nomadic bands with chiefs and councillors, as well as powerful military societies.	Semi-permanent, fortified farming villages; complex system of government based on tribal confederations.
Dwellings	Igloo or skin tent for hunting, or stone and turf den in the east. Large log dwellings in the Mackenzie Valley.	Wigwam or similar hut of birch or spruce bark, rush mats or skins.	Large multi-family plank houses with framework of logs.	Various pit dwellings, tepees, cedar bark huts.	Tepees of wooden poles, covered with buffalo hides.	Bark-covered longhouses.
Transportation	Skin-covered kayak and umiak for open water, dog sled for winter.	Birch bark canoe, snowshoes, toboggan.	Seagoing dugout canoe.	Canoe, toboggan. Some dog packs.	Dog travois.	Birch bark canoe, snowshoes, toboggans.
Food	Seal, walrus and whale, fish, caribou.	Caribou, moose, fish, bear.	Fish and seafoods of all kinds. Shellfish, oolichan oil, berries and nuts	Smoked fish, bear, deer, mountain goat, moose, caribou and waterfowl, berries, bark and nuts.	Buffalo, antelope and deer made into pemmican or jerky.	Corn, squash and beans, deer, fish and rabbit plus nuts and berries.
Food gathering	Men hunted seals and other marine mammals, caribou, migrating birds. Women and children fished, gathered berries.	Bands followed migratory patterns of animals, fish and fowl.	Women gathered clams and other shellfish, dipped fish from traps and nets. Men harpooned larger sea mammals, fished for salmon and oolichan.	When the salmon returned each year, they were dried or smoked. Birds were netted and animals trapped or shot.	The hunters followed their migrating prey; buffalo were herded into killing areas.	Men cleared the land, women and children tended the crops; fields were exhausted after a few years so the community moved on.
Tools	Harpoon, spear, hook and line, traps and snares.	Bow and arrow, spears, traps and snares for animals. Weirs for fish.	Harpoon, hook and line, gill nets and traps.	Weirs, dip nets, and spears to catch salmon; bows, knives, traps and snares.	Bow and arrow, sometimes a buffalo jump.	Crops were tended by horn and bone tools; bows, snares, blowguns and traps used for hunting.
Clothing	Double-layered parkas, pants, high boots of sealskin, caribou or other fur.	Shirts, leggings, moccasins and breech cloths made of dear, moose, caribou skins; fur robes in winter.	None in good weather; rain hats and capes of woven bark and roots; sea otter robes.	Usually deer, but sometimes moose hides made into shirts and leggings.	Light skins such as elk or deer used for shirts, leggings breech cloths; heavy bison skins became winter robes.	Similar to subarctic and plains.

land along the Yukon, Peace and Liard Rivers became ice-free. Finally, about 10 000 years ago, the glaciers began to recede, the ice age ended, and the Aboriginal peoples were able to move more freely around the continent.

Aboriginal cultures were shaped by their environment, and peoples living in close proximity to one another tended to develop similar cultures. Every geographic area demands certain specific skills and tools from people settling there. Each Aboriginal group developed special ways of using its environment for food, shelter, clothing, transportation, and all other spiritual, cultural, and religious needs. First Nations of the eastern woodlands might speak different languages, but their housing designs were often similar because they shared a similar environment. First Nations of the plains were often nomadic, skilled at hunting by chase, or corralling their prey. Peoples of the northwest coast became magnificent boat builders, and settled in large communities.

Today, we study Aboriginal development with the help of six native culture zones: Arctic; Subarctic; Northeast Woodlands; Great Plains, Plateau, and Northwest Coast. Fig 2-20 charts some of the cultural similarities and differences between First Nations settling in these regions.

Contact

The arrival of Europeans in the 1500s had dramatic impact on North American Aboriginal Peoples. Early contact consisted, for the most part, of cautious cooperation, and even curiosity as the Europeans would not have survived initially without Aboriginal help. Aboriginals and Europeans viewed one another as separate,

2-21 Native culture areas of Canada.

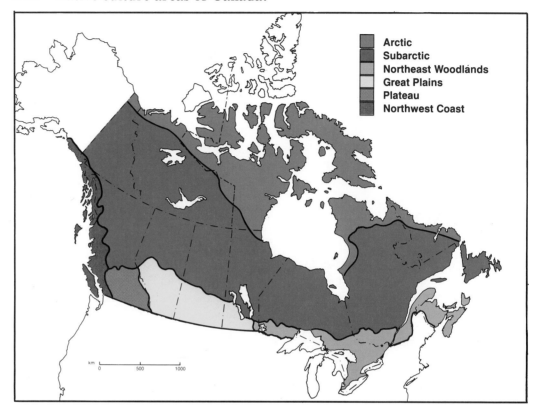

Arctic
Subarctic
Northeast Woodlands
Great Plains
Plateau
Northwest Coast

equal, and independent. Trade between them was of mutual benefit.

A Royal Proclamation of 1763 defined the relationship as follows:

Whereas it is just and reasonable, and essential to Our Interests and the Security of Our Colonies, that the several Nations or Tribes of Indians, with whom we are connected and who live under Our Protection, should not be molested or disturbed in the Possession of such parts of Our Dominions and Territories as, not having been ceded to or purchased by Us, are reserved for them, or any of them, as their Hunting Grounds.

Europeans brought guns to Canada, alcohol, and technology – and disease. Smallpox, influenza, scarlet fever, measles, typhoid, tuberculosis, chicken pox, cholera and diphtheria had enormous impact on Canada's Aboriginal Peoples. Some Nations lost an average of 25 - 50 percent of their population.

More and more Europeans settled in Canada, encroaching on Aboriginal lands. As the competition for land increased, European attitudes towards these first Canadians changed. The new settlers cleared forests, ploughed the prairie, mined the hills and altered river valleys. Many Europeans misunderstood Aboriginal peoples' concept of land because Aboriginals exhibited none of the possessiveness associated with individual land ownership. Aboriginals had no concept of individual land ownership. They believed that the land, and its creatures, and their people were one. The British Crown initiated and concluded several treaties with First Nations to remove Aboriginal title to the land. Some First Nations received cash payments for the land surrender. Others received annuities and other benefits, and were moved onto **reserves**, or land set aside specifically for their use. Sometimes these reserves were in areas where hunting and fishing was not available, and Aboriginals could no longer maintain their way of life.

The Indian Act of 1876 outlined the rights and responsibilities of Canada's Aboriginal peoples, making laws to cover many aspects of reserve life. Although revised several times (most recently in 1985), much of the act remains in use today. The Minister of Indian Affairs and Northern Development has **trust and fiduciary responsibility** with respect to all Canadian Aboriginal peoples.

The Act distinguished between **Status** or **Registered** Indians and **Non-status** Indians. The term Status Indian refers to those First Nations peoples who are registered as Indians in accordance with the Indian Act. Status Indians are eligible for government benefits and annuities which they receive as a result of various treaties and acts of government. Non-status Indians are not registered under the Indian Act.

One government goal was **assimilation**, where Aboriginal peoples would learn European customs and become "civilized." Young Aboriginal children were brought up in non-aboriginal homes or **residential schools** where their native culture was replaced by European ideas about progress and development. This policy caused many, long term, and harmful damages to Canada's First Peoples as it worked to destroy basic family units, and communities.

1. **Study Fig 2-20. How did Canada's Aboriginal peoples use their natural environment. Cite three examples.**
2. **You are a young Aboriginal meeting a European explorer for the first time.**

Write a short description of your meeting, and how you feel about these strangers.
3) **Look up the word fiduciary in your dictionary. Why do you think this particular term was used?**
4. **Describe how you might feel if you were removed from your family when you were seven years old and sent to live with another family of a different culture, speaking a different language, in a different part of the country, and attending a new school.**

New Words to Know

Aboriginal Peoples	Status Indian
Non-Status Indian	Anthropologist
Assimilation	First Nations
Inuit	Bering land bridge
Reserves	residential schools
Trust and Fiduciary responsibility	

Aboriginals in Canada Today

There were 811 400 Aboriginals in Canada in the 1996 census year. Four and one-half percent of all Canadians have Aboriginal ancestry. There are some 1000 Aboriginal reserve or settlement communities and between 60-80 Aboriginal Nations (wherein Nation is defined as "a sizeable body of Aboriginal people that possesses a shared sense of national identity and constitutes the predominant population in a certain territory or collection of territories," – Haida, Cree, Inuvialuit, Mi'kmaq, Blood, etc.).

Sixty-six percent of Canada's Aboriginal population is Canadian North American Indian (the designation includes both status and non-status Indians), 25 percent is **Métis** (mixed Aboriginal/European ancestry), and 9 percent is Inuit. Slightly under

2-23   Where the native people live.

	Status	Non-Status	Métis	Inuit	Total
Atlantic	3.9	4.4	1.5	7.9	3.7
Quebec	11.8	7.7	7.4	19.2	10.7
Ontario	24.0	34.8	12.9	4.3	22.4
Prairies	38.6	24.7	66.3	3.5	40.2
B.C.	18.5	25.5	9.1	2.0	16.8
Territories	3.2	2.9	2.8	63.1	6.2
Total	**100.0**	**100.0**	**100.0**	**100.0**	**100.0**

1. (a) Which region has the largest percentage of all native people? Which region has the smallest?
   (b) Is this the same region for each group within the native people?
2. Why might more Non-Status Indians live in Ontario?
3. Give evidence to show that native peoples are unevenly distributed by region across Canada, and by group within the regions.

2-24   Phil Fontaine Grand Chief of the Assembly of First Nations, walks with Federal Indian Affairs Minister Jane Stewart after their meeting in Vancouver at the Annual General Assembly.

200 000 respondents to the 1996 census reported an Aboriginal language as their mother tongue, with Cree spoken by 76 000 people, and Ojibway by 23 000. Two-thirds of the Inuit, or 27 000 people speak some dialect of Inuktitut, the Inuit native tongue. Some 20 000 additional persons reported learning an Aboriginal mother tongue later in life. There are more than 50 Aboriginal languages; these are usually grouped into larger language families as illustrated in Fig 2-19.

Status Indians have special rights and privileges which they have been given through various treaties and acts of government. Their personal property, when sit-uated on a reserve, for example, is exempt from tax. There were 610 874 Status Indians in 1996 (of which 331 289 lived on reserves). An **Indian band** is defined under the Indian Act as a group of First Nation people " for whose use and benefit in common, lands have been set aside" and for whom the Canadian government holds moneys in trust. Sometimes, the govern-ment will declare a First Nations group to be a legal band even if these conditions are not met. The Caldwell Potawatomi, for example, located near London, Ontario, do not have a reserve, but they are a legal band.

The Canadian Federal Government will spend some $6 billion dollars (slightly less than 6% of total federal expenditures) on Aboriginal Programs in 1997-1998 under the auspices of 13 departments including Indian Affairs and Northern Development (DIAND) which will account for $4.3 billion. Over 90% of DIAND's expenditures go towards providing basic services (such as schools, infrastructure, housing, education, social assistance) which other Canadians receive through their provincial, municipal, and territorial governments. Administration of these funds is handled by Indian band **councils**, which also have the power to zone land use and manage wildlife on the reserve, enact taxation and licensing laws, regulate reserve traffic, and oversee the construc-tion and repair of buildings.

The Canadian Constitution Act of 1982 affirmed the "existing aboriginal and treaty rights of the aboriginal peoples of Canada." This meant that existing Aboriginal land rights could not be over-ridden, although they could be given up or **extinguished** with the consent of the Aboriginal rightsholders. The Canadian Constitution also recognized an inherent right of self-government for Canada's Aboriginal peoples, and in 1995 the Canadian government began a process of negotiation with Aboriginal groups with this end in mind.

New Words to Know
Indian band
Extinguish
Inuit
Métis
council

Social Conditions

The forcible separation of Canada's Aboriginal population from the mainstream population of this country has created significant social discrepancies. The Aboriginal mortality rate is 11 per 1 000 compared to 6 for the rest of Canada, although for registered Indians the rate is 5.3. Teen suicide is significantly higher for status Indians than for the rest of Canada. Life expectancy is 69.1 for males and 76.2 for females.

One expert estimated that damage to the Aboriginal family unit as a result of the residential school system has taken over three generations (or sixty years) to correct. The average age of the 1996 Canadian Aboriginal population was 25.5 years compared to 35.4 years for the rest of Canada. Thirty-five percent of the population was children under the age of 15, and young people represent almost 20% of all Aboriginal age groups. There are 491 Aboriginal children under the age of 5 for every 1000 women of childbearing age (compared to 290 children per 1000 women for the rest of Canada). The birth rate for Status Indians is 2.7% compared to 1.3% for the rest of Canada. This fact has "sobering implications for future job needs."

Canada's First Nations have access to three educational systems: band-operated schools on reserves, federal schools on reserves, and provincially administered schools off reserves. Registered Indian and Inuit enrollment in post-secondary institutions almost doubled between 1987/8 and 1996/7. The number of band-operated elementary and high schools increased 70.2% from 262 in 1987/8 to 446 in 1996/7 as First Nations began "to create an education system that prepared children for modern-day life, while preserving their traditions. Special grants for training Aboriginal teachers, Aboriginal language classes and lessons in Aboriginal history and culture, all helped strengthen the new education system." The importance of family and of Aboriginal elders in the education process has been reaffirmed.

In 1996, as a result of the Federal government's Green Plan, 96 percent of on-reserve dwellings had water delivery systems and 91 percent had sewage systems, although "houses occupied by Aboriginal people are twice as likely to be in need of major repairs as those of other Canadians." Factors which contribute to this situation are the often extreme environmental conditions present on many reserve lands, and the sometimes minimal construction standards, lack of maintenance, and overcrowding.

Many of Canada's Aboriginal peoples are moving to build and strengthen the infrastructure of their communities. The new home building project initiated by the Gesgapegiag First Nation in Ouje-Bougoumou, Quebec (funded by government subsidy and bank credit) caused that community to be selected by the United Nations as one of 50 exemplary communities around the world. Health and social services are another area where band council control is helping resolve major discrepancies between Canadian Aboriginals on reserves and the rest of the country.

Aboriginal sense of community and family is a central part of the culture. This truth is reflected in the fact that 51 percent of First Canadian peoples 15 years and older participate in traditional Aboriginal activities, and 92 percent of them are secure in the knowledge that they have someone in their lives on whom they can call for help. These figures are particularly impressive when contrasted with those of non-Aboriginal Canadians.

The Aboriginal Business World

Canada's Aboriginal peoples comprise three percent of the Canadian labour force. They have traditionally been employed in government services, wholesale and retail trades, manufacturing, construction, and the hospitality and natural resource industries. Forty-seven percent of Status Indians living on reserves were gainfully employed on a full time basis in 1991: figures for off-reserve Aboriginals (57 percent) and **Métis** (59 percent) were marginally better, but still well below the 68 percent average for the rest of Canada. There are over 18 000 Aboriginal owned businesses in Canada. Many of these are on reserves serving local markets, but some concerns are national and international in scope. NorSask Forest Products of Meadow Lake, Saskatchewan, owned by the Meadow Lake Tribal Band, employs between 110 – 155 people, and produces lumber for construction, and high quality wood chips from white spruce, jack pine and balsam fir. First Air, the Inuit Airline, has a fleet of 75 airplanes, regular flights to the Canadian North, and employs 750 people. Advanced ThermoDynamics Corporation (30 people), a joint project with the Batchewana First Nation Reserve near Sault Ste. Marie, produces climate-control products for the trucking and transport industries, and is the only Canadian supplier of alternative heating and cooling systems. Cain Development, a business arm of the **Métis** Nation located in Happy Valley, Labrador, employs between 15 and 40 people on construction contracting work. Casino Rama on Mnijikaning First Nation (Rama) lands near Orillia, Ontario, operates 24 hours a day with celebrity performers, and employs 2 500 people. Growth areas for Aboriginal businesses include the tourist and services industries, natural resources, and health care.

During the 1990s, the Department of Indian Affairs and Northern Development launched a Procurement Strategy For Aboriginal Business Plan "to help Aboriginal firms do more contracting" with government departments and agencies. The federal government also instituted a policy of reserving all contracts worth $5000 or more serving a primarily Aboriginal population for qualified Aboriginal businesses. DIAND maintains an inventory of 3 752 Aboriginal suppliers for quotation on government contracts to facilitate this procedure.

Economic Development

Until recently, there have been limited opportunities for economic development on First Nation reserves. These lands are often located in more remote areas, with limited access to markets. Financing for development has been difficult to obtain because reserve lands are communally owned, and thus not available as mortgage or loan collateral for private individuals. Lack of infrastructure and labour force training have been additional disadvantages faced by entrepreneurs. Government programs like the Aboriginal Achievement Foundation, Northern Native Broadcast Access, Aboriginal Youth

Business Initiative, Native Internship, and Resource Access Negotiations (RAN), have been created to address these concerns. RAN helps First Nations to attract investment opportunities, strengthen business alliances, and provide jobs and community development. RAN program expenditures in 1995/6 were $3 651 477 for 118 projects.

Development in the North particularly, has often been seen as a threat to the traditional lifestyles of Canadian Aboriginal peoples. Although only slightly more than 1% of the population makes its living from fishing and trapping, hunting or harvesting, these activities have traditionally provided food and valuable supplementary income. They can be threatened by big development projects planned and engineered in the South, but erected in the North. Aboriginal communities are concerned about the development of resources on their lands without prior consultation, and without financial benefit to their communities. When large numbers of mining crews arrived at Labrador's Voisey's Bay to undertake the exploratory phase of development on Nain Hill, close to the centre of the Inuit town, residents were deeply concerned. No one had consulted with the Inuit about settlement of outstanding Aboriginal land claims in the area. What economic development does to animal habitats, water quality, and the spiritual and cultural values of Aboriginal peoples is another concern. Communities like the Walpole Island Potawatomi (near Sarnia) have successfully challenged developers and the government, holding these forces accountable for polluting the lands and waters around their reserves.

As more and more Aboriginal communities move to assume greater control of their resources, the need for capital increases intensely, along with workforce training, innovation, experience, and market development. These ingredients are vital to the successful economic growth of Aboriginal peoples in Canada.

Land Claims

Although much of Canada's land has been transferred to Crown control through treaties with ancestors of its original Aboriginal inhabitants, large areas remain where title was never surrendered. Aboriginal peoples hold **Aboriginal title** to regions of British Columbia, and selected areas throughout the rest of the country because of their ancestors' "use and long-standing occupancy of the land." **Aboriginal rights** are defined as rights to hunt, trap and fish on the ancestral lands. Additional aboriginal rights vary according to the practices, customs and traditions of each specific Aboriginal culture.

In 1973, the Canadian government, in response to a Supreme Court of Canada decision, issued the *Statement of Claims of Indian and Inuit People* recognizing two broad classes of Aboriginal claims: **Comprehensive** claims where Aboriginal title to the lands was never given up, and **Specific** claims where the government failed to fulfill lawful obligations agreed by treaty. The government is said to be in **breach** of its treaty obligations in these instances. Claims activities increased significantly after 1973, and the Office of Native Claims was established in 1974 under DIAND.

Much of the concern about land claims stems from that fact that Aboriginal peoples do not separate the land from the plants or animals, or from themselves or their culture. They look upon land not only as a resource, but as a defining element. Today, federal and provincial governments are either involved in, or investigating, over 250 Aboriginal land claims.

The first contemporary comprehensive Canadian Aboriginal land claim settlement was that of James Bay Cree and Inuit of Northern Quebec in 1975. It offered over $225 million in cash to a corporation representing the claimants, and acknowledged specific rights to a 169 902 km^2 land area, with separate agreements offering training, guaranteed incomes, and some self-government. In return, the Quebec government was allowed to proceed with Phase One of the James Bay hydro-electric project as the claimants surrendered Aboriginal title to 981 610 km^2 of land.

The Alberta **Métis** Settlement in 1990 provided that 4 856 km^2 of land, and $310 million be awarded to Alberta Metis over a period of 17 years. Alberta retained responsibility for roads and mineral rights. The Metis are able to veto the allotment of mineral rights, and to negotiate with oil and gas developers for royalty overrides.

In 1993, the Nunavut Land Claims Act was passed by Parliament, giving Inuit of the eastern and central Arctic title to 350 000 km^2 of land, mineral rights to 35 257 km^2, a cash settlement of $1.4 billion (paid out over 14 years), and a $13 million trust fund for educational training. In addition, the new federal territory of Nunavut, with over 2 000 000 km^2 of land will be established in April 1999, and the Inuit will receive a royalty share of the mineral developments on federal lands.

In 1996, the Nisga'a people of Northern British Columbia, concluded over twenty years of land claims negotiation with provincial and federal governments. The Nisga'a Treaty (released in 1998) grants the Nisga'a 2 000 km^2 of land in B.C.'s Nass River Valley, mineral rights, and a form of self-government. The Nisga'a are able to enact laws on language, employment, traffic, marriage, and land use. The Nisga'a are to provide their own health care, court system, police force, educational services and child welfare system. They have the ability to tax all people (Nisga'a or not) living on Nisga'a lands, and to hold their own elections. The treaty includes a cash payment of $190 million, funding for social programs, gravel roads, and fisheries. It must be ratified by the Nisga'a, the federal government, and the B.C. government before its provisions can be enacted.

One example of a specific claim is that of the Eel River Bar Nation in New Brunswick which charged that damming of the Eel River in 1963 by the neighbouring town of Dalhousie devastated the First Nation's "subsistence and commercial economy," causing the loss of salmon, eels, clams, and other river resources. This claim was argued in 1996 and 1997.

2-30 The mechanical rice harvester can harvest as much grain in half an hour as a hand harvester can pick in a day.

2-31   Justice Berger in Nahanni Butte in the Northwest Territories. *The commission that examined the implications of the Mackenzie Valley pipeline held hearings in the communities that would be affected.*

2-32   "The native concept of land."

It is very clear to me that it is an important
and special thing to be an Indian.
Being an Indian means being able to understand
and live with this world in a very special way.
It means living with the land, with the animals,
with the birds and fish,
as though they were your sisters and brothers.
It means saying the land is an old friend
and an old friend your father knew,
your grandfather knew,
indeed your people have always known . . .
We see our land as much, much more than the white man sees it.
To the Indian people our land really is our life.
Without our land we cannot—we could no longer exist as people.
If our land is destroyed, we too are destroyed.
If your people ever take our land you will be taking our life.

Richard Nerysoo of Fort McPherson,
quoted in *Northern Frontier, Northern Homeland*,
by Mr. Justice Thomas R. Berger, 1977.

The final report of the Indian Claims Commission rejected the claim, but allowed for an appeal to the DIAND.

A second example of a specific claim was that made by Alberta's Cold Lake Nation regarding the creation in 1954 of the Primrose Lake Air Weapons Range on 182 km^2 of land in Northern Alberta and Saskatchewan. The Indian Claims Commission released a report in 1993 recommending the claim be accepted for negotiation because Canada had breached its treaty obligations by expelling band members from much of their lands, subjecting them to "social, economic, and cultural devastation," without proper compensation. The government responded in 1995, denying any breach of responsibility, but agreeing to enter into negotiations with the band due to "the unusually severe impact" the creation of the Primrose Lake Range had on the Cold Lake and Canoe Lake Nations. Those negotiations are still underway.

In August 1998, the Supreme Court of Canada issued the Delgamuukw v. British Columbia decision which addressed the issues of Aboriginal title, its content and its protection under the Canadian Constitution, the provincial extinguishing of Aboriginal rights, and the issue of the Aboriginal right to engage in particular site-specific activities intrinsic to the practices, customs and traditions of distinctive aboriginal cultures.

The ramifications of this decision will have significant impact on Canadian land claims issues for some time.

1. (a) **Read the poem *The Native Concept of Land* by Richard Nerysoo carefully. Compare how he feels about the land with how you feel about your home.**
   (b) **Write a short poem, song, or descriptive paragraph on this topic from a variety of cultural viewpoints.**
2. **List the ways the James Bay settlement has benefited the Inuit and Cree of that region.**
3. **Suggest why you think Aboriginal land claims might be difficult to settle.**
4. **How do Canadian Aboriginal Peoples benefit from living on reserves? What are the limitations or disadvantages?**
5. **Research the Delgamuukw v. British Columbia Supreme Court decision.**

What implications do you think it will have on Aboriginal land claims in Canada.

To the Future

Canada's Aboriginal population is expected to reach 959 000 by 2006 and 1 093 400 by 2016. The Status Indian population is expected to grow by 2.3% on reserves, and 2.4% off reserves.

In 1991, the Canadian Government set up a Royal Commission to study the "relationship between Aboriginal and non-Aboriginal people in Canada, to propose practical solutions to stubborn problems" so that the future would be one of justice, peace and harmony for all peoples of this country. The seven commissioners visited 96 communities, and held over 178 days of public hearings before releasing their *People to People, Nation to Nation* report in 1996, concluding "The main policy direction, pursued for more than 150 years, first by colonial then by Canadian governments, has been wrong .... We believe Aboriginal people must be recognized as partners in the complex arrangements that make up Canada ... that Aboriginal governments are one of the three orders of government in Canada ....The three orders are autonomous within their own spheres of jurisdiction, sharing the sovereignty of Canada as a whole."

Unified Voice

Canada's Aboriginal peoples are a cultural mosaic, and it is difficult to forge a unified consensus when so many different nations and cultures are involved. Although there is wide general agreement on the nature of the problems facing Canadian Aboriginals, each First Nation, or band, or settlement, must work to preserve its own successful survival as well. The Assembly of First Nations (AFN), the Inuit Tapirisat of Canada, the **Métis** National Council, the Dene Nations and the Native Council of Canada are all national organizations devoted to the concerns of Canada's Aboriginal peoples. It is through the efforts of these lobby groups, and through the voices of Canada's Aboriginal peoples themselves – lawyers, judges, MBA's, doctors, teachers, students, singers, artists, and actors that their economic, social, cultural, and constitutional concerns will be heard.

New Words to Know
Aboriginal title/land rights
Extinguished
Comprehensive
Specific

2-33 Claim areas in northern Canada, in which four groups of Aboriginal Peoples have traditional land-use rights.

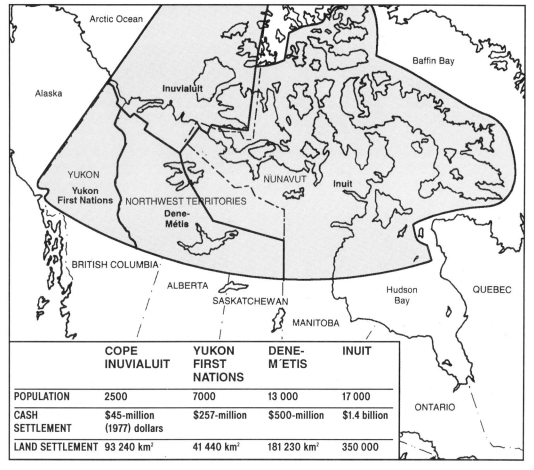

	COPE INUVIALUIT	YUKON FIRST NATIONS	DENE-M´ETIS	INUIT
POPULATION	2500	7000	13 000	17 000
CASH SETTLEMENT	$45-million (1977) dollars	$257-million	$500-million	$1.4 billion
LAND SETTLEMENT	93 240 km²	41 440 km²	181 230 km²	350 000

Migration

Canada is a sparsely populated country. Our 1997 population of over 30 286 600 people is the result of two factors. One is the birth and death rate. Until recently, the birth rate has been much higher than the death rate, contributing to Canada's population growth. In 1996, Canada's fertility rate fell to 1.59%, and as the Canadian death rate continued to rise (7.2% that same year), population or **demographic** experts estimate that Canadian birth and death rates will equal one another by the year 2020, so that any major changes in the Canadian population, its distribution and composition after the year 2020 will be the result of immigration.

In 1996, immigrants to Canada represented 17% of our population; that number is expected to rise as we enter the new millennium. Between July 1, 1996 and June 30, 1997, 223 238 people immigrated to this country, while 49 633 emigrated from it to another country. There were 364 765 births over the same period, and 216 491 deaths. The natural growth, difference between the birth and death rates, was 148 274, and the growth due to immigration was 173 605 (immigration less emigration). Thus 53% of Canadian growth during this period resulted from immigration.

Statistics Canada does not keep detailed records of those who leave this country to live in another. The Canadian government does not require exit visas or formal letters of emigration, and the emigration figures quoted here are official government estimates compiled from a variety of sources. People leave Canada for a variety of reasons – a better climate, particularly

2-34 Immigration and Emigration, 1966-1996

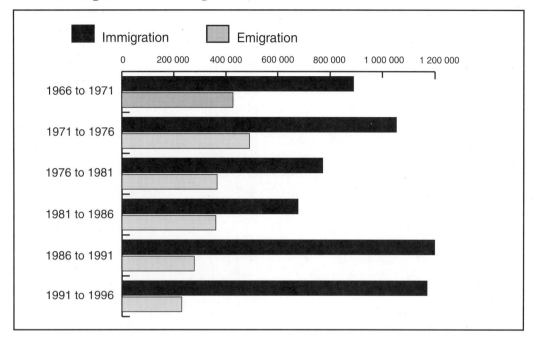

in retirement; to study abroad; for research or marriage; for work-related reasons; or even because of family. Emigration from Canada was highest from 1971-1976 with just under 500 thousand people leaving the country during that period. Since then, however, it has remained fairly steady at between 45 - 55 000 people per year.

Immigration statistics since Confederation in 1867, show that the number of immigrants to Canada has varied greatly. The time of heaviest immigration was the early part of the twentieth century, as the great vastness of our Canadian west was opened up for settlement. The number of immigrants increases significantly during economic "boom years," declining dramatically during times of economic hardship. World events have significant impact on our Canadian immigration

figures as famine and war in Somalia, the termination of Britain's 99-year lease of Hong Kong, or the strife and unrest in Eastern Europe have demonstrated.

Canada's immigration policy, which sets out the number of immigrants allowed into this country every year, and the requirements for each, has been changed and modified over the years to reflect this country's changing needs, values, and attitudes. As a result, there have been great changes in the number of immigrants, in their source countries, their occupations, backgrounds, and reasons for coming to Canada. Each new immigrant contributes in some way to this country's diverse multicultural society. Each adds to this country's social and economic growth.

Why Do People Migrate?

Each year, thousands of people make the decision to leave their homes, their countries of birth, and sometimes, their families, to come to Canada. Each individual has personal reasons for deciding to migrate (emigrate). All these reasons, however, can be described in one of two ways – as a **push factor**, or as a **pull factor**. War, poverty, unemployment, ethnic or religious persecution are examples of push factors; they are conditions within immigrants' native countries pushing them to leave. Civil war in Sri Lanka, Hong Kong's changing government, ethnic cleansing in Bosnia and Rwanda are all strong push factors. Pull factors include opportunities for education and training, better pay and social services, and the benefits of living in a peaceful democracy. Many students come to Canada to take advanced degrees in our universities; **entrepreneurs** come to Canada searching out investment opportunities; families come to Canada seeking a more prosperous lifestyle for their children, and so on. The reasons are endless.

A Profile of Canada's Immigrants

The first non-Aboriginals to settle in this country were Europeans, mainly from France and Britain. These people became farmers, fur-traders, loggers and fishermen, taking advantage of whatever the economy had to offer. Following the American Revolution in 1776, hundreds of families came to Canada from the United States. Many of these people were United Empire Loyalists who settled around the Great Lakes, and in the Atlantic Provinces. Many of these families brought their slaves with them. This was the beginning of

2-35   Russian immigrants newly arrived in Quebec City, 1911. *Many East Europeans were attracted by the free land offered by the Canadian government.*

Black settlement in Canada. During the eighteenth and nineteenth centuries, the continuous flow of British immigrants was replaced by Chinese, Italian, and Irish workers who laboured on Canada's first roads, railways, and canals. Thousands of Eastern Europeans settled the Canadian prairies in the early part of the twentieth century. Canada opened its doors to these immigrants because there was a need for cheap labour, and because the country needed to populate its empty western territories. The fact that the provinces of Alberta and Saskatchewan were created in 1905 is no coincidence.

The two World Wars, and the Great Depression had an important impact on immigration. Thousands more Europeans were forced to immigrate, many of them **displaced persons** unable or unwilling to return to their homelands. Yet Canada did not want, and did not encourage immigration during these periods. It was not, generally speaking, until the 1950s that this country began to openly encourage immigration as it had done in the earlier years of the twentieth century. Since 1990, Canada's population growth has depended upon immigration, and today, Canadian policy welcomes people from many parts of the world.

Canada's Immigration Policy

In pioneer times, Canada had no immigration policy. In fact, until 1952, Canada had no Department of Immigration. At one time, immigration was even under the Department of Agriculture. Canada's earliest immigration policies were designed to prevent certain ethnic groups from entering the country. The Chinese Immigration Act of 1923, for example, restricted entry into Canada for Asian peoples. It was not, in fact, until the 1960s that Canada finally designed an immigration system that did not discriminate on the basis of race, colour, or religion. The current Canadian Immigration Policy was enacted in 1976, with supplementary regulations added in 1978.

Canadians for immigration to Canada fall into three classes: Family class,

2-36 Immigration, 1870-1996

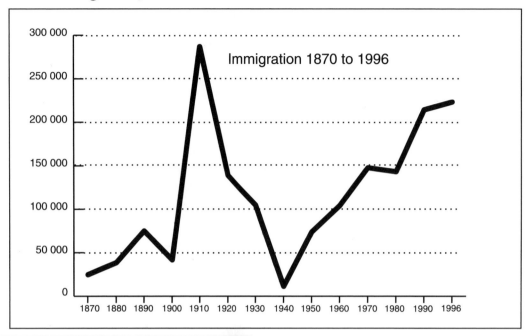

Immigration 1870 to 1996

Refugee and Humanitarian class, and Independent and other class. No matter what the class, however, all immigrants are subject to medical checks and background checks to determine whether an applicant is carrying an infectious disease (like tuberculosis), or whether he or she has a criminal record (a murder conviction). Discovery of any medical or legal impediments can cause an immigration application to be denied.

The government's Member of the Family policy attempts to make it as easy as possible for new Canadians to bring their immediate families to Canada. This class allows a wife or husband, dependent son or daughter (under 19 and single, full time student, or supported primarily by the parents after age 19), fiancee, parent or grandparent, orphaned or unmarried brother, sister, niece or nephew, or grandchild under the age of 19, or adoptee to

2-37 Hassidic Jews in Montreal. *Many immigrant groups have been able to retain their customs and culture in Canada.*

2-38 **American settlers on their way to Alberta around 1890.** *About a third of the immigrants to Canada before 1914 were Americans who were attracted by the inexpensive and plentiful land.*

enter the country as long as they are sponsored by a Canadian citizen or resident "financially capable of providing assistance" to the applicant and dependents for a specified period of time. By agreeing to become a **sponsor**, the Canadian assumes responsibility for the applicant legally, and socially, as he or she adapts to life in Canada. Being a sponsor also means making every effort to help the new immigrant make a smooth transition into Canadian society.

The independent applicant category is very different. All independent applicants must be interviewed, and are subject to a screening process through a **points test**. This process was instituted in 1967 to remove racial bias so that the screening would be equally fair to all. Points in the test are allotted for a variety of social and economic factors (see Fig 2-40 for the

Immigration Criteria in 1996. Note that it is possible to assemble more than 100 points, or a better than perfect score). The number of factors and their relative importance is constantly changing. This means that the points test in 1996 was substantially different from that in 1988. There are some exceptions and special cases built into the system, to help relatives, business applicants, and retirees. Immigrants with a relative in Canada willing to assist them can receive ten **assisted relative** points, provided they have an "undertaking of Assistance," a legal promise of support made by a Canadian relative already established in the country. The assistance is in the form of sponsorship, although the applicant does not have to be of the immediate family, and is not expected to be totally supported by the Canadian relative.

Canada has three categories of business

applicant, "investors, self-employed persons and entrepreneurs." Investors must have owned or operated a business, and have accumulated a minimum worth of $500 000 CDN (lottery winnings or inheritances are not eligible), of which a minimum of $250 000 CDN must be invested in a qualifying Canadian business or fund for a fixed, locked-in five-year period. Entrepreneurs must "show by their qualifications, experience and business plans that they intend to and are able to establish a business [which] must bring significant economic benefits to the country. Applicants in the self-employed category "need enough money to realize employment plans that contribute substantially to Canada's economy or its artistic and cultural life. They must also have enough money to support themselves and any accompanying family until their self-employment income is adequate for this purpose."

The chart in Fig. 2-41 shows Canadian immigration by major classes. The figures for 1981 (family 51 017; independent 62 622, refugees 14 979) and 1991 (family 86 378; independent 68 755; and refugee 53401) are actual. Those for 1998 are projected.

The Immigration Plan set for 1998 allows for between 200 - 225 000 immigrants and refugees to enter the country (5000 higher than the plan set for 1997). These plans are guidelines, and can be exceeded. The Ministry of Citizenship and Immigration estimated 1998 figures to include up to 192 700 immigrants, and 32 300 refugees. Within these figures, breakdowns are expected to be as follows

2-39 Immigration to Canada by source region.

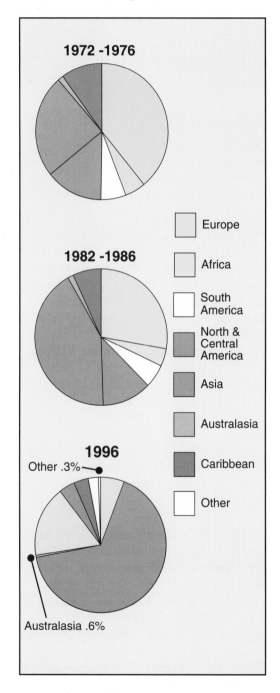

1972 -1976

1982 -1986

- Europe
- Africa
- South America
- North & Central America
- Asia
- Australasia
- Caribbean
- Other

1996

Other .3%

Australasia .6%

Immigrants	
Skilled workers	up to 106 600
Business applicants	up to 21 300
Spouses, Fiancés and Children	up to 38 300
Other	6 500
(Live-in caregivers, special categories, provincial, territorial nominees)	

Refugees	
Government-assisted	7 300
Privately-sponsored	up to 4 000
Landed in Canada	up to 18 000
Dependents abroad	up to 3 000

Normally, anyone wishing to apply for immigrant status must do so from outside the country. There are hundreds of Canadian consulates, embassies, business immigrant coordination centres, and visa offices around the world, in addition to the Department of Citizenship and Immigration's comprehensive website at http://cicnet.ci.gc.ca., all of which handle hundreds of hundreds of thousands of inquiries per annum. 225 495 people immigrated to this country in 1996, almost two-and-one-half times as many as the 1986 total of 99 219, so it is obvious that these offices are well used.

1. (a) **How do you think most Canadians feel about immigrants?**
   (b) **What are your views? Explain.**
2. **Name the classes of immigrants. List the similarities and differences among them.**
3. **In your own words, describe the changes in the size of each class of immigrant between 1981 and 1998.**

2-40 Immigration criteria, 1996.

FACTOR	POINTS	EXPLANATION
Language	10	10 for fluency in one language 15 for both French and English
Vocational Training	18	1-3 months = 3 points 1-2 years = 9 points 10+ years = 15 points
Education	16	1 point for each year of formal education
Occupation	10	determined by demand for selected occupations. If there is no demand this means status is refused
Arranged Employment	10	applicants must have a letter from a Canadian employer showing they have a job in Canada
Age	10	10 points for age 21-44 but 2 points are subtracted for each year above or below
Personal Suitability	10	based on a personal interview with an immigration officer
Demographic Factor	8	quota number set by the Minister
Experience	8	point for each year of job experience
TOTAL	**100**	
Bonus	10	for assistance from a close relative
	30	for self-employed immigrants

A number of points are given for each category in this "points method" of determining an immigrant's suitability to settle in Canada.

New Words to Know
Family reunification
Assisted relative
Refugee
Push and pull factors
Sponsor
Points test
Demographic
Entrepreneur

Where Do Immigrants Go?

Since the 1950s, over 90 percent of immigrants have listed Ontario, Quebec, British Columbia or Alberta as the province where they intend to live. Fig 2-45b illlustrates those Canadian urban areas to which immigrants were most attracted. Note what percentage of new immigrants chose Vancouver or Toronto.

This pattern continues to add to the imbalance in Canada's population distribution. We know that this country's population is located in pockets or "islands" separated from each other by huge, sparsely inhabited regions. Immigrants to Canada do not "fill in" the spaces or even expand the size of the "islands". The majority of new Canadians move to provinces where the largest proportion of Canadians already live, thereby increasing the population densities in these regions. This puts greater pressure on the urban areas and communities within these provinces.

Many cities have developed "ethnic" neighbourhoods" where immigrants with similar background dwell when they first come to Canada. These neighbourhoods often act as a magnet for people of similar background. Recent immigrants can speak with their neighbours in their native

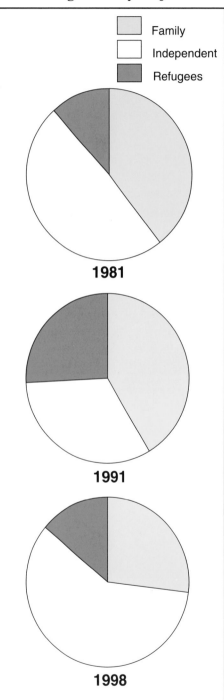

2-41 Immigration by major classes .

Family
Independent
Refugees

1981

1991

1998

2-42 Top ten intended occupations of immigrants to Canada.

	1981	1991
Entrepreneurs	293	3 615
Manager/Administrators	3 601	4 633
Managerial	3 894	8 248
Professional/Technical	13 677	15 273
Clerical	7 044	7 385
Commercial & Financial	2 151	3 203
Service & Recreational	4 361	9 518
Transportation & Communication	691	1 114
Agriculture	2 931	1 939
Mining & Quarrying	221	122
Manufacturing	12 189	16 401
Labourers	674	1 067

Total immigrants to Canada in 1991: 232 758

tongue, while receiving second language training in English or French. They can buy familiar goods, clothing, a wide selection of foods unique to their culture, and newspapers in their own language. They can establish social clubs and support agencies.

Anyone who has moved from one house to another is aware of the kind of problems that can result from relocation. Moving from house to house, however, is far less difficult than moving from one country to another.

The feelings of isolation and discomfort felt by recent immigrants to Canada frequently lead to culture shock. New people in Canada are overwhelmed by the great differences between what they were used to in their native countries and what they face here – new customs, new currency, new laws, a new way of dress and behaviour, and, often, a different language.

Large urban areas across the country have worked to counter these problems by establishing self-help centres and organiza-tions to assist immigrant populations. The West Central Community Information

Centres in Toronto, for example, offer life skills, language and housing advice/assistance in Spanish, Portuguese, Hindu, Bengali, Punjabi, Urdu and English, while the Etobicoke North Community Centre offers children's summer camps, income tax assistance, and guidance on preparing documents, affidavits, and a variety of government forms in French, Somali, Spanish, Arabic and English.

When immigrant groups assimilate into the Canadian cultural mosaic, achieving a secure economic base, they usually move away from the city into the suburbs, as other more recent immigrant clusters move in to replace them. Most urban areas are subject to similar stages of succession as different ethnic groups migrate in and out of the city core, enriching the urban mosaic with every change.

Immigrants and the Work Force

Statistics Canada figures show that between 40-50% of Canadian immigrants enter the work force; the remainder are spouses, children, students or retirees. Fig 2-42 lists the top ten intended occupations for Canadian immigrants.

Immigrants tend to be slightly better educated; those with postsecondary degrees are more likely than their Canadian counterparts to be graduates of professional programs in engineering, mathematics, and applied science. The employment percentage is about equal to that of native-born Canadians, although "immigrants with jobs are more likely to be self-employed," and immigrant men are more likely than their Canadian counterparts to have full-time jobs, while for immigrant women this is a less likely scenario.

Cultural Diversity

The energy, enthusiasm and skills that immigrants bring to Canada have made an immense contribution to the economic development of this country. Yet immigrants bring much more than business acumen or professional skills: they also bring their culture and heritage.

Holidays and festivals, traditions of dress and social customs are just a few examples of cultural attributes. Some are visible objects. A house may be remodelled to include a cantina, or cool storage area, beneath a verandah or porch. Clothing may also be associated with a particular group. Can you pick the country where one would find a tam? a poncho? a kimono? a sari? Some immigrants bring new social customs, traditions and religious beliefs. Which immigrant groups are associated with Caribana? a mosque? Rosh Hashanah? Gurdwara? Kwanzaa? These things cannot

2-43 Many new immigrants establish businesses that attract other new citizens from the same countries. *This section in the east end of Toronto offers the foods and clothing that are part of the East Indian culture.*

always be seen, but they are part of Canada's cultural diversity, and are becoming part of the Canadian heritage.

Most Canadians welcome immigrants because they add to this country's multi-culturalism. In large urban areas, it is easy to find foods that are native to many different countries, to hear music from around the world, and to listen to many different languages and dialects. Fig 2-45a illustrates the wide range of visible minorities in this country.

Cultural diversity has always character-ized Canada. Aboriginal peoples spoke many different languages, and practiced many different customs, and traditions. This diversity was not officially recog-nized until 1967, with the passing of the first discrimination-free immigration act. Bilingualism was recognized in 1969. Canada became officially multicultural in 1971, and our multicultural policies were changed into laws in 1987. In 1997, the Ministry of Citizenship and Immigration appointed a legislative review advisory group to "analyse the current legislative provisions regarding immigration and refugees to determine whether they remain sufficiently flexible and complete to provide an optimum response to emerging issues and migration trends." That report, and its recommendations, to be released late in 1998, will form the basis of Canada's immigration policies into the new millennium.

Canada's multicultural flavour is easiest to spot in our urban centres. In 1981, 48 percent of Toronto's population was of British ancestry. This percentage had dropped dramatically to the low 20s by the late 1990s. By 1996, 25 percent of the Toronto population was Chinese,

24.7 percent was South Asian, and Blacks accounted for 20 percent. In Halifax, Blacks accounted for 54 percent, Arabs and West Asians 10 percent, and South Asians 11 percent. In Saskatoon 17 percent of the population is Chinese, and 6 percent Fillipino. Latin Americans accounted for 11.6 percent of Kitchener's population mix, and 9.2 percent of the people in London, Ontario.

Canada is officially a bilingual country, and it is rapidly becoming a multilingual one. All of our large urban centres have newspapers in languages other than French or English; many stores have signs indicating the availability of multi-

language service staff. Todays' employers often give preference to job applicants who are fluent in more than their mother-tongue or first language.

1. **(a) Why is the destination of most Canadian immigrants of concern to immigration officials?**
(b) Why does this problem have no easy solution?
2. **Give examples of visible and invisible cultural attributes.**
3 **List five items you use that are not of "Canadian" origin.**
4. **Study the charts in Fig 2-45. How do you think Canadian immigration patterns have changed in recent years from those at the turn of the century?**

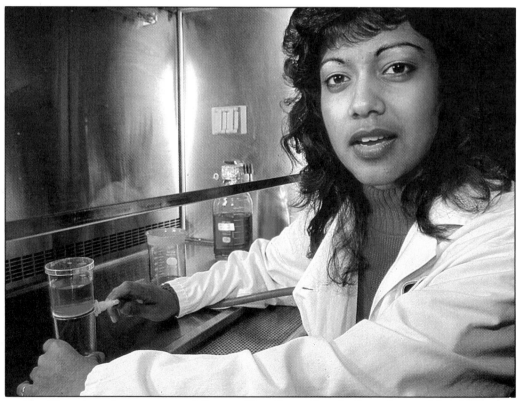

2-44   New immigrants are able to offer Canadian industries and businesses the skills and education that they acquired in their own countries.

Canadian 1996 census figures for ethnic origin indicate the top 25 most frequent responses, ranked in order, are: French, English, Chinese, German, Italian, Scottish, South Asian, Irish, Aboriginal, Ukrainian, Dutch, Polish, Portuguese, Filipino, Jewish, Greek, Jamaican, Hungarian (Magyar), Spanish, Norwegian, Russian, Swedish, Welsh, and American.

2-45a & b *Nova Scotia's Black Cultural Centre, in Dartmouth, depicts Black culture and history. The Centre's programs feature lectures, slide shows, and school and community presentations. It airs a local tv show* **Umoja Nova Scotia** *on Access cable.*

2-46 *Caribana, sponsored by the Caribana Cultural Committee, is Canada's largest annual summer festival. The Caribana parade in Toronto attracts over one million participants and viewers from all over the world, as Mas Bands compete for annual Festival King and Queen honours, and Mas Band of the Year.*

2-45c & d *Toronto has a vibrant Chinese community.*

New Words to Know

Cultural attribute	Ethnic Group
Mother tongue	Cultural mosaic

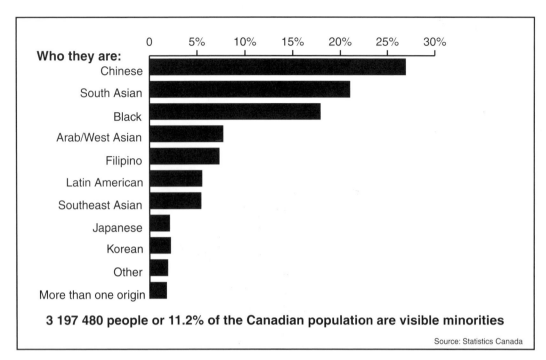

3 197 480 people or 11.2% of the Canadian population are visible minorities

2-45a   Visible minorities in Canada, 1996.

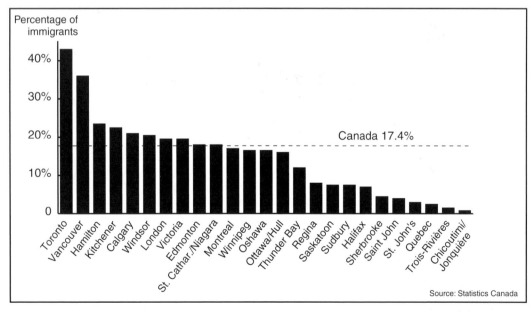

2-45b   The 25 top destinations for immigrants arriving in Canada, 1996.

The Italian-Canadian Experience in Canada

It is a well know fact that Canada's human diversity is the result of immigration. All immigrants, from the French settlers of the 1700s, to the British of the 1800s, to the varieties of peoples of the 1900s, contributed much to Canadian culture and economy. They also faced similar problems and obstacles.

All groups had to adjust to new language, foods, climate, laws, and customs. This culture shock resulted in many challenges for these new Canadians. The rate of acclimatization depended upon the specific background. British immigrants, for example, found it easier to adjust to the new Canadian way of life than did people from China, or India, Italy, or the Caribbean. Individual immigrant experiences varied. Some immigrant groups faced much more racism and discrimination than others, although, for the most part, immigrants overcame the initial challenges they faced, and integrated into Canadian society. The experiences of one Italian immigrant represent, in many ways, the typical experience of many Italians in Canada. In fact, Antonio's story is very similar to that of many immigrants who came to this country after 1950. Antonio Montopoli writes about his immigrant experience:

I came to Canada in October, 1956. I came to Canada for a better life for my family. After a two week ocean voyage, I disembarked at Halifax. From there, I took my pre-arranged train ride to Toronto, where my brother Enrico met me at the station. It was Enrico who sponsored me to Canada. He and his family had been here three years.

The house my brother rented had hot running water, a phone and television. These were new conveniences which excited me. Only rich people had these facilities in Italy. The novelty was short lived, for I quickly experienced loneliness as I missed my family very much. Sometimes I became depressed. I left behind my wife Lucia and my two children, Enio and Angelo. In the 1950s, its was typical for men to immigrate and leave their families behind.

Two days after my arrival, I began working as a construction labourer. In Italy we were farmers. The work was hard and the hours were long. We often worked 56 hours a week. The pay of 75 cents per hour was an incentive to work hard. I saved as much money as I could to bring my family to Canada. Each week I would send some money to my wife Lucia.

The first few years were very difficult. I never experienced a climate as cold as Canada's. In addition, I did not speak English and had to learn how to get around the city by bus. In 1956 we did not have health insurance, nor did we have government agencies that helped immigrants adjust to their new environment. You could only depend on relatives for support.

By 1959, I saved enough money to sponsor my family to Canada. In June my wife Lucia arrived with our two sons. My wife had a difficult time adjusting. She had left her mother and sister behind. I knew what she was going through. I too, left my parents behind.

Also in 1959, my brother Enrico combined our savings and put a down payment of $3 000 on a $15 700 home. The three bedroom house now accommodated two families – 9 people. In 1961, we sponsored my younger brother Rocco.

Can you imagine three bedrooms and one washroom for ten people? In those days it did not seem like an inconvenience. This was progress, after all, we had our own home. For most Italians it was important to have a home. We had one in the old country.

Most Italians lived in the same part of the city. Eventually this area was referred to as Little Italy. Having other Italians nearby made the adjustment easier. There were common experiences which bound us as a group. Going to church on Sunday was more than a religious experience. It was also a social experience. It was a chance to spend some time with the other Italo-Canadians. Many Italians from different parts of Italy became friends and neighbours. Many social organizations were started by people of the same region. The Abruzzesi Club from central Italy is an example. These clubs sponsored feasts to commemorate the patron saints of their hometowns. Many clubs are still active today.

Many Italians began their own businesses serving our community. These family businesses provided employment for Italians and helped them meet their needs. Stores, travel agencies, bakeries, pizzerias, butcher shops, barber shops and tailors are examples. The most interesting was the cafe. The cafe was in any ways similar to today's sports bar. It was here that men congregated in the evenings and weekends. Expresso coffee and cappuccino with homemade Italian pastries were popular. (The women were at home doing the chores).

The construction industry in Toronto employed many Italians. In Northern Ontario many were employed in the mining industry. Much of the work was done by manual labour and there were many men willing to work. Unfortunately, many suffered injuries and some even death as a result of work accidents. Working conditions were not as regulated and therefore, not as safe as they are today.

Family-owned stores carried a lot of Italian products such as salami, cheeses, and pastas. Some small family businesses grew to become large importing and exporting companies. We bought California grapes from a company called Darrigo Grade Juice. Cappola Foods, which specialized in cured Italian meats, was familiar to many. Unico brand was well known for its canned food products.

We used to make our own wine just like in the old country. Of course, of all my relatives and neighbours, I made the best wine. This was a tradition that Italian men continued in Canada. In late summer we would buy tomatoes to make our own tomato paste used for sauce. Since pasta was a popular dish, we prepared enough tomatoes to make sauce for the year. We also cured our own meats such as salami and prosciutto (delicious naturally cured ham). Looking back, these activities were undertaken because of tradition. They were also important because in the early years, there was a limited selection of Italian products in Toronto. One can say it was a case of meeting our needs.

My wife Lucia, like most other wives of immigrant men, had a full time job. She

worked for 25 years at Holiday Sportswear as a sewing machine operator. She truly earned every penny she made as she got paid by "piece work". After a hard day's work she would come home to prepare dinner and take care of the household.

In retrospect, much credit must be given to Italian-Canadian women like my wife. It was because of her household management we bought our own home in 1966. Contrary to what many Italian men say, it was the wife who controlled the family *finances and forged strong family ties.*

My sons Enio and Angelo had an easier time learning English and adapting to new ways. They quickly learned to like hot dogs and burgers. Like Canadian kids, they wanted to buy skates and play hockey. In a way, this was typical of children of Italian descent. In fact, by the mid-sixties, a few Italian Canadians from Northern Ontario were playing in the National Hockey League. One was Phil Esposito.

Enio and Angelo attended school. Within a few short years, they forgot much of the Italian language they knew. By the early sixties, a generation gap and cultural gap developed between me and the boys. They were much more Canadian than Italian. This, however, presented difficulties for me. Many arguments and shouting matches about teen-aged freedoms and independence resulted.

A turning point in my life came in 1966. It was then I took my first trip back to Italy. I was happy to see my parents who lived with my older sister and her family. When I got back to Toronto, I realized Canada was becoming my country. The following year I became a Canadian citizen with full citizen rights. I was proud to vote in elections. By the 1990s, after several trips to Italy, it had become a strange country for me. Yes I do have friends and relatives in there, but their beliefs and values are different from mine. Life in Italy had also changed. Their way of life was different. I had integrated into the Canadian mainstream – economics, politics and social life. Sure, I still make my own wine, sausages and help my wife process the tomato for sauce. But, I am now a Canadian of Italian descent. Perhaps my feelings can best be expressed by the motto once used by an Italian

Canadian newspaper, the Corriere Canadese: **fiercely Canadian; proudly Italian**.

The Italians have endured many hardships in Canada. Difficulties adjusting to a new land, new climate, new laws and even discrimination. As a rule though, I think they have adapted well. The second and third generation Italians have become directly involved in Canadian life. Many have shown leadership in politics and have become M.P.s and M.P.P.s. Many family businesses have become corporations. Other Italians have become successful in the medical, legal and other professions. The second and third generation Italians have taken advantage of the potential Canada offers.

I believe that recent immigrants to Canada, from 1980 to the present, experience many of the same problems and frustrations I experienced. However I also believe that they have a much easier time to acclimatize and integrate. One major reason is that the world has become smaller. People call it a global village now. Communication around the world by television, cell phones, faxes, e-mail, and air travel has resulted in a better informed immigrant. Upon coming to Canada, many already know what to expect. They know the support network that is available. In addition, they are not as homesick because of cheaper international travel and inexpensive global telephone network.

In addition today there are many government agencies and church organizations that provide assistance to immigrants. For example there are heritage language classes for their children, and they also receive automatic health insurance. Not having health coverage in the late 50s and

early 60s worried me a great deal.

Toronto was declared to be the most cosmopolitan city in the world in 1998. There are so many different nationalities represented here that newcomers feel at home. More important, there is a general feeling of acceptance and tolerance in Canada. Sure, discrimination and racism still exist, but not like it did 50-75 years ago. The head tax on Chinese at the turn of the century and the internment of Japanese Canadians during World War II are two of many examples. These are wrongs of the past. Today our governments spend much money promoting multiculturalism. Be it in British Columbia, Manitoba, Ontario or Nova Scotia, we are encouraged to be proud of our heritage. Because of these factors, recent immigrants like the Roma refugees from the Czech Republic will have an easier time in making Canada their adopted country.

1. **In a brief report of one hundred words or less, describe your reactions to Mr. Montopoli's comments and opinions.**
2. **Why does the motto "fiercely Canadian, proudly Italian" best describe the patriotic feelings of many Italian immigrants?**
3. **How would it be different for a person immigrating to Canada today than it was for someone, like Mr. Montopoli, who came in the 1950s?**
4. **Are your parents 'first generation' Canadian? If so, please share some of their experiences with the class.**

Refugees

Most immigrants came to this country to better their lives. Some, however, fled to this country in fear for their lives or to gain freedom from oppression. Such new Canadians are called refugees. On humanitarian grounds, they do not have to go through the regular immigration screening process.

Canada uses the United Nations definition of a refugee. According to that body, a refugee is a person who has a well-founded fear of being persecuted for any reason, and who wants to live outside the country of his or her nationality. Refugees may be unable or unwilling to return to their homeland. Since 1945, one of every ten newcomers to Canada has been a refugee.

Canada has received praise as a leading country in accepting refugees from around the world. They include 186 150 **displaced persons** from Europe following the Second World War; 37 149 Hungarians following the 1956 Hungarian Uprising; 228 Tibetans after the Dalai Lama was exiled in 1959; 11 943 Czechoslovakians after the Soviet occupation in 1968; 7069 Asian Ugandans fleeing Idi Amin; 7016 Chileans, Argentinians and Uruguayans escaping oppression in the 1960s and 1970s; 11 321 Lebanese fleeing civil war; and 60 000 Indochinese from Vietnam, Cambodia and Laos in the aftermath of the Vietnam War in the 1970s. During the 1980s, refugees flocked to this country from Poland, Sri Lanka, Iran, El Salvador, Lebanon, and Guatemala. During the 1990s, the largest percentage of refugees came to Canada from Asia and the Pacific (31%), and from Africa and the Middle East (27%), and Europe and the United

2-48 *A girl, holding her younger brother, takes a rest under a tree on the way to a United Nations refugee camp new Huay Samran, 460 kilometres northwest of Bankok. Heavy fighting between the Khmer Rouge and Cambodian forces has forced thousands to seek safety in Thailand.*

Kingdom (32%). The top ten source countries for refugees in 1996 were Bosnia-Hercegovina; Sri Lanka; Afghanistan; Iran; Iraq; India; Bangladesh; Somalia; Pakistan, Algeria, Yugoslavia, mainland China, and Ethiopia.

The Canadian government has allocated funds to be paid to refugees as settling-in expenses, and special loans are available to help a refugee adapt to Canada, and refugees can claim immediately the social benefits available to all Canadians. Many refugees, however, become self-reliant within six months, according to immigration officials, and many actually pay back the government's costs incurred while bringing them to Canada.

The number of refugees clamouring for entry into Canada is increasing, but it is just a tiny part of a huge global problem. Drought conditions on the African continent, and savage civil and ethnic wars around the world have left millions of people homeless, destitute, or starving.

Thousands of people are waiting patiently for a chance to emigrate to Canada. They must go through the regular screening process, which is often long and difficult. Refugees, however, often appear to be "jumping the queue," because they can apply from within Canada and are considered immediately for admittance when they arrive here. They only need to be in good general health, and able to adapt to life in Canada in order to gain entry. Note in Fig 2-50, the age variance in refugees coming to Canada – and the fact that more males than females are accepted here. The applicant's age, education, job skills, and knowledge of French or English are used as guides when a refugee's claim is being processed; the amount of financial or resettlement assistance required is also taken into consideration. None of these factors, on its own, is sufficient cause to refuse an applicant. Some cases, however, will be refused if they might be a security or health risk to Canadians.

It is essential to establish who is a "legitimate" refugee. The life and liberty of the applicant must be truly at stake — and this is very often difficult to prove. Take, for example, two women who arrive in Canada claiming refugee status. One has lived in Western

2-49 *A family rests as other ethnic Albanians from the Yugoslav republic of Kosovo walk in the rain along the hills around the village of Tropoja, 500 metres from the Yugoslav border, and 250 kilometres north of Tirana. Thousands of refugees are entering into the area as fighting between Serb forces and the Kosovo Liberation Front continues. Refugees say that Serb forces are burning their villages after shelling them, forcing families to leave.*

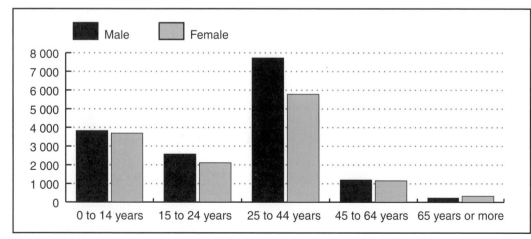

2-50 Refugee class by age and gender, 1996

Europe for two years since fleeing an oppressive regime. Another has come directly from a refugee camp. The first applicant would probably be denied refugee status since she is not being persecuted in her **country of origin**. The second would be considered a refugee, since she was held in a displaced-persons camp.

People who cannot meet immigration criteria, or do not want to wait, will sometimes claim refugee status to avoid the regular immigration process. It is difficult for the immigration officials to decide who is a refugee and who is an **illegal immigrant**. The process can take up to eight years if appeals are involved, and the Supreme Court may have to make the final decision. During this time the applicant can live and work and raise a family in Canada. To deny applicant citizenship after they have put down roots seems harsh, but it is the law. Making an appeal in a court case does not guarantee winning.

Normally only about 30 percent of those people applying for refugee status are accepted. Reasons for refusal may include health, a criminal record, or the likelihood that the applicant would be a chronic welfare case. Those people refused are detained and deported, unless they request a hearing before the Immigration and Refugee Board (IRB).

The IRB was created in 1989, by an Act of Parliament "to make well reasoned decisions on immigration and refugee matters, efficiently, fairly, and in accordance with the law." Its Convention Refugee Determination Division-CRDD (whose members are appointed for terms of up to seven years) deals exclusively

Review Policy on Citizenship, Robillard Says

Suggested not all born here should automatically be Canadians.

Canada should reconsider its policy of automatically granting citizenship to everyone born here, Immigration Minister Lucienne Robillard says.

Her comments immediately drew criticism from immigration lawyers and opposition parties. Robillard was reacting to an Ontario court decision that a Toronto woman can't be deported because the rights of her Canadian-born children were ignored.

"When we have to remove a parent, if the kid is a Canadian, we are torn between the two, and this is the problem that we have here," Robillard said yesterday

Right now, everyone born on Canadian soil is entitled to Canadian citizenship, no matter who their parents are. Robillard said that maybe only children born to Canadian citizens or permanent residents should automatically get citizenship.

"I think that we will have to raise again that question, even if it's a controversial question among Canadians," she said.

Toronto immigration lawyer Barbara Jackman condemned the suggestion. "The end result of a law like that would be that we are creating stateless children," she said.

The debate erupted after Mr. Justice Edward McNeely of the Ontario court, general division, ruled this week against deporting single mother Joyce Francis to Grenada.

To deport her, he ruled, would force her either to abandon two young Canadian-born daughters, or deprive them of their charter right to remain in Canada by taking them along....

Immigration department officials say there are no records of the number of Canadian-born children of non-Canadians. But every year more than 20 000 refugee claimants – many of them in their child-bearing years – enter Canada to seek asylum. Typically about half of those people will eventually be denied refugee status, and most of them embark on an appeal process that can stretch over several years before deportation is finally executed. The U.S. extends citizenship to everyone born on its territory. In Britain, children born to non-residents don't get citizenship automatically, but have to apply before age 18 and demonstrate a "close and continuing link with Britain." And many European countries that host large number of guest workers do not automatically give citizenship to children born to non-residents.

by Allan Thompson, with files from Donna Jean MacKinnon and Canadian Press. reprinted from the *Toronto Star* May 9, 1998

with the determination of refugee claims made within Canada (over 60% of refugee claims made in 1996 were from within Canada). Members travel across the country to hear these cases (some 25 000 each year), and they have granted refugee status to some 12 500 persons every year as a result. The two other IRB branches hear immigration appeals such as removal orders, permanent residence appeals, and Minister's appeals. All in all, the IRB is Canada's largest court; it hands down in excess of 40 000 decisions per annum.

1. How have the source countries of refugees changed since 1950?
2. What do you think about Canada's policy of extending citizenship to all persons born on Canadian soil?
3. Why might some people who intend to claim refugee status get rid of all their identification documents before landing in Canada?
4. Canadian Immigration officers can hold refugees without identification, screen them, and within 72 hours deport them from Canada. Debate the implementation of this policy from the viewpoints of immigration officials, refugee claimants, and individual Canadians who support or oppose this practice.

New Words to Know

Displaced person
Country of origin
Illegal immigrant
Port of Entry
Deportation

110

Canadian Settlement Patterns

A **settlement pattern** is a distinctive layout of roads, fields, towns, cities and other features that are created within the natural environment as a result of human occupation. There have been many settlement patterns in Canada over time, and they have differed considerably from region to region. These different settlement patterns tell much about the various cultures, economies and societies that different groups of people have created in Canada over the years.

Even the absence of a settlement pattern in the landscape is important evidence about the life and culture of Canada's aboriginal peoples. The tribes of the western plain, the eastern woodlands and the subarctic forests were nomadic or seminomadic — they followed the animals that they hunted for food, and had no permanent settlements. Even the agricultural Huron abandoned their villages every ten years or so. They moved to a new location when the soils became exhausted of minerals.

To the aboriginal peoples, the land was owned equally by humans, animals and the natural vegetation. All members of a tribe had equal ownership of all the lands in and around the tribal area. There was no legal ownership. This was very different from the European concept of land ownership. The Europeans looked on the land very differently and used it much more intensively.

When the first Europeans arrived, they brought with them European customs of land-ownership deeds, road allowances and public lands. The terms and concepts of lots, tracts, parcels, deeds and surveys were all transported across the Atlantic Ocean. Different parts of Canada were settled by different groups of European immigrants. As a result, a variety of settlement patterns occurred in different regions of the country.

In most parts of Canada, settlement practices were decided by government policy. The policy of the time dictated some or all of the following regulations: which land areas were available for settlers, where roads were to be built, the location of villages and towns, and the size and cost of parcels of land. These sets of regulations were called **survey systems** and they governed how the land in Canada was divided up.

All the land in Canada is owned by somebody. Land can be owned by individuals, corporations or groups. It always comes with a property "deed," which is a document of legal ownership. The public, through the government, can own land: such lands may be called "Crown lands," federal and provincial

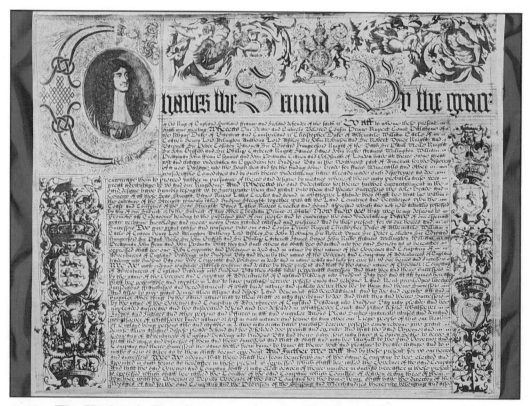

2-52   The Hudson's Bay Company charter, first page. *The charter, issued by the King of England in 1670, gave a group of merchants and courtiers the rights to trade and commerce in the regions whose waters emptied into Hudson Bay.*

parks, reserves and reservations, and so on. For example, the Manitoba Act of 1870, which created the province of Manitoba from land owned by The Hudson's Bay Company, gave all title for ungranted or waste land to the Crown, to be administered by the Canadian government. Eventually that title was transferred to the government of Manitoba. There are no unowned lands in Canada.

In some areas of Canada, the settlement pattern is a direct response to the physical environment — the linear pattern along the river valleys of British Columbia is just one example. Even in areas where there is a regular survey system, natural features — Mount Royal in Quebec, Lake Erie in Ontario and the Red River in Manitoba are three examples — have altered the local settlement pattern. It is, after all, difficult to build homes and roads on steep slopes or on water.

Some settlements result directly from the presence of forest or mineral wealth. Thompson, Manitoba, is one example. Other settlements result from a government decision to establish a settlement, or even move it, as was done with the town of Inuvik in the Northwest Territories. However, these places tend to be isolated points of settlement in largely unoccupied areas.

Many regions of Canada have a settlement pattern that was established by the original settlers. At the same time, very few large regions of Canada were settled all at once; instead, settlement occurred in waves. As a result, the original settlement patterns have been adapted to meet the changing needs of new settlers.

Understanding the original settlement patterns is important, since they have a direct effect on the kinds of development that can occur today and in the future. They can affect the price of housing, for example. A developer may have to accumulate 50 or 100 small house lots to build one apartment complex in an urban area. Each small piece of property is owned by different people and each has to be checked to make sure that the ownership is clear. This adds time and inconvenience to the developer's job.

1. The aboriginal peoples did not have permanent settlement patterns. Which features of their lifestyle made settlement patterns unnecessary?

2. The early European settlers' first job in Canada was to build a fort or defensive settlement. Which features of their lifestyle would make a permanent settlement necessary?

3. How would the physical environment affect the choice of a first settlement? List five factors you think would have been important.

4. Examine maps of your local community at a variety of different scales.
(a) Can you see a basic pattern to the streets and building lots?
(b) Explain why some areas are very different from this basic pattern.

5. In small groups, discuss the advantages and disadvantages of having title to all the Crown land in Canada.

2-53 A settlement in Upper Canada in the late 18th century.

The Atlantic Settlement Pattern

The Atlantic region was the first part of Canada to be settled by Europeans. The bare coastlines, hilly interiors and shallow soils of most of the region did not provide much arable land. The climate was generally cool and damp and the growing season short, compared to more southerly settlements. A few river valleys and narrow coastal plains and the island of Prince Edward Island did have good farmland. But most of the Atlantic region was, and still is, unfit for agriculture.

On the other hand, it was endowed with a deeply indented shoreline, and with coastal waters that teemed with fish. The many inlets, fjords, coves and protected bays gave shelter to ships large and small and some opened onto gently sloping lands suitable for building settlements.

Naturally, the early settlers built their homes along the coast. This pattern of settlement is called **peripheral** since most of the people live around the edges or "periphery" of the region. Few live in the interior. Similarly, a transportation map shows that most of the main roads, and even the major railways, parallel the coast and link those original settlements.

A second characteristic of the Atlantic provinces is that the urban settlements are **dispersed** — that is, widely separated — as, for example, the Halifax region, the area around St. John's, and the eastern section of Cape Breton Island. Between these settlements are large areas of unoccupied forest or scrubland.

Many early settlements that survived by combining farming, logging and fishing can no longer compete in a complex, high-production, modern economy. These smaller settlements are shrinking; some have been abandoned entirely. Almost 50 percent of the farms in southern Nova Scotia were abandoned in the 1960s and 1970s. When farmland is idle, the local agricultural towns and villages tend to shrink, and fewer services and facilities remain for the people who have stayed on. The effect can gather momentum and result in "ghost towns" — abandoned towns, full of peeling paint and memories.

This dispersed pattern of **nodes of settlement** makes transportation and communication difficult. Until recently, many of the smallest settlements were not accessible by road or rail. A road link between Prince Edward Island and the mainland is still only being discussed, and won't exist until the end of this century at the earliest.

Within settlements, the land-use patterns tend to be **natural**; roads curve around obstacles, each parcel of land tends to have an irregular shape, and buildings are scattered.

Not all of the Atlantic region has this kind of settlement pattern, however. The Saint John River Valley in New Brunswick, the Annapolis and Cornwallis valleys of Nova Scotia, and the gently rolling lands of Prince Edward Island, are all quite different from the region's norm. These areas are better suited to farming and from the earliest period developed large mixed farms. This strong agricultural base could support a dense, even distribution of people. As a result, a number of towns also developed. However, these more favoured areas are exceptions to the general settlement pattern of the Atlantic provinces.

2-54 **The Battery, St. John's Harbour, Newfoundland**

Margaree Valley Map Study

The Margaree Valley is located in the Cape Breton highlands of Nova Scotia. The area selected for this study shows the lower Margaree Valley where the river empties into the Gulf of St. Lawrence. The valley is wide and flat-bottomed, probably a result of erosion by a glacier a long time ago. Because the river flows slowly and with an erratic volume, the lower reaches have silted up. The water flows through winding channels called **braids**.

Many of the original farms have been abandoned, and of the remainder a large number are marginal. There is, however, a new industry developing in this rugged area. Tourism along the Cabot Trail has led to the development of museums, golf courses, motels, beach houses and fishing lodges in the Margaree Valley and along the coast. A very recent development is whale-watching tours and boat rides.

1. Use the map grid to find what is located at:
   (a) 475398 (b) 437446 (c) 430428
   (d) 478372 (e) 462411 (f) 453448.
2. In what direction does the Margaree River flow? What evidence can you provide?

2-55 Part of the Margaree Valley in Cape Breton Island, Nova Scotia.

3. Name four streams that flow into the Margaree River. What do these brooks suggest about the climate in this area?
4. Where is the highest point on the map? Give its height in feet and convert this to metres. Give its location using the map grid.
5. Locate the village of Chimney Corner in square 4139.
   (a) How far is it in a straight line to the intersection in the village of Margaree (square 4839)?
   (b) How far is it by road?

2-56 Part of Margaree topographic map 11 K/6. *Contour interval is 50 feet. For scale and key, see page 405.*

6. Draw a *contour profile* across the valley from 460390 to 490400. Label the river, the roads and the village of Margaree.
7. A new motel complex has been built near MacKays Cape and a gravel road goes from the road out to the tip of the cape.
   (a) What attraction brings tourists to this isolated location?
   (b) Suggest five other tourist attractions that might be found in the Margaree area.
8. Examine the map and find place names that are of French, English and Scots origin.
9. Read the description of the Atlantic region settlement pattern and create an organizer to see whether the Margaree Valley area is typical of Atlantic settlements.
10. Use the map and the photos on pages 114 and 116 to write a short paragraph explaining why you would, or would not, like to live in the Margaree Valley area.

New Words to Know
Braid
Contour profile

2-57 Drawing a contour profile.

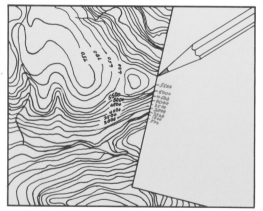

1. Place the straight edge of a piece of paper between two points on the map. Mark and label the contour lines where they meet the edge of the paper.

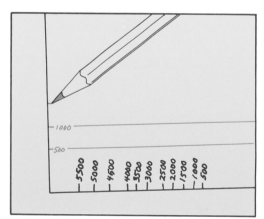

2. Determine the contour interval of the map and work out a suitable vertical scale to represent this (e.g. 50 vertical feet = 1 cm). Prepare a graph as shown above, and transfer the position of the contour lines to the base of the graph.

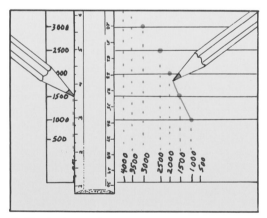

3. Directly above each contour line marked across the bottom, find the graph line for that height, and place a dot there.
4. Connect all the dots. The line passing through the dots is a cross section or contour profile of the land between point A and point B on the map.

114

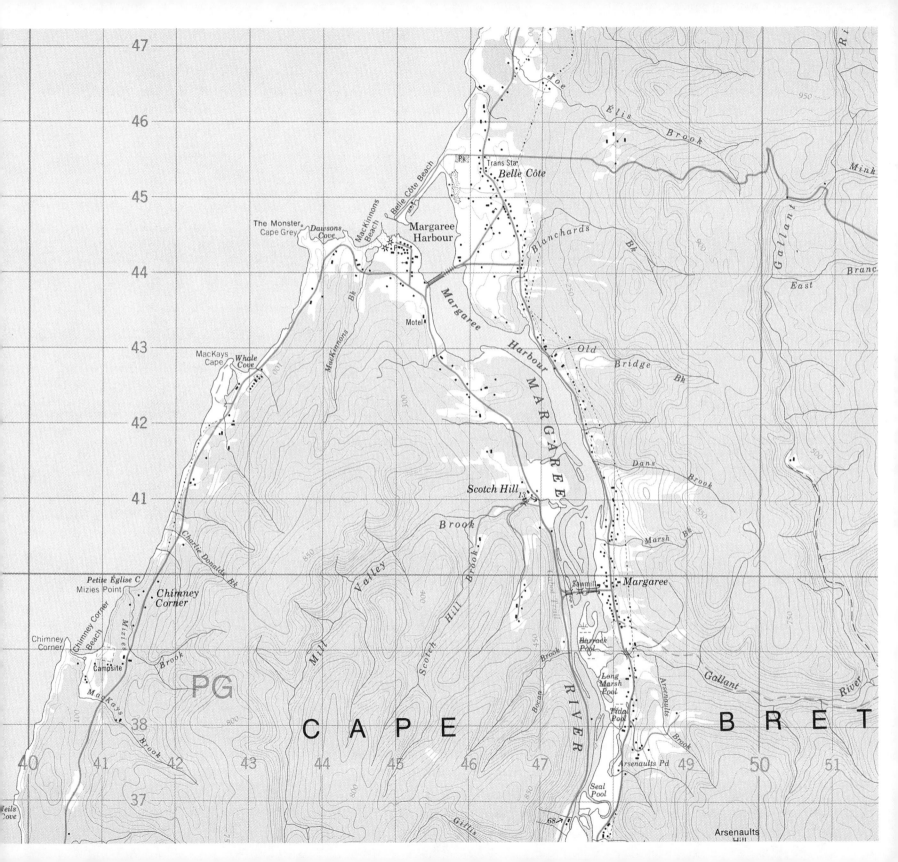

Peripheral, nodal, dispersed and natural — each of these terms describes the Atlantic settlement pattern. Each provides advantages and disadvantages.

The fragmented settlement pattern has created serious problems for the modern economy. Today's manufacturing technologies are best suited to mass production. In the absence of a large market, a diversified labour force, and varied support services, it is difficult for the Atlantic region to attract and hold manufacturing industries. The small, dispersed towns and cities of the Atlantic provinces must compete with the huge urban areas of the American northeast and the St. Lawrence River basin.

Communities are forced to duplicate many functions such as postal services, police and fire protection. They cannot buy goods and organize services on a large enough scale to lower costs.

Transportation and communication costs are greater than in southern Quebec and Ontario. Small, winding roads are picturesque, but they add to the cost of hauling goods. Ferries or air freight add time and cost on any route going to Newfoundland or Prince Edward Island. Among those island dwellers, this leads to a sense of isolation from the rest of Canada.

In Newfoundland, the provincial and federal governments have been trying to resettle people from scattered outports to larger communities. Between 1954 and 1965 over 7500 people were moved from 115 tiny communities under this scheme. By 1971 another 143 communities had been abandoned, and the program had begun to attract criticism. Many people found it difficult to re-establish themselves in larger and sometimes less friendly settlements. Some have since

moved back to their old settlements. The resettlement scheme has since been placed on hold.

But the Atlantic provinces' settlement pattern has some advantages. The sight of picturesque villages perched at the head of rocky coves, with small fishing boats leaning in the mud of a protected harbour, is something that sets tourist cameras clicking. Winding coastal roads, isolated beaches and startling seascapes draw tourists from across Canada and around the world, and have brought new economic life to some communities.

1. **What characteristics of the physical and cultural environment of the Atlantic provinces led to the development of peripheral rather than evenly distributed settlement?**
2. **Use map 2-8 of population distribution and an atlas map of the Atlantic region to investigate and name three nodal regions not mentioned in the text. Look for distinct areas of dense population around a large town or city well separated from other populated areas.**
3. **Explain why abandoned farms and fishing villages would accelerate the depopulation of an entire region.**
4. **(a) Why would the government of Newfoundland want to move people from outports to larger centres? (b) How would you react to being resettled in this way?**
5. **Create an organizer to show the advantages and disadvantages of the basic settlement pattern of the Atlantic provinces. Separate it into "Then" — for the original settlers — and "Now" — for people living there today.**

New Words to Know

Settlement pattern	Dispersed settlements
Survey system	Nodes of settlement
Peripheral	Natural land-use pattern

2-58   Whale Cove on the Cape Breton coast, near the mouth of the Margaree River.

The Long-Lot Pattern of Southern Quebec

The settlement pattern of southern Quebec was introduced in the 1600s by the first settlers from France. These *habitants*, as the French peasants were called, faced an inhospitable land and depended on each other for help. The thick forests surrounding the tiny settlements were impassable, and there were no roads. As a result, rivers and streams became the major transportation routes. Because of the importance of the rivers, the settlement pattern that developed in southern Quebec was based on rows of long, narrow river lots or *rangs*.

The *habitants* were tenants who paid dues to a *seigneur*, a feudal landlord who had been given a grant of land by the King of France. Since the riverfront land was so valuable, the *seigneur* divided his land into rows of narrow lots that extended away from the river at right angles. Each row of lots was a *rang*; each lot in a *rang* was called a *roture*. The *seigneur* charged rent on the basis of the width of each *roture*. The result was a characteristic pattern of narrow strips of farmland with a length-to-width ratio of about ten-to-one.

Along the St. Lawrence River there were approximately 240 *seigneuries*, occupying around 3.2 million km². The land was divided quite evenly. The early *habitants* could survive on their *rotures* through subsistence farming, woodcutting and the trading of furs.

For many years this long-lot pattern allowed each *habitant* access to the river for travel — by boat in summer or on the ice in winter. The river provided

2-59   The long-lot pattern of farming along the St. Lawrence River in Quebec.

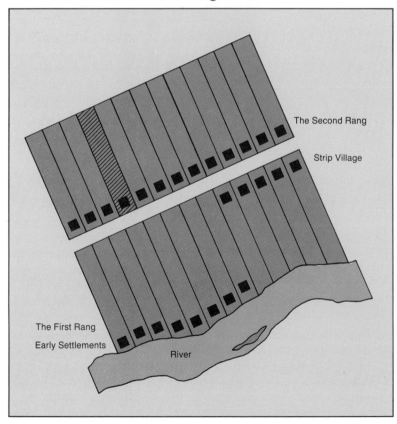

The Second Rang

Strip Village

The First Rang
Early Settlements
River

2-60   The long, narrow fields of the original French settlers can still be clearly seen in parts of southern Quebec.

both water supply and sewage disposal. The forest at the back of the lot provided lumber, fuel, maple sugar and, in the early years, furs.

However, the long, narrow lots along the river were quickly filled and soon second *rangs* had to be built behind the first. Between the first and second *rangs* was a road. Over time, this road became the focus of transportation in the community. Many of the first settlers rebuilt their homes to face the road. The pattern of farms and homes took on the appearance of a **strip village** — a continuous row of houses along the road. This pattern is still advantageous when it comes to providing services like garbage pickup and grocery delivery. It still serves a key social function.

The disadvantages of the long-lot pattern have to do with the length of the lots. Because the farms were side by side and close, it was necessary to fence them so that cattle from one farm stayed out of the crops of the neighbouring ones. Similarly, the farms often had a great variety of soils and topography. Land near the river might have been flat, silty loam; further from the river, another section might have been shallow or stony, and a third area swampy muck. In the traditional pioneer farm, this variety was an advantage, because various crops could be grown for market or for home consumption. For example, land suited for growing grain was balanced by land suited only for rough pasture. Today, this kind of variety makes it more difficult to use large machines and modern farming methods.

Long, narrow farms are very time-consuming to operate. This is especially true if the home is at one end of the *roture*. Imagine the time it would take to move animals from a barn or milk shed to the pasture at the other end of the long lot!

Adding to the problem is the way the *rotures* were later divided. Some of the original, narrow *rotures* were split lengthwise so that sons received equal shares of the land and the river frontage. In cases where the land was not subdivided, the younger or less-favoured son would be forced to leave the area.

This long-lot pattern was taken by French-Canadian settlers to parts of Nova Scotia and New Brunswick, to eastern Ontario along the St. Lawrence and Ottawa rivers, and to Manitoba along the Red and Assiniboine rivers.

Not all of Quebec, however, has this distinctive long-lot pattern. After the British conquered New France in the mid-1700s, they instituted a township survey using rectangular lots of various sizes. An example of this kind of settlement pattern is the area known as the Eastern Townships. Around Abitibi is another distinctive Quebec system of townships called a "canton." It has lots halfway in shape between the narrow long-lot and the rectangular Eastern Township lots.

The long-lot pattern was very effective when it was instituted. Today, however, the long-lots seriously hamper modern development. Farmers who want to enlarge their holdings find it difficult to acquire adjacent lots — the result is fragmentation of their farms. Farms that cannot compete are abandoned and, as a result, huge areas of former farmland are now returning to forest. Not all is bad news, however. Those picturesque

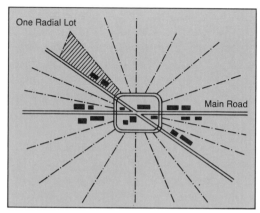

A third type of settlement pattern in Quebec was the radial village. This type of settlement was encouraged by the colonial government, but never became popular. What advantages would this kind of village have over the long-lot pattern? Why do you think the authorities tried to establish it?

strip villages of the St. Lawrence River valley are a world-class tourist attraction.

1. (a) Explain why the long-lot system was particularly well-suited to the early *habitant* settlers.
   (b) How have conditions changed?
2. Where would the long-lot pattern be found outside Quebec?
3. (a) What is a strip village?
   (b) What are the advantages and disadvantages of this kind of settlement?
4. Why would tourists visit the rural long-lots in Quebec?
5. Which advantages would there be to the township lots of the Eastern Townships?

New Words to Know

Long lot	*Roture*
Habitant	*Seigneurie*
Rang	Strip village
Seigneur	

118

The Ontario Township

European settlement in Upper Canada, as Ontario was originally called, did not begin until much later than in the Atlantic region and Quebec. The British government hoped to make the process of settlement orderly and rational by adopting a standard survey pattern based on an American design, called the Ontario Township Survey. First used along the St. Lawrence River near Kingston but later extended across the province, the survey was based on a standard unit known as a **township**. It started with the selection of a base line, usually parallel to the river or the shoreline of the Great Lakes. A road was built along this base line, and two rows of lots, called **concessions**, were surveyed back from it, each lot being about 100 acres (40 ha). At this point a second road, called the "second concession," was built parallel to the base line, another double row of lots was surveyed, back to the fourth concession — and so on. The distance between concession roads was usually a mile and a quarter (about two kilometres); at about the same distance on the east-west axis, sideroads were built along the edges of the concessions, at right angles to the base line and the concession roads. The lots were grouped into townships of various sizes; townships, in turn were grouped into larger political and administrative units called counties.

Although each township was laid out as a neat, regular grid, this logical pattern did not necessarily continue from county to county, or even from one township to the next. In a given township, the base line may be at a different angle from that of the adjacent township. The lots may have different sizes; there may be more or fewer concessions or side roads, and the township's area may be larger or smaller than that of its neighbour. A result of such mismatched survey patterns is that many country roads are forced to zigzag because the two township-survey systems did not connect. This also produces some awkwardly shaped building lots.

John Graves Simcoe, the first governor of Upper Canada, wanted each township to be settled by people of similar religion and background. He considered the township to be the appropriate size for a self-contained pioneer community, and hoped that villages or towns would be established at major road intersections within each township as a focus for community events and services.

This idea met with only limited success. Villages did not necessarily locate in the middle of the townships. Often

2-62   The Ontario township settlement pattern.

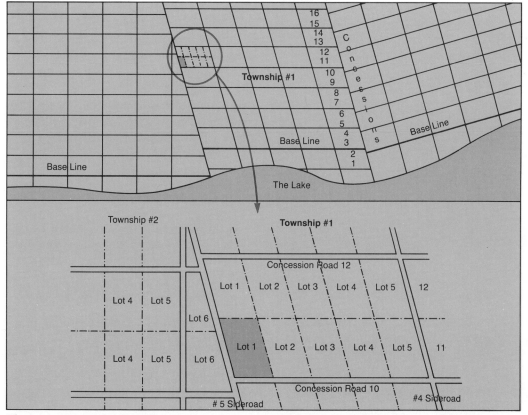

The intersection of sideroads and concession roads produces the typical Ontario grid pattern. A piece of property in Ontario is still referred to, for example, as: "lot one, concession 11, Vaughan Township in the County of York."

they grew up around water-power sites or places where streams or rivers could be crossed. The grid pattern meant that roads were built at fixed intervals whether they served any purpose or not. Often they didn't. One lasting result is the great density of roads over much of southern Ontario.

The mismatching of survey patterns led to problems later on. Many original road surveys have had to be changed in order to accommodate the modern highway network. Governments have built new routes by **expropriating**, or taking over, land from the owners. This is often necessary in order to straighten out the jogs in the original road allowances. But it has led to many court battles. Another problem is that early roads were surveyed across impassable ravines; when they were actually constructed, the routes would have to detour along the steep edge until it was possible to construct a route down and back up again. Even today, many roads simply end at natural obstacles.

It wasn't just roads that were misplaced. Because the survey pattern was so regular, lots were often broken by streams, hills or swamps. In the Bruce Peninsula, for example, whole sections of "farm lots" consisted of rock, with only occasional patches of thin soil. Many of these lots were never bought by settlers, but some were, sight unseen, and then abandoned after fruitless effort.

The problems brought on by the survey were made worse by increasing technology. By the late nineteenth century, most Ontario farmers were mechanizing. The original 100-acre lot, which had seemed so huge to the pioneer, was now too small for the efficient use of machines. Similarly, some of the early farms had survived, even if much of the lot was poor, because a small section was good farmland. When mechanization came, farms with these land conditions could no longer compete, and were abandoned.

The Ontario Township settlement pattern was successful in that it provided an even distribution of land to the pioneers. Unfortunately, Ontario's people today have to put up with extra costs for road maintenance, with unnecessary travel, and with strangely shaped building lots — all because of decisions that made sense almost 200 years ago, in a pioneer society with limited funds and simple technology.

1. **Use the description of the early settlement pattern above to create a sketch map of a typical Ontario township grid. Label the concession roads and sideroads, and shade in one of the original lots. Number your concessions and lots and give the location of your shaded lot.**
2. **(a) Use map 2-63 to suggest areas where the township settlement pattern has been greatly modified.**
   (b) Suggest problems that result from this "patchwork quilt."

New Words to Know
Township
Concession
Expropriate

2-63 Some counties of southern Ontario.

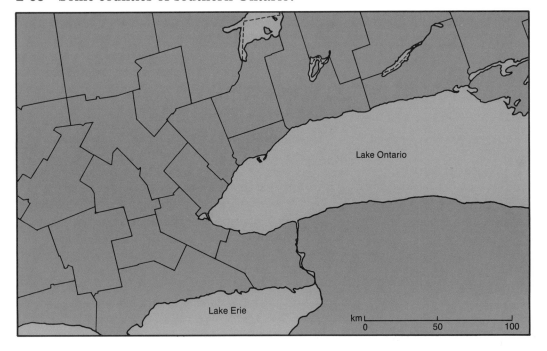

The Prairie Township

A variation on the Ontario Township survey developed in Upper Canada, the prairie settlement pattern is a rectangular survey based on parallels of latitude and meridians of longitude. Even more than the Ontario survey, which was done on a small-scale basis, the prairie survey pattern shows the desire of the government to encourage the fast and orderly settlement of unprecedented numbers of people, in an immense area of virgin land, in an almost impossibly short time period, fuelled by a transportation technology of revolutionary power. In the few short years between 1885 and 1914, between the opening of the CPR and the outbreak of World War I, over a million people flooded onto Canada's prairies and were smoothly and easily settled within the structure of the prairie survey pattern.

Unlike the Ontario survey, which was laid out township by township, the prairie survey encompassed the entire plains — from the Ontario-Manitoba border to the Rockies — in one unified system. The base line was the 49th parallel (the southern border with the United States). Every 24 miles (38.6 km) northward another base line was established. At similar intervals from east to west specific lines of longitude were designated as **principal meridians**. These created the overall grid. Within this huge area bounded by longitude and latitude lines, townships with six miles (9.66 km) to each side were laid out. Inside each township were 36 **sections** of one square mile (2.6 km²).

Unfortunately, meridians of longitude converge as they get closer to the poles. This meant that every 24 miles "correc-

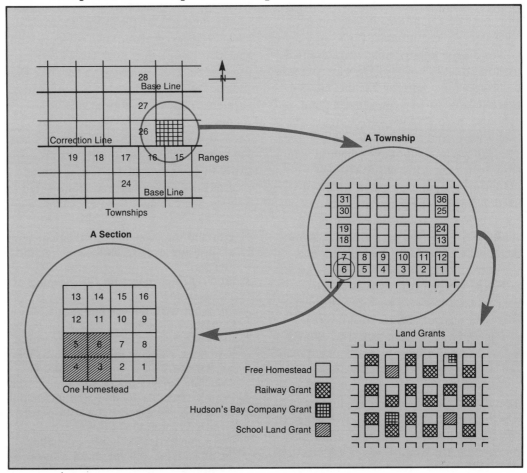

2-64   The prairie township settlement pattern.

tion lines" had to be inserted. All the north-south survey lines jog a little westward along a correction line. Roads were surveyed every mile (1.6 km) from east to west and every two miles (3.2 km) from north to south.

The townships were numbered from south to north beginning with number one at the Canada-U.S. border. The east-west divisions, called **ranges**, are numbered consecutively from east to west beginning at each principal meri-

dian. The result is a neat checkerboard of townships and ranges. With this system, it is easy to locate any place on the prairies by its township, range and section numbers.

A square-mile section of 640 acres (259 ha) was considered too large for a pioneer farmer to handle, so each section was divided into 16 equal **subdivisions**. A prairie **homesteader**, or first settler, could buy four subdivisions, i.e. a quarter section of 160 acres (64.8 ha), for the princely sum of ten dollars. The homestead shaded on diagram 2-64 would be the southwest quadrant of section 6, township 26 and range 15.

Not all the subdivisions were granted as homesteads, however. It was common to leave whole sections vacant, either to be sold later to pay for railways and schools, or to provide for churches. Some were sold as compensation to the Hudson's Bay Company, which at one time had owned the entire region.

Not all the homesteaders followed the survey pattern. In parts of southern Saskatchewan, Mennonites and Hutterites have formed agricultural communities that are centred on villages and the communal ownership of land. In Alberta, Dukhobors have also rearranged the prairie township pattern to suit their special culture and heritage.

The early settlement pattern resulted in many small towns and villages spread out among the townships. Many of these communities have shrunk, and even been abandoned, because their businesses cannot draw enough customers from the surrounding farms. Mechanization has meant that the number of farms has decreased while their size has increased. The result is that there are now even fewer farm families in a township, and that the average distance between farms has greatly increased. Many families have moved to towns or cities.

The huge prairie farms of today are a result of the consolidation of many pioneer homesteads. The original 160-acre (64.8-ha) homestead has almost disappeared. Today's prairie farm may cover hundreds of hectares. The average size of a Saskatchewan farm is now over 350 hectares, for example. An Alberta beef ranch may have as many as 4000 hectares. It is generally true that large, highly mechanized farms are more efficient because of **economies of scale**. But this also means that only larger farms are competitive. In order to survive, the prairie farmer has had to buy out other farmers. This takes a great deal of money and only a few farmers have been able to continue on the land. The land that comes available is not necessarily in the adjacent section, so some prairie farms have become fragmented as they have grown.

1. **Why did the Canadian government want a "fast and orderly settlement" of the prairies?**
2. **(a) Prepare a report on the meaning of "homestead."**
   (b) What were the original terms of a homestead contract?
3. **Calculate how many homesteads would have been available in the 25 townships on diagram 2-64 if only alternate sections had been available for homesteads.**
4. **Draw a township to show the ownership patterns if the average size of a prairie farm is 777 ha (1920 acres).**
5. **Use a comparison organizer detailing the advantages and the disadvantages of the prairie township settlement pattern.**

New Words to Know
Principal meridian
Section
Range
Subdivision
Homestead
Economies of scale

Isolated Settlements

British Columbia and northern Canada have no regular pattern of settlement. This is because of the restrictions set by their physical environments. In both regions, settlements are scattered and occur only where there is a specific reason for people to live and work. Sometimes the location itself provides a resource — like gold or fish. Other times the location serves as an administrative centre — it might, for example, be the site of a hospital serving a number of scattered settlements. Such settlements in the north and west tend to be isolated from each other by distance, rugged country, or both. They are often connected to the rest of Canada only by fragile transportation links.

Almost two-thirds of the population of British Columbia lives in just two CMAs — Vancouver and Victoria. The remaining third live in much smaller settlements sprinkled across the rugged landscape. Almost 90 percent of British Columbia is unsuitable for any kind of settlement: it is too steep, too rocky, too isolated or simply lacking in any resources that would support a settlement. However, if there is some resource that can be extracted economically, a town will develop. Over one million British Columbians live in resource-based centres of fewer than 10 000 people — isolated fishing ports, mining towns, forest-industry towns, processing towns, or towns which service small agricultural regions.

The North has many resource-based settlements centred on some **extractive industry**. Norman Wells in Alberta (oil), Dawson in the Yukon (gold), and Nanisivik on Baffin Island in the Northwest

2-65　Yellowknife, Northwest Territories.

Yellowknife in the Northwest Territories was founded as a gold-mining settlement in the 1930s. Since then, it has developed into a service and administrative centre, providing a variety of goods and services for a huge area around the actual city. Yellowknife is the site of the territorial government — the licensing and administrative offices are there, as well as the court and police station for much of the Northwest Territories. Almost 18 000 people live in Yellowknife. It has movie theatres, a high school and all the other amenities of a small city. Because of the vastness and emptiness of the region it serves, it is more important to its area than its size might suggest. No southern city of the same size is as important to its region.

1. (a) **Name three examples of isolated communities that occur in British Columbia.**
   (b) **For each one named give the primary resource that made urban development possible.**
2. (a) **Use the population-distribution map of Canada (Map 2-8) to locate the areas of dense population in British Columbia.**
   (b) **What advantages do these areas have in terms of their location?**
3. **Why is distance not always relevant when talking about travel in British Columbia or in the North?**
4. **Why would the DEW line settlements have both positive and negative effects on the aboriginal people in the area?**
5. **What makes Yellowknife a "multi-function settlement"?**

New Words to Know
Extractive industry
DEW line

Territories (lead and zinc) are three examples.

There are two other important kinds of isolated settlements in Canada's Arctic regions. One is former military installations such as the **DEW line** radar stations. Because of technical advances, these stations are now obsolete. Most are now being dismantled. While in operation, however, some of these sites lured aboriginal people to settle permanently nearby because of the presence of medical and other service personnel. Spence Bay and Rankin Inlet are just two examples of this kind of military-based settlement.

Mind Benders and Extenders

1. "I said to him, 'What is your disability?' He said, 'It is that people think I have one.'" This quote is from the Federal Task Report on Disability Issues. What does it tell you about how people with disabilities have been perceived? What does it tell you about how they perceive themselves? What major recommendations does the Task Force make?

2. Prepare a two-to-three page research report outlining the kinds of jobs held by Canadian women today. How have these changed over the years?

3. Create a census questionnaire for your school. You can survey a class, a grade, all the students or the whole school body (including staff). You should discuss which kinds of information are and are not appropriate. No question should be embarrassing. Once the data is collected, use the aggregate data to analyze such things as age and sex distributions, favourite recreation, transportation, house type, out-of-school job, left-handedness and the like. A variety of maps, graphs and charts should be used to display the results in the school hallway. The school yearbook is another possible place to record your results.

4. Script a radio or television commercial, or a newspaper advertisement, to convince the following groups that they should take part in an upcoming census. You should emphasize the positive results of "counting themselves in" to the Canadian Census.
   (a) Young people
   (b) Working mothers
   (c) New Canadians
   (d) Aboriginal peoples

5. (a) In small groups, discuss why you think the lifestyles of Canada's Aboriginal peoples and the pioneer settlers were in direct conflict.
   (b) Create an organizer in which these conflicts could be described.

6. Status Indians are entitled to government support and special privileges. What are the advantages of these programs? The disadvantages?

7. (a) What are the kinds of issues involved in Aboriginal land claims? Why do they take so long to resolve?
   (b) Why do you think Canada's First Nations have lands all across this country while those in the United States are based primarily in the west?

8. Read the chapter sections about immigration. In groups of three to five, decide on each of the following:
   (a) Should the number of immigrants be increased or decreased? Explain.
   (b) Should Canada keep the points system?
   (c) Would your group make any changes to the system?
   (d) How would your group deal with refugees?
   (e) How would your group deal with families?
   (f) What new facilities should there be for immigrants?
   For each answer, prepare a short explanation of why your group decided to make changes.

9. Prepare a questionnaire to ask people who have immigrated to Canada about the push and pull factors they experienced. In small groups, discuss the questions and organize a questionnaire. Use your form to interview at least two immigrants — who should be of different ages and from different source countries — about their push-pull factors. Write a report on your findings.

10. (a) What are the advantages and disadvantages of the Ontario township settlement pattern?
    (b) Discuss this pattern in terms of the early settlers. Then compare the advantages and disadvantages it offered them with those for people living there today.

11. Write a brief biography stressing the importance of one of these men: Captain John Palliser, H.Y. Hind, Sir Sanford Fleming, Crowfoot. Explain how the man you chose helped shape the settlement of the Canadian West.

12. Tour your local community by bus or on foot. While on the tour, keep records about each of the following:
    (a) The distance between main roads (both north-south and east-west) and the number and kind of smaller roads.
    (b) How large is the average lot? What are its dimensions?
    (c) Are there distinct styles of architecture, of building size or of location of the building on the lot?
    (d) Can you identify the date of original settlement or the date that the land was subdivided?
    Use the tour information to compare your findings with the major settlement patterns described in this chapter. Which one seems to fit best? Create a comparison organizer between the information you found on the tour and the basic settlement pattern you chose as the best fit.

13. On your local area map, pinpoint all the churches, temples or other religious institutions that are located within a kilometre of your school. Is there any pattern to their location?

CANADA
EXPLORING NEW DIRECTIONS

Urban Canada

"The metropolis today is a classroom."

Marshall McLuhan

Urbanization in Canada and the World

Chances are that as you read these words you are living in an urban area. Statistics Canada defines an urban area as a place with 1000 or more people and a population density of 400 or more per square kilometre. Many large urban areas are classified as cities. Cities have a number of things that attract people: jobs, stores, schools, universities, sports facilities and the like. Many people also come to visit relatives and friends.

Large urban centres have always attracted people. Most of the world's important events occur in cities, and most jobs are found in them. In every part of the world, people have been moving to cities in increasing numbers. Canada is not an exception to this.

Many young people learn about certain cities because they are home to their favourite sports teams. Teams such as Montreal's Canadiens and Expos, and Toronto's Maple Leafs and Blue Jays, attract national attention to their cities. Many cities are great sports rivals – for example, Toronto and Montreal (hockey), Winnipeg and Regina (football), and Edmonton and Calgary (hockey *and* football).

The Olympic Games have been held every four years since 1896, except when a world war has intervened. They are always associated with a host city. Calgary, Alberta, hosted the Winter Games in 1988; Barcelona, Spain, and Albertville, France hosted the Summer and Winter Games in 1992. In 1994, Lillehammer, Norway hosted the Winter Games and in 1996 the Summer Games were hosted by Atlanta, U.S.A. The 1998 Winter Games were held in Nagano, Japan. Other great Olympics have been held in Los Angeles (1984), Mexico City (1968), Montreal (1976) and Tokyo (1964). Besides the Olympics, there are World's Fairs at regular intervals. Vancouver hosted one in 1986, and Montreal in 1967, Canada's Centennial year. Other recent ones have been held in Seville, Spain (1992), and Lisbon, Portugal (1998).

The world's wealth is concentrated in cities — in banks, stock exchanges and insurance companies. Most financial companies are headquartered in great cities such as New York, Tokyo, London and Toronto.

3-1 Montreal's Place Jacques-Cartier (*in Old Montreal*) *is a popular gathering place for natives and tourists alike.*

Urban Growth

The first cities appeared in the Tigris-Euphrates river basin (now part of Iraq) between 4000 and 3000 B.C. Early cities contained only small numbers of people. However, their influence in the growth of trade, industry, art, science and culture was out of all proportion to their small size. Some cities such as Babylon, Rome, and Constantinople grew to great size as the centres of mighty empires. However, for thousands of years most cities remained small, and by far the greatest proportion of the population lived in the countryside as farmers.

In Europe, many cities virtually disappeared for hundreds of years following the fall of the Roman Empire. As towns and cities revived, so did trade and culture. But though towns and cities grew steadily over the centuries, most people still lived in the countryside until as recently as 200 years ago. It was only the coming of the **Industrial Revolution** in the nineteenth century that brought explosive growth in the urban population of Europe.

During the Industrial Revolution, people harnessed improved energy sources and developed new manufacturing technologies. Waterwheels, windmills, steam engines, and electricity provided power for great factories in vastly expanded cities. As the new factories grew, so did the cities; cities were where the work was. Urban areas began to dominate every nation's economy and culture.

In the early 1800s, 20 percent of Europe's population was urban; by the 1980s the figure was 70 percent. In Canada, urban growth was even more dramatic. Until the 1700s there were no

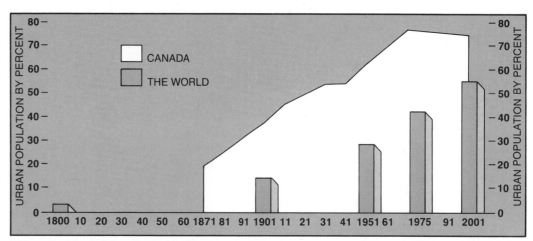

The proportion of the population living in cities has increased steadily in both Canada and the world since the early 1800s. Canada's rate of urbanization has been greater than the world average. What has happened to Canada's rate of urbanization in recent years?

3-3 Populations of largest Census Metropolitan Areas, 1971-1996

CMA	POPULATIONS			PERCENT CHANGE		
	1971	1981	1996	1971-81	1981-96	1971-96
TORONTO	2 628 043	3 130 392	4 263 757	19.1	36.2	62.2
MONTREAL	2 743 208	2 862 286	3 326 510	4.3	16.2	21.3
VANCOUVER	1 082 352	1 268 183	1 831 665	17.1	44.4	69.2
OTTAWA-HULL	602 510	743 821	1 010 498	23.5	23.3	67.7
EDMONTON	495 702	740 882	862 597	49.5	16.4	74.0
CALGARY	403 319	625 966	821 628	55.2	31.3	103.7
WINNIPEG	540 262	592 061	667 209	9.6	12.7	23.5
QUEBEC	480 502	583 820	671 889	21.5	15.1	39.8
HAMILTON	498 523	542 095	624 360	8.7	15.2	25.2
LONDON	286 011	326 817	398 616	14.3	22.0	39.4
KITCHENER	226 846	287 801	382 940	26.9	33.1	68.8
ST. CATHERINES-NIAGARA FALLS	303 429	342 645	372 406	12.9	8.7	22.7
HALIFAX	222 637	277 727	332 518	24.7	19.7	49.4
CANADA	21 568 000	24 343 000	28 846 761	12.5	18.5	33.7

1. Pick the CMA that is closest to your community and make two bar graphs to show the percentage population change between 1971 and 1981, and between 1981 and 1996.

2. List the cities by population size for 1971 and compare this with 1996. Explain the differences.

3. For 1996, calculate the percentages of Canadians who live in:
(a) the three largest urban areas. (b) the three smallest urban areas. (c) Toronto. (d) your community. Design a chart to organize and present your findings.

4. (a) List the four cities that have grown most over the 25 years 1971-1996. Suggest as many reasons as you can for the growth of each. (b) Why do you think Montreal has shown the least growth?

large urban settlements in Canada. In the 1800s, there was some growth around sites where power was available (Montreal, Trois-Rivières), fur-trade centres (Winnipeg, Edmonton, Thunder Bay), defence sites (Halifax, Quebec City), major ports (Halifax, Vancouver) and places where minerals could be found (Sudbury). Canada's urban population was growing faster and faster by 1900.

In 1871, only 19 percent of Canada's people were city dwellers. By 1930 the figure was over 50 percent. This continued to rise until 1976, when 76 percent were urban dwellers. Since then the percentage has stayed the same.

A significant trend in Canada is the increasing dominance in national life of a few ultra-large urban centres. These are often called census metropolitan areas. A CMA is, basically, any place that (a) has a large urbanized core, and (b) is surrounded by urban or rural areas that are closely integrated with the core. The integration can be economic or social. A CMA must have a population of at least 100 000.

Today there are 25 CMAs in Canada. Chart 3-3 and map 3-4 show the largest. They have been the largest for the past ten years, and are all continuing to grow. The western CMAs are growing most quickly, Edmonton and Calgary in particular: both are moving up the list. CMAs are especially dominant in Ontario.

3-4 Map of Canada's CMAs.

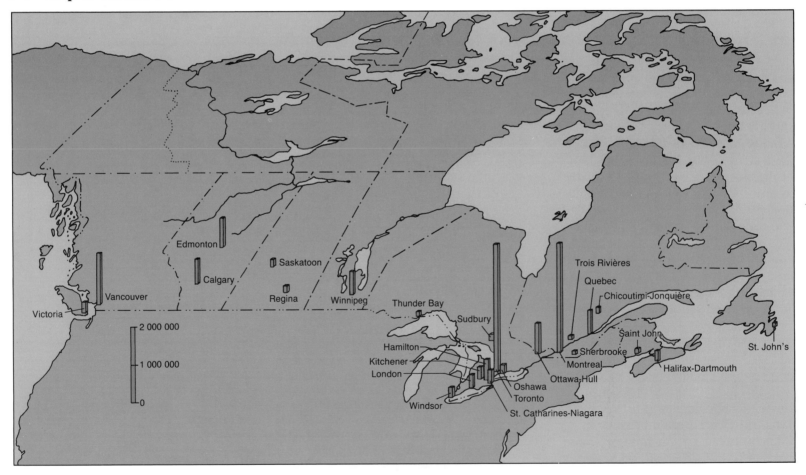

3-5 The world's ten most populous urban areas

1	TOKYO-YOKOHAMA (JAPAN)	27 856 000
2	SAO PAULO (BRAZIL)	16 417 000
3	NEW YORK CITY (USA)	16 329 000
4	MEXICO CITY (MEXICO)	15 643 000
5	BOMBAY (MUMBAI) (INDIA)	15 093 000
6	SHANGHAI (CHINA)	15 082 000
7	LOS ANGELES (USA)	12 410 000
8	BEIJING (CHINA)	12 362 000
9	CALCUTTA (INDIA)	11 673 000
10	SEOUL (S. KOREA)	11 641 000

Half of Canada's population lives in the country's 20 largest urban areas; one-third of Canadians live in or very near Montreal, Toronto or Vancouver.

Urban centres are very important to Canada. Yet only two rank in the world's top 100 in terms of population: Toronto (61st) and Montreal (66th).

1. **Why do you think some people would prefer to live in a small town rather than a large metropolitan area?**
2. **Do you expect to live in a city after you complete your education? Explain why, or why not.**
3. **List at least five Canadian towns and cities, noting their most famous professional or amateur sports teams. Compare your list with those of your classmates.**
4. **List the Canadian CMAs that you and your classmates have visited or lived in during the past two years. Note two things about each that you like, and two you do not like.**
5. **Using an atlas, plot the location of the world's 20 largest cities on a map. Describe their distribution, keeping in mind the charts you completed earlier about Canadian cities.**

Regional Urban Development

The Atlantic Region Many of the cities and other urban settlements in Atlantic Canada have been influenced by the sea. Most are on the coast, and the sea has provided a livelihood for many of the people. The largest cities in each of the four provinces, St. John's, Charlottetown, Saint John and Halifax, all have excellent natural harbours. Three of the four are the capitals of their provinces. The exception is Saint John, for Fredericton, an inland city on the Saint John River, is New Brunswick's capital.

Halifax is the only **metropolis** of the region, with over 330 000 people. Its site was determined by the huge, deep harbour where the British established a fortified naval base in 1749. The city has been a military bastion ever since, and remains the major base for Canada's navy. Halifax is the busiest port in Atlantic Canada, and has developed into the major industrial centre in the region. A large oil refinery and a car-assembly plant (Volvo) are only two of the many industries spawned by the port. It has also become the commercial, financial and cultural centre of the region.

St. John's is the second largest city of the region, with over 174 000 people. It serves as the political, commercial and cultural centre for Newfoundland, and has some secondary industry, mostly based on fish processing and ship repair. Saint John, with a population of about 125 000, is a major industrial centre, with two large pulp-and-paper plants, and Canada's largest oil refinery. Charlottetown is small, with 57 000 people, but it is the capital and regional centre for P.E.I.

Atlantic Canada's cities were, historically, among Canada's first major settlements. Today, however, their relative

3-6 St. John's, Newfoundland. *The harbour seen from Signal Hill.*

size has declined, as the younger cities in central and western Canada have grown more rapidly, particularly in this century. Since 1900, the immigrants who flooded into Canada have largely bypassed Atlantic Canada. The westward shift in population and economic activity is a major trend in Canadian history, which continues to this day.

The Central Region The Great Lakes-St. Lawrence waterway is a natural transportation corridor which has encouraged the development of settlement in this region. The wide St. Lawrence enabled the early French explorers to reach deep into the continent and establish settlements. The first was Quebec City, founded in 1608 at the strategic point where a mighty cliff guards a narrowing in the river. From this base in the St. Lawrence valley the French pushed westward as far as the interior plains, establishing scores of forts and fur-trading forts. Many of these have developed into towns and cities, from Trois-Rivières, Kingston, Sault Ste. Marie and The Pas in Canada, to Detroit, Pittsburgh, New Orleans and others in what is now the United States. Canada's two largest cities, Montreal and Toronto, were both first settled by the French as military forts and fur-trading posts.

The lakes and rivers remained important as settlers moved into central and western Canada, since travel by land was so difficult. One of the earliest industries was logging, and the rivers were essential for floating the logs to the ports for shipment overseas. Our national capital first began as Bytown, a lumber centre on the Ottawa River. In

 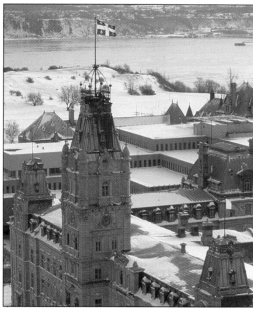

3-7 *Quebec City has been an important commercial and cultural centre from the earliest days of European settlement because of its spendid strategic location atop a high bluff overlooking a narrow point on the St. Lawrence. Capital of Quebec province, and seat of the Quebec legislature, the city is also an important symbol for Francophones, many of whom feel very strongly that Quebec province's true destiny lies in its becoming a sovereign state.*

agricultural areas of the south, scores of towns sprang up to service the farms and farmers. The coming of the railways had a considerable effect on all these. Settlements on the railway had a chance for development; others were generally doomed to little or no growth. The great transcontinental railways, beginning with the Canadian Pacific Railway in the 1880s, also opened up vast new areas in the West and the North. Mining towns were particularly dependent on the railways to move their bulky product. Some of them, such as Sudbury, were actually the result of discoveries made as the railways were built. The railways gave a big boost to both mining and the pulp-and-paper industry.

Overall, the cities of southern Ontario and Quebec have attracted industry and immigrants, thanks to their varied resource base, excellent transportation facilities, energy supply, large domestic market and proximity to the huge American market. Both Montreal and Toronto had excellent locations for their development as key transportation and communication centres, expanding also in manufacturing and commerce. Some places had particular location advantages for manufacturing and developed into major industrial cities. Niagara Falls, Welland and St. Catharines were favoured by nearby hydro power and the Welland Canal; Hamilton and Sault Ste. Marie had good harbours and easy access to the iron ore and coal needed for steel production; Windsor was across the river from the auto industry of Detroit; Chicoutimi, Shawinigan and Jonquière had inexpensive hydro power for their aluminum industries, and forests for pulp and paper. Stratford and Kitchener had nearby hardwood forests for their furniture industries.

Unlike many regions, the central provinces do not have one leading metropolis, but two, Toronto and Montreal. This reflects the bilingual and bicultural nature of Canada. For most of Canada's history, Montreal was the dominant urban centre; in recent decades, however, Toronto has become the nation's leading industrial, commercial and financial centre. Montreal, however, continues to be a vital, growing city, and retains its special position as the great urban centre of French Canada — the second largest French-speaking city in the world. The population of these two cities alone includes one out of every four Canadians.

The Prairie Provinces Several cities on the prairies, including two of the largest, Edmonton and Winnipeg, began as fur trading posts. As in other regions, suitable location on major rivers was the key factor. The coming of the railway was particularly important in stimulating urban growth in the West. Both Edmonton and Winnipeg, but also Calgary and Moose Jaw, greatly expanded when the railways chose them as divisional centres. The establishment of government administrative centres also encouraged urban growth. Regina, still Saskatchewan's capital, began as the headquarters for the North West

Mounted Police (today the R.C.M.P.) and administrative centre for the entire North West Territories.

With the huge influx of settlers at the end of the 1800s, agriculture boomed; scores of towns were founded to serve the farmers and market their grain. Saskatoon is one city that began as such an agricultural centre. Most of these towns, of course, never grew beyond small, local centres; in recent years, in fact, the tendency has been for many to shrink in size, or even to disappear, as people move to the cities. In this century, the discovery of oil in Turner Valley made Calgary Canada's oil capital. Following World War Two, another

major oil find near Edmonton gave that city an oil boom, too, though Calgary remains the financial and administrative centre of the industry. Less dramatic mineral discoveries in the shield also resulted in the establishment of mining towns such as Flin Flon and Thompson.

In 1900, Winnipeg, "Gateway to the West," was the most important city on the prairies. Today, its has a population of 1 142 000, and one of the lowest unemployment rates (6.6 percent) in Canada. Key industrial sectors of the city include financial services, manufacturing, agri-food processing, transportation, distribution, information technologies and telecommunications, film production, and health industries.

3-8 *Winnipeg, Manitoba. The city grew up around the fur trading posts at the point where the Assiniboine River flowed into the Red River. Winnipeg accounts for 80 percent of Manitoba's total economic activity, and 65 percent of provincial employment and retail trade.*

The Pacific Province The location of urban centres in British Columbia is primarily related to its resource industries. Transportation routes are the second major factor. After the fur trade, the gold rush of the late 1850s was the first event that attracted European settlers to British Columbia in any significant number. The goldfields were in the interior plateau, so that most early settlements were there, or in the Fraser River valley, along the gold rush route.

Surprisingly, the site of Vancouver — now Canada's third largest city — remained almost uninhabited until 1885, when it was chosen as the western terminus of the CPR. The city's excellent harbour became Canada's major port for Pacific trade. As the prairie grain trade grew, so did Vancouver's importance as a port. By the 1920s, it had become the primary metropolitan centre in western Canada, a position once held by Winnipeg. Canada's number one port, it continues to grow as Canada's trade with the Far East increases; wheat and potash increasingly flow from the prairies, and coal, minerals and forest products flow from British Columbia. Vancouver provides the financial and commercial base for the mining, forestry and fishing industries, as well as processing their products.

British Columbia's second city, Victoria, another early fur trading post and government centre, today remains as the provincial capital. It is also Canada's chief west coast military base. Most other cities owe their existence to resource industries, such as Prince Rupert (fishing), Kitimat (hydro power), Port Alberni (forestry) and Trail (mining).

1. **Prepare a brief report on the characteristics of any two urban areas in two different Canadian regions.**
2. **What have been the main factors promoting or limiting urban growth in any two of Canada's regions?**
3. **Set up a chart similar to the following and fill it in, using the data provided in the text, and an atlas where necessary.**

New Words to Know
Industrial Revolution   Metropolis

3-9 Victoria, British Columbia. *Victoria's magnificent harbour first attracted the British navy and colonial officials. The city has remained B.C.'s capital, and Canada's main west-coast naval base, despite the shift of population and economic activity to the mainland.*

3-10 Comparative chart of Canada's largest CMAs.

CMA	PROVINCE	POP. 1996	PHYSICAL FEATURES OF SITE	LOCATION	IMPORTANCE (ECONOMIC POLITICAL, ETC.)
TORONTO					
MONTREAL					
VANCOUVER					
OTTAWA-HULL					
EDMONTON					
CALGARY					
QUEBEC CITY					
WINNIPEG					
HAMILTON					
LONDON					
KITCHENER					
ST-CATHARINES – NIAGARA FALLS					
HALIFAX					

Fill in the information in chart 3-10.
How does your community compare to those listed above?

Toronto: "The Carrying Place"

Toronto has come a long way since Upper Canada's first lieutenant-governor, John Graves Simcoe, declared in 1792 that "the city's site was better calculated for a frog pond or a beaver meadow than for the residence of human beings." Few would view Toronto in this way today!

The Huron called Toronto's present site "the carrying place" because it was the centre of the area's fur trade. The French and later the British built a series of forts in the area. The first, Fort Rouillé (sometimes also called Fort Toronto), was built in 1750. In 1788 the British purchased control of the Toronto region from the Mississauga for 146 barrels of trade goods — blankets, axes, bolts of cloth and the like. Today that land is worth scores of billions of dollars, and is home to well over three million people.

Toronto grew quickly in part because of its location — it was central within its own province, as well as near the U.S.A. Two other factors, however, gave it advantages over other settlements in the same area. First, it had a large, sheltered harbour. And second, from 1795, it was chosen as the capital, and thus also the political, administrative, and legal centre, of Upper Canada. By 1818 more than 1000 people lived there. Originally named York, the town was incorporated as a city in 1834, and renamed Toronto.

Improved transportation added to Toronto's importance in the nineteenth century. The building of the Erie Canal through New York State improved Toronto's access to New York City. As shipping costs fell, Toronto quickly expanded its trade with U.S. cities.

By the 1850s, Toronto was also the **hub**, or centre, of the major roads in Upper Canada. The coming of the railways in the mid-1800s reinforced its position as the transportation centre of the province. The new railways gave Toronto vastly improved links with the rest of Canada and the northeastern U.S.A. They also encouraged manufacturing. The result was thousands of new jobs in the city. Roads, railways, waterways: all gave Toronto a significant advantage.

Between 1850 and 1950 Toronto dominated the commerce and industry of Ontario. In recent decades it has become the dominant commercial, financial and cultural centre within Canada as a whole.

In 1998 the Ontario Government created a new political region called the City of Toronto to replace the old Metropolitan Toronto. The new "Megacity" is composed of the former cities of Toronto, Etobicoke, North York, Scarborough and York and the borough of East York. With a population of almost 2.5 million people, one mayor, 57 councillors, 27 000 civic employees and a 1998 budget of $5.9 billion, Toronto is the largest city in Canada and one of the largest in North America. Home to 90% of Canada's foreign banks, Toronto is the financial centre of Canada. Other service activities include Canada's top law firms and advertising agencies and 80% of Canada's top public accountants and high-tech companies. The new City of Toronto has 75 000 businesses.

Despite its size, the City of Toronto is considered to be one of the safest and most liveable areas on the globe, according to international studies. It is also the most multi-cultural city-region in the entire world. At this time the details of the

3-11 *Saturdays in St. Lawrence Market are an energetic, exciting, and colourful shopping experience.*

organization and the divisions of power and responsibility for Toronto are still under discussion. Plans are for the new city to run public transit, education, public safety (fire, police and ambulance), roads, water and sewage treatment, solid waste management, libraries, social and health services, recreation and all aspects of planning. The "downloading" of these services to the city, along with a new market-value property tax re-assessment mean huge changes to the budget. The amalgamation is intended to increase economies of scale (cost savings estimated over $850 million), to increase efficiency by reducing duplication (4 500 fewer civil service jobs), and to simplify decision-making. However, there are already signs that the huge region may be too cumbersome. Even a simple decision such as where to hold meetings is proving to be a concern for the new city administrators. The division of power between the City of Toronto and the community and neighbourhood councils is another area of potential conflict. It will take time before all the changes due to amalgamation are completed. Making this process of change even more complex is the fact that the new City of Toronto is the central anchor for the even larger Greater Toronto Area (GTA).

The GTA is a massive region encompassing some 30 municipalities and 5 regional governments within one administrative unit. This new city-region contains 4.5 million people and covers 7 200 square kilometres. This means the GTA alone holds more than 42 percent of Ontario's population and produces more than half its wealth. It has the highest concentration of industrial space in Canada.

This whole urban region needs to have co-ordinated planning so the government of Ontario intends to create the GTA Services Board to run GO Transit, develop a regional waste disposal system, and cooperate on regional road, sewer and water projects. There has also been discussion of co-ordinated funding for services like assisted housing and welfare payments. However, many of the mayors of the municipalities within the GTA are resistant to these changes. They are concerned about paying for services they will never use. As of 1998 even the membership of this organization is still unknown.

Across the GTA a number of significant issues are emerging:
• curbing urban sprawl
• preserving the environment
• fair and equitable taxation
• co-ordinated planning
• funding for parks and recreation activities

• delivering social services
• meeting transportation needs
• public safety concerns
• sustainable growth and development

Toronto's Physical Site

A number of physical features helped attract early settlers to Toronto. There was a large, natural harbour sheltered by a peninsula which later became the Toronto Islands; areas of flat land made building easy; and the Don and Humber Rivers gave access to the hinterland.

Many of Toronto's surface features show the effects of glaciation. One thing the glaciers left behind was a steep **bluff** 20 to 25 m high. It marks the old shoreline of Lake Iroquois, an ancient lake that shrank and turned into present-day Lake Ontario. The bluff has interfered with the construction of many north-

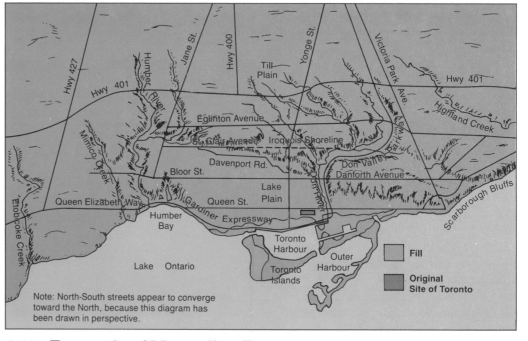

3-12 **Topography of Metropolitan Toronto.**

south roads. Many of Toronto's roads end at its base; Yonge Street, Toronto's main north-south artery, is one of the few that doesn't. Because of the bluff, Toronto grew more quickly east to west than it did north to south. A number of major east-west roads were built below the bluff before any were built above it. Davenport Road runs along the base of the old shoreline.

Toronto also developed along the river valleys that break the old shoreline. It was usually easy to build roads along the riverbanks. Later, railways were built alongside the roads. Toronto grew out from its low-lying harbour to the north, east and west in ribbons.

The city is also crisscrossed by ravines. These have tended to restrict development to the land in between them. Bridges to span them are expensive and difficult to build, and there still aren't enough of them. As a result, traffic congestion occurs daily on the roads over Highland Creek and the Don and Humber rivers. Some ravines, however, such as the valley of the Don River, can be used as transportation corridors. The ravines also provide extensive parkland.

South of the bluff is a level plain extending to Lake Ontario. This flatness made construction easy, so Toronto's central core developed there. Railways were built along the lakeshore. This pattern is still visible today.

Northern Toronto, above the bluff, is a gently rolling till plain. As development spread outward, many subdivisions filled the land between the rivers and streams.

Toronto Islands   Originally the Toronto Islands were an extension of the mainland — a spit of land composed of sand, clay and gravel. In 1858 a major storm separated the spit from the mainland, creating the Toronto Islands. The islands' hooked shape forms a natural harbour. Today most of the islands are used for recreation: there are yacht clubs, beaches, children's playgrounds and large areas of parkland drawing millions of people every year. At the islands' west end is an airport used by thousands of small planes.

People still live on the Toronto Islands. There has been a community there for over a century, though it is much smaller now than it was a few decades ago. Some politicians want the island homes **expropriated**, so that more of the islands can be used for recreation. The islanders have been fighting Metro for years, and the issue still isn't settled.

Scarborough Bluffs   These steep, 30-m bluffs are an extension of the old shoreline mentioned earlier. They are found east of the downtown core, on Lake Ontario. Parkland and subdivisions share the top of the bluffs. The bluffs are being constantly eroded by Lake Ontario and sometimes valuable property is washed into the lake. The City of Scarborough has tried to stop some of the erosion by building special wave deflectors, but the geology of this area is very complex: even if the wave erosion is stopped, the bluffs will probably erode over time into a sloping shoreline.

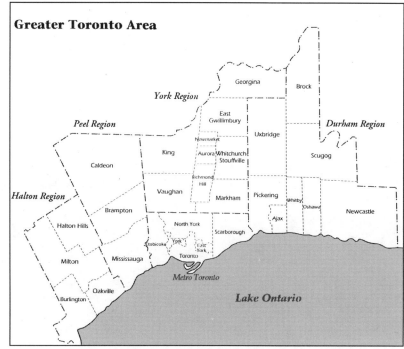

Greater Toronto Area

Georgina
Brock
York Region
East Gwillimbury
Uxbridge
Durham Region
Peel Region
Newmarket
King
Aurora
Whitchurch-Stouffville
Scugog
Caldeon
Richmond Hill
Vaughan
Markham
Pickering
Whitby
Oshawa
Halton Region
Ajax
Newcastle
Brampton
North York
Scarborough
Halton Hills
Etobicoke
York
East York
Milton
Mississauga
Toronto
Metro Toronto
Oakville
Burlington
Lake Ontario

3-13   Toronto and the Greater Toronto Area (GTA)

Climate Toronto is located in the Temperate East Coast climate region. This region generally has cold winters with heavy snowfalls, and hot, humid summers. However, Toronto's climate is moderated by its lake; its winters are far less severe than those farther inland. The monthly mean temperature for January, the coldest month, is $-4.4°C$. There are often thaws and winter rains between the cold snaps. Toronto rarely gets more than 10 cm of snow at a time; towns farther inland often receive 20 or 30 cm.

In the summer, the temperature averages 20 to 21°C. However, heat waves each summer can send the temperature into the 30s. Again, Lake Ontario moderates the summer weather; Toronto summers tend to be cooler than those farther inland. Precipitation is moderate, and occurs year round, though most falls in the summer.

1. Which features first attracted settlers to Toronto?
2. Explain how each of the following shaped the growth of Toronto: Toronto Islands, Scarborough Bluffs, the old shoreline.
3. Compare your community's climate with Toronto's.
4. List three advantages of metropolitan government.

New Words to Know
Hub
Bluff
Expropriate
Amalgamate

Population of the Toronto region

The City of Toronto is home to 2 385 421 people, with 2 059 279 more in the GTA (Greater Toronto Area), for a total of 4 444 700. This is close to 15% of the entire Canadian population, and 40% of the population of Ontario. The GTA, in addition to the City of Toronto, includes the regions of Peel, York, Halton, and Durham. Most of the population growth in the GTA over recent years has occurred in the outlying regions as Toronto City's population has seen only single digit growth while the areas of York, Peel, and Durham have been growing at annual rates in excess of 20%. One reason for this growth is that land is cheaper in the surrounding regions, and many Canadians want to own single family houses on larger lots, and option not available to many in the city proper.

It is expected that growth will continue to be greater in the regions so that by the year 2011, 53% of the GTA population will live outside the city of Toronto, compared with todays 46%. The populations of York and Peel, particularly are increasing rapidly.

Toronto is Ontario's capital and administrative centre. There are over 42 000 government jobs in the GTA – 30 000 provincial and 12 000 federal. Toronto's economy is strong and highly diversified. It has one of the highest average incomes in the country.

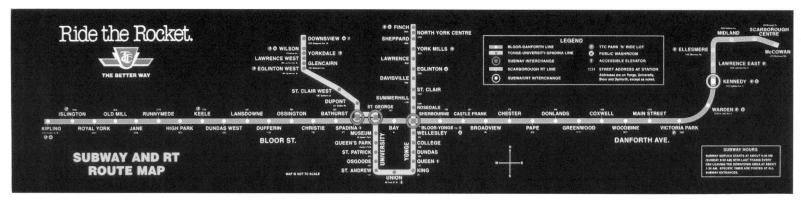

Fig 3-14 Toronto Transit Commission Route Map.

3-15 Population projections for the Toronto region.

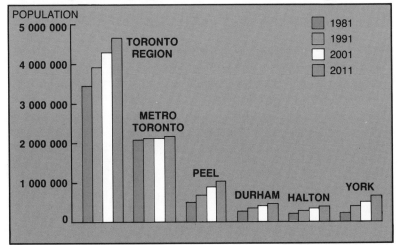

3-16 Visible minorities in Toronto, 1996.

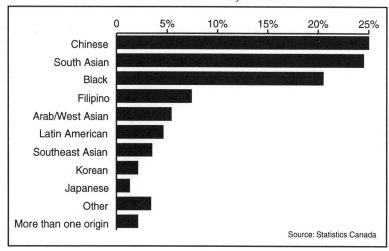

Source: Statistics Canada

Age The GTA's population – like that of the country — is aging. The number of seniors will more than double by the end of this century. The proportion of children is expected to stay about the same. However, more of these will be living in the surrounding regions rather than in Toronto itself, because young couples tend to move away from the central city to raise their families. Fifteen percent of all families in Metro are headed by a single parent — usually the mother.

The population of the City of Toronto is expected to grow little, although significant new housing projects have been developed in the downtown core on the site of the old Massey-Ferguson plant, and the former railway lands bordering Lake Ontario.

''Baby-boomers'' — those born between 1945 and 1964 — make up 36 percent of the region's population. More and more, these young adults are moving into older neighbourhoods such as Cabbagetown and the Annex. They **renovate** (repair and alter) the old homes in those inner-city neighbourhoods.

The many changes in Toronto's population can be seen in population graphs 3-15 and 3-16. Clearly, the percentage of young people has declined since 1971, while the number of seniors has grown.

Ethnic Groups Toronto is a very diverse city ethnically. A walk through almost any neighbourhood illustrates this fact. The population mix has changed greatly since the Second World War. In the 1950s, three-quarters of the city's population was British or of British origin, and as recently as 1981, the figure was 48%. Today, 25% of Toronto's population is Chinese origin, 24.7% is South Asian, and 20.5% is Black. Figure 3-16 depicts the ethnic variety of Toronto's poplation. One reason for this, is that Toronto is the preferred destination for many immigrants to Canada. Compare the charts in Fig 3-17 to discover how immigration to Toronto has changed from the years before 1961 to 1996.

Toronto is a very **cosmopolitan** city. Almost every type of food is available there, in the markets, and in its more than

5000 restaurants. Most immigrants have retained at least some of their native country's lifestyle, creating a rich multi-cultural texture for their new place of residence.

Almost every country in the world is represented in Toronto's population. When the World Cup Soccer playoffs occurred in the summer of 1998, one Toronto newspaper proudly noted the fact that Toronto had more fans for each competing soccer country than almost any other city in the world.

1. **Why are the regions of Durham, Peel, Halton and York increasing in population more rapidly than Toronto?**
2. **In which ways does the variety of ethnic groups add to the attractions of Toronto?**

New Words to Know
Diversified
Renovate
Cosmopolitan

Toronto's Transportation System

Large urban centres need many ways to move people and goods — by air, rail, road and water.

Road transporation Most of Toronto's main roads, or arteries, follow a basic grid pattern. The major expressway linking Toronto with other cities in southern Ontario is the MacDonald-Cartier Freeway, commonly known as Highway 401. Originally planned as a bypass around the north of the city, 401 has now become the major east-west artery within the GTA. Other major expressways in Toronto are the Don Valley Parkway, Highway 404, Highway 403, Highway 427, and the new 407 Express Toll Route (ETR).

There are more than 1.2 million automobiles in Toronto. Another million cars are driven into the city every day, as the average commute for workers to the city is 60 minutes. This puts great pressure on the city's streets. Traffic jams are common, and parking is at a premium, when it can be found. Although many people feel solutions to this problem can be achieved with the building of new roads, other people feel that moneys should be spent instead on public transit. Toronto, like other Canadian cities, has bus and commuter lanes reserved on its major arteries during rush hour. Transit buses use these lanes, as do commuter cars with two or more people. Flex hours within the workplace, where people come to work and leave during non-rush hour times is another potential aid to reducing the traffic on Toronto streets, as is the increased use of the work from home syndrome, wherein employees communicate with their offices through e-mail.

3-17 Immigrated to Toronto before 1961.

	Immigrated to Toronto before 1961	Percent
Italy	67 665	25.6
United Kingdom	58 630	22.2
Germany	21 700	8.2
Poland	15 490	5.9
Greece	10 110	3.8
Netherlands	9 345	3.5
Hungary	8 335	3.1
Ukraine	6 390	2.4
United States	5 430	2.1
Austria	4 375	1.7
Total	**264 630**	**100.0**

3-18 Toronto Recent Immigrants 1991-1996

	Toronto Recent Immigrants 1991-1996	Percent
Hong Kong	48 535	25.6
Sri Lanka	36 735	22.2
China	35 330	8.2
Philippines	33 210	5.9
India	33 185	3.8
Poland	18 605	3.5
Jamaica	16 780	3.1
Guyana	13 195	2.4
Vietnam	12 290	2.1
Trinidad & Tobago	11 375	1.7
Total	**441 035**	**100.0**

Urban Transit Systems Only 22% of Toronto commuters use the Toronto Transit Commission (TTC), but that is the highest percentage anywhere in Canada. The TTC is famous around the world as clean, efficient, and safe. Its lines form a vast network of routes throughout the city's core and to its perimeters, where TTC links to surburban transit operations allow millions of commuters to travel the wide range of the GTA daily.

The TTC's system is a fully integrated mixture of buses, trolley buses, streetcars, and subways. There are over 2600 pieces of equipment running on 97 km of track, carrying more than 372 million passengers a year. Wheel-Trans, the service for people with disabilities, accounted for 1 317 388 passenger trips in 1996.

The first subway in Canada was built under Yonge Street in Toronto in 1954. This subway system now has three main lines. As well, the TTC was the first system to use the Canadian-built **Light Rapid Transit** (LRT) system. This connects Scarborough Town Centre to the Bloor/ Danforth subway. The TTC is now making use of **transportation corridors,** a multilaned expressway combined with a separate **rapid-transit** system running down the centre.

Impact of Public Transit It has been estimated that traffic jams add billions of dollars a year to business costs in Toronto.

One way Toronto is attempting to meet its transportation needs is by expanding public transit. Studies have shown that one rapid-transit line can handle as many people as six expressway

3-19 Toronto's Street Car Sheds on Queen Street in the city's east end.
The City of Toronto has a total of 238 streetcars, affectionally called Red Rockets. 196 of these vehicles (known as Canadian Light Rail Vehicles – CLRVs) are regular sized or 15.24 m long; 52 of them are reticulated (with an accordian linkage between the two parts of the vehicle). These larger cars, known as ALRVs, measure 23.2 m in length.

lanes. The policy of the TTC is to have public transit service available within 600 m of 95 percent of Toronto residents.

Land values increase greatly when public transit is improved. The opening of a new mass-transit line attracts heavy investment in new business and multi-family residences. In the 30 years following the opening of the Yonge Street subway, 50 percent of new apartment construction in Toronto was built within walking distance of the line. In the same period 90 percent of all new office construction occurred near the major subway stations.

The same effect can be seen today, as major developments have sprung up beside GO stations in Burlington, Oakville, and Mississauga, as well as beside rapid transit connections in Etobicoke, Scarborough, and North York.

3-20 The TTC has a total of 1 468 buses which operated more than 93 million kilometres in 1997.

3-21 A GO train crosses the Don Valley Parkway. *The Canadian-designed doubledecker coaches can seat approximately 160 passengers, almost twice the capacity of single-level coaches.*

Toronto plans to spend billions more on its transit system in the next decades. Major projects planned include new subway lines and an LRT line for the harbour area, connecting Union Station with Harbourfront and the SkyDome. The LRT is meant to encourage suburbanites to use their cars less when coming to the SkyDome. If they don't, traffic in the dome area will be chaotic for hours before and after events.

1. **Write a brief description of the public-transit route you would take to go from your community to its CBD or to the nearest large town.**
2. **(a) List five reasons why you should take public transit.**
   (b) List five reasons why you do not like public transit.
3. **Make a map of the new TTC subway route along Sheppard Avenue. What connections can travellers make. What other areas of TTC expansion can you predict. Explain your reasons.**

New Words to Know

Light Rapid Transit	Commuter
Transportation corridor	Trolley bus
Rapid transit	

3-22 Thirty years of GO Transit.

	1967	1997
RAIL NETWORK	96.9 km	361 km
BUS NETWORK	0	1231 km
TRAIN STATIONS	15	49
BUS TERMINALS	0	13
PASSENGERS	2.5 million	34 million
BUS FLEET	0	174
RAIL CAR FLEET	49	329
LOCOMOTIVE FLEET	8	45
ONE-WAY FARE, OAKVILLE TO UNION STATION	95¢	$4.60

Rail transportation Canadian National (CN) and Canadian Pacific (CP) link Toronto to every part of Canada, the U.S.A. and Mexico. VIA Rail, this country's passenger network, offers inter-city, transcontinental and international connections. The Ontario Northland Railway connects Toronto with communities in northern Ontario, as far north as Moosonee.

GO Transit Commuters from GTA communities, use the GO Transit system. The GO network uses trains and buses to serve an area of 8000 km² with some four million inhabitants. The system links communities along the Lake Ontario shoreline between Hamilton and Oshawa, and inland centres such as Guelph and Barrie.

Union Station in the City of Toronto is the hub of the GO system, where GO links with the TTC. The GO Transit system now handles 34 million passengers yearly.

Air Transportation Toronto's main airport is Lester B. Pearson International Airport, located 27 kilometers north-west of Toronto. Pearson was opened in 1939, its first terminal a converted farmhouse. Today, Pearson handles over 26 million passengers and over 375 000 aircraft movements to rank it 25th in the world One third of Canada's air passenger traffic and 40 percent of its cargo passes through Pearson each year. Some 61 airlines fly direct to 140 destinations in 45 countries. As Pearson nears its present capacity of 28 million, there is currently a $2.8 billion 10-year redevelopment program underway at the airport to help it meet demands of the future.

The smaller Toronto Island Airport, close to Toronto's CBD, is used by private planes and commuter aircraft, linking Toronto with Ottawa and Montreal. The Toronto Childrens' Hospital runs its emergency helicopters out of this airport.

The Port of Toronto The GTA's industrial areas are served by Canada's largest inland port. The city's sheltered harbour is about 3 km long and 1.6 km wide. It must be dredged constantly to remain deep enough to handle all the vessels using the St. Lawrence Seaway.

The harbour is well-equipped with the latest cargo-transfer systems, including special heavy-duty cranes for loading extremely bulky and heavy shipments. Toronto's harbour is a fully integrated, intermodal transportation site, which means that different types of transport systems can easily exchange cargo.

3-23 *Pearson Airport's Trillium Terminal was constructed in the 1990s.*

Ports such as Halifax and Vancouver can stay open all year. The Port of Toronto's navigation season is only eight to nine months; the rest of the time the harbour, or the St. Lawrence Seaway, is ice-bound. The port regularly handles bulk cargoes such as salt, oil, sugar, wheat, soya beans, cement, sand and gravel. It also handles container and general cargoes, such as automotive parts, food, liquor, furniture, electronic components and many types of consumer products.

Land Use in Toronto

As a city grows, its neighbourhoods tend to take on specific characteristics, and to be used for specific purposes. The **land-use pattern** refers to the dominant activities taking place in different areas. Most cities contain the following land-use areas: commercial, retail, industrial, residential, institutional, transportation, parks.

Central Business District The heart of any city is its downtown, or CBD.

The focus of Toronto's CBD is the financial district, centred on the intersection of Bay and King Streets. Toronto is the leading financial city in Canada, and the greatest part of that activity takes place within a radius of, at most, half a dozen blocks of that intersection. The Toronto Stock Exchange, the eighth largest in the world, is located nearby.

In broader terms, Toronto's downtown extends south to the railway lands, north to Bloor Street, west to University Avenue, and east to Jarvis Street.

3-24 Two views of Toronto skyline: above, 1967; below, 1998.

3-25 *Trizec-Hahn's Bay-Adelaide Centre, in the heart of Toronto's financial district, will be the first new office building to be built in the downtown core in ten years. 50 stories high, this 1 200 million square foot office tower is scheduled to begin construction in Fall 1999. This image is a computer generated architectural rendering.*

141

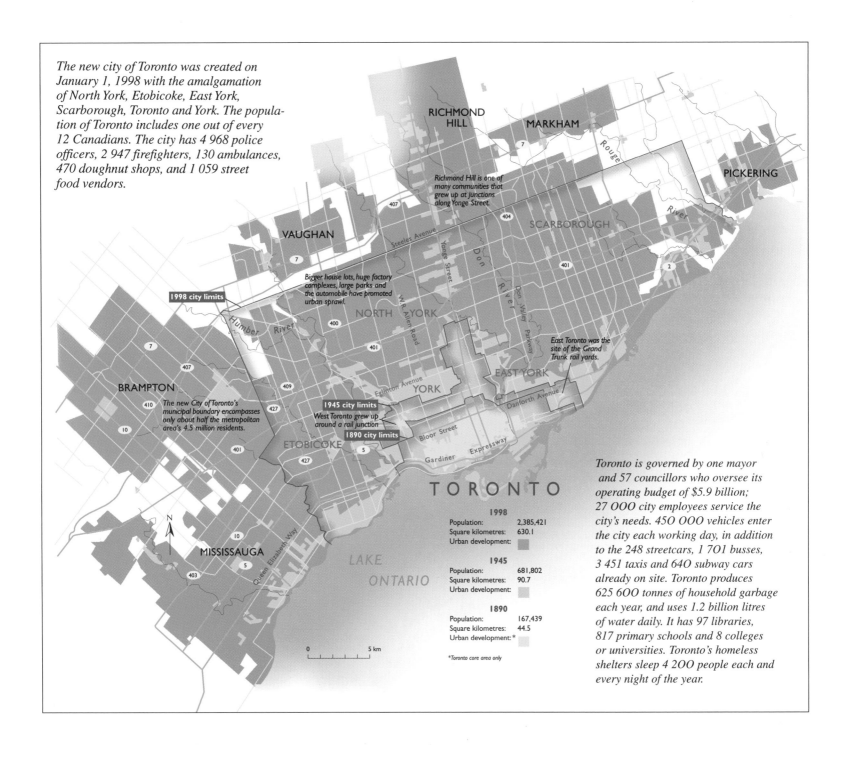

The new city of Toronto was created on January 1, 1998 with the amalgamation of North York, Etobicoke, East York, Scarborough, Toronto and York. The population of Toronto includes one out of every 12 Canadians. The city has 4 968 police officers, 2 947 firefighters, 130 ambulances, 470 doughnut shops, and 1 059 street food vendors.

RICHMOND HILL

MARKHAM

PICKERING

Richmond Hill is one of many communities that grew up at junctions along Yonge Street.

SCARBOROUGH

VAUGHAN

Bigger house lots, huge factory complexes, large parks and the automobile have promoted urban sprawl.

NORTH YORK

East Toronto was the site of the Grand Trunk rail yards.

EAST YORK

1998 city limits

BRAMPTON

YORK

The new City of Toronto's municipal boundary encompasses only about half the metropolitan area's 4.5 million residents.

1945 city limits

West Toronto grew up around a rail junction

1890 city limits

ETOBICOKE

Bloor Street

Danforth Avenue

Gardiner Expressway

TORONTO

MISSISSAUGA

LAKE ONTARIO

1998

Population:	2,385,421
Square kilometres:	630.1
Urban development:	

1945

Population:	681,802
Square kilometres:	90.7
Urban development:	

1890

Population:	167,439
Square kilometres:	44.5
Urban development: *	

0   5 km

*Toronto core area only

Toronto is governed by one mayor and 57 councillors who oversee its operating budget of $5.9 billion; 27 000 city employees service the city's needs. 450 000 vehicles enter the city each working day, in addition to the 248 streetcars, 1 701 busses, 3 451 taxis and 640 subway cars already on site. Toronto produces 625 600 tonnes of household garbage each year, and uses 1.2 billion litres of water daily. It has 97 libraries, 817 primary schools and 8 colleges or universities. Toronto's homeless shelters sleep 4 200 people each and every night of the year.

In a large city, the CBD always has the greatest concentration of tall buildings. And the nearer a building is to the CBD, the higher it is likely to be — that is where land is most expensive. Office buildings dominate much of Toronto's CBD. The amount of office space in the GTA for 1996 was 112 million square feet, with a vacancy rate of 18.1%. The largest concentration of space, as expected, exists in the City of Toronto downtown core, where many **head offices** are located. Toronto has more head offices than any other city in Canada.

The **highrise** office towers in the Toronto CBD rise on average over 30 stories. The tallest office building in Toronto in 1989 was First Canadian Place, which is 72 stories high and rises 290 m. It contains thousands of legal, brokerage, banking and other business offices. These people need to work in close proximity, for meetings and other business activities.

Toronto's City Hall and municipal offices are in the CBD. So are the legislative assembly for Ontario, most provincial government offices, the headquarters of the police and fire departments, and many courts of law. So is the Trade and Convention Centre.

International trade links Toronto with the world. Over 70 foreign consulates and trade commissions have offices in Toronto, many of them located in the CBD.

Toronto's CBD is the hub of many transit routes. The subway lines all converge on the CBD, which is also the centre of a huge regional transportation system: the main bus terminal is on Bay Street near Dundas; the terminus for commuter trains and VIA Rail is Union Station on the CBD's south edge.

Toronto's CBD is also an entertainment centre. In the area are live theatres such as the Royal Alexandra, the Princess of Wales Theatre, and the Hummingbird Centre ... and sports arenas like Maple Leaf Gardens, the SkyDome, and the soon-to-be completed Air Canada Centre.

Harbourfront and the Railway Lands

Exciting new commercial, entertainment and residential areas have developed south of the CBD. These have given the city new contact with the harbour and waterfront areas, which have largely been cut off from the rest of the city by the railway tracks and the Gardiner Expressway.

Harbourfront is a non-profit Crown Corporation established on federally owned land in 1972. Originally an industrial area of warehouses and factories, it has been turned into a "people place" with art and craft galleries, shops and restaurants, marinas and condominiums. The developers planned at the outset to have people living full-time in the area, but so many highrise apartment buildings have been built that some people feel these have become another barrier between the city and its waterfront.

Between Harbourfront and the CBD, there have been substantial new developments in the area of prime real estate formerly known as the railway lands. This area was formerly a railyard for assem-

3-27　Part of Harbourfront. *Both on land and on water the harbour area is being used more and more for recreation and leisure activities.*

3-28 The rise of large suburban superstore malls is a comparatively recent phenomenon. *Each store in one of these complexes is usually not less than 1500 square metres, rising to be as large as 4000 square metres. Many of the "box stores" as they are called are built by large Canadian or American retail chains, including Price Costco, Canadian Tire, Business Depot, Home Depot, Chapters, Moores, Michaels, and others. Smaller independent retailers are finding it increasingly difficult to compete with the retail muscle of these outlets.*

3-29 *Downsview Station on the Yonge-University subway line is one of five key stations equipped with elevators making the system more accessible to persons with disabilities.*

bling freight trains. The area is very close to the CBD and very attractive for urban development. The CN Tower and the SkyDome have already been built there. The CN Tower is the world's tallest free-standing structure (533 m). The SkyDome is the domed, retractable roof, stadium that is home to the Toronto Blue Jays, and the Argonauts sports teams. Other areas of the 80 hectare site have been developed as a series of office buildings, condominiums and/or apartments (for 15,000 people) and shopping, entertainment venues. The project, largest in Toronto's history, is scheduled for completion by the year 2010.

Suburban Business Districts The boroughs of Etobicoke, Scarborough, and North York have have modern business districts, built along subway lines. These urban subcentres consist of modern office buildings, auditoriums, public library outlets, shopping centres and various municipal offices. Development of these suburban business centres encourages new growth within the City of Toronto spreading economic development more evenly around the city, removing pressure from the downtown core.

Retail Land Use About five percent of all the land in Toronto's CMA is devoted to commerce — both stores and

offices. The greatest concentration of retail activities is in the CBD, but as the city grew, retail activities followed the major roads. This type of development is called **strip-commercial** because the small stores, offices, gas stations and so on, line both sides of the streets in a ribbon pattern. Arteries such as Yonge Street, Bloor Street and Eglinton Avenue East are excellent examples of this. Where two major arterial roads intersect, there is often a major plaza.

Large, modern commercial developments are often called, simply, "shopping centres." They are of various kinds, from small neighbourhood plazas to huge, enclosed regional malls. The

3-30 The North York Metropolitan Subcentre along Yonge Street.

3-31 Inside the Eaton Centre.
Despite the growth of CBDs in suburban municipalities, development in the downtown core continues to surge.

neighbourhood plazas serve local areas of a few thousand people and generally have a food store, a variety store and a few other small outlets. Regional shopping centres, on the other hand, will serve tens of thousands of customers throughout a large region, and are located with that in mind.

Toronto has 9 regional shopping centres, each with an area of at least 90,000 square feet and between 100-200 stores – and anchored by a major department store chain. These centres offer free parking and a wide retail variety – both in chain and specialty stores. Recently, the rise of big box stores, and warehouse clubs has altered the suburban mix, and increasingly,

in the Toronto area new shopping centres are becoming clusters of big box stores, each a stand alone outlet, rather than the earlier concept of numerous smaller stores under one giant enclosure.

Eaton Centre and Yonge Street

Malls which span several city blocks in CBDs, like the Eaton Centre with its over 300 shops, cinemas, and restaurants on four climate-controlled levels, have changed the downtown core as well, although Yonge Street itself attracts millions of visitors every year. Yonge is a very dynamic street and appeals to shoppers young and old alike. It has thousands of interesting stores, ranging

from the popular Sam the Record Man to the Art Shoppe, which sells expensive furniture. Many people prefer the noise, the intense activity and the jammed sidewalks of Yonge Street to the more sedate atmosphere of the neighbouring Eaton Centre.

1. Why is commercial land often found along major arteries and near cloverleafs?
2. (a) List the major activities in the CBD in the largest community located near your home.
   (b) Compare your list to the activities found in Toronto's CBD.
3. What are the advantages of locating new subdivisions on rapid transit lines?

Residential Land Use In some cities, including Toronto, over 80 percent of the land is zoned as "residential." Most of Toronto's residential land is in the suburbs. It is estimated that Canadians spend 75 percent of their time in the area of their home. Housing in Toronto can mean anything from a single-family detached house to an apartment in a highrise, from a huge estate to a high-rise penthouse, from a townhouse to a rooming house or a flat above a store. Four of the most common residential types are single-family houses, multi-family houses, apartments and **condominiums**.

Unlike some cities, Toronto has a surprising number of residential buildings in or near the downtown. Most of these are highrise apartment buildings, and many of these highrises are condominiums which are owned, rather than leased by their occupants. Many of these condominium buildings have been developed for a combination retail-residential use, with retail or business outlets on the lower floors, and residential suites above. These residential developments mean that there are people in Toronto's CBD 24 hours a day. This helps create a safer, more liveable downtown area.

For most Canadians, the preferred type of accommodation is probably a detached house in the suburbs. In the Toronto area, however, this is becoming less and less possible for many people. By choice or necessity, more and more people are living in apartments. Young professionals often prefer to live in apartments near the CBD. University students live where they can afford to rent — often in flats above stores or rooms in older houses.

More than any other factor, income dictates where people live. Everywhere in Canada, the cost of housing is fluctuating.

There have been rapid increases in Vancouver and Calgary. In Toronto the average price of a home in 1998 was $225 323 up from $215 638 in 1997. There were 5 410 single family dwellings sold in June 1998 compared with 5 048 in June 1997. Average price paid in the 29 western districts of the Toronto Real Estate Board for June 1998 was $208 476, in the 14 Central districts $342 709, $232 180 in the 23 Northern districts, and $173 943 in the 21 Eastern districts. Compare these number with those in Fig 3-34.

In general, Toronto's most expensive homes are located north of the CBD, near ravines and south of Highway 401. The most expensive districts of all are the Bridle Path, Rosedale and Forest Hill. The former is in North York, the latter two just north of the CBD.

The high cost of housing creates problems for families moving into Metro. Some people even turn down a promotion or a better job if it means moving to Toronto, because they can't or won't buy into Metro's housing market.

In Toronto the cost of land raises the price of housing more than any other factor. The more popular the location, the higher the land prices. Consequently, only 31 percent of Toronto's households are single-detached; almost 45 percent are apartments or condos. About half of Torontonians own their own homes, the other half rent. The trend is now toward renting, however, because of the high cost of buying.

New Words to Know

Land-use pattern	Boutique
Head office	Strip-commercial
Highrise	Condominium
City centre	

3-32 Highrise apartment buildings have been built in or near many traditional residential neighbourhoods.

3-33 Comparative costs for an average detached bungalow

	1987	1997
Vancouver	123 000	450 000
Calgary	97 000	167 000
Edmonton	84 000	114 000
Regina	91 000	100 000
Winnipeg	105 000	100 000
Toronto	179 000	408 000
Ottawa	115 000	137 000
Montreal	100 000	106 000
Charlottetown	88 000	100 000
Saint John	78 500	92 000
Halifax	132 000	136 000
St. John's	89 200	93 000

survey Royal LePage

Industrial Land Use Toronto has the largest labour force of any city in Canada — 14 percent of the national total. One in five employees in Metro works in manufacturing or warehousing. The CMA has over 7000 manufacturing companies — 21 percent of the national total. It is a major producer of almost everything an industrial nation needs: food, beverages, furniture, automotive parts, aircraft components, chemicals, books and other printed goods, building materials, communications equipment, appliances, textiles, plastics, pharmaceuticals, electronics . . . the list goes on almost forever.

Between 10 and 15 percent of Toronto's land is industrial. Industries first located on the waterfront and near the mouth of the Don River. Later they followed the railway lines northeast and northwest of the CBD. There are still factories in those areas.

In recent decades, most companies have moved out to the suburbs; certainly, almost all new factories are being built there. Many are in **industrial parks**. In this section, an older industrial area will be compared to a newer industrial park.

3-34 Median house prices for Metro Toronto, by area.

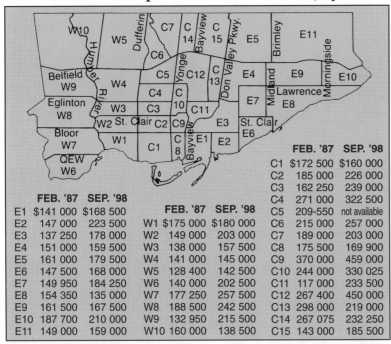

	FEB. '87	SEP. '98
E1	$141 000	$168 500
E2	147 000	223 500
E3	137 250	178 000
E4	151 000	159 500
E5	161 000	179 500
E6	147 500	168 000
E7	149 950	184 250
E8	154 350	135 000
E9	161 500	167 500
E10	187 700	210 000
E11	149 000	159 000

	FEB. '87	SEP. '98
W1	$175 000	$180 000
W2	149 000	203 000
W3	138 000	157 500
W4	141 000	145 000
W5	128 400	142 500
W6	140 000	202 500
W7	177 250	257 500
W8	188 500	242 500
W9	132 950	215 500
W10	160 000	138 500

	FEB. '87	SEP. '98
C1	$172 500	$160 000
C2	185 000	226 000
C3	162 250	239 000
C4	271 000	322 500
C5	209-550	not available
C6	215 000	257 000
C7	189 000	203 000
C8	175 500	169 900
C9	370 000	459 000
C10	244 000	330 025
C11	117 000	233 500
C12	267 400	450 000
C13	298 000	219 000
C14	267 075	232 250
C15	143 000	185 500

Figures such as these become obsolete almost before they are printed; however, they indicate comparative prices for different areas. According to the data, what is (a) the most, and (b) the least desirable part of Toronto to live in? What might explain this?

3-35 *One indicator often cited by analysts studying and predicting economic growth is new housing starts. According to the Canada Mortgage and Housing Agency (CMHC), new housing starts in the Toronto metropolitan census area were 38 791 in 1988, 25 574 in 1997, and 8 235 through April 1998. Work stoppages, including strikes, by drywallers, concrete and drain personnel, concrete forming specialists, framing carpenters, and highrise trim carpenters will have a significant impact on the 1998 year end figures.*

Don/Harbour Area Industry came to this area for a number of reasons. Food processing was an important industry in early Toronto, and the raw materials for it usually came by water. That made the harbour a convenient place for flour mills, sugar refineries and other food plants. In the 1800s, railway lines were connected to the docking facilities, the better to transport finished goods to market.

As well, the Don/Harbour area was flat, which made it easy to build on. It was also close to workers' houses, which was important, since there was no public transit and the work hours were much longer than they are now.

Gradually, other companies built factories in the Don/Harbour area. Today, in addition to flour and sugar mills, there are plants producing or refining cement, metal parts, chemicals, whiskey malt, carpets, meat and plastics. There is also warehousing nearby.

Many of these factories created smoke, noise and foul odours. They were usually large, cavernous buildings, difficult to adapt to the needs of businesses today, and with very little room for expansion. Pollution was not a concern at the time they were built, and some plants used to dump their wastes directly into Toronto Harbour. When modern laws limiting pollution were enacted, many of these manufacturers were unable to comply economically. The hugh capital costs required to upgrade their facilities were not justifiable in a business sense, and so

the old factories were destroyed, or sold or abandoned, as the businesses within them moved away or shut down. Toronto's old heavy industrial sector was a desolate area of the city in the 1980s, like those of many other cities.

Although large and out-of-date, many of the old factory buildings in the Don/Harbour area remain magnificent examples of industrial architecture. As the concepts of mixed use, and urban preservation/restoration began to become popular in the late 1980s and 1990s, many of these

building were re-evaluated for other purposes. Some became theatres, or museums, or restaurants, or office buildings or professional offices or residental/retail complexes. The St. Lawrence neighbourhood was created with housing ranging from low income to high income within the same project. Parks were developed, and new housing plans call for a population of 12 000 within the area, as long term use for the harbour land is to return it to the people for recreation and residency.

3-36 One of the earliest industrial areas in Toronto. *Originally, industries located near the mouth of the Don River and the harbour. Nowadays, the narrow streets and high density of land use cause congestion and other problems.*

Skyway Industrial Park

This industrial park was built some twenty years ago. An industrial park is generally an area where one owner – an individual, consortium, or a local government – has developed a large section of land for industrial or commercial use. The area is fully planned ahead of time so that many of the problems found in ad hoc industrial areas do not occur.

Skyway Park is designed to create a parklike setting. The roads are wide enough for big trucks and built in long curves. Housing subdivisions are kept well separated from the park, which makes for fewer conflicts with home-owners.

The industrial properties are well-landscaped and maintained, because research has shown that attractive work areas make for happier and more productive employees. Companies often provide picnic tables and landscaped rest areas, and parking lots well away from the trucking areas.

The plants in these newer industrial areas are usually one storey in height, making for more efficient production. The plants are built near major **arterial** roads and expressways, since most manufactured goods in Toronto are now shipped by truck. The Skyway Industrial Park, for example, was built near Highways 401 and 427 — both are multi-laned expressways. It is also near Pearson International Airport.

The Skyway area is zoned "industrial and commercial." Businesses commonly found in industrial parks are called "light-secondary." Data-processing companies, insurance offices, service and repair companies, warehouses, office supply companies, graphics companies, and packaging, light-assembly and fabricating plants are typical.

Today, many large secondary industries such as breweries and chemical plants are finding industrial parks a convenient business location. Industrial parks also attract smaller businesses which serve the larger businesses – sandwich shops, copy shops, car leasing agencies, security agencies, insurance companies, employment agencies, and so on.

Other secondary industries, like flour mills and steel mills, must be near a harbour because the cheapest method of shipping in their required raw materials is by water.

3-37 A view of Skyway Industrial Park. *The open areas are pleasing esthetically, and minimize traffic congestion.*

1. Set up a chart to compare older industrial areas and industrial parks, using the following criteria: transportation systems, landscaping, environmental impact, room for expansion.

New Words to Know
Industrial park
Heavy industry
Arterial road

Vancouver: Canada's Pacific Gateway

Vancouver has a site of great physical beauty. The Pacific Ocean is at its doorstep, while snow-capped mountains form a magnificent backdrop. It is no wonder that Captain George Vancouver was impressed with the site when he first saw it in 1792. Still, no one settled in the area until the Fraser River Gold Rush in 1856-57.

The area was originally named Granville. In 1870, sawmilling was its major industry. The town grew slowly until 1886, when the Canadian Pacific transcontinental railway was completed, and Granville was chosen as the final destination. That year, the little sawmill town was renamed Vancouver, after Captain George Vancouver, and it began a spectacular growth which continues to this day.

Vancouver's growing industries were now able to ship their merchandise by train to markets in both Canada and the U.S.A. By 1901 Vancouver's population was 30 000 and it was becoming a major transportation and distribution centre. In 1914 the Panama Canal was opened. Ships from Vancouver now had swifter and safer access to Europe and South America. Even before the beginning of the twentieth century, Vancouver was also linked to the Pacific Rim countries such as China and Japan.

Vancouver also became the focal point for the numerous fishing and forestry operations on the British Columbia coast. By the mid-1920s Vancouver had become the dominant city in western Canada. It was the centre for transportation, commerce, finance, manufacturing and trade. A solid base for the city's metropolitan development had been established. Vancouver was and remains a city with one foot planted firmly on a strong resource base – fish, lumber, materials – and the other anchored on an excellent transportation system, the financial and communication industries, tourism, film-making, and computers.

The Physical Site

Vancouver is built on the south shore of Burrard Inlet. To the north and east are the Coastal Mountains, which rise to over 3000 m. The southern shore of Burrard Inlet, however, is part of the Fraser River delta, and is easy to build on. The rich, alluvial soil is flat and fertile.

Vancouver has water on three sides. To the west is the Pacific Ocean, to the north Burrard Inlet, and to the south the Fraser River. The natural environment abounds with wildlife. Bald eagles are commonly seen. Bears and cougars occasionally wander into the suburbs!

3-38 Burrard Inlet and downtown Vancouver from the north shore.
Vancouver harbour is a natural inlet of the sea, sheltered by Stanley Park peninsula.

Seals and killer whales are sometimes spotted in nearby inlets. The area is world-famous for sports fishing — particularly for Pacific salmon. Vancouver has taken full advantage of its picturesque location by developing first-class outdoor recreation facilities.

Vancouver's Climate The Vancouver area is an outdoor enthusiast's paradise. The moderate climate permits year-round golf, tennis, hiking, swimming and sailing. During the winter it is possible to play golf in the morning, sail in the afternoon, and ski at night.

3-39 Comparison of Immigrants to Vancouver prior to 1961.

	Immigrated to Vancouver before 1961	Percent
United Kingdom	27 485	31.8
Germany	9 890	11.4
Italy	6 655	7.7
Netherlands	6 250	7.2
China	5 545	6.4
Poland	2 775	3.2
Hungary	2 670	3.1
United States	2 565	3.0
Denmark	2 445	2.8
Austria	1 445	1.7
Total	**86410**	**100.0**

Vancouver Recent Immigrants 1991-1996

Vancouver Recent Immigrants 1991-1996		Percent
Hong Kong	44 715	23.6
China	27 005	14.2
Taiwan	22 315	11.8
India	16 185	8.5
Philippines	13 610	7.2
South Korea	6 335	3.3
Iran	4 640	2.4
United Kingdom	4 040	2.1
Vietnam	3 855	2.0
United States	3 640	1.9
Total	**189 660**	**100.0**

Monthly average temperatures are above freezing all year. The lowest is 3°C in January. The Pacific Ocean moderates the climate, keeping summer temperatures around a pleasant 18°C. Rainy days are common, however, especially in winter, and total annual precipitation is over 1000 mm. Decembers receive an average of 164 mm of rain. Summer rainfall is low, averaging 26 to 35 mm in July and August. Occasionally Vancouver receives a snowfall that melts quickly. The year-round moderate temperatures and the absence of harsh winters, make Vancouver one of the more attractive places in Canada to live.

Vancouverites: The People of Vancouver

Vancouver's metropolitan area is Canada's third-largest in population.

The 1996 census reported that the CMA of Vancouver contained 1 831 665 people. Metropolitan Vancouver, which consists of 17 member communities, includes large centres such as Burnaby,

3-40
The University of British Columbia's Museum of Anthropology, overlooking Howe Sound and the Georgia Strait, is home to one of the world's most impressive collections of Haida carvings, specifically totem poles.

Coquitlam, Delta, New Westminster, North Vancouver, West Vancouver and Richmond. Surrey, a suburban area, is growing at a much faster rate than the City of Vancouver itself.

Traditionally, people of British and U.S. origin have dominated in the region, but Vancouver is now one of the most diverse, multicultural cities in Canada. Its Chinese community (in excess of 49.4% of the total population) is the second largest in North America; South Asians represent 21.3% of Vancouver's population, and Filipinos 7.2%. People of Indo-Pakistani descent are another powerful ethnic force in the city. Vancouver's cultural mosaic encompasses over 100 languages. 50% of the city's school children are enrolled in some type of second language program, the extra costs of which often place a severe strain on the financial resources of the urban region. Many traditional shopping districts now have a strong Asian influence which can have both positive and negative effects.

A large percentage of Vancouver residents are between the ages of 20 and 35. This probably reflects the desire of many young Canadians to go west for a new start in life, and it certainly has been impacted by the increase in Asian immigration, which accounted for over 60% of Vancouver's immigrants over the 1991-1996 period. Vancouver is now the most Oriental of any non-Asian city in the world, a fact that is causing major adjustments for its earlier inhabitants.

Land-Use Patterns and Development

The physical site of Vancouver greatly affects how the city can expand. Its

151

growth is blocked by the Pacific to the west and the Coastal Mountains to the north. It can only grow south and east, along the Fraser River delta.

Industrial Development   Most of the city's heavy industries — flour mills, oil refineries, chemical plants, shipyards and repair docks — are concentrated on Burrard Inlet.

There are also major industries along the Fraser River and in the New West-minster, Surrey, Coquitlam and Delta areas. In each case, industrial development occurred because there was flat land, access to fresh water, and excellent rail conditions.

Vancouver companies manufacture wood products, metal parts, machinery, electronic goods, canned and frozen foods – especially fish, submersibles, computer software, and construction equipment. About 14% of the city's workforce is involved in manufacturing, much of which is based on British Columbia's natural resources – its forests, mines, fishing grounds, agriculture, and increasingly on the well-educated workforce.

Commercial Development Vancouver's CBD is squeezed between Burrard Inlet and False Creek. Most offices and retailing activities are concentrated here as well as regional head offices for some of Canada's leading financial institutions. One distinctive skyline silhouette is Canada Place, built for Expo 86, which provides facilities for hotel and convention service. Another, the 323,000 square foot main branch of the Vancouver Public Library, designed by Canadian architect Moshe Safdie, opened in May 1995.

Some of the older commercial districts have been renovated into **specialty shopping** and tourist areas. Gastown, once a rundown hotel and bar area, is now highly tourist-oriented and a mandatory stop for the over 700 000 people from cruise ships who come to visit Vancouver each year. Exclusive stores are found in the Sinclair Centre on Hastings Street, and in the subterranean Pacific Centre below the downtown core.

Along Granville Street, and around the Hotel Vancouver area are newly developing areas catering to Asian customers, as the old Chinatown is being challenged by suburban malls such as the Yaohan and Aberdeen Centres in Richmond. Multinational brand-name retail stores are moving in along Robson Street, and its counterpart older European sections.

The rest of Vancouver's commercial development follows the major arterial roads leading into the CBD – Hastings Street and Marine Drive are two of the

3-41   Land use in Vancouver.

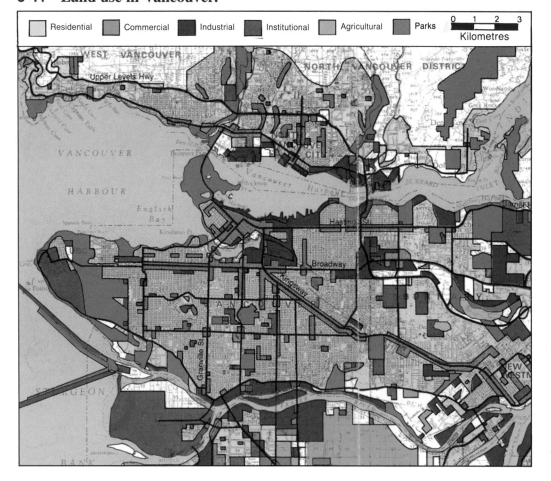

Residential   Commercial   Industrial   Institutional   Agricultural   Parks

0 1 2 3
Kilometres

main drives. Regional shopping centres like Oakridge and Park Royal are located at major intersections.

A unique shopping area has been developed on a small island south of Vancouver's CBD. Granville Island was once an industrial and warehousing district. Today it is an exciting combination of **light industry**, restaurants, live theatres, craft shops, art schools and a public market that is one of the island's main attractions. Fresh fish, vegetables and baked goods are the mainstays of the market. The island, which can be reached by car, bus, foot, boat or mini-ferry, draws locals and world tourists.

Residential Development Single-family detached homes dominate in Vancouver. This produces a low population density because of the large amount of land required by single-family houses. It is a paradox, because one of Vancouver's biggest problems is lack of space for growth which is creating traffic grid-lock.

The residential land-use pattern is typical of other Canadian metropolitan areas. There is a small inner core of high-

3-42   Street scene in Chinatown.

density apartment buildings around the CBD, then a ring of older single-family houses, and finally newer homes in the suburbs. Asian buyers have driven up the prices of homes. Some people from Hong Kong have built huge mega-houses in older residential areas creating resentment in some neighbourhoods.

Some of Vancouver's most expensive homes are in Shaughnessy Heights, south of the CBD, and in West Vancouver, which is north of Burrard Inlet. The West Vancouver homes have beautiful views of English Bay and Stanley Park, so some of the most expensive real estate in Canada is located in this area.

3-43   Granville Island.

More and more apartments are being built in newer suburbs such as North Vancouver, Burnaby, Richmond and New Westminster. One of the most popular apartment districts is Vancouver's West End between the CBD and Stanley Park. People who live there are within minutes of their downtown jobs and just as close to recreation. The West End contains about nine percent of the city's population on less than two percent of its area. This gives it one of the highest population densities in Canada.

Vancouver's apartments are home mainly to young singles, young, childless couples, retired couples and, to an ever-increasing extent, single-parent families. Many apartment buildings are old, although the trend is increasingly toward highrise apartment blocks.

Another interesting residential sector is the False Creek redevelopment. Once a heavily industrialized area containing shipyards, railways and sawmills, today False Creek provides housing for more than 3000 people. The development offers a great variety of housing for all income levels. It includes senior-citizen housing, luxury condominiums, low-income housing, low- and medium-density apartments, townhouses and even floating homes. All residents have access to parkland and walkways along the water. The main access is a shoreline path that connects all the dwellings. This discourages cars from using the district. The False Creek project includes a marina and an elementary school. The project is a fresh approach to solving urban housing needs and a model for other urban areas across Canada.

Recreation Areas   The city's irregular shoreline makes for many scenic coves, bays and inlets. There are many public beaches within minutes of the CBD, close to the densely populated west end. Stanley Park is one of North America's largest urban parks. The peninsula on which the park is situated was set aside by the British government as a military reserve. In 1888 the federal government deeded the land to the city for "the use and enjoyment of people of all colours, creeds and customs for all times." It still belongs to the federal government.

3-44   *A view from Stanley Park's Prospect Point showing the Lions Gate Bridge.*

3-45   *A cruise ship at Canada Place.*

154

Today the 405-ha park is used by millions and is a major tourist attraction. It is heavily forested with many species of trees, including some 800-year-old Douglas firs. This wilderness setting is one of Stanley Park's major attractions. It is used heavily by picnickers, hikers and nature-lovers. Picnic tables dot the park, and dozens of interesting nature trails crisscross it. It is encircled by a 9-km seawall with walking paths and bikeways. The park has tennis, lawn-bowling, cricket and fishing facilities, and many interesting historic sites including the Vancouver aquarium, with its killer whales.

Transportation

Vancouver's International Airport Authority is one of Canada's busiest. The airport is located on the Fraser River delta just south of the city. Air cargo is an increasingly large proportion of its business. Vancouver is served by 23 airlines and has over 475,000 flights each year. The airport has a $200 million impact on the region with the new $360 million expansion. This expansion enhances the airport capacity as well as sea-air cargo efficiency. Increased passenger flow adds to the Port of Vancouver's cruise-ship business and increases the air connections to Asia. Vancouver is 1600 kilometers closer to Hong Kong than Los Angeles, therefore planes and ships can carry more cargo from Vancouver because they carry less fuel. The airport is only about 25 minutes from Vancouver's CBD and from the port's inner harbour facilities – including the two cruise ship terminals.

CN and CP link Vancouver by rail to other cities and regions of Canada. The British Columbia Railway runs northward into the interior. The Burlington Northern

3-46 *Transit B.C.'s Seabus travels the Burrard Inlet between downtown and North Vancouver's Lonsdale Quay. The trip lasts 12 minutes.*

Santa Fe – a U.S. railway – provides a rail connection directly to Seattle, Washington. From there, Vancouver's products are shipped throughout the U.S.A. and Mexico. All railroads provide modern double-stack, **intermodal** service. This has enabled Vancouver to compete with American Pacific ports.

Within the Vancouver region, a public-transit system connects the suburbs with the CBD. Buses, trains and trolley buses are the mainstay of this system to the suburbs. However, the newest transit mode in Vancouver is the Canadian-designed Sky Train. This is one of the most

advanced rapid-transit systems in the world. It has over 42 km of route, some elevated, some underground. The computer-run trains are separate from ground-level traffic at all times. The Sky Train has triggered new office, hotel and apartment projects along its route. Property values have tripled in some neighbourhoods since the system's completion.

Sky Train links the CBD with B.C. Place Stadium (Canada's first completed domed stadium), with new developments on the Expo 86 site, and with the nearby communities of New Westminster, Burnaby and Surrey. Vancouver is linked with North Vancouver by means of the Sea Bus, an ultramodern ferry connecting the Waterfront Sky Train station with the other side of the harbour. Sky Train technology is being sold to Far Eastern cities such as Singapore.

The Port of Vancouver Vancouver is Canada's largest full-service port. It has facilities to handle all types of cargo. It is one of the world's top-ten ports in terms of export tonnage. The port is open all year round and has a fully protected deep-water harbour with no depth restrictions. The huge harbour area allows dozens of ships to anchor safely while waiting to load or unload. Close to 2,000 vessels use Vancouver's facilities every year. It is the busiest port in the Americas and the most competitive. Over 85 percent of its shipments are bulk goods handled through the port's 17 ultra-modern bulk terminals using SPARCS – a fully computerized cargo control system.

The Port also operates two container terminals. Greater Vancouver is a hub for British Columbia's ferry services. These ferries link isolated communities along

British Columbia's coast using rail barges, car transporters and passenger ferries.

Increasingly, the port is being used by cruise ships plying the waters from Vancouver to Alaska. In 1997 cruise ships carried over 700 000 passengers to Vancouver. They create thousands of jobs in the two main cruise-ship terminals and in nearby suppliers. In 1998 the Port of Vancouver provided 10 700 full-time equivalent jobs.

Canada Place, built for Expo 86, is the main cruise-ship terminal. It has a trade-and-convention centre, restaurants, an Imax theatre (which uses a huge screen and high-tech projectors) and a major hotel. It is a focal point for B.C.'s rapidly growing tourist industry.

The Greater Vancouver Regional District (GVRD)

In the early 1970s regional government came to the Vancouver area. The area is now guided by a federation of the 17 member communities which make up the GVRD.

The GVRD is not as powerful as its equivalent in the Toronto GTA. Its main responsibilities are water supply, regional parks, regional planning and transportation. One example of its successes is the efficiently run water system. Most cities get their water from wells or nearby lakes. Vancouver, however, uses water from the mountains of the Capilano, Coquitlam and Seymour watersheds. The water is gathered in six storage lakes, then fed to the region by large underground water mains.

This system serves 1.8 million people — about 50 percent of the province's population. It is essential that the source of this water be protected, so the water authority has planted over 2.5 million seedlings. Healthy forest areas protect the water supplies because the foliage and roots prevent soil erosion. It is soil that filters, cleans and stores the water.

3-47 Greater Vancouver water system.

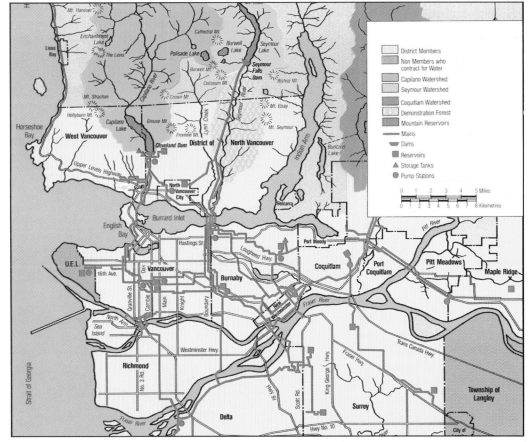

Because it comes directly from rainfall, Vancouver's water is so pure it is fit to drink without any purification treatment whatever.

1. **What makes Vancouver one of North America's most attractive cities?**
2. **What are the origins of Stanley Park? Outline the advantages of having Stanley Park so close to the centre of Vancouver.**
3. **Give four reasons why Vancouver was chosen as the site for a World's Fair.**
4. **Where might each of these people decide to live in Vancouver?**
   (a) **an artist or musician**
   (b) **a young lawyer**
   (c) **a young family of three**
   (d) **a wealthy retired couple**

New Words to Know
intermodal

Calgary: The Booming Boom-and-Bust City

Calgary is an adolescent in a world full of mature cities. Some European cities are over 1000 years old; Calgary is only 114. The city came to the world's attention in 1988, when it hosted the Winter Olympics. The Calgary Games were seen by 1.5 billion television viewers around the world.

Calgary's first building was a North West Mounted Police post built at the **confluence** of the Bow and Elbow Rivers to stop American whisky traders working out of Fort Whoop-Up (present-day Lethbridge). It was an excellent place for a city — there was plenty of wood for construction, a flat plain to build on and plenty of clean, fresh water.

In the beginning Calgary was a small ranching centre. By 1881 the population was only 75. It was not until the arrival of the railway in the 1880s that it began to grow. The site provided the CPR with an ideal bridging point across the Bow River. The railway boosted Calgary from a frontier outpost to a regional centre for the pioneers settling in the Far West.

Early industries were flour milling and meat packing. There were also railway repair shops and shipping companies. In 1914, oil was discovered in the nearby Turner Valley. This provided the biggest boost yet to Calgary's development. By the late 1920s and 1930s, however, severe droughts, declining world markets and the Great Depression had sent Calgary into a bust. Calgary did not suffer as much as many single-industry towns, because it had a broader economic base. However, its population still decreased between 1931 and 1936, during the worst years of the world depression.

After the Second World War, Calgary boomed again, with the discovery of the Leduc oil fields. Since then, Calgary has been the undisputed centre of the Canadian oil-and-gas industry.

The Foothills City

Calgary stands at the junction of two rivers draining the Rocky Mountains: the wide Bow and fast-flowing Elbow. The CBD is spread out within a wide **meander** of the Bow River. Flat-topped bluffs rise above the river floodplains to

3-48   Aerial view of Calgary, looking toward the Rockies.

the north and northwest, across from the city centre.

Early growth naturally followed the river plains and flat-topped bluffs where building was easiest. Beyond the bluffs to the north, small, barren, wind-blown hills dominate. The most prominent of these is called Nose Hill. To the west, the **foothills** of the Rockies provide a dramatic backdrop to the city. The hills to the southwest provide the city with water from the Glenmore Reservoir.

Climate

Most prairie cities, because they are so far inland, are known for extremes in temperature. Calgary, however, may have the most moderate conditions of any major city on the prairies.

Calgary is in the semiarid prairie climate region. This means low precipitation, long, cold winters and short, hot summers. Calgary receives about 443 mm of precipitation every year. Fortunately for the region's farmers, most of this falls as rain during the growing season.

The most spectacular — and unusual — thing about Calgary's climate is the **chinook** wind. Imagine a cold February day with the temperature at −20°C and a snow cover of 5 cm. Within an hour the winds swing around to the west-southwest, gusting up to 90 km an hour. The temperature jumps by 20°C and more, and the fresh snow melts.

This happens on average 25 times a year. It is a strong, blustery, dry, warming wind which provides relief from the cold. However, it causes severe headaches and sleeplessness for some people, and can dry out the soil and lead to erosion.

3-49 Climate severity indexes for selected CMAs.

VICTORIA	13	WINNIPEG	51
VANCOUVER	18	TORONTO	35
CALGARY	34	MONTREAL	44
EDMONTON	37	SAINT JOHN	48
SASKATOON	42	HALIFAX	43
REGINA	47	ST. JOHN'S	56

The lower the index the more desirable the climate. Severity is generally measured in terms of temperature ranges and extremes, humidity, and wind velocity.

The People of Calgary

In 1996, Calgary's population was 821 628, up nine percent from 1991, and 31 percent from 1981. Much of the city's growth is due to migration which accounted for close to 75 percent of the increase between 1976 and 1996. 52.8 percent of Calgary's population is youthful; only 7.1 percent of its residents are over 65, while 58 percent are under 35. The city's density is 161 persons per every square kilometer.

The city's **labour force participation rate** of 76.1 percent is the highest in Canada, and Calgary's percentage of women working outside the home also ranks among the highest in the nation.

Calgary's Economy

The Energy Connection Among Canadian cities, only Toronto and Montreal have more corporate head offices than Calgary. Over 80 percent of this country's oil-and-gas producers, 86 percent of its geophysical supply companies, and 75 percent of its coal companies have their head offices in Calgary. Every major oil company in the world has an office in Calgary and

3-50 *The Calgary Stampede, first held in 1912 (when it attracted 14 000 people) has developed into an eleven day celebration which attracted 1 124 271 visitors in 1998.*

14 international banks have offices there.

Here are found the thousands of highly educated specialists needed to run Canada's energy business. Nearly 50 percent of Calgary's labour force is employed in the production, exploration and development of gas and oil. As well, more people are employed in the mining industry in Calgary than in any other large city in Canada. A vast pipeline network extends outward from Calgary to many parts of North America. This, along with the railways, makes possible the development of oil refineries and gas-processing plants in Calgary. But though the energy industry is clearly the major basis of Calgary's wealth, it is also the cause of its boom-and-bust economy. The energy industry is highly competitive and variable. Shifts in the international supply and price of oil and gas affect Calgary immediately. Calgary's fortunes are more affected by the world economy than those of almost any other city in Canada.

The Food Connection Alberta is a major agricultural producer. Cattle-raising is the most important activity. Many Canadians still identify Calgarians with the famous ten-gallon cowboy hats. However, the climate and rich soil around Calgary also yield a wide variety of field crops.

Calgary is the **agribusiness** centre for much of western Canada. It is involved in all types of food processing and boasts the largest stockyard in western Canada. It is the beef capital of Canada; food processing employs one out of every five manufacturing workers.

The Variety Connection Calgary has diversified its economy. Reliance on food and energy production has exposed the city to boom-and-bust cycles for too long. To prevent another bust, local entrepreneurs have invested billions of dollars in new industries and businesses.

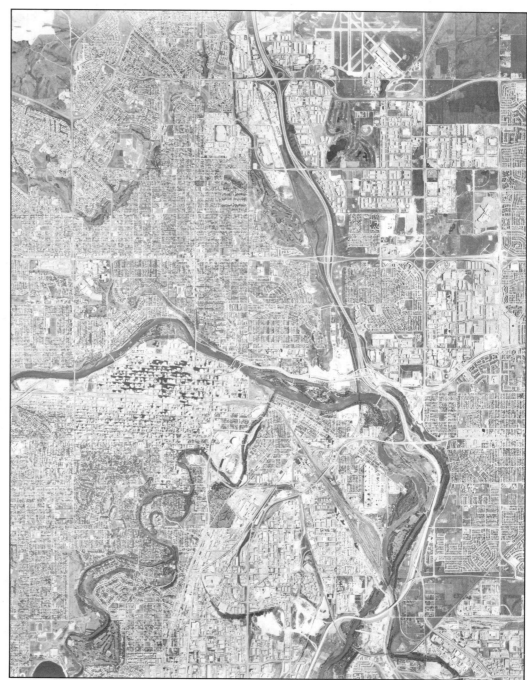

3-51 An air photo of Calgary. *Note the location of the central business district, the airport, and major industrial areas.*

There are about 30 industrial parks in Calgary. Concentrated in the city's south and east, they are easily accessible by major highways, by railway and by Calgary's international airport. Diversification in manufacturing means that Calgary produces everything from all-terrain vehicles (ATV's) to computer software.

Research and Development Calgary boasts the largest concentration of science, mathematics and engineering workers in Canada with 44 employees per 1 000 population in these technical industries (the national average is 22 per 1 000). Almost 50 percent of Calgary's population has some post-secondary education – the highest percentage of any major Canadian city. This **brain trust** reflects the city's high number of research facilities.

More than two-thirds of Alberta's advanced-technology companies are located in Calgary, many of them in new industrial parks. The University of Calgary has acquired a worldwide reputation in engineering, geophysics and related fields. The University and mining companies work closely together to develop new technologies for the energy industry.

Some 15 percent of the world's **seismic** industry operates from Calgary.

The Tourism Connection Calgary is located only 40 minutes from the mountains, and Canada's oldest National Park, Banff. The city's world class recreational facilities, many of which were installed for the 1988 Olympics, include a bobsled/luge track, North America's first fully enclosed speed-skating oval, and an alpine centre. The Glenbow Museum, one of Canada's most respected repositories of western Canadian history is in Calgary, while Spruce Meadows, the popular horseracing track is six kilometers away.

Calgary is the site of Canada's second largest zoo, and numerous other parks. Heritage Park, on the city's outskirts, shows what the city was like in its earlier days. Sports fans follow the Flames (hockey) and the Stampeders (football) in the Calgary Saddledome sports stadium, a city landmark.

3-52 *The town of Banff has long attracted tourists from around the world because of its proximity to the 6 641 km2 Banff National Park, and because the town itself is a mecca for artistic and cultural events like the Banff Festival of the Arts, or the Banff Publishing School. In 1998, the federal government, working in conjunction with Banff officials, moved to develop long term planning strategies for Banff, and for six other Canadian towns adjacent to National Parks, so that in future, commercial development interests can grow without adversely impacting on the ecological integrity of the surrounding parks. Banff National Park currently receives 5 million visitors per year. This amount is expected to reach 19 million by the year 2020.*

Transportation in Calgary

Calgary is trying to achieve a balanced urban transportation system, especially in its CBD. Traffic flow downtown is intense and concentrated. It is made worse by the fact that many residential neighbourhoods are on the city's west side, many industries on the east. A great deal of rush-hour traffic must pass through or near the CBD.

Calgary developed one of the first LRT systems in Canada, called the C-Train. Its 22 km of double track serve the major suburban areas and carry commuters to the CBD, where one-third of Calgarians work.

While downtown, people can move easily from place to place using the unique "Plus-15" system. This is a network of elevated, glass-enclosed pedestrian bridges linking the major downtown buildings — the largest system of its kind in the world. All told, there are 41 bridges in the system. They link the LRT stations with various retail malls, office blocks and "parkades" — small indoor parks with fountains. The elevated walkways give pedestrians views of the mountains while protecting them from winter's cold. Between them the LRT and the Plus-15 have encouraged the development of new highrise projects, attracted people downtown, and helped ease traffic congestion.

1. **Why did the coming of the transcontinental railways have a major impact on Calgary?**
2. **How do the names of Calgary's two professional sports teams reflect that city's major economic activities?**
3. **Why has Calgary gone through so many boom-and-bust cycles?**
4. **Name five major advantages for Calgarians of the Plus-15 system.**

New Words to Know

Confluence	Agribusiness
Meander	Brain trust
Foothills	Seismic
Chinook	
Labour force participation rate	

3-54a *The Calgary Saddledome*

3-54b *Oxford Properties Group's new 23 story Millennium Tower, with its 38 000 m² of office space and 1 900 m² of retail area will help alleviate the tight office real estate market in Calgary where the vacancy rate was 5.3 percent at the end of 1997. Migration into the city remained steady at 1 200 people per month throughout the year. Retail sales were up 11 percent over 1996.*

Halifax: "Warden of the North"

Halifax has one of the finest natural harbours in the world. The sea has always been the reason for this city's existence.

Halifax was founded by the British in 1749, as a military base. The port's defences were anchored by Citadel Hill, where a vast fort was built. Other, smaller forts were built on small islands in the harbour. Because of these strong defences, no foreign army ever succeeded in capturing the city. The British made it their leading military and administrative centre in this hemisphere. This encouraged its rapid growth.

Halifax played an important role in the two world wars. The huge double-harbour — Halifax Harbour itself and the even larger Bedford Basin beyond it — were assembly areas for the large ship **convoys** bound for Britain. The wars gave a dramatic boost to the industrial and commercial life of Halifax.

Halifax has always been more dependent on national and global events than on its own hinterland. It has always been Canada's most important east coast city. It is still the Atlantic region's only real metropolis.

3-55 Halifax and Halifax Harbour, looking toward the cloud-covered Atlantic.

The Site of Halifax

Halifax occupies a peninsula jutting into a large, sheltered bay of the Atlantic Ocean. There is salt water on three sides: Halifax Harbour to the east, Bedford Basin to the northwest, and the Northwest Arm to the southwest. The latter is a beautiful, fjord-like bay that is used as a yacht basin. Many expensive, private estates dot its shoreline.

Halifax Harbour is the second-largest natural harbour in the world. It is also one of the safest and most sheltered. The outer harbour extends 9 km inland from the Atlantic. It is linked by a narrow passage to Bedford Basin, the inner harbour, which is exceptionally deep.

Across the harbour is Dartmouth, the sister city of Halifax. Together they comprise the Halifax/Dartmouth CMA. The cities are linked by two suspension bridges and a passenger ferry.

The Halifax area has a cool, temperate, maritime climate. Its annual mean temperature is 7°C. Its monthly temperature ranges from 18°C in summer to −4°C in February. Average winter temperatures are just above freezing. The average annual precipitation is 1381 mm; the average snowfall is 205 cm.

Halifax has the lowest snowfall and highest winter temperatures of any Canadian east coast port.

The weather is damp and often foggy, because the warm Gulf Stream and the cold Labrador Current meet and mix nearby. These currents also help create one of the world's most important fishing grounds — the Grand Banks.

Haligonians: The People of Halifax

Halifax was founded and settled by the British; its people are still primarily of English, Irish and Scottish descent. Germans, Americans and French came in later waves. Today Metropolitan Halifax is home to 332 518 people. Well over 200 000 more live within 160 km of the city.

Halifax has grown by over eight percent in the past five years. Its current growth rate is 14 percent per decade. It has a higher percentage of people in the 18-25 age bracket than any other large Canadian city. This is because it is the major educational centre in Atlantic Canada, with eight large colleges, universities and research institutes.

Halifax is also the largest military centre in Canada. The federal government's plans to strengthen this country's naval forces will add to Halifax's population and development.

Haligonians are employed as follows: community, business and services, 33 percent; public administration and defence, 18 percent; transportation and communications, 8 percent.

Close to half of all working Haligonians are employed in port activities. The port is the focal point of the city's economy.

3-56 Historic Properties *This area of restored factories and warehouses attracts many tourists and shoppers to the harbour area. This former fishery building now houses boutiques and specialty shops featuring Nova Scotia handicrafts.*

The Economy of Halifax

Halifax is headquarters for the Canadian Armed Forces' Maritime Command, the Canadian Navy and the Coast Guard. As capital of Nova Scotia, it is the province's judicial and administrative centre. It is also the province's educational and research centre. Halifax is home to Dalhousie University, St. Mary's University and Mount St. Vincent University.

It is also home to the world-famous Bedford Institute of Oceanography. This is the second-largest centre for marine studies in the world, after Woods Hole in Massachusetts. The institute carries out research in marine biology, sea-bottom charting, pollution and seismic studies. The institute is on the northeastern shore of Bedford Basin.

Conventions and Tourism Halifax is the Atlantic region's major retail, entertainment and cultural centre. It also serves as a major tourism and convention focus. The Halifax World Trade and Convention Centre was built in 1984, on the city's waterfront. It can accommodate over 3 500 guests. In addition, there are over 2 500 first class hotel rooms within a five-minute walk of the CBD.

The Historic Properties, a group of carefully restored waterfront buildings from the eighteenth century, now a wide variety of retail boutiques, is a popular tourist

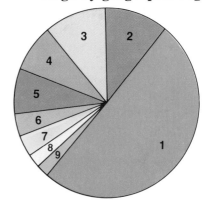

3-58 Port of Halifax container cargo by geographic region.

1 U.K./CONTINENT 51.2%
2 FAR EAST & SOUTHEAST ASIA 10.7%
3 SCANDINAVIA 10.5%
4 INTERCOASTAL 8.0%
5 MEDITERRANEAN 7.5%
6 AUSTRALIA/NEW ZEALAND 3.9%
7 OTHER 3.7%
8 CARIBBEAN 2.3%
9 MIDDLE EAST 2.2%

3-57 Bedford Institute of Oceanography.

attraction. The Halifax Citadel, completed in 1856, is another. The area abounds with restaurants, museums, live theatres and open terraces.

Nearby is the *Bluenose II*, an exact replica of the famous schooner *Bluenose*, which is featured on the back of the Canadian dime. *Bluenose* was the fastest sail-powered fishing boat in the western world.

Halifax is a major cruise-ship centre. Thousands of people are employed in provisioning and manning the ships.

Commerce and Industry Halifax is Atlantic Canada's regional head office for many large financial corporations. It is also a leading centre for insurance companies and brokerage firms. It is not, however, a major industrial centre by Canadian standards. Traditionally,

Halifax has concentrated on fish and food products, oil refining, shipbuilding and repair, and metal fabrication.

Halifax has expanded its industrial base. Its industries produce tires, electronic components, aerospace parts, oil-rig components, high-tech exploration equipment, and heavy engineering products. The city's business leaders are linking new industries to the extensive research and educational facilities within the metro area. Many local industries have international connections.

Woodside Ocean Industrial Park is unique in that it combines serviced lots with dock facilities. Its purpose is to foster all forms of marine and ocean-related industry. The development of the Hibernia oil field off Newfoundland sparked significant new activity in the area.

The Halifax Regional Municipality has 13 business/ industrial parks which are home to 1 500 companies.

Halifax Harbour

Halifax is on the **Great Circle Route**. It is the closest North American port to Europe, Africa, the Mediterranean and much of South America. This advantage is not obvious on a flat map. It *is* on a globe. For ships, Halifax is at least one day closer to Rotterdam, London, Rio de Janeiro, Capetown and Rome than are other ports of Eastern North America. One result is that Halifax is a major importer and exporter of intercontinental shipments.

With at least 21 m of water at low tide, Halifax is one of the world's deepest harbours, capable of handling

3-59 The World Trade and Convention Centre in Halifax.

3-60 *Semi-submersible oil rig at Common User Dock, Woodside Ocean Industries Park.*

165

the world's biggest ships. The harbour is free of navigational obstructions and dangerous currents. There are no bridges, no shallows, no narrow channels and only minimal **tidal fluctuations**. All this means that ships can arrive in Halifax safely, even at low tide. It is also ice-free, and roomy enough for ships to manoeuvre. That is why so many shipping companies have located in the city.

Around the harbour are industries such as shipbuilding, ship repair and **offshore fabrication**. The harbour is home to Canada's east coast naval fleet. NATO ships also use Halifax, giving the city the largest concentration of warships in Canada. The harbour also services fishing vessels from many countries.

Container traffic has also given a big boost to Halifax Harbour. Halifax has two container terminals. The city suc-

cessfully competes with New York City, Quebec City, Montreal and Toronto for container traffic. This is because of its position on the Great Circle Route, its greater security, and the transcontinental links offered by the Canadian railway network. It is often faster for a ship to unload onto rail in Halifax than to sail all the way to Montreal.

1. Using well-labelled sketch maps, give ten major reasons why Halifax developed into an important port.
2. Using a piece of string or narrow ribbon on a globe, prove that the Great Circle Route for North American-European ship traffic gives Halifax an advantage over ports of the eastern U.S.A.
3. How has Halifax diversified its industrial base? Why has it done so?
4. Draw and label a sketch map showing all the major transportation routes serving the Halifax metropolitan area. Use your atlas for reference.
5. (a) Plot on a map of the world all the ports shown on chart 3-60.
   (b) Refer to the transit times listed in the same chart. Note on your map the number of days it takes a container ship to go from each port on its route to the next port.
6. Why is it easier to pilot a ship into Halifax Harbour than into Montreal Harbour?
7. Why would a Dutch exporter of expensive cheeses prefer a container vessel to a general-cargo vessel?

New Words to Know
Convoy
Great Circle Route
Tidal fluctuations
Offshore fabrication

3-61 The Naval Dockyard. *The ships of Canada's east coast fleet in the foreground, with the historic fortress on Citadel Hill in the background, testify to Halifax's founding, and its continuing tradition as a military centre.*

166

Single-Industry Towns

Canada has many settlements involved in primary economic activities. The result is that it has a great many **single-industry** towns. Such towns depend almost entirely on the exploitation of one local natural resource — minerals, fishing or timber. Such towns often shrink, or even become **ghost towns**, when the local resource is exhausted, or the market for that resource disappears.

For instance, Cumberland, British Columbia, had 13 000 people in the early 1900s. The town's main employers were the local coal mines. However, with the increasing use of petroleum, the demand for coal fell; by 1958, only 2000 people were left. Ireland's Eye, Newfoundland, was once a thriving fishing village. It no longer exists. Farm-based towns such as McConnell, Manitoba, and Expanse, Saskatchewan, almost disappeared because of the ravages of drought and the Great Depression. Schefferville, Quebec, where iron ore was mined, suffered a similar fate in 1982, when its mine was closed.

Many people enjoy living in single-industry towns. They often provide ready access to nature and the outdoors. Hunting, fishing, snowmobiling and boating are common activities. The companies that build the towns usually include extensive recreation facilities. Well-organized local sports programs attract hundreds of participants. There is often a strong sense of community. And the work often pays very well — much better than equivalent work in the South.

But such communities are prone to boom-and-bust cycles. These swings

3-62 The boom-and-bust cycle.

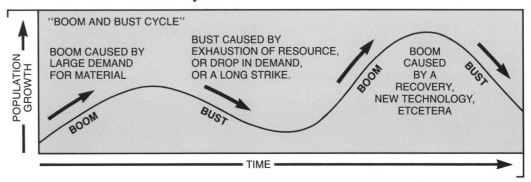

create severe problems for the companies that exploit the resource, but even greater problems for the townspeople. The companies may be forced to "write off" millions of dollars worth of buildings and equipment, but ordinary people may have their lives shattered. Sudden unemployment, loss of investment in businesses and houses, dislocation of friends, expensive relocation costs — all are common when the "bust" is on.

Single-industry towns are often isolated. Even during boom times the social, educational and employment opportunities for women and teenagers are limited. The result is that people don't tend to stay long. Young people often leave for larger cities, where there are more jobs and a chance of better education.

Single-industry towns generally have a short lifespan, which means that people who live in them often face more pressures. Family stress, drug and alcohol abuse, petty crime, and child and spousal abuse are often more prevalent. It is often women and children who

suffer most during the *bust* part of the cycle. These towns seldom provide much employment or social support for women. Recently a number of companies in single-industry towns have been encouraging women to do the same work as the men. This helps.

Some of these towns have tried to widen their economic base by developing a tourist industry or by attracting other industries. Such diversification, if it succeeds, creates long-term employment and economic stability. After all, if one industry shuts down, the town has others to fall back on. Port-Cartier, an isolated town in Quebec, is one example: it had suffered when the nearby iron mines collapsed. The federal government helped it to diversify by building a maximum-security prison there in 1988. This has created hundreds of permanent jobs. However, some have argued that it is a poor location for a prison, as it is much too far from the families of most of the inmates.

For single-industry towns to be suc-

167

cessful, they must base their development on sound **planning principles**. Good planning protects the environment and provides the residents with a pleasant place to live. It reduces conflicts between the industrial and the residential areas by locating the town itself away from mines and refineries. Well-planned towns have parks, shopping and recreation areas near where the people live.

The following case studies illustrate some of the problems of single-industry towns, and some of the attempts that have been made to overcome them.

Thetford Mines, Quebec

Asbestos, the "miracle fibre," has been part of people's lives for hundreds of years. It has a tremendous number of uses: it is found in the Space Shuttle and Formula One racing cars, in home roofing and oil-drilling equipment. Asbestos is a tough fibre which can withstand extremely high temperatures without burning, and resist the corrosive power of many chemicals. Its high tensile strength (that is, its ability to resist tearing or parting) makes it a very versatile material. The word *asbestos* comes from a Greek word meaning "indestructible."

In 1876 the free world's largest asbestos deposits were found in southeastern Quebec. They led to the founding of one of Canada's most famous single-industry towns — Thetford Mines.

This town is dominated by mining companies. The Thetford Mines region is the world's largest exporter of asbestos. The town was founded in 1899 and had a population of 2000 by 1937. As more uses for asbestos were discovered, Thetford Mines experienced a

3-63 The town of Thetford Mines is located virtually on top of the mine.

3-64 Tailings from the mining operations are almost in people's backyards.

168

boom. Its population peaked at over 22 000 in 1971.

During the 1970s and 1980s, medical studies indicated that prolonged exposure to asbestos could cause serious health problems. Many mine and mill workers had died over the years from lung cancer, asbestosis (a lung disease) and other related respiratory diseases. These serious health problems had long been a constant plague in the asbestos-mining communities of the region. As the demand for asbestos plunged, the town began to experience problems.

In the 1980s the town entered a major bust period. Shipments of asbestos had declined by almost 60 percent in the 1970s and 1980s. In the Thetford Mines area the number of asbestos miners has declined from over 4000 in 1980 to about 1200 today. The town's population has decreased by almost 15 percent since 1971. As miners were laid off and moved out of town, other service workers, such as clerks, electricians, merchants, lawyers and teachers also left.

To make things worse, Thetford Mines was not a planned town. As the mines developed, the town grew without any zoning restrictions. The resulting land-use conflicts could have been avoided. The open-pit mines were right beside the houses and stores! As the mines expanded, houses were relocated elsewhere. In a nearby town, Asbestos, some houses actually toppled into the open-pit mines. The bust cycle made this deterioration worse, as there was no money to improve conditions.

Old **tailings** (piles of mine waste) cover thousands of hectares. Many homes are right beside these mounds and in danger of being buried by earth-slides. Dust and dirt blow off the mounds into the nearby homes. On top of that, severe air and noise pollution devalues properties nearby and harms the health of the townspeople. Asbestos dust coats buildings and cars in the downtown area. Blasting in the mines can be heard all over the town.

Community leaders are working to prevent a further decline in Thetford Mines. They are striving to diversify the economy. Lumber, textile, metal fabrication and plastics industries have been attracted to the town. Tourism is increasing. Retail outlets are expanding. The town is developing a more secure and permanent economic base. Recent studies have shown that with proper controls and regulations, asbestos can be used safely. All in all, Thetford Mines' future is looking promising again.

1. **List five conflicts in land use at Thetford Mines.**
2. **Describe four things about Thetford Mines that you find unattractive.**
3. **Why did the town's population drop in the 1970s and 1980s?**

New Words to Know
Single-industry town
Ghost town
Planning principles
Tailings

3-65 Asbestos, Quebec, 1974. *Fifty families were threatened and three houses destroyed when the rim of this mine collapsed.*

Thompson, Manitoba

INCO — the International Nickel Company — started to develop Thompson in 1956. It built the town from scratch, mostly at its own expense. Nickel ore had been discovered nearby many years before. For two years, workers battled the severe winters, the blackflies, the deep, sticky mud and the rugged terrain of the Canadian Shield to build a town. Thompson, which had been completely planned even before the construction began, opened in 1958. The same year an open-pit nickel mine began production.

Today Thompson is the site of an **integrated industrial complex**, one of the largest mining/smelting operations in the world. At Thompson, INCO mines, mills, smelts and refines nickel ore for shipment to markets around the world. There are also research facilities in the town.

The open-pit mines and mill sites were all located well away from the town, which overlooks the beautiful Burntwood River. The town's population has almost doubled in the past 20 years. It now stands at almost 15 000. Over 80 percent of the population is under the age of 45. Less than two percent is over 65.

Aerial photo 3-66 and land-use map 3-69 show that Thompson was as thoroughly planned as Thetford Mines was not. INCO took great care in providing all of Thompson's residents with well-designed and well-located housing, recreational facilities and centrally located shopping. Homes, parks, schools and churches ring the CBD. Residential areas are bypassed by

3-66 The open-pit mine and smelter at Thompson, Manitoba.

3-67 The townsite at Thompson. *It is carefully planned with separate areas for different land uses. Note the greenbelt surrounding the town.*

170

arterial roads. This increases the subdivisions' privacy and greatly reduces traffic within the neighbourhoods.

The mine sites are located downwind from the town, many kilometres away. This greatly reduces air-pollution problems for the residents. Extensive recreation areas have been set up along the Burntwood River. The sewage-treatment plant is downstream from the town. All the industrial activities — mine, mill, smelter — have been placed to avoid conflict with the townspeople. Even the railyards are located well away from the townsite.

To encourage more diversification, heavy and light industrial areas were established at sites separate from the residential areas. Areas for future expansion are also set aside in the town-

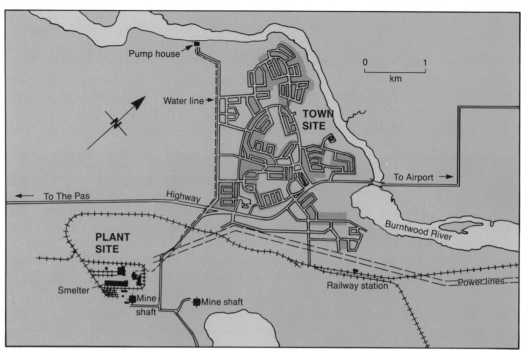

3-69 Mining operations and the town are located in separate areas.
Can you tell what direction the two photos were taken from by referring to the plan?

3-68 Population Statistics, Thompson, Manitoba, 1991

AGE	MALE	FEMALE
0-4	760	755
5-9	720	725
10-14	695	685
15-19	775	650
20-24	680	685
25-29	800	810
30-34	695	660
35-39	595	655
40-44	655	580
45-49	550	415
50-54	405	275
55-59	225	145
60-64	110	85
65-74	70	75
75 and over	25	40
TOTAL	7760	7240
TOTAL POPULATION	15 000	

site proper. The opening of the new Thompson Open Pit Mine ensures employment into the next century.

There is less worker turnover in Thompson than in most single-industry towns. In fact, many employees have 30 or more years of Service at Thompson. INCO continues to donate money and staff time to all facets of the town's life. The main union, the United Steel Workers of America, is also actively involved in improving life in Thompson for the workers and their families. The company has contributed to the building of a ski club, recreation centres and a museum. INCO also subsidizes health care in the community by paying part of the salaries of special health-care professionals such as physiotherapists. Thompson has

served as a model for many other new single-industry towns in Canada and other countries.

1. Why can there be special family problems in a rapidly expanding single-industry town?
2. If you were a teenager in a single-industry town, would you plan to stay there or move to a larger place? Explain why?
3. Why should there be more employment opportunities for women in new single-industry towns?
4. List five reasons why economic diversification is a good thing.

New Words to Know
Integrated industrial complex

Toyota in Cambridge: A Case Study in Planning the Impact of a New Factory

In 1985, the Toyota Motor Corporation began a search for locations for two new automobile plants, one in Canada and the other in the United States. Toyota is the fourth largest automobile manufacturer in the world, after GM, Ford, and Daimler-Chrysler. Following its own motto, Toyota realised that

3-70 "Good Thinking, Good Products." *These Japanese characters express the Toyota company's motto.*

much of the success of the new plant in Canada would be determined by its location. Therefore, it made a careful study of several possible locations, using a specific set of locative factors or criteria to help make the choice.

For the Canadian plant Toyota's criteria were as follows:

- a southern Ontario location near the Canadian market and existing suppliers
- a site within 50 to 130 km of Toronto, ensuring good access to services, markets and airport facilities
- a suitable labour force within the **commutershed** of the proposed plant

- a suitable climate, in order to minimize delays for supplies or shipments due to snow etc.
- a site of approximately 150 ha, with adequate room for expansion
- a high, relatively level site
- an essentially square site
- lands with minimal environmental constraints
- good highway access via arterial roads
- rail service
- availability of water, sewers, drainage, natural gas and electrical power
- separation from residential areas
- location within an established urban community

3-71 The Toyota site in Cambridge.

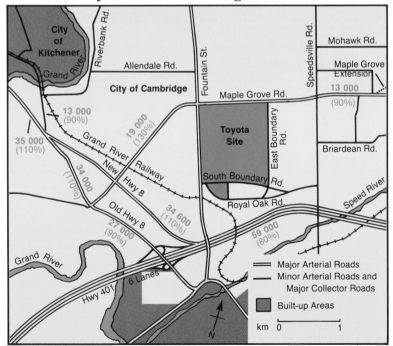

The map shows details of the location in north Cambridge, plus access roads. The statistics show the estimated traffic volume and the estimated capacity of each road in use by 1996.

3-72 Some tri-cities automotive-related manufacturers

COMPANY	MAJOR AUTO PRODUCTS	EMPLOYEES
CAMBRIDGE		
BUTLER METALS	METAL STAMPINGS	550
CANPARTS AUTO	BRAKE MATERIALS, PADS	243
ELECTROHOME MOTOR	MOTORS, FANS, BLOWERS	350
S & H FABRICATING	AUTO TUBES AND HOSES	102
WALKER EXHAUSTS	EXHAUST SYSTEMS	425
BUNDY OF CANADA	AUTO TUBING	222
KRALINATOR FILTERS	AIR, OIL, GAS FILTERS	321
LONG MFG. LTD.	OIL COOLERS, RADIATOR PARTS	200
CANADIAN GENERAL TOWER	VINYL FOR AUTO INTERIORS	446
KITCHENER		
APEX METALS	METAL STAMPINGS, ASSEMBLIES	100
B & W HEAT TREATING	HEAT TREATING FOR METALS	115
BECKERS LAY-TECH	INTERIOR TRIM & PRODUCTS	447
BUDD CANADA	STAMPINGS, CHASSIS & FRAMES	1639
EPTON INDUSTRIES	MOULDINGS, HOSES, GASKETS, SEALS	488
LEAR SIEGLER	AUTO SEATS	922
M.T.D. PRODUCTS	HOSE STAMPINGS & CLAMPS	530
B.F. GOODRICH	TIRES	970
PEBRA INC.	AUTO TRIM	190
UNIROYAL	TIRES	1670
KUNTZ ELECTROPLATING	CHROME PLATING	350
WATERLOO		
BAUER INDUSTRIES	FELT PRODUCTS	252
CUSTOM TRIM	AUTO TRIM & LEATHER WRAPPING	615
VAN DRESSER	AUTO HEADLINERS	160

How might this information have influenced Toyota's decision to locate where it did?

The Impact of the Toyota Plant The site chosen was a parcel of land of approximately 160 ha in north Cambridge, Ontario. This land was part of a larger 485-ha parcel which would require $32 million to service so that factories could be built. The Province of Ontario assumed 87 percent of this cost, with the Regional Municipality of Waterloo and the City of Cambridge assuming the rest, approximately $2 million each. The provincial government, however, agreed to pay for all servicing of the Toyota site. Obviously all three governments felt that it was worth millions of dollars to ''capture'' the Toyota plant.

Adding an auto plant with a capacity of 50 000 cars and 1 000 employees had a great impact on the community. The Cambridge Toyota plant impacted areas well beyond the site where the factory was actually built, specifically the twin cities of Kitchener-Waterloo – just north of the plant, and the city of Cambridge itself. Auto assembly plants also draw parts from plants hundreds of kilometers away. Many of those plants increased their production and work forces dramatically. Some new plants had to be built.

Changes also went far beyond the automobile industry. When hundreds of workers are added to a labour force, hundreds of workers from other industries are attracted to the region, causing employment in other areas of the community. Most of these additional workers were in the service industries as the influx of automotive workers and their families required more teachers, dentists, doctors, real estate agents, more retail clerks in the stores, more cooks in more restaurants, etcetera. Economists and geographers refer to this as the

3-73 Estimated new employment in the Toyota plant and related local industries.

	TOYOTA PLANT	OTHER AUTO PLANTS	SERVICE AND RETAIL	TOTAL JOBS
1988	600			600
1989	800	100	600	1500
1990	1000	200	900	2100
1991	1000	300	1200	2500
1992	1000	400	1300	2700
1993	1000	450	1400	2850
1994	1000	500	1500	3000

multiplier effect. It is also sometimes referred to as the **ripple effect**, because it is similar to the dropping of a pebble in a pond. The ripples go out in ever-increasing circles. The factory, in this case, is the pebble, and the ripples are the effects felt in industries, stores, schools, hospitals, churches and institutions throughout the regional pond.

Obviously a large factory like the Toyota plant causes great changes in any local community. It is important for people to anticipate what these changes will be prior to new plant establishment – so that plans can be formulated to meet and adapt to them. Most communities prepare an **impact study** of large new developments, and publish a report outlining the probable changes that will result. In the case of the Toyota plant, the Regional Municipality of Waterloo published a book entitled *Toyota-North Cambridge Development Impact Report*. Charts and a map from this report are included here to give an idea of the effects of the new plant, and the kinds of concerns that local communities need to study.

Toyota Continues to Grow
The Toyota plant has continued to change and expand since 1988 when it produced

its first cars. Employment followed the predicted growth rate (see Fig 3-73), and in 1997, a large addition required ever more workers. Employment in 1998 was 2 500, and the original 1988 assembly line was producing about 20 000 Solaras per year. The new assembly line's annual production is 150 000 Corollas. The Solara, which is made only in Canada, was designed in Calfornia, and engineered in Michigan for Canadian production.

1. **Consider carefully Toyota's location criteria and explain why your community would, or would not, have been a good location for the plant. Which criteria were most favourable to your community, and which were most unfavourable?**
2. **Which benefits do you think the Ontario government expected to get from the Toyota plant that made it spend almost $30 million?**
   Explain why you think this was a good, or a bad, use of public money.
3. **Explain in your own words what is meant by the term ''multiplier effect,'' and show how it applies to Toyota.**
4. **Explain at least three reasons why the companies in chart 3-72 would have a good chance to become suppliers to Toyota.**
5. **Using map 3-71, describe the relative location of the Toyota plant in terms of urban, transportation and physical features.**
6. **Of the 1000 new employees at Toyota, it was expected that about 400 would be newcomers from outside the region. In fact, the number of outsiders was much smaller. Speculate about possible reasons for this.**

New Words to Know

Commutershed	Impact study
Multiplier effect	Basic growth
Ripple effect	Non-basic growth

Residential Areas

An Older Neighbourhood and a Newer Suburb in London

Between the 1950s and the 1980s most new residential growth in this country was in the suburbs. There are big differences between residential neighbourhoods in older sections of towns and cities, and newer, suburban residential areas. For example, anyone living in a suburb almost has to own a car — some families need two. Also, suburbs usually offer larger homes on larger lots. As well, suburbs are usually quieter and less congested than older downtown neighbourhoods.

This section compares two residential areas in the southern Ontario city of London: one is older and closer to downtown, the other is a newer suburb nearer the edge of the city.

London is a city of about 326 000 people. It has a strong, diversified economy. No one industry dominates its economic base. Most of the work force is employed in office, commercial and service industries. Over 10 000 people alone work in the city's five major hospitals. Approximately 5 600 are employed by the University of Western Ontario. London is the headquarters of a number of financial and insurance companies, including Canada Trust (2 500 employees) and London Life (1 650 employees). In the manufacturing sector, London has major factories – General Motors (diesel locomotives), Siemens Electric, and 3M Canada (adhesives). Just those manufacturers employ over 5000 people.

3-74   Aerial view of Whiteoaks. *Whiteoaks is a planned subdivision with each neighbourhood centred on an elementary school. The streets form a curvilinear pattern.*

3-75   A general view of Whiteoaks. *Large open areas of greenbelt have been left throughout Whiteoaks, linking the school sites with residential areas. Previously the area was farmland, and trees have not had time to grow to maturity.*

3-76 Aerial view of Old South London. *This 100-year-old residential area is just minutes from London's CBD.*

3-77 Street scene, Old South London. *This area has a variety of land uses intermingled. Here several old houses have been converted to commercial purposes.*

The older residential area will be referred to as Old South London. The new subdivision is Whiteoaks. Both are middle to upper-middle income areas.

Old South London is centred at 805570 on the topographic map on page 179, and extends for about five blocks in all directions. Whiteoaks is centred at 810530 and includes the subdivisions to the immediate north and south. Compare the map with aerial photos 3-74 and 3-76.

Residential types The homes in Old South London are generally two to three stories high, and built of brick or stone. Over one-third are single-family dwellings. An increasing number have small apartments in the upper floors. This helps meet the demand for inexpensive apartments near London's CBD. Older five- to ten-storey apartment buildings are located along major arterials, and in many places single-family houses exist right beside these apartment buildings.

There is no green, open space separating the houses from the apartment buildings. A few older houses have been demolished to make room for townhouses and condominiums. This accounts for most of the new home construction in the area.

In Whiteoaks, homes are generally one storey in height and built of brick. Zoning regulations do not permit the building of apartments where houses exist. Instead, apartment buildings are

175

clustered near major intersections, where higher densities are permitted, well away from the single-family housing area.

Well over half the homes in Whiteoaks are duplexes, triplexes, townhouses and "condos". There are more of these **multi-family** dwellings being built because land prices are so much higher than they were 60 years ago. Many of these multi-family structures are grouped together, sharing open space and play areas for the children. Most of the lots are wider than the lots in older, more established areas. In Whiteoaks there are more split-level and ranch-style homes.

Population Characteristics Both Whiteoaks and Old South London are growing in population, but the former is growing five times faster, because there is more vacant land available. Old South London, however, has a population density at least three times higher, because lots are smaller and there is less parkland and open space.

Almost one-third of the residents of Whiteoaks are children under 14. In Old South London, only 12 percent fall into that age group, whereas one resident in five is a senior citizen. In Whiteoaks the ratio is about one in 27.

In Whiteoaks, 81 percent of the families have two parents, 18 percent are led by a single parent — usually the mother. In Old South London, 88 percent of the families have two parents, only 12 percent are led by a single parent. The proportion of single people is greater in the older neighbourhood. Singles tend to prefer older residential neighbourhoods because they are closer to downtown and rents are cheaper.

Street Design and Pattern The streets in Old South London are in a grid pattern. Most of them intersect at

3-78 Typical houses in Old South London. *Solidly-built, two-storey, single-family brick homes predominate here.*

3-79 Newly-built detached houses in Whiteoaks. *Suburban families prefer two-car garages and modern architecture.*

176

3-80 Comparative population characteristics, 1986.

| | OLD SOUTH LONDON | | WHITEOAKS | |
	TOTAL	PERCENT OF TOTAL	TOTAL	PERCENT OF TOTAL
POPULATION	6623	—	3940	—
CHANGE 1981–86	4.9%	—	23%	—
AREA	1.35 km²	—	2.46 km²	—
POPULATION DENSITY/km²	4906	—	1601	—
CHILDREN (0–14)	855	12.9%	1115	28.3%
UNMARRIED (15+)	1760	26.5%	700	17.8%
SENIORS (65+)	1235	18.6	145	3.7%
SENIORS (75+)	610	9.2%	35	less than 1%
SINGLE DETACHED HOUSES	1145	34%	200	14.2%
APARTMENTS (5 OR MORE STOREYS)	935	27.7%	435	30.9%
DUPLEXES, TOWNHOUSES, CONDOMINIUMS	1280	38%	765	54%
NUMBER OF FAMILIES	1635	—	1045	—
TWO-PARENT FAMILIES	1435	88%	855	82%
SINGLE-PARENT FAMILIES	200	12%	190	18%
AVERAGE FAMILY INCOME	$39 908	—	$39 887	—

right angles to each other. There are many stop streets. The streets in Whiteoaks are often curved and winding. Others come to dead ends, or **cul-de-sacs**. Whiteoaks is bordered by major arterial roads, which cut through Old South London in several places. Traffic is heavier in the older neighbourhood, and noise and congestion are worse.

Old South London is better served by public transit, because of its grid pattern and population density. Emergency vehicles and delivery trucks find it easier to locate houses in the older neighbourhood. This is because the streets run in a more predictable pattern.

Land-Use Mix Since Whiteoaks was totally planned before construction, it has very few land-use conflicts. All parts of the subdivision are zoned for one

land use and are clearly separated from other land uses by open space. Old South London has a great mixture of land uses. Some houses have been converted to variety stores, video stores, beauty salons and professional offices. These activities can cause noise and congestion, but they are also convenient for the local residents. In Whiteoaks, retail activities are in strictly zoned areas — generally in malls near major intersections.

Aesthetic Factors Old South London is considered by many to be more attractive, because the streets are lined by many tall, old trees. Many of the yards are well-landscaped, with great varieties of flowers, bushes and mature trees. This vegetation increases property values by beautifying the area. It also

increases privacy, cleans the air of pollutants, provides shade for outdoor activities, muffles sound and conserves energy in the houses.

Whiteoaks is not old enough for its trees to have matured. It will eventually be as attractive as Old South London. It does, however, have more open space, parks and recreation land.

Old South London has a neighbourhood shopping district similar to a small village. Along some streets there are restaurants, antique stores, book stores, gift shops and boutiques that create a unique atmosphere in the middle of a residential area. On the other hand, Whiteoaks is close to one of the largest regional shopping centres in southwestern Ontario, Whiteoaks Mall, which contains hundreds of stores in one climate-controlled building. Much of Old South London is within walking distance of London's CBD. Whiteoaks is 10 to 15 minutes from downtown by car.

Both Whiteoaks and Old South London are fairly typical residential neighbourhoods. Both are attractive and desirable areas in which to live. Both offer a lifestyle that is typical of modern Canadian urban life.

1. (a) **Why do some people prefer to live in a new subdivision?**
   (b) **Why do some people prefer to live in an older, more established residential area?**
2. **Where is it safer to ride a bike — in a subdivision or in an older residential area?**
3. **Explain the importance of mature vegetation to a residential area.**

New Words to Know
Multi-family   Cul-de-sac

South London Map Study

With a metropolitan population approaching 400 000, London, Ontario is Canada's tenth largest urban centre. Located midway between Toronto and Windsor/Detroit, London lies at the heart of southwestern Ontario. It is the economic, social and cultural centre of the region, and its main transportation hub. It is surrounded by a rich agricultural region, with significant amounts of prime farmland. Dairy products, beef, hogs, poultry, grain, fruit and vegetables supply many food-processing plants in the region. The map shows the London CBD, the southern suburbs, and the adjacent rural-urban fringe.

1. Use the map grid to find what is located at:
   (a) 825520 (b) 817530 (c) 755504
   (d) 840490 (e) 803572 (f) 818562.
2. Locate the main intersection in Lambeth (755506). Measure the distance by road from there to the intersection at 789541.
3. Where are the lowest and highest points on the map? Record their locations using the map grid.
4. List the grid references for one example of each of the following: (a) senior citizens' home (b) chemical plant (c) high school (d) communications tower (e) swamp (f) golf course.

3-81 Suburban residential area, south London.

5. Name and give the grid references of at least five tourist attractions.
6. Explain the major advantages of the location of the Wilton Grove Industrial Park (845530) in South London.
7. Why would Westminster Ponds (822550) make life more enjoyable for Londoners?
8. Count the number of churches in Old South London and compare this with the Whiteoaks area. Why is there such a difference in numbers?
9. Why would some business people convert an older house in Old South London to a boutique or office rather than locate in a mall?

3-84 Part of St. Thomas topographic map 40 I/14. *Contour interval is 10 m. For scale and key, see page 405.*

10. Plan a 15-km bicycle tour with a friend using this topographic map and air photos. Begin and end your route at the same intersection. Describe what you and your friend would see at the five- and ten-kilometre points on your route. Pick a good place to stop for a picnic lunch.
11. Read the text on page 180 to find out about rural-urban fringe.
    Locate the rural-urban fringe on this map. List ten urban activities that are located in the rural-urban fringe, and give the grid reference for an example of each.
12. Locate and explain at least one possible land-use conflict in the rural-urban fringe.
13. Locate two examples of shopping malls in the rural-urban fringe. Why would a company build a shopping mall there?
14. Why would industry likely locate near highways 402 and 401?
15. You are a recreational planner charged with the job of selecting a site for a new theme park that will feature Olde English Culture. Choose a site of approximately 50 ha and explain the reasons for your choice.

3-82 Rural-urban fringe, south of London.

3-83 Rural-urban fringe, south of London.

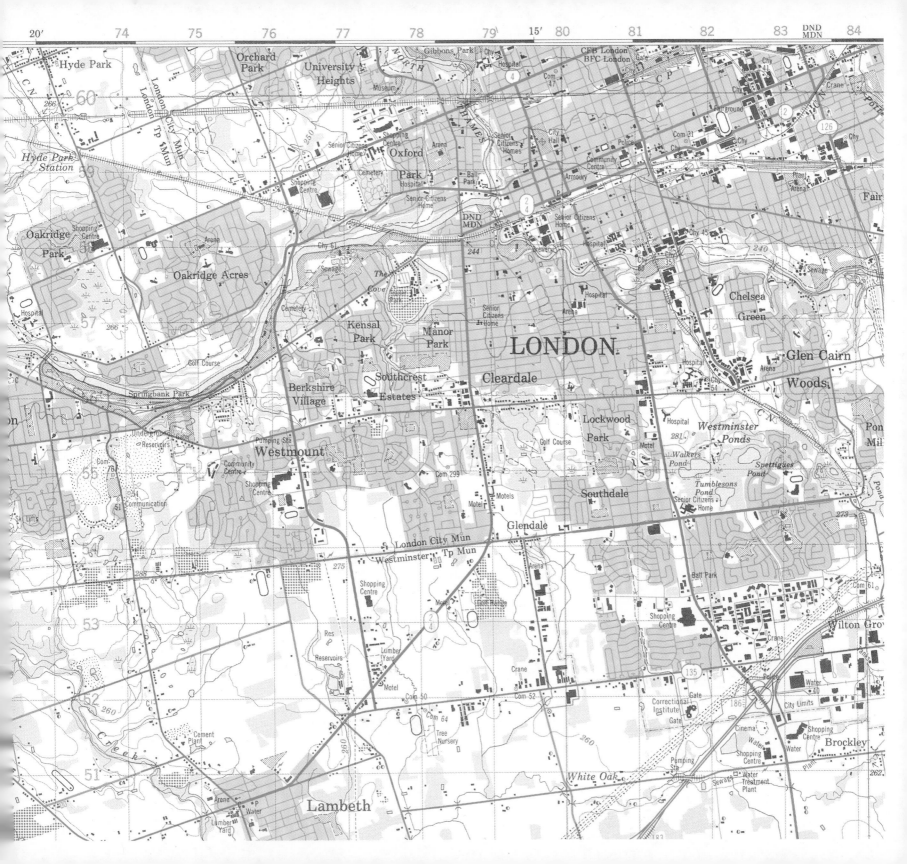

The Rural-Urban Fringe

Beyond the suburbs of most cities is a special area known as the **rural-urban fringe**. These areas are created mainly by the outward expansion of cities. They adjoin the cities and are neither totally urban nor totally rural. They have characteristics of both.

The typical rural-urban fringe is a mixed-land-use area. Some of it will still be farmed. The commercial activities include plazas, drive-in theatres, lumberyards, motels, restaurants, gas stations, nurseries, auto wreckers . . . in fact, just about anything and everything. The industrial activities tend to be just as varied. There will also be residential areas: scattered housing, subdivisions, mobile homes, residential estates and senior citizens' homes. The rural-urban fringe is also used for recreation — particularly of the type that requires a lot of land. Golf courses, theme parks and conservation areas are usually found just outside cities. Finally, there are things such as airports, **landfill sites** and sewage treatment plants. These are considered "public" land uses.

The growth of Canada's cities is steadily taking away some of this country's best **food land**. Urban dwellers sometimes forget that they are dependent upon farmers for food. Also, many Canadians have the mistaken impression that Canada has unlimited farmland. In reality, just over 11 percent of Canada's total area is suitable for any kind of agriculture. And only 0.5 percent is "prime agricultural land." Farmers using this land — which is usually called Class I land — can produce four times the food that can be produced on poorer-quality Class IV land. In the past 50 years, close to one-third of Ontario's prime agricultural land has been lost, most of it through the expansion of cities and towns.

Over half of Canada's Class I land is in southern Ontario. From the top of

3-85   The rural-urban fringe.

Familiarize yourself with the photos in this section, which illustrate typical land-use scenes found in a rural-urban fringe area, and answer the following questions:

Toronto's CN tower, it is possible to see 37 percent of that land. But it is this very land that is being lost most rapidly to urban growth. Some experts suggest that new industries should be encouraged to locate on Classes III and IV land, to help preserve the best food land for crops.

The rural-urban fringe usually has a variety of farm types, including traditional dairying and field crops, but also more specialized operations such as market gardens, "pick-your-own" farms, orchards and other types that rely on the closeness of the city. Some farms are idle because the original owners have sold them, and other farmers cannot afford to buy them. Usually the idle land is owned by **land speculators**, who hold on until the price increases, then sell it without ever having used it.

Sometimes urban developments "jump over" expensive vacant land and locate instead far out into the rural areas. This is called **leapfrogging**. Leapfrogging and land speculation often result in areas of irregular and "premature" land use. A subdivision that has been built before sewers, roads and other urban services are available is one example of premature development.

The rural-urban fringe is a transition area. Urban activities compete with — and often overwhelm — rural ones. Eventually rural land use will give way entirely. This can cause conflicts, because many types of land use are incompatible with each other.

The biggest conflicts are between farmers and non-farmers. Non-farmers move to the rural-urban fringe expecting to find rustic, picturesque countryside. Instead they discover that farms have noisy machinery, dust and "smelly" animals. They are offended by the odours from manure and compost piles. They complain to the farmers and to the township councillors.

At the same time, farmers may be angry. The newcomers damage fences, allowing farm animals to run loose. Urban noise disturbs the animals. Increased traffic congestion disrupts the farm activities. Farmers believe their traditional way of life is being disrupted and some may leave their farms as a result. This takes more food land out of production and encourages **urban sprawl.**

C

D

1. **Record the number of the photo(s) which show a conflict between:**
(a) a business and a farm.
(b) a farm and a residential area.
(c) a farm and transportation.
(d) a business and a residential area.
(e) a government activity and a farm or residential area.

2. **Explain in two or three sentences the nature of the conflict. Consider such things as noise, privacy, smells, congestion, image of a business, costs**

Without good planning, other conflicts may occur in the rural-urban fringe. For example, a hotel may be located beside a meat-packing plant, or a scrap-metal yard may be located near a new subdivision. A church may find a go-kart track springing up next to it. These land-use conflicts create complaints, anger and declining property values. In cities, zoning regulations usually prevent these conflicts from happening. In rural areas, such conflicts are more common because those areas have often been developed before there was any zoning. Government cannot close down a scrap yard just because the area is now zoned for residential use.

Annexation: "For" and "Against"

In order to expand, cities have to take land from neighbouring **municipalities**. This process is called **annexation**. Cities can only do this with the permission of the provincial government. Plans to annex may create major conflicts between a city and the neighbouring townships. A city often wants to expand because:

- there is little vacant land in the city.
- it needs more land to attract large new industries such as automotive and food-processing plants.
- it has solved its problems by annexation before, and believes it should be allowed to do so again.
- uncontrolled industrial and commercial growth in the outlying municipality may cause urban sprawl, as the townships are often ill-equipped to handle urban developments.
- urban growth along its borders conflicts with the needs of those residents who live just inside the city's boundaries.
- it needs more tax revenue.
- nearby municipalities will be competing with the city for new industry, shopping centres and entertainment attractions.

E

F

of production, real-estate values, stress, impact on the environment.
3. Briefly outline possible solutions to any of the conflicts noted.
4. What other possible conflicts can you see?

5. Locate each photo according to where you think it most likely would be on the map, and record the grid reference. Explain the reasons why you located each picture where you did.

- larger communities can provide more efficient services.

Townships are generally against annexation because they believe:

- the city is exaggerating its need for more land.
- they have as much right as the city to develop new industrial and commercial growth.

- they can plan properly and prevent urban sprawl.
- the cities could increase their population density, thereby cutting down the need for more land.
- annexation will cause them to lose industrial and commercial taxes.
- annexation will increase residential taxes in the townships.
- the city will only want to annex more land in the future, so they should be stopped now.
- the city's population is aging and the birth rate will decrease, therefore they will not need more land.

Some townships become towns in the belief that this may protect them from city's annexation bids. For instance, Sarnia Township, near Sarnia, Ontario, is now the Town of Clearwater. Westminster Township south of London, Ontario, became the Town of Westminster in 1988. Time will tell if new towns such as these will alter the make-up of the rural-urban fringe, or affect the process of annexation.

1. Why might farmers dislike city dwellers moving into the rural-urban fringe to live or start a business?
2. Explain how you would feel if your family moved into the rural-urban fringe. If you already live in a rural-urban fringe area, how would you feel if you moved to the city?
3. Create an imaginary map of a rural-urban fringe area, marking on it all the land-use types mentioned in this section, along with others you think might be found in the fringe.
4. Discuss in two paragraphs whether a large city should or should not be permitted to use land in the rural-urban fringe for a landfill site for dumping the city's garbage.

New Words to Know
Rural-urban fringe
Landfill site
Food land
Land speculator
Leapfrogging
Urban sprawl
Municipality
Annexation

3-86 **The expansion of London, Ontario, through annexation, 1840 to 1993.**

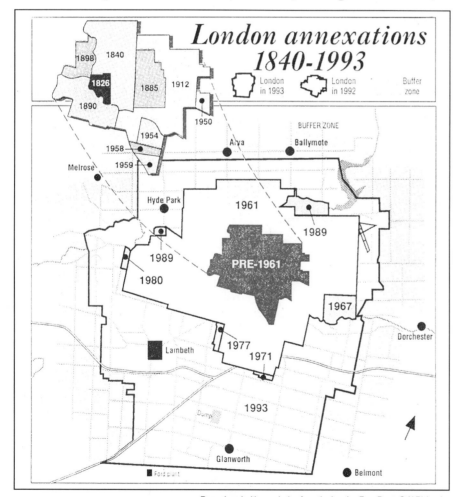

London annexations 1840-1993

Reproduced with permission from the London Free Press © Al Richards.

Mind Benders and Extenders

1. Many people feel that Toronto would be an ideal site for the 2008 Summer Olympics and that either Quebec City or Calgary would make an excellent host for future Winter Olympics.
   (a) What are the pros and cons for each of these cities as an Olympic host?
   (b) Montreal and Vancouver have each hosted a World's Fair. What other Canadian cities should be considered for this event? Why?

2. Research your own community. Why was it founded where it was?

3. Interview a local business person to discover why he or she located in your area.

4. Prepare a three-minute oral presentation explaining why you would, or would not, like to live in a large city ten years from now.

5. In your own words, are suburban shopping malls better than downtown areas for shopping? List what you see as the advantages and disadvantages for each.

6. Would you rather work in Toronto's CBD or in a small community outside Metro Toronto?

7. By referring to the photos of the different types of transit vehicles, and drawing on your own experience, answer the following:
   You and your classmates have been hired by the TTC to plan to increase the number of passengers using their system. Create a report for the TTC board of directors outlining how you would attract more commuters to public transit. Make sure your team discusses the following issues: cost to the government, cost to the commuter, time, safety, impact on the environment, effect on the growth of the city. Present your findings to the class.

8. Compare and contrast the harbours of Halifax and Vancouver under the following headings: physical size and site, major transportation facilities, major areas served.

9. Write a brief news report for the National TV News describing the arrival of a strong chinook the day after a blizzard dumped 35 cm of snow on Calgary.

10. Compare Thetford Mines and Thompson under the headings: population, distance from major cities, environmental problems, examples of good planning, future potential.

11. If you had to live in either Thompson or Thetford Mines, which would you choose? Why?

12. Name at least five other single-industry towns in Canada, and name the basis of the economy for each.

13. The map in the Toyota study was created in 1985–6 as a "future" map. It attempted to predict new traffic flow after the Toyota factory was built. What techniques and sources of data might have been used to make these predictions ten years into the future?

14. Study the map and suggest where traffic tie-ups will likely occur. Which improvements will be needed to avoid them?

15. Write an essay of 500-600 words suggesting what impact the new Toyota plant had on Cambridge.
    Give both positive and negative aspects of the impact. Pool your ideas with others in the class. How could you go about finding out if your ideas came true?

16. Working from the aerial photos of South London neighbourhoods, sketch two maps showing the road patterns and major land uses.

17. Write two real estate ads for the local newspaper. In one, sell a house in Old South London. In the other, sell a house in Whiteoaks.

18. (a) Help plan a class discussion to explore the annexation issue further. The class might be divided into two groups representing the city's and the township's point of view. Each group would be expected to prepare an oral presentation of at least three minutes in length, outlining the reasons for its position on annexation. It would be advisable if each group developed at least three arguments in addition to the ones in the text.
    (b) After all groups have presented their positions, have a full-class debate on the advantages and disadvantages of annexation. Record a vote on the issue and compare it with the results taken in other classes studying this issue.

19. (a) Find out whether there is an annexation proposal under active consideration in your area. Research the reasons behind the plan and relate them to the arguments presented in this chapter.
    (b) Invite local politicians or planners from the city and neighbouring areas to discuss the advantages and disadvantages of annexation with your class.

20. Working in groups of two or three, develop an organizer to compare the advantages and disadvantages of living in your community compared to any other community mentioned in the book.

CANADA
EXPLORING NEW DIRECTIONS

Natural Resources

*"For the next century, Canada must become
the world's "smartest" natural resource developer;
the most high tech; the most environmentally friendly;
the most socially responsible; the most productive."*

Ralph Goodale
Minister of Natural Resources

Forestry

Forests are the most prominent feature of Canada's landscape. From Newfoundland in the east to British Columbia in the west, forests cover nearly half of Canada's total area. On a global scale, this represents ten percent of the world's forests. Only the Russian Republics and Brazil have more.

One of every ten Canadians works in the forest industry. Forest products are among Canada's leading exports and thus play an important role in the economic well-being of every Canadian. On top of that, millions of Canadians and foreign tourists use the forests for recreation — for sightseeing, hikes, camping holidays, and canoe trips. Forests are also vital to the quality of Canada's environment. The root mass of the trees, which binds the soil together, collects rainwater and spring runoff. This protects watersheds from erosion and flooding. Through their life processes, forests provide much of the earth's oxygen while removing unwanted carbon dioxide from the atmosphere. Finally, forests provide a habitat for a wide range of plants and animals.

Canada's Forest Regions

Graph 4-1 shows the proportion of forested land in Canada; however, it does not show the tremendous variations in forest types across the country. Map 1-81 shows that there are six major forest regions, each with its own unique characteristics. The variations are mainly the result of climate conditions. When the climatic regions on map 1-76 are compared with the forest regions, a close relationship is obvious. For exam-

ple, the Boreal forest region almost matches the Subarctic climate region. The West Coast forest region closely matches the Temperate West Coast climate region. Note that not all of Canada is forested: significant areas of the North and of the southern prairies have no natural forests.

The largest forest region is the boreal forest, a broad, continuous band of coniferous trees running from Labrador

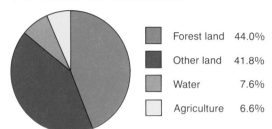

4-1 Land use in Canada.

Forest land   44.0%

Other land   41.8%

Water   7.6%

Agriculture   6.6%

Forests are the dominant feature of the landscape.

4-2 *Boreal forest. In 1995, 1 011 328 hectares of forest were harvested, of which 866 435 was clearcut. 66 900 persons were employed in the logging and forestry industries in 1997.*

to the Yukon. Coniferous trees, or soft-woods, are well-suited to the region's severe climate. In general, the trees of the boreal forest become smaller and more scattered as the latitude increases. White and black spruce, jack pine and balsam fir abound, and provide the backbone for the pulp-and-paper and lumber industries.

In contrast, the smallest forest region is the deciduous forest of southern Ontario. In this temperate zone, conifers are still common but deciduous trees, or hardwoods such as maple, oak, hickory, birch, poplar, aspen and beech dominate. Although hardwoods represent only about 10 percent of the trees cut each year, their lumber is very valuable. It is used in furniture, flooring and panelling.

The mixed forest region is the transition zone between the deciduous trees of the South and the coniferous trees of the North, and has trees of both types. It is the home of the fabled white pine. This tree, along with animal furs, provided much of Canada's wealth during the nineteenth century. The Acadian forest of the Atlantic region is quite similar to the mixed forest. It does, however, have a greater percentage of conifers. This is because of the cooler climate.

The forests along the Pacific coast are truly unique. Because of the long growing season and the heavy precipitation, this region's coniferous trees — Douglas fir, western red cedar, sitka spruce and western hemlock — grow to immense size in dense stands. Douglas firs can grow to almost 100 m — they are the tallest trees in Canada. The high-quality **plywood** from this region is world-famous.

4-3 Major tree species of Canada by area.

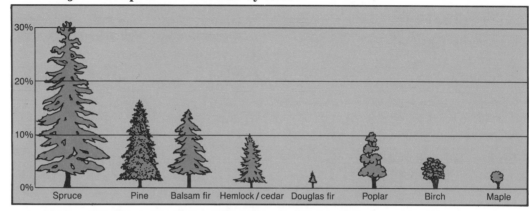

Compare the area covered by softwoods with that covered by hardwoods.

The forest cover of the B.C. interior is considerably more complex than that of the coastal forest. Because of the cooler, drier climate, trees in the interior are smaller and grow more slowly. Lodgepole pine and spruce make up over 50 percent of the interior's forest and the rest is a mixture of Douglas fir, cedar and hemlock.

Fig. 4-4 Merchantable volume.

PROVINCE AND TERRITORY	SOFT-WOODS	HARD-WOODS (million M3)	TOTAL
NEWFOUNDLAND	488	39	527
PRINCE EDWARD ISLAND	16	10	26
NOVA SCOTIA	153	101	254
NEW BRUNSWICK	434	212	646
QUEBEC	2 951	1 292	4 243
ONTARIO	2 320	1 302	3 622
MANITOBA	609	302	911
SASKATCHEWAN	434	393	827
ALBERTA	1 709	974	2 683
BRITISH COLUMBIA	9 245	691	9 936
YUKON TERRITORY	567	65	632
NORTHWEST TERRITORIES	315	131	446
Total	**19 541**	**5 512**	**22 753**

Merchantable volume is the amount of wood in a tree that is considered useful.

Graph 4-3 shows that softwoods are by far the predominant type of trees in Canada. Spruce, pine and fir are the most common. Of hardwoods, aspen and poplar are the main species. In the past, these two species had little economic value; increasingly, however, they are being used to manufacture **waferboard**. Waferboard is made of large, thin chips, or "wafers," of aspen and poplar that are bonded together under intense heat and pressure. It can often be used instead of plywood, which is more expensive.

Forests cover an immense area of Canada, but not all of them are productive. **Productive forests** are those that can be harvested to make a profit. Nonproductive forests may be in extremely rugged areas; they may be too far from a mill to profitably cut and transport; they may be in a region where the climate limits tree growth; the trees may not be suitable for processing. In Canada, 44 percent of the land is forested; 61 percent of that forest land is productive.

4-5a Destinations of Canada's Forest Exports

UNITED STATES	71%
JAPAN	12%
EUROPE	9%
OTHER	8%

4-5b Forest Land and Value of Forest Product Exports

	Land Area (mill ha)	Forest Land (mill ha)	Value of Exports (mill $)
Newfoundland	37.2	22.5 (61%)	497
Prince Edward Island	0.57	0.59 (51%)	1.4
Nova Scotia	5.3	3.9 (74%)	483
New Brunswick	7.2	6.1 (85%)	1700
Quebec	135.7	83.9 (62%)	7800
Ontario	89.1	58.0 (65%)	5900
Manitoba	54.8	26.3 (48%)	278
Saskatchewan	57.1	28.8 (50%)	305
Alberta	64.4	38.2 (59%)	1300
British Columbia	93.0	60.6 (65%)	14100
Yukon Territory	47.9	27.5 (57%)	3.0
Northwest Territories	329.3	61.4 (17%)	0.05
TOTAL	**921.5**	**417.6 (45%)**	**32400**

© *Canadian Council of Forest Ministers/National Forestry Database*

1. **Name three specific regions of Canada that are not forested, and explain why this is so.**
2. **What common tree species would you find in every province in Canada?**
3. **(a) Using table 4-4, list the four top provinces in terms of merchantable volume of wood.**
   (b) Which province ranks fourth? Does it surprise you? Why?
   (c) Note the significant difference between hardwood and softwood volume. Can you find out the difference in market value of these two groups?
4. **(a) Both the Yukon and the Northwest Territories have a low amount of productive forest compared with their** total forest. Explain why.
   (b) Why do B.C. and Prince Edward Island have very high amounts?

New Words to Know

Plywood	Productive forests
Waferboard	Merchantable volume

Green Gold

Canada's vast forest resource is both **renewable** and **biodegradable**. It is often referred to as "green gold." The forest industry is profoundly important to Canada's economy; it is Canada's largest, directly employing almost 270 000 people in harvesting and processing. Another 470 000 workers are *indirectly* employed by it; that is, employed in supplying forest companies and their workers with equipment and consumer goods. From coast to coast, the forest industry is the dominant employer in 345 communities! Without a healthy forest industry, many of these communities would suffer severe unemployment and social distress. Prince George, B.C., with a population of over 70 000, is one such community. It is in the B.C. interior on the Fraser River, and is that province's largest single-industry town. Over 75 percent of its workers are directly or indirectly employed in the forest industry. Its pulp, lumber and other wood products are exported to over 20 countries throughout the world. One Prince George company produces three million chopsticks a day for export to Japan. No wonder Prince George has been dubbed "the Chopstick Capital of the World."

Another community that relies heavily on the forest industry is Pine Falls, Manitoba. It has a population of nearly 1000, and is located 130 km northeast of Winnipeg. Pine Falls is also a company town, built in the late 1920s by Abitibi-Price. Today over 90 percent of the town's workers are employed by the company. Many work in its newsprint mill, which is the lifeblood of the town. The company also maintains or supports the schools, recreational facilities and fire department. The lives of the town's inhabitants would be seriously affected if the mill had to be closed because of a world depression, prohibitive government regulations, or the loss of export markets.

Canada is a trading nation, and much of this trade involves forest products. Over 50 percent of its forest products are exported. Canada is the world's largest exporter, and top producer, of pulp-and-paper products. Canada is also the world's largest exporter of softwood lumber. B.C.'s share *alone* of the world's softwood-lumber market is greater than that of any other country in the world. The province of Quebec holds the same position in terms of newsprint exports.

The U.S. is this country's largest market for forest products. There are several reasons why. Clearly, Canada's geographic location is one of them: nearness to a market reduces transportation costs. A well-developed transportation system between the two countries allows easy access, especially to the large urban areas in the central and northeastern regions. Also the domestic forest industry in the U.S.A. can't supply all of the demand in that country. Canada's popu-

Region	1993 Volume		1994 Volume	
	million m2	%	million m2	%
Canada	42.9	50.3	44.8	52.6
USA	5.4	6.3	4.90	5.4
Europe	25.9	30.4	29.50	32.4
Oceania	0.9	1.2	1.04	1.2
Former USSR	7.1	8.4	7.60	8.4
World	85.3	100	91.10	100

Source: Food & Agriculture Organization

4-5b Global softwood lumber exports

Changes in the Forest Industry

Long gone are the days when teams of oxen or horses and a gang of six men spent a day cutting, squaring and transporting a single tree to a river. However, 100 years ago in the Ottawa Valley — which was then the centre of the Canadian lumber industry — this was the normal procedure. At that time only enormous red and white pines were cut. A tree had to be free of imperfections for at least 10.5 m, with its narrow end large enough to cut down into a 30-cm

lation is a tenth that of the U.S.A., but its forest industry is much bigger than the U.S. one. As a result, this country has a great surplus of wood products, and the U.S.A. buys most of it — about 80 percent of this country's two major wood-export products, newsprint and lumber. Europe and, increasingly, Japan consume most of the rest.

1. Suppose you live in a single-industry town such as Prince George. Name at least three direct forms and three indirect forms of employment in the forest industry.
2. Which new products used in house construction contribute to a reduction in the amount of "solid" wood being used?
3. Why is the U.S.A. the largest buyer of Canadian wood exports?
4. In 100 words or more explain why Canada's forests are often called "Green Gold."

New Words to Know
Renewable
Biodegradable

4-6 A feller-buncher. *Large, specialized machines are used to harvest the forest.*

squared log. After a tree was cut down, it was shaped into a square by a gang of six men with broadaxes. This usually took a full day's work! It was extremely wasteful, since only about half the tree's length could be used and the outer third was lost in squaring. However, in those days, the forests seemed limitless.

Nowadays most muscle power has been replaced by machine power. For instance, in the B.C. interior over half the felling is done by a type of mechanical harvester called a **feller-buncher**, which shears off the trees at ground level and lays them down in rows. **Skidders** then drag the trees to a central loading area, where they are delimbed, topped and loaded onto trucks for transport to the mills or to ports on the coast.

In many parts of eastern Canada much the same process takes place. The smaller trees in this region allow for the use of a **multi-function harvesting machine**, which delimbs and tops trees as they stand, and cuts them off at ground level. Then it cuts them into uniform lengths and stacks them. One machine, operated by one person, can do all this to one tree in less than a minute. It is possible for a tree to be made into newsprint without ever having been touched by a human hand. Many operations in Canada are not this fully mechanized, but as the international market grows more competitive so will the use of automated equipment spread.

Logging in coastal B.C. requires different techniques, because of the mountainous terrain. Trees are felled, delimbed, topped and **bucked** (cut into transportable lengths) using chainsaws.

Sometimes helicopters are used to transport logs from steep slopes. The logs are then dragged to a central loading point by a series of steel cables pulled by a winch mounted on a **mobile spar**. Trucks transport them to the coast or a riverbank, where they are tied into rafts, called **booms**, and floated to the mills. Sometimes special barges are used. At the mill, one side of the barge is deliberately flooded and the logs are unloaded by sliding down the slanted deck.

1. **(a) How has the life of a logger been made easier in recent years?**

4-7a Aftermath of clearcutting.

(b) How have these changes affected the forest industry economically?
2. **Explain, using examples, why different logging techniques must be used in different regions of Canada.**
3. **What advantages does the self-loading/dumping barge have over the traditional methods of rafting and booming?**

New Words to Know
Feller-buncher
Skidder
Multi-function harvesting machine
Buck
Mobile spar
Boom

4-7b Canada is a world leader in manufacturing recycled paper.
Since 1989, Canadian pulp and paper industries have invested $1.5 billion in recycling research and new equipment. In 1996, there were 60 Canadian paper and paperboard mills. They used 4.5 tonnes of recycled paper.

Managing the Resource

The management of Canada's forest resources is a complex activity. Both natural forces (fires, insects and storms) and human forces (cutting systems, over-cutting, pollution and economics) place stress on the forest environment. Forest management attempts to reduce or eliminate these stresses. The three main activities of forest management are harvesting and processing, **silviculture** (the art and science of growing trees), and resource protection.

Harvesting the Crop By far the most favoured **crop-cutting system** presently used in Canada is **clear-cutting**. This system simply removes all of the trees in a given area of forest. Over 75 percent of Canada's forests are logged in this fashion. Sometimes, the **slash**, or tree

4-8 Maturity of Canada's forests.

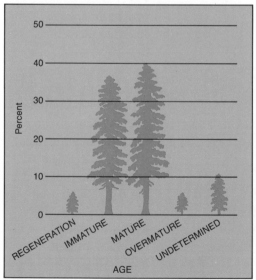

What future problems does the graph indicate for the forestry industry?

debris, is left to rot. This provides nutrients for new growth. Frequently, however, forest managers burn the slash in what is called a **prescribed burn**. This rids the forest floor of the unusable portions of the harvested trees, along with any dead and decaying trees that constitute a fire hazard. It prepares the site for **regeneration** and prevents unwanted trees, brush and shrubs from growing before the desirable species become established. Regeneration can be either by nature (from the seeds of the uncut forest) or by **reforestation** (replanting by hand).

Two variations of the clear-cut system are **strip-cutting** and the **seed-tree system.** In strip-cutting, trees are cut in strips rather than in patches or blocks. The width of the strip is usually equal to the height of the adjacent trees, and at a right angle to the prevailing wind. The uncut forest provides seeds for natural regeneration, as well as shelter for the young **seedlings**. Once regeneration is well-established, the uncut trees are logged.

In the seed-tree system, some mature individual trees or clumps of trees are left standing, while the rest of the area is clear-cut. The remaining trees provide the seeds for regrowth, and are cut after regeneration has taken place.

Clear-cutting is used mainly in the boreal and West Coast forests. Typically, these forests grow in large stands of trees of the same age and species. Most of these mature **virgin forests** originate after a forested area suffers some form of natural disaster such as fire, disease or pests.

The deciduous and mixed forests of southern Canada and the B.C. interior tend to contain a greater mix of tree species and wider variations in tree ages. As a result, **selective cutting** is used. This removes small mature stands or individual trees, and leaves the remaining trees to fill in the resulting open space. As more trees mature, the process continues.

Silviculture Silviculture involves improving and regenerating existing forests, and the creating of new forests by planting nursery-grown seedlings. In 1994, in Canada, close to $2.5 billion was spent on silviculture. It may seem unusual to spend so much money on a resource that is both large and renewable. However, it is clear that in the past, Canadians did not take adequate care of their forests. In some parts of the country there has been over-cutting. This happens when the volume of the harvest exceeds the annual growth of the forest.

There is no danger of Canada running out of trees, but some regions may be running low on trees worth cutting. A perfectly managed forest would have an even distribution of seedlings, immature trees and mature trees. Insufficient replanting, along with the destruction caused by fires and insects, has created large areas where the age distribution of the forest cover is unbalanced. Nearly half the forests are classified as regeneration (or immature) forests, which are of little use to the forest industry. Of equal concern is the low percentage of Canadian forest in seedlings — the trees of the future.

Silviculture involves more than just planting seedlings. It means using intensive forest-management techniques to care for a forest from the seedling stage to the mature-tree stage. This can be compared with growing vegetables in a garden. A row of carrots planted as seeds in the spring, then watered, fertilized, weeded and thinned throughout the growing season, will produce far superior carrots to a row left unattended. A forest, of course, may take up to 60 years to reach maturity, covers a much larger area, and is much more expensive and difficult to manage and maintain.

1. **Outline the differences between clear-cutting and selective cutting.**
2. **Which environmental problems might arise in a clear-cut section of forest?**
3. **Why might clear-cutting be considered a wasteful method in the Deciduous Forest region?**
4. **You are the manager of a harvesting operation in a virgin forest along the coast of B.C. Which cutting system and equipment would you use to move the logs from the forest to a mill many kilometres away?**
5. **Why has silviculture become a necessary part of forest management?**

New Words to Know

Silviculture	Reforestation
Crop-cutting system	Strip-cutting
Clear-cutting	Seed-tree system
Slash	Seedlings
Prescribed burn	Virgin forest
Regeneration	Selective cutting

Forest Protection   Each year millions of hectares of forest are destroyed or damaged by fire, insects and disease. Together, these factors destroy almost as many trees as the forest industry harvests in a year. This is not only an economic loss of millions of dollars, but it also hurts other Canadians who use the forests for recreation. Protecting the forests from these threats has become an integral part of forest management.

Forest fires grab people's attention more than any other forest enemy. Every summer the television news shows huge, uncontrolled fires sweeping through forested areas, leaving behind a trail of smoldering trunks and dead wildlife. Often lodges and cottages are burned, roads and recreational areas are closed, and whole communities must be evacuated.

The number of forest fires, and the number of hectares burned have increased over the years since the first records in 1918. In 1989 over 12 000 fires burned 7 000 000 ha. As of August 1998, there were 1 709 fires in B.C. alone, and fires in the Northwest Territories had destroyed over 800 000 ha. The annual average seems to be 9-10 000 fires which destroy some 3 000 000 ha.

Knowing how forest fires start is important: it allows forest managers to focus on prevention, rather than on much more costly fire suppression. Lightning — the only natural cause of forest fires — was responsible for only 16 percent of all fires. All other fires were started by people.

The key to controlling *any* fire — not just forest fires — is quick detection and response. Modern technology has made detection much easier than it used to be, especially in remote areas. Sophisticated computers can analyze changing weather conditions to predict where lightning storms area most likely to occur. Several provinces operate lightning-detection networks, which can locate where lightning has actually struck the ground within minutes after it occurs. Where the fire danger is rated high or extreme, aerial surveillance is carried out daily. In less remote areas, many fires are reported by the general public, or by private and

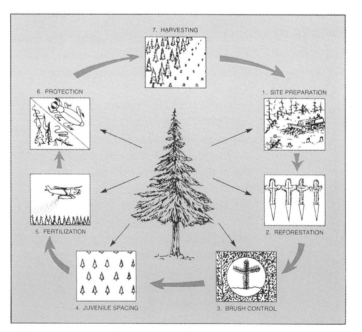

4-9   *The cycle of intensive forest management follows a tree from seedling to harvest.*

commercial pilots.

Although forest fires are dramatic and start quickly, the number of trees destroyed each year by disease and insects is often greater. The most destructive forest pest in Canada is the **spruce budworm**. When its eggs hatch in the spring they develop into caterpillars which feed on the buds and new **foliage** on the trees, damaging new growth, causing the trees to be stunted and deformed. If the infestation is severe enough, trees die, but even if a tree is not killed, it may be left in such a weakened state that other insects or diseases can kill it. In 1994, insects were responsible for destroying nearly 12 million ha of forest which is double the amount lost due to forest fires.

Pests can't be totally eradicated: many of them are simply part of a healthy forest ecosystem. However, periodically their numbers grow to epidemic proportions, at which time some form of control is necessary. The most widespread method is aerial spraying with a bacterial **insecticide**. The one used most often these days is Bt. Unlike the insecticides used in the past, which harmed or killed other insects, birds and fish, and in some cases even humans, Bt is intended to destroy only budworms. It has little effect on other forest creatures.

Forest managers don't attempt to spray the entire forest and protect every tree. Instead, they choose certain areas — usually prime timber, recreational and tourist areas. By concentrating on those, valuable forest is kept alive and healthy until the pests decrease naturally to acceptable levels.

As well as aerial spraying, researchers are now experimenting with parasites, viruses and predators that attack and destroy the budworm. Also, there is ongoing research to develop pest-resistant trees.

1. **How do forest fires indirectly affect all Canadians?**
2. **How can small caterpillars such as the spruce budworm kill a large tree?**
3. **Why is the total eradication of the spruce budworm impossible and perhaps inadvisable?**
4. **What dangers are there in using insecticides?**
5. **What risks are associated with the introduction of parasites, viruses and predators? Research examples of species introduction and the impact this practice has had.**

New Words to Know
Spruce budworm   Insecticide
Foliage

4-10   Forest fires in Canada.

Number of fires

Area burned (million ha)

Explain the discrepancy between the number of fires and the area actually burned in different years.

4-11   Aerial spraying. *Bacterial insecticides help control the spread of insect pests.*

193

British Columbia: Lumber

Canada's vast forests were once considered a hindrance to settlement. Consequently, great stands of forest were simply cut down and burned. In the nineteenth century, England emerged as the world's major industrial power, creating an international demand for

4-12 A large coastal log.

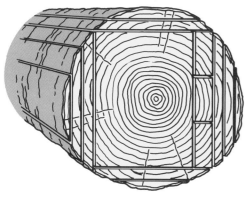

Cuts are carefully planned for each log to yield the maximum amount of usable lumber.

4-13 A smaller interior log.

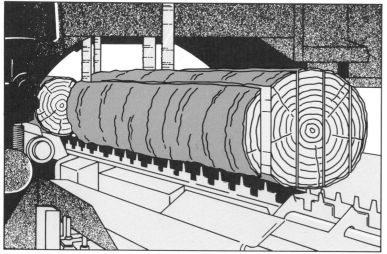

Quad saws quickly cut logs into standard dimension lumber.

Canadian timber. Most of this trade was in the form of squared timber from the Maritimes and Quebec. By the beginning of the twentieth century, trade with Britain was declining and markets for sawn lumber were opening in the U.S.A. By the mid-1980s, Canada was the source of over half the world's exports of softwood lumber.

In the past, eastern Canada dominated the industry; now B.C. does. Its industry lacks the romance associated with eastern Canada, because it is a product of the twentieth century. Until the completion of the Canadian Pacific Railway in 1885, B.C. was isolated from the major markets of eastern Canada and the northeastern U.S.A. By 1995, B.C. sawmills were accounting for 57 percent of all the lumber produced in Canada.

Two distinctive lumber industries have developed there — one on the coast, the other in the interior. The coastal industry harvests very large trees and produces lumber of varying qualities, or "grades," from a single log. Each log is individually cut, making numerous, specially-planned passes through the saw to extract the maximum amount of "clear" or knot-free lumber. The British Columbia Forest Products mill in Victoria processes about 13 000 logs a month in this manner. Many coastal factories are **integrated sawmills**, with pulp-and-paper operations included. This simplifies using waste from the sawmill, such as sawdust and chips, as a raw material for making paper.

The interior forests produce smaller trees, primarily of the lower construction grade. The trees, however, are generally consistent in quality, which allows rapid, automated, high-volume processing. Logs are often sawn by a **quad saw**, which has four blades operating at the same time. In the British Columbia Forest Products mill at Mackenzie, some

4-14 *The harvesting of Canadian forests has sometimes attracted considerable controversy and protest, particularly when virgin or old growth forests are at risk. Several enviromental groups, including Greepeace, the Western Canada Wilderness Committee, and Canadian Parks & Wilderness Society (CPAWS) have actively protested Canadian logging.*

220 000 logs can be cut into standard construction lengths and sizes per month. Interior mills produce 70 percent of the province's lumber, with the remainder coming from coastal mills.

Over half the lumber exported from B.C. is destined for the U.S.A., mostly for the home-building industry. Because it depends so heavily on exporting to the U.S.A., B.C. is often at the mercy of that country's economy. When the U.S. economy is on an upswing, and/or mortgage rates are low, B.C.'s lumber output peaks, as it did in the mid-1980s. On the other hand, lumber production declined drastically in 1982 when U.S. housing starts fell during a recession and mortgage rates climbed. The lumber industry, like most resource industries, is cyclical in nature.

1. **List the characteristics of lumber production on the B.C. coast and in the interior.**
2. **The lumber industry is cyclical in nature. Explain.**
3. **Why would B.C. lumber producers be less concerned with a drop in housing starts in Canada than in the U.S.A.?**

New Words to Know
Integrated sawmill
Quad saw

Quebec: Newsprint

What the lumber industry is to British Columbia, the pulp-and-paper industry — especially newsprint — is to Quebec. Canada produces almost one-third of the world's newsprint — more than any other country. Quebec's share of this is 44 percent, making that province the third-largest producer in the world, after

the rest of Canada and the United States. Quebec is also the world's greatest exporter of newsprint.

Quebec's pulp-and-paper mills have a number of features in common. All are in the southern half of the province, in the boreal or the mixed forest, to ensure that they are near an adequate supply of softwoods — mainly black and white spruce. These two species are favoured for making paper because of their long fibres.

A second common feature is that all the mills are on rivers or lakes. Water in enormous quantities is needed for paper production, to carry the wood fibres through the various stages of the process. It is also used to clean and debark the logs, to cool equipment and, in the form of steam, for heating. Newsprint machines, which may be as long as

4-15 Trade of newsprint

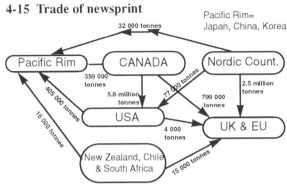

Pacific Rim = Japan, China, Korea

Source: Food & Agriculture Organizations

a football field, use vast amounts of electric power. Pulp-and-paper mills consume one-eighth of the electric power produced in Canada, and much of this is hydro-generated. Finally, water is often used to transport logs to the mill.

The basis of paper-making is to extract cellulose fibre from wood. This can be done by grinding the logs

4-16 Major pulp and paper mills in Quebec.

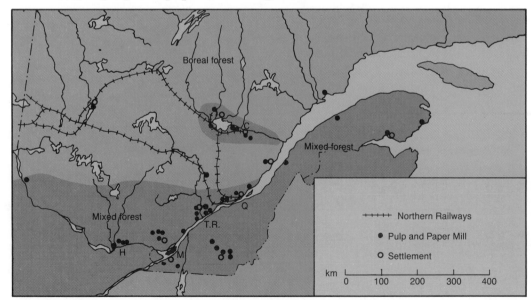

Quebec's output of pulp and paper is greater than that of any other province.

195

mechanically into pulp, or by cooking woodchips in chemicals. Both processes break down the wood into individual fibres, which are mixed with water and collected on a flat screen. The wet wood fibres stick to one another. Then, using heat and pressure, the paper machine gradually removes the water from the pulp to form paper.

Quebec has vast forest and water resources and an abundant supply of cheap power. Its forestry industry spent over $1.2 billion in public and private funds, upgrading and modernizing its plants. This put state-of-the-art computerized equipment in place, and increased productivity by 5 percent.

Competition from other parts of the world is a cause of concern to Canada's, and especially Quebec's, newsprint industry. In 1960, Canada's share of world exports was 76 percent. By 1994 the figure had dropped to 57 percent. Sweden, Brazil and Chile have made inroads into the newsprint market. Brazil, in particular, is a future threat. Its forest cover is greater than Canada's, and its trees reach maturity in less than half the time. Brazil also has enormous hydro potential and cheaper labour.

1. **List the key site factors common to paper mills in Quebec.**
2. **Why is water an important ingredient in making paper?**
3. **What upgrades would make a paper mill more productive?**
4. **Why do Latin American countries pose a future threat to Quebec's dominance as a newsprint exporter?**

Forestry Issues

In recent years a number of groups have challenged forestry companies and governments on their management of the forests. Environmentalists have criticized forestry companies for their cutting systems — especially clear-cutting — and for the pollution they cause. Preservationists have pressured governments to set aside large areas of virgin forest for parks and wildlife reserves. Aboriginal groups argue that governments have given logging rights to companies on lands they consider sacred, or essential to their way of life. The tourist industry wants more forest land opened and set aside for its use. In the past, conflicts over the multiple use of forests were rare: this country's forests were too vast and isolated for the average citizen to be overly concerned. However, as the forest industry gets ready to enter the next century, and competition for forest lands constantly increases, many of these issues are demanding attention.

Issue: Clear-cutting and Prescribed Burn

One forestry practice which has come under severe criticism in B.C. is the system of clear-cutting and prescribed burn, or "cut-and-burn."

For

"Clear-cutting is a good, efficient way of harvesting trees and renewing the forest resource."

"It is used in many other countries, such as Sweden, with good results."

"Clear-cutting is needed when there are stands of trees that are all of the same mature age."

"Many new trees cannot grow in the shade of other, bigger trees; they need full sunlight in open areas."

"Avalanches and silting are rare; they are mistakes, and can be prevented with proper care and research."

"The effects of clear-cutting are only temporary; they quickly give way to new forests and other vegetation."

"Clear-cutting reduces the habitat of some species, but actually increases food and habitat for moose and deer."

"Slash burning actually reduces the likelihood of wild fires, and creates areas for new trees to grow."

Against

"Clear-cut logging destroys the land. It leads to massive erosion, climate change and degradation of the land for the future."

"Clear-cutting causes avalanches and silting of streams and lakes."

"Trees absorb the force of rainfall, allow the land to absorb water, help prevent flooding."

"Roots bind the soil and help prevent erosion."

"Forests transpire water into the air, which forms clouds and produces rain."

"Forests stabilize the climate and recycle carbon, oxygen, nitrogen and water."

"Trees provide a natural habitat for many animals. Clear-cutting slaughters the entire forest community."

"Slash fires burn directly on top of the earth. Often they are so hot they destroy organic matter in the soil."

Agriculture

"Farming is a business, it always has been and always will be. The challenge for my generation is to make it more efficient and profitable." Paul Hunt, Toronto *Globe and Mail*.

At the time of Confederation "the business of farming" would have had great meaning to most of the Canadian population. Agriculture was the backbone of the economy; nearly 80 percent of the labour force was engaged in it. By the beginning of the twentieth century, farms occupied 25 million hectares, and grain exports, mainly to Britain, had reached 270 000 t. This made Canada one of the world's great food producers. It is still one of the world's major grain exporters, but since that time there has been a shift from labour-intensive farming to machinery-intensive farming, which uses high-tech equipment and scientific methods. This has resulted in enormous changes. The number of Canadian farms has declined drastically, to less than 300 000 from a high of over 700 000. On the other hand, the average farm size has more than quadrupled. Accompanying these changes has been a sharp drop in the farm labour force; it is now only three percent of the total.

Today, Canada's farms are more efficient and highly mechanized, and produce crop yields unheard of at the time of Confederation. Modern agriculture relies heavily on a variety of chemicals to obtain these high yields: fertilizers to increase the yield, **pesticides** to kill crop-damaging pests, and **herbicides** to destroy weeds and unwanted vegetation, are all used in increasing quantities. Agricultural scientists have also developed new genetic varieties or **strains** of crops which are hardier, grow faster and produce higher yields than those of the past. These innovations, along with mechanization and improved cultivation techniques, are responsible for the high yields. An Ontario farmer at the turn of the century was able to produce enough food for 12 people; by the end of the 1980s that figure was nearly 100 people!

In 1995 the amount of farmland under cultivation was 68 million hectares, and wheat exports were almost 21 million tonnes. This made Canada the world's tenth-largest producer, and second-largest exporter of wheat. Only the U.S.A. exported more. As graph 4-18 illustrates, the market for Canadian wheat is worldwide with a strong emphasis on developing countries in Asia, Africa, and the Middle East.

4-17 Farming in Canada

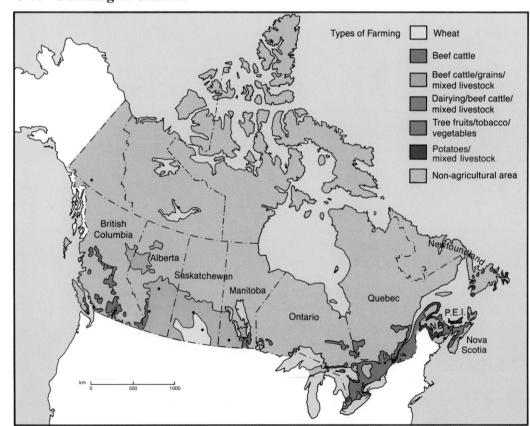

Types of Farming
- Wheat
- Beef cattle
- Beef cattle/grains/mixed livestock
- Dairying/beef cattle/mixed livestock
- Tree fruits/tobacco/vegetables
- Potatoes/mixed livestock
- Non-agricultural area

1. What type of farming occupies (a) the largest and (b) the smallest land area?
2. Determine what type of farming is found: (a) in south-central Saskatchewan (b) near Chatham, Ontario (c) on Prince Edward Island (d) near Calgary, Alberta (e) in the Annapolis Valley.

4-18 Canadian wheat exports, 1994-95 crop year.

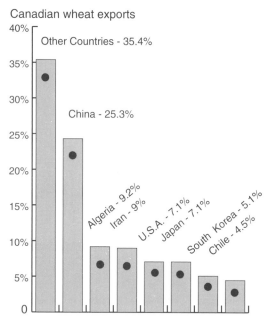

Canadian wheat exports

Other Countries - 35.4%
China - 25.3%
Algeria - 9.2%
Iran - 9%
U.S.A. - 7.1%
Japan - 7.1%
South Korea - 5.1%
Chile - 4.5%

Until the early 1960s, western Europe received the bulk of Canadian wheat exports. Now countries in eastern Europe and Asia have become large importers of Canadian wheat. Account for this change.

It is true that wheat and other grains such as barley, rye and oats, represent 59 percent of this country's agricultural exports by value; but Canada produces a wide range of crops and livestock for domestic and export markets.

As with most exports, the U.S.A. is the leading market for many non-grain farm products. There are a number of reasons why. A major one is Canada's close proximity to the U.S. northeast. The rail and road systems connecting the two countries are well-developed, which is important, since many products are perishable. As well, Canadians and Americans share similar eating patterns, and grading and packaging systems; this makes Canadian farm products easily acceptable in the U.S.A.

Although agriculture is no longer the backbone of the Canadian economy, it is still an indispensable industry because it provides a basic necessity of life — food. Many other businesses are directly dependent on farming. These are called collectively the agribusiness sector. They include manufacturers of farm equipment, fertilizers and pesticides, along with food-processing plants such as flour

4-19 *A farm worker harvesting cranberries in British Columbia's Fraser River Valley. There are over 1 100 hectares of cranberries under cultivation in the province for an annual crop of $28 million. British Columbia is the third-largest producer of cranberries in the world.*

4-20 Selected agricultural exports to the United States

COMMODITY	PERCENT
SUGAR	95
LIVE ANIMALS	89
RED MEATS	89
FRUITS AND NUTS	70
POULTRY AND EGGS	64
POTATOES	49
TOBACCO	35
GRAINS	3

mills, canneries and meat-packing plants. About 15 percent of Canadian workers are in the agriculture-supply and food-processing industries.

1. **What basic changes have taken place in Canadian farm populations, numbers, farm sizes and production since Confederation?**
2. **What methods have farmers used to dramatically increase crop yields?**
3. **Using graph 4-18, list the regions of the world which imported the majority of Canadian wheat in 1995.**
4. **Give three reasons why the U.S.A. is Canada's main market for most agricultural products, except grains.**

New Words to Know
Pesticide
Herbicide
Strain

Patterns of Farming

Only a little over six percent of Canada's land area is suitable for crop agriculture. An equal amount of land is covered by native grasses that can be used for livestock pasture.

Most of Canada's farms are in the South; nearly all of them are less than 500 km from the U.S. border. The northern regions are simply too dry and cold, and have too short a growing sea-son for successful crop production. In other regions, such as the Maritimes and British Columbia, the lack of level land restricts agriculture almost entirely to river valleys and coastal lowlands. The Precambrian Shield, which includes most of northern Quebec and Ontario, and much of the Prairie provinces, is unsuitable for agriculture, except for a few small pockets.

Although agriculture is an important industry in every province of Canada, it is much more important in some than in others. For example, Newfoundland has only a few pockets of farmland while Saskatchewan has almost half its area in farms. The Prairie provinces, along with Ontario and Quebec, have over 90 percent of the productive farmland in Canada. It is also worth noting that there is significant diversity from one region to the next. Wheat farms are concentrated in Saskatchewan, cattle ranches in the B.C. interior, dairy farms in Ontario and Quebec, and potato farms in Prince Edward Island and New Brunswick. At the same time, **mixed farms**, which produce both crops and

4-21 Canadian Farm Statistics

	Number 1991	1996	Average farm size hectares
Newfoundland	725	731	65.6
Prince Edward Island	2 361	2 200	109.7
Nova Scotia	3 980	4 021	99.8
New Brunswick	3 252	3 206	115.6
Quebec	38 076	35 716	90.2
Ontario	68 633	67 118	79.5
Manitoba	25 706	24 341	300.8
Saskatchewan	60 840	56 979	441.9
Alberta	57 245	58 990	363.8
British Columbia	19 225	21 653	124.6
	280 043	274 955	242.1

livestock, are found all across Canada. However, a mixed farm in Nova Scotia will differ in size, the specific crops and livestock raised, the degree of mechanization, and the income it produces, from a mixed farm in Manitoba.

Farm sizes vary considerably from region to region. Prairie farms average four times the size of those in other regions. In fact, in the drier parts of the prairie wheat belt, some farms are as large as 6400 ha. Farms this size require considerable equipment and minimal amounts of labour, and produce relatively low yields per hectare. This type of farming is called **extensive farming**. In the rest of Canada, farms average around 80 to 100 ha. Nearly 20 percent of Ontario's farms are less than 30 ha. Many of these smaller farms specialize in fruits and vegetables, or poultry. Though they are small, they produce high yields of high-value products. They generally require large amounts of labour, machinery, irrigation and fertilizer to operate efficiently. This type of farming is called **intensive farming**.

1. **Why is so little of Canada's land mass suitable for farming?**
2. **The Precambrian Shield is Canada's largest physical region, yet it contains only small pockets of farmland. Explain why.**
3. **What are the restrictions on farming in both the Maritimes and B.C.?**
4. **What are the differences between extensive and intensive farming?**

New Words to Know
Mixed farming
Extensive farming
Intensive farming

The Atlantic Provinces

Farming in the Atlantic provinces is widespread only in Prince Edward Island. Large parts of Nova Scotia and New Brunswick and most of Newfoundland are not utilized for agriculture. The region's climate is generally suitable, but the lack of level land and the thin, stony, infertile soils limit agriculture. As well, the Maritimes are relatively isolated from Central Canada; production is mainly for the small regional market. Only potatoes and apples are grown in sufficient quantities for export to other parts of Canada and abroad.

Maritime farming is primarily located in river valleys and coastal lowlands. Both tend to have flat to gently rolling terrain, as well as deep, fertile soils. Mixed farming dominates, with an emphasis on dairy cattle, hogs, poultry and vegetables — particularly potatoes.

In Nova Scotia, the main agricultural areas are along the Bay of Fundy, the Northumberland Strait and in the Annapolis Valley, which is famous for its apples. The valley of the Saint John River is the most important farming area in New Brunswick; produced there are dairy products, poultry and vegetables for the local urban markets of Saint John and Moncton, along with potatoes for export.

Prince Edward Island is unlike the rest of the Maritimes — agriculture is its leading industry. The gentle rolling landscape and light, deep, fertile, stone-free soils, combined with a moderate and moist climate, are ideal for potatoes. As well, because the province is an island, it is isolated from many of the pests and diseases which routinely attack potatoes in other parts of the Maritimes. All of the above help make P.E.I. this country's largest producer of high-quality table potatoes. Since the island has a small population; most of the 1 262 887 mt grown annually are exported, mainly to other parts of Canada and to the U.S.A. Many end up as "fries" in fast-food outlets. P.E.I. is also the world's second-largest exporter of disease-free seed potatoes. Exports of seed potatoes alone are about 43 218 mt. Roughly one-quarter goes to the U.S.A.; the rest are shipped worldwide.

1. Why is Maritime agriculture concentrated in river valleys and coastal lowlands?
2. Why does mixed farming dominate agriculture in the Maritimes?
3. "In terms of agriculture, Prince Edward Island is quite different from the rest of the Maritimes." Explain.
4. Traditionally, part-time farming has been common in the Atlantic Provinces. Why has this been the case?

4-22 *Nova Scotia is Canada's leading blueberry producer, with an annual harvest in excess of 12 million kilograms, worth over $12 millon. Sven Robinson, fourth generation of an Oxford, N.S. blueberry dynasty, proudly displays one day's harvest. Blueberry patches are found throughout the province, but concentrated in the western sector.*

4-23 The rich soil, rolling landscape, and cool climate of P.E.I. allow it to grow high quality potatoes.

The Central Provinces

By far the most productive area of the central provinces is the Great Lakes-St. Lawrence Lowlands. Flat land, fertile soils, moist and temperate climate, and a long growing season support a thriving agricultural industry with a wide variety of livestock and crops. **Forage crops**, to feed livestock, take up most of the crop land, so mixed farms are the most common type. The region has a huge urban population, and this has also encouraged specialized intensive farming. An excellent network of highways and freeways gives the farming areas direct and rapid access to their markets. This is a key consideration in growing vegetables, grapes and peaches, and dairy products. All of these are perishable.

Ontario and Quebec produce more hogs, poultry, dairy products, eggs, sheep and lambs, fruits, vegetables, corn, soyabeans, flowers, nursery stock and tobacco than any other region. They are also a close second to the prairies in the raising of cattle. The bulk of the products are consumed fresh, but some are used as raw materials for other products. Canneries, wineries, cheese factories and meat packers are scattered throughout the region. Altogether, this region contributes nearly 50 percent by value of Canadian agricultural production. Two-thirds of the output is consumed within the region; the remainder is exported, mainly to the U.S.A.

Within this region are small areas producing **horticultural crops**, which are not found at all — or at best in limited quantities — in the rest of Canada. These include certain fruits, vegetables, berries, flowers and nursery products. Most are grown on very small farms averaging less than 50 ha. The crops they produce show the highest yields per hectare in the country, as well as the highest retail values. For example, a bushel of peaches, apples, strawberries, tomatoes or carrots all sell at several times the price of a bushel of wheat. The Holland Marsh just north of Toronto and the Leamington area in southwestern Ontario, together grow most of Canada's tomatoes, carrots, sweet corn and onions. Also in southwestern Ontario, along the shore of Lake Erie, almost all of Canada's tobacco is grown. Tobacco is an **industrial crop**, since it is intended for a manufacturing industry and is not consumed as a food. Further east is Canada's most important **tender tree fruit** and grape-growing area, the Niagara Fruit Belt.

1. **What physical characteristics have allowed the Great Lakes-St. Lawrence Lowlands to become a major agricultural area?**
2. **(a) Why is proximity to markets and transportation routes essential for vegetable- and fruit-growers?**
   (b) Why is it generally necessary for dairy farms to be close to their markets?
   (c) How can some dairy farms be successful hundreds of kilometres from their markets?
3. **Why is most of the agricultural output of the Great Lakes-St. Lawrence Lowlands consumed locally, rather than exported, as is prairie wheat?**
4. **Excluding tobacco, list two industrial crops, and their uses and source regions, grown by farmers anywhere in the world.**

New Words to Know

Forage crop	Industrial crop
Horticultural crop	Tender tree fruit

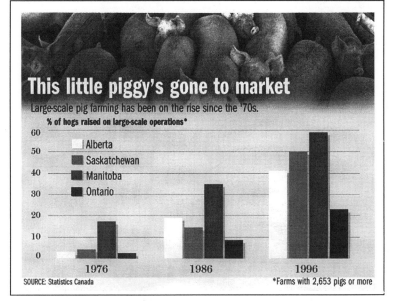

This little piggy's gone to market

Large-scale pig farming has been on the rise since the '70s.

% of hogs raised on large-scale operations*

- Alberta
- Saskatchewan
- Manitoba
- Ontario

1976   1986   1996

SOURCE: Statistics Canada   *Farms with 2,653 pigs or more

4-24 *Beef and poultry have traditionally been strong Canadian staples. Pork is a relatively recent addition.*

4-25 The Niagara fruit belt.

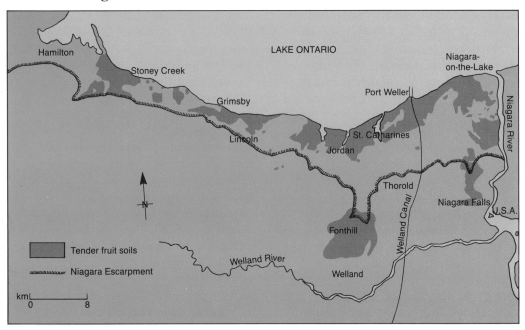

Tender fruit soils
Niagara Escarpment
km 0 — 8

Niagara Fruit Belt

The Niagara Fruit Belt produces 90 percent of the nation's grapes, 70 percent of its peaches and over 40 percent of its other tender tree fruits. What makes this area so productive is its combination of climate, physical features and soils.

Climate The Niagara Fruit Belt is in the southernmost part of Canada. It has the longest growing season and frost-free period in the country. The length of the frost-free period is very important for tender fruits: frost can easily destroy the delicate blossoms and developing buds in the spring, or damage the tender skin of the fruit in the fall.

Lake Ontario moderates the area's climate. In spring, the lake's cold water cools the area, which delays blossoming several weeks, compared to inland locations. This prevents premature blossoming during a warm spell and saves the area from damages which might result from a late frost. Since the lake does not freeze over, it warms the region in the winter. This helps to protect the trees from severe winter temperatures.

Topography Most of the orchards are on a narrow plain between Lake Ontario and the Niagara Escarpment. The latter is a high ridge or cliff which runs from the Niagara River to Hamilton and then swings north to Georgian Bay. When cold air from the north descends on the area, it tends to flow over the escarpment and, sinking low to the ground, continues down the sloping plain to the lakes. The gradually sloping plain prevents any build-up of cold air, which could kill or damage the blossoms or trees. This is called **air drainage**.

Soils The lake plain is the remnant of a glacial lakebed. Its soils are mainly sands and gravels, which drain well and are ideal for peach, apricot and cherry trees. In other parts of the region, on the lake plain and above the escarpment, are clay soils used primarily for vineyards.

The region is less than an hour's drive from Toronto, the heart of Canada's largest urban market, in the Hamilton-Toronto-Oshawa "golden horseshoe." Unfortunately, this advantage is also a problem for the Niagara Fruit Belt. The characteristics that have made the region ideal for fruit growing have also made it a magnet for urban growth.

1. **What physical advantages does the Niagara Fruit Belt have that allow it to grow tender tree fruits?**
2. **What advantages are there for the fruit growers in being located on a major expressway linking Hamilton and Toronto?**

New Words to Know
Air drainage

Loss of Agricultural Land

Over the years, agricultural output, farm sizes and farm incomes have all increased. At the same time, the actual *amount* of land used for farming has decreased, by over 2.5 million hectares since 1951. This trend is expected to continue well into the next century as the pressures of housing, industry, transportation and recreation grow.

There are still 67 million hectares of farmland in Canada. However, just ten

4-26 Cold air drainage.

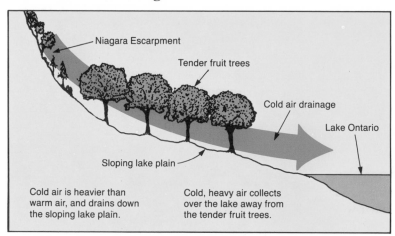

Niagara Escarpment

Tender fruit trees

Cold air drainage

Lake Ontario

Sloping lake plain

Cold air is heavier than warm air, and drains down the sloping lake plain.

Cold, heavy air collects over the lake away from the tender fruit trees.

4-27 Much of the work in a vineyard is done by hand.

percent of the country is suitable for farming, so any loss of agricultural land is significant. Once farmland becomes urban land it is usually lost forever. Moreover, only one percent of the total land area of Canada is classified as **prime farmland**. Unfortunately, a great deal of the farmland under pressure from urbanization is this prime farmland, Canada's best.

Nowhere has the loss of farmland, particularly prime farmland, been more evident than in Southern Ontario. About 40 percent of Ontario's prime farmland is within 160 km of Toronto, and for every 1 000 people who move to the urban metropolis, 46 ha. of farmland is lost. In recent years over 2 000 ha. has been lost annually due to Toronto's growth. In Ontario as a whole, between 1976 and 1986 nearly 40 000 ha. of farmland was taken out of production. Between 1986 and 1996, 5 637 farms ceased to exist. In Niagara County, of the Niagara Fruit Belt, over the same period, the number of farms dropped over 15 percent.

The problem is not restricted to Ontario. The same thing is happening, to some degree, in every province. Various organizations have been formed to combat this problem. They all suggest that if prime farmland is allowed to keep disappearing, a number of problems could result:

- *More expensive food, because farmers are being forced onto less productive land where it costs the same per hectare to produce less food than on prime farmland.*
- *Food laced with chemicals, because more pesticides and fertilizers will be needed to maintain levels of production on a smaller and smaller land base.*
- *Greater reliance on an unstable supply of increasingly expensive imports, because other major food-producing regions in the world are also losing their farmland and facing such problems as water shortages.*

Toronto Star

Unfortunately, urbanization is difficult to control and almost impossible to stop. Many municipalities would rather use farmland within their boundaries for purposes that generate more taxes. As well, some farmers support urbanization, because they can sell their land at very high prices — almost double the price per hectare when the land is sold for housing or industry rather than for agriculture.

1. Why is the loss of prime farmland a matter of concern for Canada?
2. Why is the Niagara Fruit Belt such a desirable area for the construction of houses, factories, malls and recreational facilities?
3. (a) What negative effects could result if the loss of farmland is not controlled?
   (b) In small groups, discuss the concerns raised by the farm organizations eager to preserve farmland.
   (c) Would consumer groups have similar feelings?

New Words to Know
Prime farmland

The Grape Industry in Peril

While the area of land in tree-fruits has dwindled over the years, the area in vineyards has increased. First, grapes can be grown on heavier clay soils, which are not suitable for tender tree-fruits and are less in demand for non-agricultural uses. Second, Canadians are drinking much more wine. The total grape purchase in Ontario for 1996 was 28 700 metric tonnes. Wine produced was 37 million litres, for a retail sales total of $275 million. Ontario has 800 winery operations, including 550 grape growers, with over 10 000 ha. of land devoted exclusively to vineyards. Wine-related industry employment is 4 000 people. Some 300 000 tourists visit the Niagara vineyards every year.

Vineyards are typical intensive farms. They are very small by Canadian standards, averaging only 11 ha. Grapes are harvested by machine in the fall, but most other steps in the process — pruning the deadwood in winter, tying the vines to supports and thinning the flowers in spring — must be done by hand.

Grape growers face many of the same problems as other farmers: Hail; too much rain at harvest; too little moisture in the growing season; a late-spring or fall frost; migrating birds.

However, the critical threat to Niagara grape growers comes not from nature but from competition with the U.S.A. and Europe. To help develop and promote Ontario's young wine industry, the provincial government imposed a tariff so that imported wines cost, on average, 65 percent more than local wines. In 1988, European wineries complained to the international commission named **General Agreement on Tarriffs and Trade** (GATT) that the tarriff policy was unfair. They won their case, and as a result, tarriffs were reduced and imported European and American wines increased their share of the Ontario market.

Ontario **vintners** moved to protect themselves and their industry by establishing the Vintners Quality Alliance (VQA) in 1988 to designate specific wine-producing areas and enforce basic wine-making standards. In 1991, one Ontario vintner won the Grand Prix d'Honneur at the French VinExpo, with the result that the Canadian wine-producing industry began to gain international recognition. In 1996, the VQA established National Wine Standards. The Cold Climate Oenology and Viticulture Institute was also established in that year to provide education opportunities and research to the Canadian wine making industry.

1. **Why have vineyards expanded while tender tree fruit orchards have declined?**
2. **Explain why grape growing can be classified as intensive farming.**
3. **Why did the Ontario government impose high tariffs on imported wines? Do you think this was a good idea?**
4. **Why do you think winning an international award would be important to Canadian vintners?**

New Words to Know
General Agreement on Tarriffs and Trade (GATT)
Vintner

Prairies

To many people, farming in Canada means the Prairie provinces, and farming there means wheat. They have an image of wheat fields stretching from Manitoba to the Rocky Mountains. In fact, over 80 percent of the nation's farmland *is* on the prairies. However, while wheat is by far the region's most important crop, the prairies are far more than one vast wheat field. Other grains such as barley, rye and oats are also grown there, mainly as livestock feed. Non-grains such as canola (used to make margarine and cooking oils) and flax (a

4-28 Sources of farm cash receipts from farming operations, 1996.

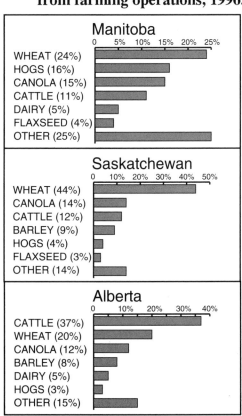

Direct program payments (D.P.P.) are subsidies from various federal and provincial agencies.

source of oil and fibre) occupy significant amounts of land. Livestock are also important, especially in Alberta. Together, the Prairie provinces raise over two-thirds of all the cattle in Canada, and more than a third of the hogs and sheep. Cattle ranches are located mainly in southwestern Saskatchewan and the Alberta foothills where the land is too dry and rugged for crops.

Saskatchewan Wheat

Saskatchewan is the centre of wheat production in Canada, with 60 percent of the total. The three Prairie provinces together grow over 95 percent of Canada's wheat. What has made the Prairie provinces the breadbasket of Canada and one of the world's best wheat-growing areas? It is a combination of climate, soil and topography.

Wheat is a member of the grass family, and does not require a lot of moisture. This is fortunate, since precipitation is lower on the prairies than in most other agricultural areas of Canada. About 30 percent falls as snow, which melts into the ground in spring and provides early moisture for the seeds and young plants. Much of the rest of the precipitation comes in the spring and early summer, when it is needed for early crop growth. The hot prairie summers provide heat, which is the driving force necessary for wheat to grow and mature. Later in the year, the warm days and cool nights of fall are ideal for ripening it. Most varieties mature in around 100 days; this is within the region's frost-free period of 125 days.

The soil cover also lends itself to wheat. The clay subsoil retains moisture for the plants' roots during the long, dry summer. The topsoils, which range from 25 cm to 50 cm thick, are among the most fertile in the world. They contain essential nutrients, and are deep enough to allow plants to develop mature root systems.

The prairie allows farmers to cultivate huge expanses of land with large-scale mechanized equipment. This is important in extensive-farming areas, where yields and crop values per hectare are lower than in intensive-farming areas.

Two main types of wheat are grown in Saskatchewan. As its name implies, **spring wheat** is planted in the spring. It is the main type grown, and is ideally suited to the hot, dry summer. It is high in protein, and used mainly to make bread. It forms the bulk of Canada's wheat exports. About 15 percent of the

4-29   The extensive wheat farms of the prairies require large-scale mechanized equipment to harvest their crops.

crop is **durum wheat**. This is also a spring wheat, but it contains less protein. It is grown to make pasta products such as spaghetti and macaroni.

Problems To combat insects and weeds, farmers use a variety of insecticides and herbicides. To control wheat rust and other diseases, resistant strains of wheat have been developed. Also, farmers now practise **crop diversification** — growing several crops rather than just one. This lessens the impact of a major disease outbreak in one crop.

Prairie weather is highly variable; the area often experiences hailstorms, frosts and periodic droughts. In the average year, hail destroys about four percent of Saskatchewan's crops. An early frost or snowfall can also damage or destroy a mature crop. In early days of settlement, farmers relied on a few European varieties of wheat that were not well-suited to the region's shorter growing season. Since that time, new strains of wheat have been developed which mature faster.

Of all the weather hazards, drought is the most common and critical. **Drought** is defined as a period of dry weather of sufficient length and severity to cause at least a partial crop failure. If a drought extends over several years, as during the mid-1980s, severe crop damage may result. Farmers have developed a number of techniques to conserve moisture and soil. **Summer fallow** has traditionally been the main method. Each year, farmers leave a different portion of their land, usually one-third, unseeded. This allows the moisture in the soil to recover and accumulate. Another method is to leave the stalks of the harvested wheat

or **stubble** standing in the fields over the winter. This holds snow on the fields rather than letting it blow off.

1. **Why is it inaccurate to think of prairie agriculture as simply wheat farming? Refer to graph 4-28.**
2. **(a) Which aspects of climate favour wheat growing?**
   (b) How is the prairie soil suited for growing wheat?
   (c) Why is the flat terrain an advantage for Prairie wheat farming in particular?
3. **In a paragraph of about 100 words, explain the techniques farmers are using to reduce the hazards of wheat farming.**

New Words to Know

Subsidy	Drought
Spring wheat	Summer fallow
Durum wheat	Stubble
Crop diversification	

4-30 Drought conditions afflicted the prairies during the 1930s. *Large areas of the prairies still face the problem of lack of moisture if rain and snowfall are less than needed to replace the moisture used by crops.*

British Columbia

Due to the mountainous terrain and variable climate of British Columbia, only two percent of the province is suitable for agriculture. The agricultural areas that do exist are located in scattered pockets — mainly in river valleys. The exception to this is the Peace River district in B.C.'s northeast corner, which is an extension of the prairies.

The province's geography has also created a wide variety of growing conditions that allow B.C. to produce a greater diversity of agricultural products than any other province. B.C. can be divided into five distinct agricultural regions, each with its own unique characteristics.

The most important and intensively cultivated is the south coastal region, especially the Fraser River delta. This has a mild climate, abundant precipitation, very fertile soils and flat terrain. Most farms are small, and specialize in dairy and poultry products, vegetables, berries and fruits for the nearby urban areas. Parts of the delta and the lower Fraser Valley are being lost to urban development. The B.C. government has imposed restrictions to slow down the conversion of agricultural land, but the rapidly increasing population poses a constant threat.

Further inland are the Thompson-Cariboo and central regions. Both are areas of mountain ranges and rolling plateaus cut by deep river valleys. The landscape is dry and rugged, with extensive natural grasslands. Cattle ranching dominates agriculture in both regions. Crop production is limited to irrigated parts of the river valleys, and to native meadows, almost all devoted to forage crops, to feed livestock in winter.

The Peace River region is similar in climate and topography to the mixed-farming areas of the Prairie provinces. Here are grown wheat and other grains, oilseeds, forage crops and cattle.

In the southeast is the Okanagan Valley. From here come almost all of the grapes and tender tree fruits grown outside the Niagara Fruit Belt. Apples from the Okanagan, especially the B.C. red and golden delicious, are world-famous. Also, all the grapes for B.C.'s wine industry — the second largest in Canada — come from there. It has a less favourable climate for orchard crops than the Niagara Fruit Belt, since it is drier and colder. However, mountain lakes provide water for irrigation, without which the Okanagan would be unable to produce tree crops, and would be of little use except for grazing. In fact, the lower valley resembles a desert. There are few orchards on the valley floor, where cold air collects and frost damage could occur. Most are on the valley slopes to take advantage of cold-air drainage.

1. **How is the geography of B.C. both an advantage and a disadvantage for agriculture?**
2. **Which conditions have allowed the Lower Fraser River valley to concentrate on intensive agriculture?**

4-31   British Columbia's main farming areas.

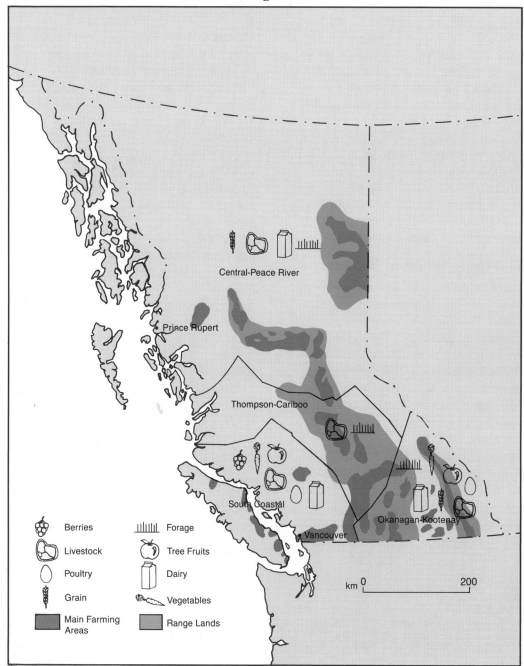

The varied terrain and climate in B.C. allow for a wide variety of agricultural products.

207

Mining

"To discover certain islands and lands where it is said that a great quantity of gold and other precious things are to be found." This was the command that François I gave to the French explorer Jacques Cartier as he set out for the New World in 1534. Cartier did not disappoint his king; he returned to France with a ship loaded with rocks containing a pale-yellow mineral. Unfortunately, what Cartier thought was gold turned out to be iron pyrite, which is often called "fool's gold." It contains iron and sulphur — but no gold.

The Importance of Mining

Later explorers and modern prospectors have been much more successful than Cartier was. In 1995, Canada was the third-largest mineral producer in the world, surpassed only by the Russian Republics and the U.S.A. The value of these minerals was a staggering $20 billion, making mineral production one of the mainstays of the Canadian economy.

Canada is well-endowed with minerals. During 1995 roughly 300 mines throughout the country produced about 60 different mineral products. Canada leads the world in the production of potash, zinc and uranium, and ranks second in nickel, asbestos and cadmium. It is a major producer of gold, silver, platinum, molybdenum, gypsum, lead and copper.

Canada is the world's greatest exporter of minerals. Nearly 80 percent of its production is sent to foreign markets. This accounts for 20 percent of Canada's export earnings. The U.S.A. is by far the most important customer, taking 68 percent of its **crude** (unprocessed) and **semi-fabricated** (semi-processed) mineral exports. Japan and Western Europe imported another 18 percent.

The degree of dependence on exports varies from one mineral to another. Almost the entire output of nickel, asbestos, copper, potash and gold is exported. For others the export dependency is less. Large quantities of aggregate (sand and gravel used to make concrete), limestone (which is pulverized to make cement) and stone are mined, but very little is exported, since these are common throughout most of the world. Most of them are used domestically, primarily in road-building and construction.

Mining and mineral-processing:
- account for about half of the total value of rail traffic and one-third of domestic shipping of all kinds;
- are among the largest consumers of energy in Canada;
- require large amounts of specialized equipment, machinery and haulage vehicles;
- require many highly-trained specialists in their labour force;
- use enormous quantities of construction materials for mine and mill development, and chemicals for mineral-processing;
- have led to the opening and development of remote areas, not only for mining, but for other activities such as agriculture, forestry and recreation.

Mining areas are found throughout Canada. Over 120 communities, from Newfoundland to British Columbia and up into the Far North, depend primarily on mining for their existence. Almost 200 000 Canadians earn a living directly from mining and mineral-processing. Although they employ only 1.5 percent directly, it is estimated that mining companies support nearly eight percent of the Canadian labour force. They spend over $2 billion annually, mostly in Canada, on a multitude of products and services from other industries. The employment-multiplier effect creates many more jobs. Hundreds of thousands of people, both in and out of mining regions, are indirectly linked to mining.

1. **Why would Jacques Cartier be amazed at the Canadian mining industry of today?**
2. **How does Canada rank in the world as a producer of minerals?**
3. **Account for the fact that some of this country's minerals are in greater demand outside Canada than others.**

New Words to Know
Crude mineral
Semi-fabricated mineral

Location and Types of Minerals

Canada's minerals can be broken down into three major categories: **metallic, non-metallic** and **structural**. Each type is created under certain geological conditions and, as a result, occurs in specific geological regions.

Most metallic minerals, or metals, are associated with igneous rock, particularly the Precambrian Shield, which is often called "Canada's Mineral Storehouse."

The shield covers much of Quebec and Ontario; consequently, these provinces produce nearly 50 percent of this country's metallic minerals. They are particularly important as sources of iron ore, nickel, uranium, **base metals** and **precious metals**. Base metals include copper, lead and zinc, which are "inferior" in value to precious metals and are used for commercial and industrial purposes. Precious metals —

mainly gold, silver and platinum — are valued for other than commercial and industrial purposes.

The Appalachian and Cordilleran regions are also significant sources of metallic minerals, though only where there are intrusions of igneous rock into the sedimentary rocks that largely make up these mountains. B.C. produces over 40 percent of Canada's copper, and New Brunswick contributes nearly 25 percent of its zinc.

Metals make up the most valuable category of minerals mined in Canada.

However, non-metallic minerals, or non-metals, are also significant. The most important are gypsum, potash, salt and asbestos. They are mainly used in industrial and manufacturing processes in their natural state, and are frequently referred to as **industrial minerals**. Non-metals are generally found in the sedimentary rocks of Nova Scotia, New Brunswick, Saskatchewan and southern Ontario. They owe their origin to the ancient saltwater seas that once covered these regions. As the seas slowly evaporated, the salts in the water were deposited on the ocean floor in thick layers. Asbestos is an exception: it is contained in igneous rock which has been superheated, twisted, shredded and

4-32   Principal mining areas of Canada.

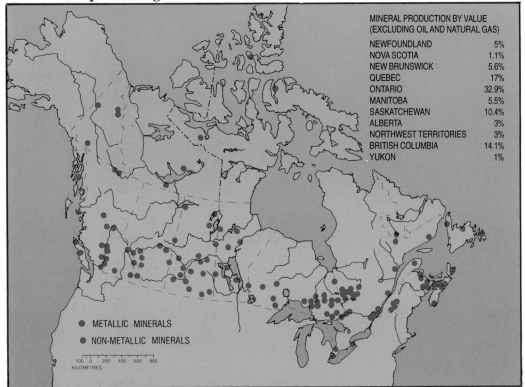

MINERAL PRODUCTION BY VALUE
(EXCLUDING OIL AND NATURAL GAS)

NEWFOUNDLAND	5%
NOVA SCOTIA	1.1%
NEW BRUNSWICK	5.6%
QUEBEC	17%
ONTARIO	32.9%
MANITOBA	5.5%
SASKATCHEWAN	10.4%
ALBERTA	3%
NORTHWEST TERRITORIES	3%
BRITISH COLUMBIA	14.1%
YUKON	1%

● METALLIC MINERALS
● NON-METALLIC MINERALS

100 0   200   400   600   800
KILOMETRES

1. **Which two provinces dominate mineral production by value?**
2. **In which physiographic region are these provinces located?**
3. **What type of mineral is most important in these regions?**
4. **What type of mineral dominates on the prairies and in the southern parts of Ontario and Quebec?**
5. **Which physiographic region has both mineral types?**
6. **Draw a graph to illustrate the value of mineral production by province.**

4-33 *Production and value of leading mineral materials, 1996 ($millions).*

	Production	1996 Value	1990 Value
Metals			
Gold (kilogramme)	164 136	2 803.0	2 407.6
Copper (kilotonnes)	656	2 037.2	2 428.9
Nickel (kilotonnes)	185	1 958.2	2 027.9
Zinc (kilotonnes)	1 188	1 652.3	2 272.6
Iron ore (kilotonnes)	36	1 310.5	1 258.8
Uranium (tonnes Uranium)	11 448	645.8	887.9
Silver (tonnes)	1 228	280.5	249.7
Platinum group (kilogramme)	14 234	146.2	189.4
Cobalt (kilotonnes)	2	168.4	49.6
Lead (kilotonnes)	246	261.6	279.3
Non-metals			
Potash (kilotonnes)	8 165	1 263.8	964.9
Salt (kilotonnes)	12 289	316.2	240.9
Asbestos (kilotonnes)	521	238.1	272.1
Peat (kilotonnes)	783	128.9	n/a
Sulphur (kilotonnes)	8 131	95.6	368.9
Structural Materials			
Cement (kilotonnes)	11 050	931.5	991.4
Sand and gravel (kilotonnes)	217 898	778.3	817.3
Stone (kilotonnes)	86 057	552.6	663.4
Lime (kilotonnes)	2 491	212.3	n/a
Clay products (kilotonnes)	..	117.1	n/a
Mineral fuels			
Petroleum, crude (thousands cubic metres)	116 832	19 008.5	13 103.4
Natural gas (millions cubic metres)	152 985	8 718.9	5 692.0
Natural gas by-products (thousands cubic metres)	25 882	2 456.5	2 370.8
Coal (kilotonnes)	75 950	1 943.1	1 823.7

** figures not available   Source: Statistics Canada, Catalogue no. 26-202-XPB.

squeezed to create a metamorphic mineral with fibrous characteristics.

Structural or building minerals include limestone, **aggregate** and clay. These rarely receive the glamourous attention accorded to gold or silver, yet as a group they are almost as valuable as non-metals, because they are the basic materials for the construction of most roads and buildings. When aggregate — sand and gravel — is mixed with tar, the result is asphalt; when mixed with cement, it becomes concrete. Deposits of aggregate and clay are found in every geological region of Canada, but are most numerous in areas where there was a lot of glacial deposition. Ontario — particularly southern Ontario, because of its geology and large urban and industrial base — dominates Canada's production and consumption of structural minerals.

1. **For the three mineral categories name:**
   (a) **the physiographic regions in which they are found**
   (b) **the rock types**
   (c) **three specific examples of minerals.**

New Words to Know
Metallic minerals
Non-metallic minerals
Structural materials
Base metals
Precious metals
Industrial minerals
Aggregate

From Mine to Marketplace

Canada has some 300 mines producing a vast array of minerals. However, before any of these mines actually reach the point where a mineral or rock can be sold profitably, a number of activities must take place. These include exploration, extraction/production and refining. These activities require enormous amounts of capital, which is defined as money for investment. The process is always a lengthy one; for example, the Thompson North mine at Thompson, Manitoba was opened in 1986 after five years of development, at a cost of $100 million! It is estimated that the average base-metal mine in Canada needs between $50 and $150 million, and between eight and ten years of exploration and development, before it actually goes into production.

Exploration

For many people, hunting for minerals brings to mind a scruffy prospector struggling through the woods with a pick, shovel and packsack. In the past, many of Canada's greatest mineral discoveries were made by such individuals as they examined exposed surface rocks for signs of mineralization. But despite their successes, the early prospectors were unable to detect most of Canada's mineral wealth, which lies far below the surface.

Modern exploration relies heavily on **geophysics**, a branch of science that combines geology, the study of rocks, with physics, the study of matter and energy. Geophysical surveys are conducted on the ground, or by aircraft towing complex measuring instruments over thousands of square kilometres of wilderness. The instruments measure and collect magnetic, radiation and gravitational data about the rocks beneath the surface. One of the most widely used instruments is the **magnetometer**, which

4-34   Core samples obtained by diamond drilling are examined for their mineral content.

measures changes in the earth's magnetic field. Geologists look for differences in the regular pattern of the rocks. These may indicate the presence of a mineral deposit.

Once a possible deposit has been located, a detailed study is made to determine its exact location, size and mineral composition. Unless the mineral is exposed on the surface, a drilling crew must be sent in to take thousands of **core samples**, or "rock worms," of the underlying rock. Core samples are cylindrical pieces of rock extracted with a hollow drill. Only after a close examination and an **assay** (scientific testing) of the core samples can the geologist begin to know what lies buried in the rock. In most cases, the assay indicates that the valuable-mineral content is non-existent, or so low that mining would not be profitable. On average, only one of a

4-35   Underground in a New Brunswick potash mine.

thousand exploration prospects ever develops into a mine!

1. **Explain, in one paragraph, how modern exploration methods differ from prospecting during earlier times.**
2. **Which role would the following people play in deciding whether or not to develop a mine?**
   (a) **a geologist**
   (b) **a marketing specialist**
   (c) **a transportation expert**
   (d) **a town planner**
   (e) **an environmental-protection expert**
   (f) **a government official**

New Words to Know

Geophysics   Assay
Magnetometer   Ore body
Core sample   Grade

To Mine or Not to Mine?

Even after a mineral deposit has been discovered, a mining company must still decide whether or not it is an **ore body** — a mineral deposit from which one or more minerals can be extracted *at a profit*. Obviously, no company is going to spend perhaps $100 million on a mine, however rich the deposit might be, if other factors make it unprofitable. A number of other considerations must be taken into account.

Quality of the resource   Geologists must map the mineral deposit and determine its size, extent and depth, as well as **grade** of the mineral content. From this information, projections can be made as to the amount of valuable minerals that could be obtained, the life of the mine, and the cost and the mining method required to extract the ore. The mining company may decide the deposit is not worth mining. Many mineral ores remain buried because they cannot be exploited economically.

Demand and value   The present and future demand for a mineral, as well as its value in world markets, must be assessed. For example, in 1988 the world price of nickel moved from $11,000 (U.S.) a tonne to $22,500 (U.S.) a tonne. International Nickel (INCO) re-opened one mine and developed another. The price of nickel in 1998 was $5710 (U.S.) a tonne and INCO laid off 1 275 employees between November 1997 and February 1998.

Transportation   New mine sites are often isolated and inaccessible. There must be a way to bring in the tonnes of equipment required to open and operate the mine, and to ship the mineral to market. A road or railway may have to be built through rough and rugged terrain. This extra cost may make the project too expensive. The high cost of building railways in the B.C. interior has limited that region's mineral development.

Labour force   Without workers, a mine cannot function. Unfortunately, most new mines are in areas where there are few towns. This may mean that a brand-new town must be created.

   Those are only a few of the major concerns a company must face. Others are:
• Is there an available source of power?

Timmins Map Study

Timmins is located approximately 700 km north of Toronto in the Canadian Shield. This city, along with many other communities in the surrounding area, owes its existence to the rich mineral deposits found in the igneous rocks of the shield. In past decades, the gold from such mines as Hollinger, McIntyre, and Dome provided much of the area's wealth. Today, the Kidd Creek zinc/copper/silver mine is the city's biggest employer.

1. What features indicate that this area is in the Canadian Shield?
2. Find features which show the presence of mining. Locate a place name which indicates a major mineral found in the area.
3. Explain what is meant by mine waste. What is another name for mine waste?
4. Find as many mine sites as possible on the map. How many of these are abandoned, compared to those that are still active?
5. Name and give the map references of two forestry-related activities.
6. What structures have been built at 776679 and 781684? On what physical feature have they been built? Why is this a good location for such structures?
7. What utilities and transportation and communication facilities have been constructed to serve this community?
8. Explain how and why the road system of Mountjoy and Melrose Gardens is different from that in the core of the city.
9. If you had your choice, where would you build a fast-food restaurant? Give three reasons for your choice.
10. Suppose you became lost at 731629. How far is it to the nearest main road leading to Timmins and what obstacles would you face in walking to it? If you decided to walk to the radio tower at 715660, what direction would this be on your compass?
11. Compare the landscape of this map with that on the map in chapter one.

4-36 Road, housing, and rail contruction in the mining town of Tumbler Ridge, B.C.

4-37 Part of Timmins topographic map 42 A/6. *Contour interval is 50 feet. For scale and key, see page 405.*

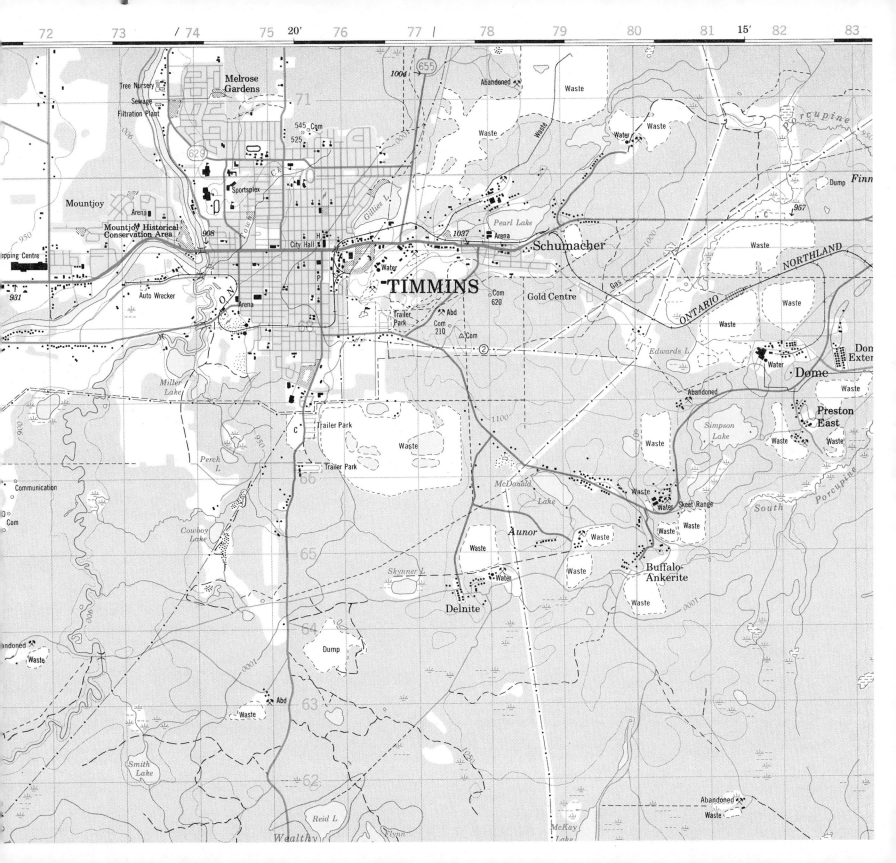

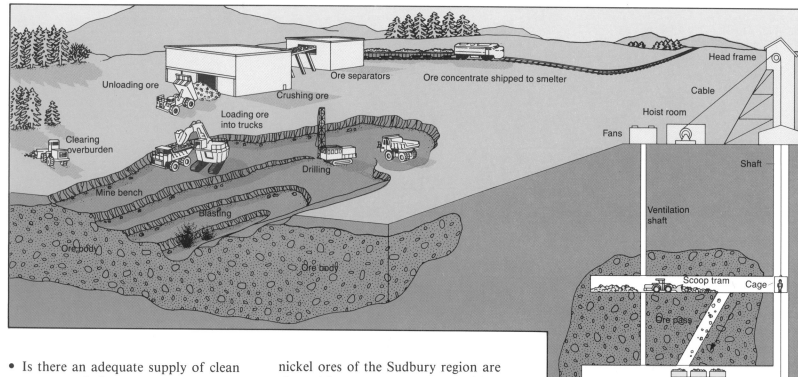

Labels in figure 4-38: Unloading ore; Crushing ore; Ore separators; Ore concentrate shipped to smelter; Loading ore into trucks; Clearing overburden; Drilling; Mine bench; Blasting; Ore body; Ore body

Labels in figure 4-39: Head frame; Cable; Hoist room; Fans; Shaft; Ventilation shaft; Scoop tram; Cage; Ore pass; Ore body; Skip; Ore body

- Is there an adequate supply of clean water nearby?
- Which environmental problems might arise?
- How will government taxes and subsidies affect development?
- How will the company finance the new mine?

From Ore to Metal

Very rarely will an ore extracted from the earth have a high enough valuable-mineral content that it can be used without further processing. This is particularly true of metallic ores, which require considerably more processing than other minerals. Most metallic ores have an extremely small percentage of valuable minerals — the rest is waste. The rich nickel ores of the Sudbury region are about 96 percent waste; most gold ores have less than six grams of gold per tonne.

A variety of processes have been developed to recover the minerals profitably. In the **mill**, which is usually close to the mine, the ore is crushed and ground into particles the size of grains of sand. Depending on the physical and chemical properties of the valuable minerals, the ore may be washed, sorted, passed over a magnetic field or treated with chemicals. Whichever process is used, the end result is that the valuable minerals are collected in a form called **concentrate** and the waste, or **tailings**, is disposed of.

In Sudbury's mills, the concentrate is a mixture of nickel, copper, gold, silver,

4-40   From ore to metal.

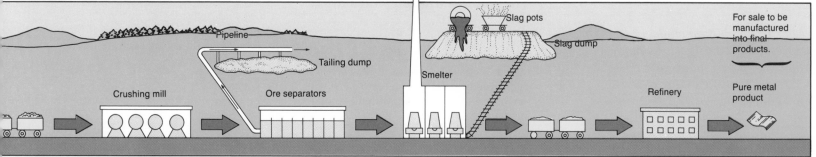

The illustrations on this spread show the two basic approaches to mining, combined with the process of refining the ore into usable metals. **Open pit mining** is used when an ore body lies at or near the surface. First, **overburden**, which is the rock and soil lying on top of the ore, is cleared away. Then crews begin to drill into the ore, blasting it loose and loading it into huge trucks, which haul it away by means of a series of steps, or **benches**, which spiral up the wall of the pit to the surface. **Shaft mining** is used when, as often happens in Canada, the ore is located far below the surface. Above the vertical main shaft is a building called a **headframe**. This contains a hoist, which is used to lower workers, supplies and equipment into the mine in a big elevator,

or **cage**, and to lift the ore to the surface in a large bucket called a **skip**. The shaft also carries water, power, and air to the mine. Horizontal tunnels lead off the shaft, some to large openings called **stopes**, where the ore is drilled and blasted from the rock face. Small rail cars, conveyor belts, or special vehicles called **scoop trams**, carry the ore to a collection point where, usually, it is crushed before being loaded into the skip. Ores are usually concentrated at or near most mines, but shipped elsewhere for processing into metal. Smelters and refineries are huge, expensive complexes, and are built only in conjunction with large mines that have vast ore deposits.

platinum and cobalt, along with a variety of other minerals and impurities. At the **smelter** the nickel-rich concentrate is heated to high temperatures, in combination with other chemicals. The heat drives off water and many of the impurities, leaving a metal of about 50 percent purity. Finally, it must go to a **refinery**, where it is further purified to almost 100 percent nickel, and is ready for sale to the industries that use it.

Most other metal-bearing ores follow a similar path from mine to refinery. Some refining processes are simpler than others; all, however, involve a series of steps. Most non-metallic and structural minerals require far less processing, since they occur naturally pure or almost pure. Crushing, washing, screening and sorting are the most common methods of processing these minerals.

1. **What determines whether an open-pit or a shaft mine is used?**
2. **Why is an open-pit mine generally safer and cheaper to operate than a shaft mine?**
3. **Why would large mines need to have additional shafts for ventilation?**
4. **Why is it necessary to send metallic ores to a mill, smelter and refinery?**
5. **What is the purpose of producing a concentrate?**
6. **Why do most non-metallic and structural minerals require less processing?**

New Words to Know

Open-pit mining	Stope
Shaft mining	Scoop tram
Overburden	Mill
Mine bench	Concentrate
Headframe	Tailings
Mine cage	Smelter
Skip	Refinery

The Future

Until 1993, Voisey Bay up near Goose Bay on the coast of Labrador was a caribou hunting ground and home to a few Native coastal settlements. Today, geologists, drillers and prospectors are working around the clock to unearth one of the richest nickel, copper and cobalt deposits discovered in Canada.

The Voisey Bay discovery holds much promise. It is offering up immense amounts of ore, much of which is twice as pure as the ore found in other operating mines. The ore is also close to the surface, which allows open-pit mining at a lower cost, and, because Voisey Bay is near the ocean, this promised reduced transportation costs.

For Canada, this is good news. During the mid-1990s, international demand for nickel outpaced supply. Canada is the world's second-largest producer of nickel. The Voisey Bay site alone may produce as much as 15 percent of current world production. Canada's share of world nickel production was 17.2 % in 1994.

For nearby Newfoundland, Voisey Bay has meant more jobs, and has injected new vigor into the economy. In 1995 alone, prospectors poured $55 million into the province, standing in line for hours to register their mineral claims.

Analysts expected the Voisey Bay Mine to be in full operation by 1999, however Aboriginal land claims, environmental issues, unresolved conflicts between the province and the developer, and a downturn in the mineral market set this schedule back.

1. Why has there been a decline in the world demand for minerals and metals?
2. If the present trends continue, which problems might the Canadian mining industry face?
3. Which measures could Canadian mining companies take to regain, or at least retain, their share of the world market?
4. A mining company has discovered a rich metallic-mineral deposit near your community. It wants to build an open-pit mine, mill, smelter and refinery, and will present its plans at a public meeting.
   (a) Compose five questions you would ask the company regarding its policies on the environment.
   (b) A real-estate agent, a road contractor, the owner of a golf course, an unemployed mechanic and a homeowner living next to the mine are also at the meeting. Make up one question that each of these people might ask.

The Mining Industry and the Environment

The mining industry, rightly or wrongly, has often been accused of having little regard for the environment. Below is a list of environmental problems associated with the mining industry, along with their possible remedies.

Problem:
• *Abandoned and worked-out open-pit mines, and mountains of overburden and tailings, scar the terrain leaving an ugly sometimes dangerous landscape.*
• *Pollutants in the overburden/tailings escape into nearby rivers and lakes, and seep into the ground water.*
Remedy:
• Smaller open-pit mines are partially refilled with the overburden. Others are made into ponds for recreation.
• Modern mines build containment dams around mill tailings to prevent run off into nearby water bodies.

• Tailing dumps are reshaped and landscaped, and covered with trees and grass.
• Mine or pit openings are fenced off for the safety of children and adults (particularly snowmobilers).

Problem:
• *Water used in the various processing stages contains harmful chemicals, is disposed of in nearby lakes and rivers.*
Remedy:
• Installation of recycling processes or improved water-treatment facilities.

Problem:
• *Smelters are notorious air polluters, creating moonlike landscapes for many kilometres around.*
Remedy:
• Stringent government controls, along with improved recovery technology recovery, reduced sulphur dioxide emissions. The INCO in Sudbury, Ontario has made considerable progress in this area.

4-38b *This lush landscape outside Stouffville, Ontario, was once a thriving sand and gravel pit supplying builders and contractors throughout the area. Once the aggregate had been removed, the land was graded, and reclaimed so that it could be used for residential and farming purposes. Land recycling is an excellent example of sustainable development.*

Fishing

Canada's fishing industry has three sectors: the Atlantic, the Pacific and freshwater fisheries. They are considerably different in many aspects: the size of the catch; the species caught; the number of workers involved; the methods of harvesting; the markets they sell to. Together these three fisheries landed 1.1 million tonnes of fish and shellfish in 1995. This sounds like a lot but it is actually small compared with the ten million tonnes caught annually by both Japan and Russia. Canada, however, leads the world in the *export* of fish and fish products. As with so much else, the U.S.A. is the primary export market — 60 percent of the total — followed by Japan and the U.K.

The Atlantic Fishery

Armed Canadian Fisheries Officers Board US Dragger

French Naval Tug Seizes Newfoundland Trawler

Spanish Trawlers Arrested in Canadian Waters

Canadian Salmon Fishermen Blockade Alaskan Ferry.

These headlines are typical of many that Canadians read in 1996-97 – about disputes between friendly neighbours and allies — over fish! The incidents took place in the northwest Atlantic, for centuries the mainstay of the Canadian fishing industry. Not only is this area important to the Atlantic provinces, but it also attracts foreign fishing vessels from around the world. It may seem strange that ships come from as far

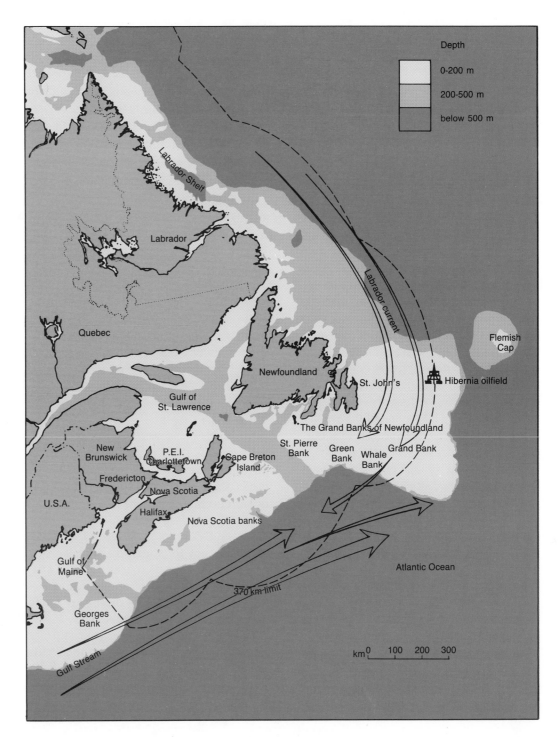

Depth

- 0-200 m
- 200-500 m
- below 500 m

Labrador Shelf

Labrador

Labrador current

Quebec

Flemish Cap

Newfoundland

St. John's

Hibernia oilfield

Gulf of St. Lawrence

The Grand Banks of Newfoundland

New Brunswick

P.E.I. Charlottetown

Cape Breton Island

St. Pierre Bank

Green Bank

Whale Bank

Grand Bank

Fredericton

Nova Scotia

U.S.A.

Halifax

Nova Scotia banks

Atlantic Ocean

Gulf of Maine

370 km limit

Georges Bank

km 0 100 200 300

Gulf Stream

away as Japan, South Korea and Russia to fish in or near Canadian waters, when there are millions of square kilometres of ocean throughout the world. However, only certain small areas of the oceans contain fish in numbers large enough to support a fishing industry. As people in that industry often say, "90 percent of the fish are in ten percent of the water." The waters off Canada's Atlantic provinces are part of this ten percent.

The Atlantic region dominates Canada's fishing industry. It has 70 percent of the total catch and 67 percent of the value of fish caught. For centuries cod was "king". Along with other **groundfish** (fish that live on or near the ocean floor) these species made up over 60 percent of the catch. However in the early 1990's a moratorium was placed on many groundfish to preserve the remaining stock and today they represent only 20 percent of the catch. **Pelagics**, or surface feeders, such as herring and mackerel, and **shellfish** make up 80 percent of the catch. Some seafoods are more valuable than others. For example, lobsters represent only six percent of the volume of landings, but a remarkable 36 percent of the dollar value. They are quite limited in numbers and are considered a gourmet food. Most are trucked, or flown live, to restaurants and seafood houses in Canada, the U.S.A. and Europe. Lobsters, together with other shellfish, are the most valuable part of the Atlantic fishery.

By the end of the 1980s, some 80 000 people were employed in fishing or fish-processing in 1300 communities throughout the region. Almost half of these towns and villages depended solely on the industry. Fishing, particularly for cod, was the backbone of Newfoundland's economy and a major source of income in the other Atlantic provinces as well.

The waters of the Atlantic region provided excellent fishing grounds because there was plenty of food for fish. Unique physical characteristics of that area created an ideal environment for microscopic plants called **phytoplankton**, which are the beginning of the food chain in the ocean.

Extending out underwater from the continental land mass is a shallow sea floor called the continental shelf. In the Atlantic region it is exceptionally wide. On the shelf are even shallower areas called **fishing banks**. The largest and most famous of these is the Grand Banks. Since the water is shallow, sun-

4-42 Canada's fish catch by region.

	value %	weight %
Atlantic	67	70
Pacific	30	28
Freshwater	3	2

light, one of the vital ingredients for phytoplankton growth, reaches to the ocean floor. Over this shallow water two ocean currents meet and mix. As the cold Labrador Current sinks under the warm Gulf Stream, it stirs up the water, spreading out the nutrients, which would otherwise sink to the ocean floor. This perfect combination of sunlight and nutrients allows phytoplankton to thrive and multiply in the billions. They become the food for larger (but still microscopic) animals called **zooplankton**, which are eaten by small fish and shellfish, which in turn are caught and eaten by humans.

But in the early 1990s, disaster struck Newfoundland and the Atlantic fisheries. The northern cod stock, which in 1988 represented almost 60 percent of the total Atlantic catch, was nearly depleted. The near extinction of the fish was caused by overfishing — done mostly by large offshore stern trawlers (see next page). To protect and ensure the survival of the remaining fish, the Canadian government in July 1992 called a stoppage — or a **moratorium** — on the fishing of cod off the northern banks of Newfoundland. The cod, however, did not return in significant numbers. In 1994, therefore, the Department of Fisheries extended the moratorium to include the closure of all cod fisheries in Atlantic Canada. An industry that had for centuries been the mainstay of the Atlantic economy was no longer. Thirty thousand Atlantic fishermen lost their jobs. Unemployment rates in Newfoundland soared, reaching 18 percent, more than twice the national rate, in 1995. The moratorium remains in effect today.

1. **Why do other countries come to the waters off Atlantic Canada to fish?**
2. **(a) Which are the three major fish or shellfish caught by weight and value? (b) Why are there wide differences between the weight and the value?**
3. **Briefly describe the food chain in the ocean.**
4. **Why are the Grand Banks an ideal breeding ground for phytoplankton?**

New Words to Know

Groundfish	Fishing bank
Pelagic	Zooplankton
Shellfish	Phytoplankton
Moratorium	

Offshore Fishing

The Atlantic fishery is characterized by two types of fishing, inshore and offshore. The inshore fishery is, largely small-scale traditional, while the offshore fishery represents a large-scale, more modern approach. The offshore fishery concentrates on groundfish and accounts for 40 percent of the catch, yet it involves less than 20 percent of the workers and five percent of the boats. It operates year round, in most weather conditions, using large **stern trawlers**. These are large fishing boats, averaging 35 to 45 m in length. They take their name from the large net, or trawl, that is let out and hauled in at the stern of the boat. They have the latest in sophisticated electronic fish-finding and navigational equipment and often roam the ocean hundreds of kilometres away from their home port. They carry a crew of about 15 and generally stay at sea up to two weeks at a time.

The large trawl nets, shaped like a cone, are dragged along the ocean floor and are capable of scooping up thousands of kilograms of fish in a single haul. They are often referred to as "vacuum cleaners of the sea." The catch is cleaned while the ship is still at sea, then iced or refrigerated and stored in the hold, which can take several hundred tonnes of fish. Because of their high cost, few stern trawlers are owned by individuals. Most are part of fleets owned by large **vertically integrated companies**. These firms have the crews process and package the catch and market it worldwide.

4-43   A stern trawler can catch thousands of fish in a single haul.

Fishery Products International: A Case Study of the Offshore Fishery

FPI is a global seafood enterprise comprised of a processing division and a trading division. From being a completely vertically integrated company in 1987, FPI has had to rationalize its processing activities in response to the dramatic decline in the groundfish resource on the Grand Banks of Newfoundland that occurred in the late 1980s and early 1990s.

In 1996 it owned and operated a fleet of eight deep sea trawlers, working out of Newfoundland, that harvested 9 million kilograms of groundfish. It also procured from European and Russian suppliers 22 million kilograms of groundfish to complement its own harvest. From its Nova Scotia plant, 5 offshore trawlers landed nearly 1 million kilograms of scallops. Its freezer trawler, the Newfoundland Otter, harvested 4 million kilograms of cold water shrimp, a large percentage of which was cooked on board. In addition, it purchased and processed millions of kilograms of groundfish, shrimp and crab from independent inshore fishermen.

FPI owns 10 processing plants, 8 in Newfoundland and one each in Nova Scotia and the USA. These plants manufacture numerous seafood products for several markets. In 1995 and 1996, over 40 new products have been developed for the North American market place alone.

The company's trading division purchases a multitude of seafood products from suppliers in Asia, South America, Central America and North America and sells them to customers throughout the world.

FPI has 6 sales offices in Canada, 2 in the USA and 1 each in England and Germany. It is constantly searching for new markets and new processing methods. In addition, it is developing new harvesting methods for its trawlers that have conservation of the resource in the forefront. FPI employs 3 000 people worldwide.

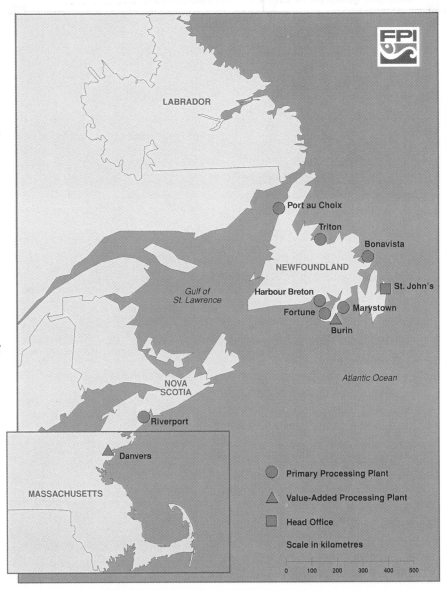

4-44   FPI's plants in Atlantic Canada.

1. Why are the plants scattered around the coast of Newfoundland?
2. What advantages might there be to locating plants in the United States?

Inshore Fishing

The inshore fishery is much smaller-scale than the offshore. In it, thousands of independent inshore fishermen harvest 60 percent of the region's fish, using small boats. Over 90 percent of the boats in the Atlantic fishery are under 15 m in length. Most of them are owner-operated and fish shallow waters close to their home ports. Since these boats lack the storage facilities and sophisticated equipment of the stern trawlers, they must return to port each day. The crews sell their catch to private processing plants or to **co-operatives** (which are plants owned by the fishermen themselves). These co-operatives, or "co-ops," process, pack and market fish like other processing plants, and the individual members of the organization share in the profits (if any).

The inshore fishery tends to be more seasonal and more **labour-intensive**, and to use a greater variety of harvesting techniques, than offshore fishing. The fishing method used depends on the season, on the characteristics of the sea bottom and shoreline and on the type of fish sought. The major species caught are cod, herring, lobster, crab and scallops.

1. (a) **What is meant by a vertically integrated fishing company?**
   (b) **Which are the three advantages of this type of company?**
2. **Which are the advantages to an inshore harvester of being a member of a fishing co-operative?**
3. (a) **Why would the introduction of factory freezer trawlers greatly concern both inshore fishermen and processing-plant workers?**
   (b) **Why is it very difficult for the inshore fishing sector to use up-to-date technology?**

New Words to Know
Stern trawler
Vertically integrated company
Factory freezer trawler
Co-operative
Labour-intensive

4-47 Selected species landed in the Atlantic fishery, 1991 and 1995

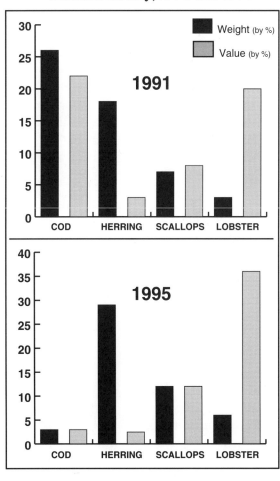

4-46 Inshore fishing boats usually return to port daily.

The Pacific Fishery

Canada's Pacific fishery lands far fewer fish than the Atlantic (15 percent by weight), but kilogram for kilogram, its fish are more valuable — nearly 30 percent of the total value.

Whereas "King Cod" once ruled the Atlantic fishery, salmon still thrives in the Pacific. This species makes up almost 50 percent of the total weight of landings, and 65 percent of the value. Most salmon is sold whole — fresh or frozen — and the remainder is canned. Herring (which are caught for their **roe**, or eggs), groundfish such as halibut and rockfish, and shellfish — mainly clams, crabs and shrimp — form the rest of the catch. Canadian Pacific salmon is prized as a delicacy in many parts of the world, and has a large international market that is less dependent on the U.S. than is the Atlantic catch. Only 18 percent of Pacific exports are destined for the U.S.A. The other 82 percent goes mainly to Japan, the U.K., France and Australia.

The Pacific seabed has a very different topography from the Atlantic. There is only a narrow continental shelf, and no banks to attract and concentrate the fish. Ocean currents don't mix there, as they do on the Atlantic coast — in fact, they move away from each other. However, the habitat is ideally suited for salmon.

Along the coastline, numerous freshwater rivers flow into the ocean. These are essential in the life cycle of salmon, which can only **spawn** (lay their eggs) in the shallow gravel beds at the headwaters of these freshwater rivers and streams. When the eggs hatch, the small fish spend some time developing in the fresh water, but then migrate to the ocean for their adult life. After 3 to 5 years, depending on the species, the adult salmon make the return migration, always in the fall, and always to the same gravel beds in which they were born. There they lay their eggs and die.

Salmon spend most of their adult life roaming the Pacific in search of herring and shrimp. Here they are harvested throughout most of the year by nets, or by trolling, which involves dragging a number of hooks through the water. Great numbers are also caught at the mouths of rivers and streams as they return to their spawning grounds.

Because of the higher value of the fish, and the restrictions on the number of licences, incomes are higher in the Pacific fishery than in the Atlantic. This is especially true of the inshore sector. Higher incomes mean that fewer workers require a second job or unemployment insurance to supplement their incomes, as many do in the Atlantic provinces.

4-48 A purse seiner on the west coast. *The net is laid out in a huge circle, then pulled tight at the bottom with the catch inside.*

The Inland Fishery

It may come as a surprise to many that Canada even has an inland fishery. It was a $63 million industry in 1992. By 1995 this figure had risen to $75 million with 5 500 people employed. The inland or freshwater fishery is concentrated in the Great Lakes, and in the larger lakes and rivers of north-western Ontario, the Prairies, and the Northwest Territories. It provides an important source of income, particularly for small native communities in the North.

On Lake Erie and Lake Winnipeg the fishing fleet consists of large trawlers. Smaller boats, using gill nets, are used elsewhere. The catch includes pickerel, perch, pike, whitefish, smelt and lake trout. Most of the catch is sold fresh or frozen to the U.S.A., Japan and France.

Protecting the Fisheries

In the past, Canadians gave little thought to protecting and managing their fish resources; fish stocks seemed inexhaustible. This perception changed in the 1970, when it became clear there were problems associated with overfishing, habitat destruction, foreign competition, outdated fishing fishing methods and competing user-groups.

Overfishing Over the years a number of different species – Atlantic and Pacific salmon, herring, lobster, bluefin tuna and cod – have gone through periods when their population levels dwindled due to overfishing. This occurs when more fish are caught than are being reproduced.

For example, the total catch, domestic and foreign, of Atlantic cod averaged 400 000 t per year until the 1950s, when it climbed to 900 000 t under increasing foreign and domestic pressure. This pressure increased in the 1960s when the U.S.S.R. began sending fleets of trawlers to the Grand Banks, accompanied by mother factory ships, sometimes called "floating processing plants." The trawlers caught the fish and then transferred them to the mother factory ships for processing and freezing. Other countries followed the U.S.S.R.'s lead. By 1968, the catch of cod had leapt to almost 2 million tonnes; but by 1992, there were few cod to catch and the Canadian government was forced to close the entire Atlantic cod industry. Other species have suffered a similar fate, and both Canadian and foreign ships have been guilty of overfishing.

In 1977 Canada declared an extension of its **territorial waters** from 22 km to 370 km. In effect, this gave Canada control over the continental shelf on both coasts, as well as most of the banks of the Atlantic provinces. This control allowed federal fisheries officials to set and enforce restriction on both foreign and Canadian fishing vessels. These include **quotas** or limits on the number of fish caught, and restrictions on when, where, and how they can be caught. As well, licenses are used to limit the number of Canadian and foreign fishing vessels. Foreign vessels may still fish in Canadian waters, but only with a license, and only for fish stocks deemed surplus to Canadian needs. This is why the Spanish and US ships mentioned earlier were boarded, and their captains arrested. Neither of these ships had a license to fish in Canadian waters; both were eventually fined and had their catches confiscated.

Boundary Disputes One particular case was the boundary dispute between Canada and France off the coast of Newfoundland. For hundreds of years France has owned the tiny islands of St. Pierre and Miquelon just 20 km off the south coast of Newfoundland. After Canada declared a 200-nautical-mile (370 km) boundary off its coasts in 1977, France counterclaimed with the same boundary for its possessions. As a result, both countries laid claim to the same rich fishing area. Canada contented that international law only allowed such islands a 19.2 km limit within the boundaries of Canada; therefore, these were Canadian waters, and Canada had the right to give out licenses and set quotas. France, of course, disagreed.

The dispute came to a head in the spring of 1988, when the Canadian government suspended French fishing rights in Canadian waters. When a trawler from St. Pierre and Miquelon challenged the new restrictions, its crew was arrested and spent two days in jail before being released on bail. France retaliated by arresting a Canadian trawler and imposing the same punishment.

In March 1989, the two countries agreed to submit the dispute to a five-member international tribunal, which was given three years to study the matter before making a decision on who owns the sea off southern Newfoundland. France agreed not to increase its fishing in the area during the dispute; in return, Canada allowed France to catch almost twice as much cod in areas north of

Newfoundland during the three years (conclusion of dispute to come).

This is only one example of a boundary and fishing dispute. Others still exist between Canada and the U.S.A., on the northern and southern coasts of B.C. and in the Beaufort Sea.

1. Why is the presence of a large number of freshwater rivers and streams emptying into the ocean important to the salmon fishery?
2. Why is the inland fishery still of special importance despite its small size?
3. How does the extension of Canada's territorial waters act as a management and conservation tool?
4. "The arrival of mother factory ships from foreign countries contributes to the collapse of the cod fishery." Explain.
5. (a) Why would some inshore harvesters be angry with the restrictions placed on them by the federal government fisheries officials?
   (b) Why are these restrictions needed?

New Words to Know

Roe	Territorial waters
Spawn	Quota
Mother factory ship	Aquaculture

4-49b *Workers at a salmon farm in Charlotte County, New Brunswick, clear seaweed away from the salmon cages.*

Aquaculture: Future Fisheries

While the fishing industry in general has suffered dwindling catches and volatile markets over the past years, one sector of the fishery, aquaculture, has not. Aquaculture is the raising of animals and plants in salt or fresh water; it is often called "fish farming." It is a rapidly growing industry. This growth is the result of an increasing demand by North Americans for low-cholesterol, high-protein foods even while native stocks of fish decrease.

4-49 **Aquaculture salmon farm.**

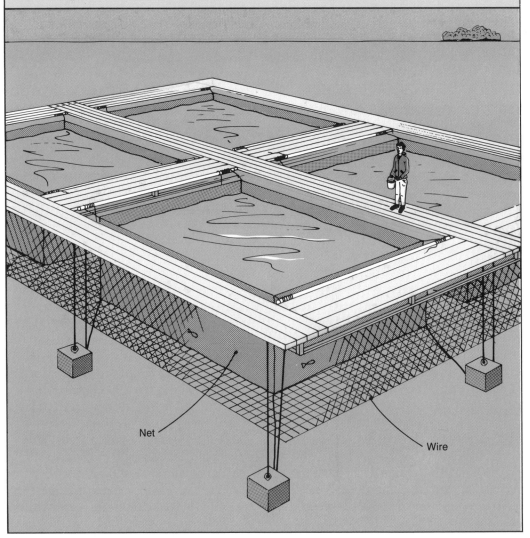

Net

Wire

Energy

"Energy: it powers our industries, provides human comforts and conveniences, fuels private and public transportation, generates jobs directly and indirectly; it underpins Canada's prosperity and is a valuable chip in the world marketplace."

The Energy Question

This statement is a reminder of just how vital and necessary an abundant supply of energy is to the daily lives of Canadians. Without it, this country's present high standard of living would not exist. The lifestyle changes would be huge if Canadians lost access to electricity, gasoline, heating oil and natural gas. There would be no television, no radio, no automobiles and no way to heat houses or water for a shower. Yet Canadians are so "hooked up" to energy that they often forget this, and take for granted that it will be readily available at the flick of a switch or the turn of a key. It generally only takes an electric power failure of just a few hours in the winter to bring Canadians back to reality.

Production and consumption of energy have skyrocketed in every country of the world, particularly in this century. Canada is not an exception and has become a major producer and consumer of energy. In fact, Canada has a particularly large appetite for energy. Canadians consume more of it on a per capita basis than any other country in the world! One reason for this is the climate: Canada has long, cold, dark winters, so large quantities of fuel are consumed to heat and light houses and other buildings. Another reason is this country's vast size: goods, services and people must travel vast distances from one end of the country to the other, and transportation consumes a great deal of energy. On top of that, many of Canada's primary industries are energy-intensive. A final factor is Canada's high standard of living: most Canadians use a lot of energy-consuming products.

In the past hundred years there have been enormous changes in the *types* of energy consumed in Canada. In 1890 coal and wood supplied almost all of the country's energy. They were used in transportation, in homes for heating and cooking, and in industry to power steam engines. Today, coal and wood are hardly used; Canada's large appetite for energy is now satisfied mainly by petroleum, natural gas, hydro-electricity and uranium.

In both quantity and variety, few countries in the world are so well-endowed with energy resources as Canada. As a result, Canada is a large producer and exporter of energy.

1. **Which factors contribute to Canada's large consumption of energy?**
2. **(a) Construct a bar graph to illustrate the sources of energy used in Canada in 1890 and in 1995. How has the energy mix changed?**
   (b) List five common uses of energy today which did not exist, or were of minor importance, in the nineteenth century.

4-50   Domestic demand for primary energy.

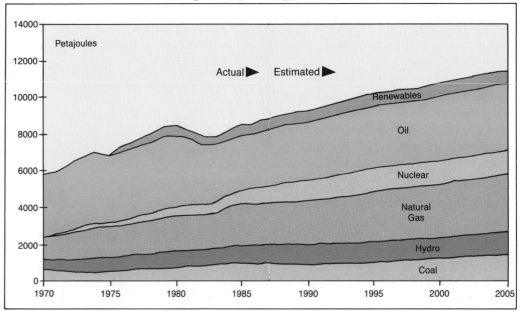

1. **Which primary energy source was the most important in 1970? What changes are projected for this source by the year 2005?**
2. **Which energy source is expected to be almost as important as oil in 2005?**
3. **What energy sources were almost unused in 1970? How is this projected to change in the future?**

Electricity: "White Gold"

Electricity is a **secondary energy resource**. This means that it is manufactured from other sources known as **primary energy sources**. Canada's primary energy sources include water, coal, uranium, oil and natural gas. Canada is considered to have an abundance of all these.

Canada's electrical capacity is one of the largest in the world. In 1995, Canada generated over 534 869 gigawatt hours of electricity (a gigawatt equals one billion watts) of which over 90 percent (or 498 896 gigawatts) was consumed nationally. The surplus was sold to the United States for $1 185 billion dollars. Quebec is the largest source of electricity exports, (42 percent in 1995) although Alberta, British Columbia, Ontario and Manitoba are also active in this area. One Alberta company, TransAlta Utilities was the 13th largest power marketer in the U.S. in 1996. New York State remains the largest single importer of Canadian electricity.

Most of Canada's electricity comes from **hydro-electric** or **thermal-electric generating stations**. In a hydro-electric station, the force of falling water is used to power the **generators** which produce electricity. In a thermal-electric station, fuel like coal, uranium, oil or natural gas is used to produce steam, which in turn, drives the generators. A **nuclear-powered station** uses the heat released from uranium as fuel.

Almost 62 percent of this country's electricity comes from hydro-electric stations, 21 percent from thermal-electric stations, 17 percent from nuclear-powered stations. A fractional amount comes from stations using **tidal power**, or the power of the ocean's waves.

Hydro-electricity is the cheapest form of power. Suitable sites and water sources are not evenly distributed across the country, however, so each province uses a different combination of energy resources to produce electricity. Newfoundland, Quebec, Manitoba, and B.C. rely almost exclusively on hydro-electric power. They all have large volumes of water available year-round, large, fast-flowing rivers, and steep topography.

Phase One of Quebec's James Bay project, alone, diverted seven rivers to double the flow of the La Grande River, creating over 200 dams and dikes to flood 10 000 plus km^2 of land. Phase two of the same project, generates over 15 500 megawatts of electricity per year. Manitoba has constructed similar hydro-electric projects along its Upper and Lower Nelson Rivers. Other massive hydro-electric development projects have been proposed, however Aboriginal, environ-

4-51 Electricity Prices Residential Sector – 1996.

	US Kw/hr
Sao Paulo, Brazil	16.47
Brussels, Belgium	15.98
Madrid, Spain	15.21
Boston, U.S.A.	12.60
Paris, France	11.69
Detroit, U.S.A.	9.65
Taipei, Taiwan	9.62
Houston, U.S.A.	8.51
Minneapolis, U.S.A.	7.98
Toronto, Canada	7.55
Ottawa, Canada	5.78
Portland, U.S.A.	5.76
Sydney, Australia	5.56
Calgary, Canada	5.40
Montreal, Canada	5.18
Vancouver, Canada	5.10
Winnipeg, Canada	5.01
Seattle, U.S.A.	4.42

Source: Natural Resources Canada & Canadian Electrical Association

4-52 Canadian Electrical Production/Consumption (gigawatt hours) 1995

	Hydro	Coal	Oil	Gas	Nuclear	Production	Consumption
Newfoundland	36 277	0	1 625	0	0	37 902	11 181
Prince Edward Island	0	0	16	0	0	16	931
Nova Scotia	917	7 053	1 414	0	0	9 551	10 032
New Brunswick	2 659	5 238	2 882	0	1 579	12 663	14 300
Quebec	166 903	0	216	25	4 511	171 630	175 201
Ontario	38 239	16 136	168	5 165	86 216	146 193	138 898
Manitoba	29 013	145	0	17	0	29 225	19 551
Saskatchewan	4 118	11 321	44	695	0	16 356	16 212
Alberta	2 213	43 465	80	5 374	0	52 313	51 337
British Columbia	49 829	0	1 347	5 347	0	58 017	60 430
Yukon	319	0	71	0	0	390	390
North West Territories	203	0	311	99	0	613	613
Total	330 690	83 358	8 174	16 697	92 306	534 869	498 876

(1960 Consumption was 109 304)

Source: Natural Resources Canada & Canadian Electrical Association

4-53 Canadian Electricity Use by Sector

	1995		1960	
Residential	134 967	(27%)	20 397	(19%)
Commercial	116 191	(23%)	12 632	(12%)
Industrial	214 673	(43%)	66 353	(60%)
Line Losses	33 145	(7%)	9 920	(9%)
Total	498 976		109 302	

Source: Natural Resources Canada & Canadian Electrical Association

1. What is the difference between a primary and secondary energy resource?
2. Why would the northeastern U.S.A. be the main market for Quebec's surplus electricity?

mental, political, and economic concerns have suspended many of them.

As Canada moves into the new millennium, federal, provincial and municipal authorities have re-examined the country's electrical utilities – de-regulating and restructuring many of them. This process, which is still ongoing, should make Canada's electrical energy more competitive within this country, and to markets outside the country. Some restructuring proposals suggest Canadian ultilities be privatized. Others propose alliances with private industry as a significant opportunity to generate new volume.

New Words to Know
Secondary energy resource
Primary energy resource
Hydro-electric generating station
Generator
Thermal-electric generating station
Nuclear power station

Hydro-electricity

Producing power from moving water is not new. Early in Canada's history sawmills and grist mills were built beside fast-flowing rivers or at waterfalls. The force of falling water was used to spin a paddle wheel, which in turn rotated a saw blade or grindstone. The same principle is now used to produce electricity in modern hydro-electric stations.

Generally, two kinds of sites are considered ideal for a hydro-electric power

1. Which type of electrical generation shows a steady increase in growth? Which has remained relatively constant?
2. Why might hydro production decline in a particular year?
3. State two reasons why it is necessary to purchase small amounts of electricity from outside the province. Which provinces are the likely source of these purchases?
4. Which types of electrical production are the least expensive, and the most expensive? Explain why.

4-54 A hydro-electric generating station.

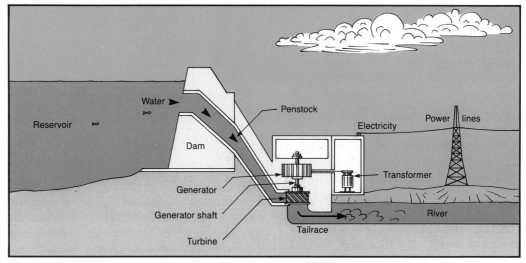

The majority of electricity used by Canadians is produced in hydro generating plants.

station: the base of a high waterfall, or the base of a dam built across a river. In the case of damming a river, the water is backed up behind the dam to create a reservoir or artificial lake. Water drops from the top of the dam through pipes called **penstocks** to the bottom of the dam, where the power station is located. In effect, an artificial waterfall has been created. As the water falls through the penstocks, it builds up tremendous force. This force causes a **turbine**, the modern equivalent of a paddle wheel, to spin. The turbine in turn spins a generator. The generator transforms the mechanical energy of the falling water into electrical energy. After the water has passed through the turbine it is released back into the river through the **tailrace**.

If a natural drop or waterfall is present, a dam is not required. Penstocks simply lead the water from the top of the waterfall to a power station that has been built on the riverbank.

Thermal-electricity

Regions which don't have the right resources to generate hydro-electricity must build thermal-electric stations, which are more costly. A thermal-electric station must be near a large supply of water for steam production and cooling; otherwise, they can be built almost anywhere. Most are built at or near their markets, in much less time than it takes to construct a hydro-power complex. The fuels burned, although they are non-renewable resources, are generally abundant in Canada. Coal, the dominant thermal fuel, is found throughout western Canada and the Maritimes in huge quantities.

The burning of fuels to generate electricity has its drawbacks. The gases and ash emitted from the smokestacks — primarily at coal-burning plants — are a major source of air pollution. Sulphur dioxide, one of the main gases emitted, is one of the major sources of acid rain. In recent years, new emission-control technology has significantly reduced the amount of sulphur dioxide and ash emissions into the atmosphere. Another disadvantage is cost. Water is basically free; fuel costs money to buy, transport and store. Thermal-electricity costs about four times as much as hydro-electricity.

1. Why is hydro-electricity considered "clean, safe and economical"?
2. Which concerns would you have if a large hydro-electric project were to be built near your community?
3. Compare the advantages and disadvantages of hydro-electricity and thermal-electricity.

New Words to Know

Penstock   Tailrace
Turbine

4-55 Revelstoke dam and power station in British Columbia.
A hydro-electric power station is non-polluting, safe, and uses water, a renewable resource, as the primary energy source. If a reservoir is created it can often be used for recreation and irrigation.

4-56 A thermal-electric generating station.

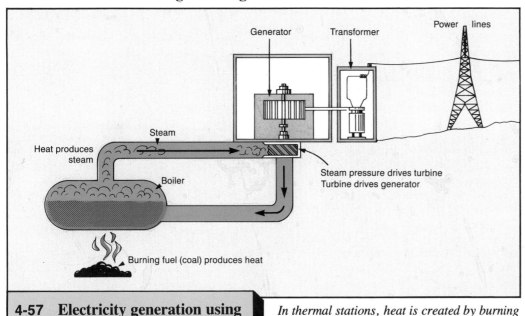

Generator   Transformer   Power lines

Steam

Heat produces steam

Boiler

Steam pressure drives turbine
Turbine drives generator

Burning fuel (coal) produces heat

4-57 Electricity generation using uranium as fuel.

Steam

Heat produces steam

Boiler

Coolant transfers heat to boiler

Uranium atoms produce heat

Nuclear power stations produce electricity from heat created by nuclear fission using uranium as a fuel.

In thermal stations, heat is created by burning a fuel to convert water into steam to drive the turbines.

Nuclear Power

Ontario, New Brunswick and Quebec generate electricity from nuclear power, with Ontario producing 90 percent of the total. Nuclear stations generated about 60 percent of Ontario's electricity in 1996! This is spectacular growth, since the first nuclear station in the province only started generating power in 1962. The province's nuclear-power industry has grown to the point where Ontario Hydro is recognized as a world leader.

Essentially, a nuclear station operates the same way as a thermal station: water is heated to create steam, which is then piped to turbines and generators that produce electricity. The main difference is that a **nuclear reactor** provides the heat that produces the steam.

Nuclear reactors produce heat by splitting uranium atoms. The reactor is loaded with thousands of **fuel bundles** containing concentrated uranium. Each bundle is about 10 cm in diameter and 50 cm long and weighs 25 kg. Each one provides an amazing amount of power — as much electricity as 380 t of coal or 800 barrels of oil.

Darlington Nuclear Generating Station is one of North America's largest energy projects and was completed in 1992. It took $8.8 billion and 14 years to build and can provide over 3 500 000 kw of electricity. This is enough to serve a city of three million people at half the cost of thermal-electricity. Darlington is less than an hour's drive from Toronto. The site was chosen because it was close to its urban market, and close to Lake Ontario, which supplies it with cooling water. It is also downwind from Metropolitan Toronto — an important safety consideration.

Nukes — Yes or No?

Nuclear energy is controversial. In an opinion poll conducted in 1988, about half of all Canadians expressed some fears about it. Some environmental and citizens' groups contend that the dangers associated with nuclear energy outweigh any advantages. They argue that technology has yet to find a completely safe way to dispose of radioactive wastes, which can be deadly. They are also concerned about the danger of a nuclear accident. In 1986, at Chernobyl in the Ukraine, an explosion and fire in a nuclear reactor killed 31 workers and released vast quantities of radioactivity into the atmosphere. The worldwide

229

effects of Chernobyl will not be known for decades. However, some scientists conducted research on death rates in the U.S.A. during the summer of 1986. They found that wherever the winds concentrated the radioactivity, death rates increased by as much as five percent. There is no absolute proof that radioactive fallout was the sole cause of these increases, but the suspicion is there.

Finally, some people raise questions about the costs of such plants. They suggest the money might be better spent on researching and on developing alternative sources of energy, or on developing and promoting energy conservation. Certainly, conservation would considerably reduce the need for more and more thermal plants, as well as the dangers they represent to the environment.

Those who favour nuclear energy present a different set of arguments. Some are given in the following newspaper article.

1. **Your provincial electrical utility has chosen a location near your community to build a nuclear power station. A local meeting is planned to discuss the project.**
   (a) As spokesperson for the utility, prepare a report, for distribution at the meeting, in support of the plant.
   (b) As spokesperson for a local citizen's group, prepare a report to oppose the project.
2. **Organize a class debate on this issue, with half the class on each side.**

New Words to Know

Nuclear reactor   Fuel bundle

Crisis at Hydro: CANDU Falls on Hard Times

Problems have plagued Ontario's nuclear program

CANDU, Canada's much-touted nuclear power generator, has long been advertised as the best in the world. But it has fallen on hard times in Ontario.

Ontario Hydro, whose nuclear program has been plagued by leaks, malfunctions and management problems, has decided to cut its losses and close the older seven of its nuclear generators-about one-third of its nuclear power-producing capacity - likely for good.

One other nuclear generating unit at the Bruce station was "laid up" or closed last year without fanfare because it needed costly repairs.

The utility has announced it plans to spend up to $ billion to replace the power lost by closing these units and to refurbish its newer generating units at the Pickering, Darlington and Bruce nuclear stations.

While Hydro has been saying it can provide 65 per cent of Ontario's power needs from nuclear sources, it has actually provided only about 50 per cent in recent years because reactors have been out of service for maintenance and retooling, as well as because of leaks and other problems.

The Pickering nuclear plant, with all eight reactor units running in top order - something that hasn't happened much in the last decade - could provide 20 per cent of Ontario's power needs; that's enough power to light all of Toronto.

The CANDU system is considered by many to be safer than the U.S. system, which uses enriched uranium and ordinary water as a coolant.

The Pickering nuclear station has had a number of radioactive leaks and spills in recent years that have caught the attention of the federal nuclear regulatory agency.

But, most experts agree, there has been no danger to Hydro employees or the public.

In fact, they say, if you look at nuclear power production as an industry, compared to the auto industry or coal mining, it has a much better worker and public safety record.

But the problems that have forced Hydro to reduce its nuclear capability drastically were not expected to occur so soon.

Atomic Energy of Canada Ltd., which sold Hydro the units built over the last 27 years, told the utility they would have a 30-year lifespan.

However, Tom Adams of Energy Probe says that the oldest of the Pickering units - Unit 1 in the A plant - is not far off the 30-year mark, at 26 years of operation.

4-58   Ontario Hydro's Darlington nuclear power station has the capacity to provide enough electricity to serve a city of three million people. *Its site on the shore of Lake Ontario near Bowmanville was chosen because of its proximity to the energy markets of Ontario, good transportation access, an abundant supply of cooling water, relative isolation, and excellent bedrock for the station's foundation.*

Adams said that, in 1982, Hydro was faced with a major reactor building program at Pickering and Darlington, and arbitrarily boosted the life expectancy of the units to 40 years as an accounting move.

Cracks appeared prematurely in reactor pressure tubes at Pickering in 1983, resulting in $1 billion in repairs to the A plant there.

The same retubing was due to be done at the Bruce station in Kincardine, but now has been put off indefinitely.

Ontario Hydro president Allan Kupcis resigned at the start of the utility's board meeting on Tuesday, just before the critical report on the utility's nuclear division was tabled.

That report, authored by Carl Andognini, the utility's chief nuclear officer, found the Ontario nuclear stations' performance to be "minimally acceptable."

But it also said that they would be allowed to continue operating in this condition in the United States.

The Atomic Energy Control Board, Canada's nuclear regulatory agency, criticized "human errors and operational failures" when it granted Hydro a limited nine-month license renewal for Pickering in June.

A major embarrassment for Hydro lately was the revelation in The Star recently, that 1,000 tonnes of copper and zinc washings from the brass heat condensers at Pickering and other provincial power-producing facilities had washed into Lake Ontario over a 10-year period, without notification to the proper environmental authorities.

Rather than continuing to pump money into the aging and failing nuclear units, Hydro now has decided to close them down and concentrate its financial resources on upgrading the remaining newer operating nuclear generators.

Meanwhile, the provincial officials have said that Hydro's monopoly as the province's supplier of electricity likely will end in the near future as the North American power grid is thrown open to competition.

Previously, other potential power suppliers, such as Ajax Hydro which wants to produce local power from a steam plant, have been thwarted by Hydro's enforcement of its monopoly on power production.

Likewise, a private firm that wanted to produce power at a refurbished private dam on the Trent River just north of Trenton also has had a long and frustrating battle with Hydro.

Nuclear watchdog groups, such as Durham Nuclear Awareness, say one reason Hydro is cutting its nuclear losses now is that power produced in this way will not be competitive when the electricity market is thrown open in competition from private producers.

The Toronto Star, August 1997

Petroleum and Natural Gas

Petroleum and natural gas play the most important role in meeting Canada's energy needs. **Petroleum** — which is crude, unrefined oil — along with coal and natural gas are **fossil fuels**. They can all be burned to produce energy.

Most Canadians are well aware how vital oil and gas are, as fuels for homes, schools, factories and transportation. But in addition, these resources provide us with a multitude of other very useful products.

Crude oil has few uses until it has been processed in a refinery. Modern refineries are complex, highly sophisticated and fully automated. They transform crude oil into various fuels, and into other products for the **petrochemical** industry. Petrochemicals are chemicals derived from the processing of oil and gas. Plastics, synthetic fibres, synthetic rubber, cosmetics, medicines, paints, insecticides, fertilizers and many other things, are all made of petrochemicals.

Hibernia

4-59 Hibernia

The Hibernia oil field, discovered in 1979, is located 315 km southeast of St. John's, Newfoundland in 80 meters of water. The total field is estimated to contain 3 billion barrels of oil, of which some 615 million barrels are recoverable. The field also contains some 3.5 trillion cubic feet of natural gas. Hibernia is a joint project of Mobil Oil Canada, Chevron Canada Resources, Petro-Canada, Murphy Oil, Norsk Hydro, and the Canada Hibernia Holding Corporation. Through March 1997, Hibernia has spent $5.4 billion in exploratory, construction and drilling costs. 5 800 people were employed at the Hibernia Bull Arm project during peak construction. This illustration depicts the Hibernia platform or Gravity Base Structure (GBS). It weighes 1.2 million tons and is 224 meters high. The platform is used for drilling, production, storage and living quarters. Approximately 185 people work on board. Estimated oil recovery from the Hibernia resevoirs (Hibernia and Avalon) is 185 barrels per day, yielding a projected field life for this platform of 18 years from December 1997, the date of first oil recovery.

The Origins of Oil and Gas

No one is sure how oil and natural gas were created. Geologists believe that oil and gas may have originated from organic matter, both plant and animal. About 500 million years ago, on the Canadian prairies, a vast sea existed which swarmed with billions of tiny plants and animals. When they died they sank to the seabed, and turned into a thick layer of organic material. Over the eons the organic material became buried by layers of mud, sand and shells.

Over many millions of years, these sediments and organic materials changed. The sediments became compressed and turned into sedimentary rock: mud became shale, sand became sandstone and the shells became limestone. The organic material was converted by pressure, heat and bacterial action into **hydrocarbons**, the basic elements of oil and gas.

As the earth buckled and shifted, the hydrocarbons and water were squeezed out of the rock in which they were formed, and migrated into **permeable** rock, which is porous. This means that fluids and gases can pass through it. Oil and gas are held in porous rock in much the same way as a sponge holds water. They move through the rock until their flow is blocked by impermeable rock, which does not allow the passage of fluids and gases.

A porous rock formation which contains concentrations of hydrocarbons and water is called a reservoir. This does not mean that the oil exists in lakes under the earth. An oil reservoir actually contains billions of tiny droplets of oil in the "pores" of the rock.

Finding an oil reservoir buried deep in the earth's crust is not easy. Sometimes oil seeps through cracks to the surface where it can be seen, but this is rare. To find oil, geologists must usually rely on more sophisticated means. Aerial photo-graphs, surface rock samples and instruments to measure the gravity and magnetic properties of the underlying rocks are commonly used. The most useful tool is a **seismic reflection study**. In this procedure, sound waves are used to produce pictures of the underground rock formations. This enables geologists to identify potential traps for oil and gas.

But generally the only way to *really* know what is below the surface is to drill a well. Drilling an exploration or "wildcat" well in unproven territory is a risky and expensive gamble. Often millions of dollars are spent on drilling only to discover a dry hole, a well that cannot produce oil or gas in commercial quantities.

4-61 Diagram of a seismic reflection study

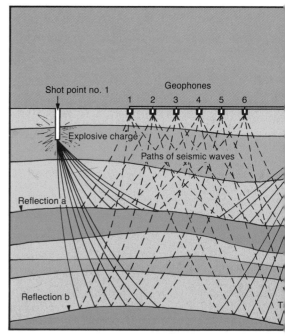

In practice, the wave produced by each blast would reflect off every stratum. The data

4-60 Common oil and gas traps.

(a) The stratigraphic trap.

(b) The anticlinal trap.

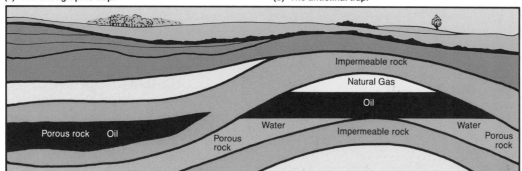

In this trap porous rock tapers off in the impermeable capping rock. This form of trap is common in Saskatchewan and Alberta.

An upward arching of rock strata is called an **anticline**. *Oil and gas can collect in the permeable rock beneath the dome of impermeable rock. Oil floats above the denser water, which is often salty. Gas collects above the oil.*

Thousands of oil and gas wells are drilled annually across Canada. Less than half strike oil and less than a quarter find gas.

1. **Besides providing fuel, how has petroleum become a necessary part of modern society?**
2. **What makes it possible for oil companies to drill hundreds of dry wells, and not go bankrupt?**
3. **Geologists refer to oil as being "trapped" in the earth's crust. Explain what they mean.**
4. **Why is it incorrect to think of an oil reservoir as an underground lake?**

New Words to Know

Petroleum	Permeable rock
Fossil fuels	Seismic reflection study
Petrochemicals	Anticline
Hydrocarbons	

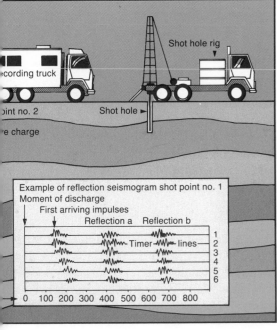

recorded can be very complex and difficult to interpret.

4-62 Oil and gas resources.

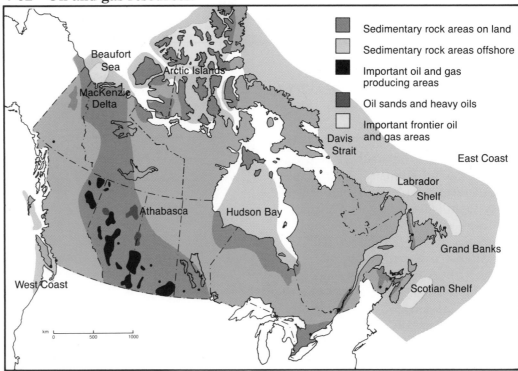

Oil and natural gas are found in areas known as sedimentary basins.

Location of Canada's Oil and Natural Gas Resources

Geologists have extended their search for new oil and gas fields to all regions of Canada; however, the heart of the industry is still the western sedimentary basin in Alberta and Saskatchewan. In 1995 these two provinces together produced 96 percent of Canada's crude oil. Alberta alone produced almost 80 percent. Outside of western Canada, very little crude oil is produced: there is none at all east of Ontario. A similar situation exists for natural gas. In 1995, Alberta produced about 87 percent of the country's natural gas.

For many decades Canada has exported oil, primarily to the USA; in 1996 that country purchased 53 percent of Canada's crude oil, for over $9.8 billion. Oddly enough, at the same time Canada *imported* over $6.6 billion worth of crude oil! There are several reasons why. For one thing, it is more profitable for western Canada to export crude oil to the western and central U.S.A., than to supply Canadian markets east of Montreal. For the Maritimes and eastern Quebec, it is cheaper to import oil from abroad than to transport western Canadian oil there. Also, much of what is exported is heavy oil, a thick, sticky form of crude oil with a high asphalt and sulphur content. When refined, it produces lower quantities of lighter, higher-demand products such as gasoline. Canada has only limited refining facilities and demand for this type of oil.

How Much Oil and Natural Gas Does Canada Have?

Canada's reserves of oil and gas are buried beneath the earth's surface. However, using various methods, geologists can make careful estimates of how much is left of them. In 1988 the National Energy Board concluded that Canada's total reserves, from both conventional and frontier areas, were substantial. **Conventional areas** are the Prairie provinces, B.C., southwestern Ontario, the southern Yukon and the Northwest Territories, areas where oil and gas are easiest and cheapest to find and develop using ordinary drilling methods. **Frontier areas** are northern and offshore areas in the Beaufort Sea, the Mackenzie Delta, the Arctic Islands and off the eastern shores of Newfoundland and Nova Scotia. Frontier reserves have remained largely undeveloped, because exploration, drilling and transportation there are very expensive, difficult and dangerous.

Canada is self-sufficient in natural gas. In 1996 about 50 percent of production was exported to the USA, which is Canada's only export market for gas. That market is growing, however and by 2000, 13 percent of the natural gas used by Americans will likely be Canadian.

Natural gas reserves continue to grow in both conventional and frontier areas, and should be large enough to meet demand well into the next century. On

4-63 A drilling rig in Alberta.
Drilling for oil is a costly and risky undertaking. However, it is the only sure method of verifying the presence and quantity of oil or gas.

the other hand, reserves of light crude oil from conventional areas — the most desirable type — are declining. At current levels of demand, Canada's supply will run out by the end of the century. Unless more is discovered in conventional areas and frontier areas are developed, Canada will become increasingly dependent on imports.

1. **Why is Canada both an importer and exporter of crude oil?**
2. **Explain the phrase "Canada is presently self-sufficient in oil and gas."**
3. **Why does heavy oil have a limited market in Canada?**

New Words to Know
Conventional oil and gas areas
Frontier oil and gas areas

Non-Conventional Resources

Canada is fortunate in having alternative sources of oil for the future. These are the heavy-oil deposits of Alberta and Saskatchewan, the Alberta tar sands, and frontier deposits.

Heavy Oil Heavy-oil deposits were discovered in the 1930s by oil-exploration crews searching for ordinary light oil. Geologists estimate that there are millions of cubic metres of this molasses-like oil buried in the heavy-oil belt surrounding Lloydminster in Alberta and Saskatchewan. In 1995 about 30 percent of Canada's crude-oil production was heavy oil. However, only a small portion of the heavy oil is thin enough to pump to the surface by conventional methods. **Enhanced recovery techniques** — particularly steam injection, which heats and thins the oil — have increased production, but they are costly. Also, sand sticks to the oil, clogging well bores and production equipment, so that they must constantly be shut down for cleaning.

After many years of research and planning an upgrading plant has been built in the Lloydminster area. This multi-billion dollar plant processes heavy oil to a point where it can be used in Canadian refineries to produce more valuable products.

Alberta Tar Sands Large amounts of hydrocarbons occur in in the form of tar sands, a thick, sticky mixture of sand, clay, water and a tar-like oil called **bitumen**. The tar sands of northern

Alberta are among the largest known reserves of oil in the world. At Canada's present rate of consumption, they represent a 250-year supply.

Recovering bitumen from tar sands is complex and difficult. It is too thick and sticky to be recovered by conventional oil wells. Tar sands near the surface — those with less than 45 m of overburden — are mined by the open-pit method. Mammoth earth-moving machines strip off the overburden; then a huge bucketwheel conveyor-belt system carries the oil sands to a nearby extraction plant. There the bitumen is separated from the sand; however, it is still too thick, and too contaminated with sulphur and nitrogen, to be used in conventional refineries. It must be upgraded into a lighter, purer, thinner oil, which is called **synthetic light crude**.

In 1996 the oil-sands mining plants of Syncrude and Suncor produced the bulk of Canada's synthetic crude. These two plants, both located near Fort McMurray, accounted for about 20 percent of Canada's total light crude oil. Unfortunately, most tar sands deposits are too far below the surface to be mined. Deeper deposits require costly techniques similar to those used in heavy-oil production. Until more economical methods are found, most of Canada's synthetic crude oil will be obtained by surface mining.

Frontier Resources Since the 1960s, substantial reserves of oil and gas have been found in Canada's frontier regions. The major discoveries have been made in the Mackenzie Delta/ Beaufort Sea area, the Hibernia oil field off Newfoundland, and the Sable Offshore

4-64 Mining the tar sands. *Giant bucketwheel excavators are used in Syncrude's open-pit mine at Fort McMurray. They move millions of tonnes of tar sand to extraction and upgrading plants, where the sand is removed and synthetic crude oil is produced.*

natural gas project off the coast of Nova Scotia. Development of these resources has been hindered by the extremely high development costs and the harsh environment.

These must be considered production areas for the next century. They will require new and special production techniques and transportation facilities. Hibernia's development costs are estimated at over $5 billion, the Beaufort Sea's at $4 to $5 billion, and Sable Offshore at $3 billion. These huge capital costs make it difficult for Canada to produce oil and compete with other low-cost world producers.

Coal

Like petroleum and natural gas, coal is a fossil fuel. This solid, black mineral was formed millions of years ago in ancient swamps from thick deposits of decayed vegetation. Later, inland seas flooded and covered these swamps with sediments. With the passage of time, the heat and pressure of the overlying sediments compressed and changed the vegetative matter into coal.

Canada has substantial coal deposits, 98 percent of which are in western and northern regions, far from the major markets in Canada. In 1996, Canada produced less than ten percent of the world's output, and nearly half of this was exported, mainly to Japan and South Korea. Most of this output came from Alberta (47 percent), British Columbia (33 percent), and Saskatchewan (14 percent) with the remaining 6 percent from Nova Scotia and New Brunswick.

Until the 1950s, coal was the major source of energy in Canada. Since that time it has been largely replaced by oil and gas to heat homes and to power factories and various modes of transportation. Currently its main uses are for generating electricity in thermal plants, and making iron and steel.

1. **What are tar sands?**
2. **Why is open-pit mining and upgrading used in the tar sands rather than the conventional well-and-pump techniques?**

New Words to Know
Enhanced recovery techniques
Bitumen
Synthetic light crude oil

Water

Fresh water is essential for life. The human body is 70 percent water and requires two litres every day just to function properly. Much of our food is comprised of water; a typical fast-food meal of hamburger, french fries and a cola contains at least 80 percent water. Although not essential for survival, our North American lifestyle of daily showers, flush toilets, dishwaters, green lawns and car washings is highly dependent on large quantities of water. The average Canadian uses between 250 and 300 L daily. In comparison, the average Briton consumes 180 L each day, a rural villager in India, only 25. No wonder immigrants are amazed and sometimes shocked at the seeming waste and misuse of this precious resource.

Of all the water on earth, 97 percent is salty and of little use to humans except for transportation. Only three percent is fresh, and most of this is

4-65   The world's water resources.

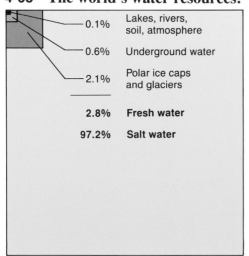

0.1%	Lakes, rivers, soil, atmosphere
0.6%	Underground water
2.1%	Polar ice caps and glaciers
2.8%	**Fresh water**
97.2%	**Salt water**

4-66   The hydrologic cycle.

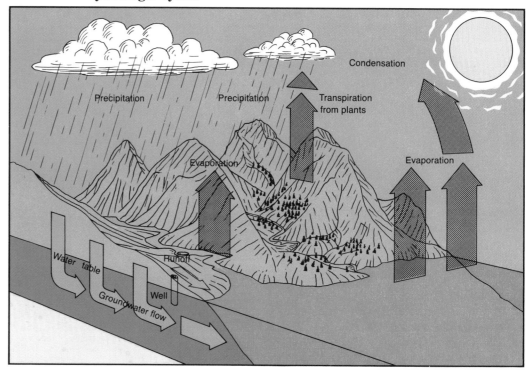

The hydrologic cycle is the constant changing or recycling of water. **Transpiration** *is the action of trees and plants giving off water vapour through their leaves.*

buried deep beneath the earth's surface or frozen in the polar icecaps. Less than one-hundredth of one percent of the world's water is accessible for human use.

Canada is very fortunate. Nearly eight percent of the country's surface is covered by lakes, rivers and streams, another 12 percent in **wet-lands** such as swamps, marshes and bogs. Canada has 20 percent of the world's surface fresh water. Most Canadians live close to a source of it. Perhaps that is why they too often take it for granted that water will be available in limitless quantities whenever it is needed. Unfortunately, not all of this

country's water is usable or accessible. Much of it is permanently frozen in the polar regions, or flows north to the Arctic Ocean away from the population centres. Nevertheless, Canada's fresh water supply is the envy of virtually every other country in the world.

1. **"Water is the liquid of life." In your own words, explain this statement.**
2. **(a) Why is Canada very fortunate in terms of its supply of fresh water? (b) Why is Canada less fortunate in the geographic distribution of its water?**
3. **How does the hydrologic cycle make water a renewable resource?**
4. **Some environmentalists fear that the**

destruction of tropical rain forests of South America may disrupt the existing hydrologic cycle. How might this destruction affect the cycle?

New Words to Know
Wetlands
Hydrologic cycle
Transpiration
Ground water

Lake Patterns

Canada has more lake area than any other country in the world. The Great Lakes, which Canada shares with the U.S.A., contain 20 percent of this planet's surface fresh water. Lakes are important because they collect and store water during times of heavy **runoff** and high precipitation. This reduces the problems of downstream flooding, and maintains the flow in rivers and streams in drier periods. Lakes are so useful that dams have been built to create artificial ones. Such lakes have many uses — they store water for use in times of drought, prevent flooding and provide water for irrigation, power and other uses.

Map 4-67 shows only Canada's largest lakes. There are many other large lakes besides the Great Lakes. For example, Great Bear Lake and Great Slave Lake are actually larger in surface area than either Lake Ontario or Lake Erie. Besides these huge lakes, the landscape is dotted with hundreds of thousands of smaller lakes and ponds from coast to coast.

The lakes on the outer edge of the Precambrian Shield were formed when the melt water from retreating ice sheets was trapped in large depressions shaped

4-67 Canada's largest lakes.

All lakes shown in colour are more than 1000 km² in size. On an outline map of Canada, locate and label these lakes.

by the glaciers. Many thousands of smaller lakes follow this pattern. Except for the Great Lakes and Lake Winnipeg, most of them are in the North.

There are few large lakes in the southern Prairie provinces and British Columbia. B.C. is too mountainous, the prairies too flat, though many small ponds or **sloughs** exist on the latter.

Runoff Patterns

Hydrologists, scientists who study surface and ground water, consider runoff a good indication of the supply of fresh water. They estimate that half of Canada's precipitation ends up as runoff.

Generally, areas of very high precipitation are also areas of very high runoff. Coastal B.C. is a perfect example: high precipitation, steep, barren mountain slopes and low temperatures at higher elevations give this region the highest runoff rates in Canada. At the other end of the scale, the southern prairies have low annual precipitation, high summer temperatures, a flat landscape and a heavy cover of soil and vegetation. All of this makes for low runoff rates. In years of exceptionally low precipitation — such as the Dust Bowl of the 1930s or the drought of 1988 — many parts of the southwest prairies had no recorded runoff! When

237

this happens, it can have disastrous effects on agriculture.

The Far North is a curious region. Total precipitation is much lower there than in most other parts of Canada; at the same time the amount of precipitation which becomes runoff is relatively high. There are a number of reasons for this. The low temperatures reduce the rate of evaporation, and also restrict the growth of vegetation so that less transpiration takes place. As well, less water is absorbed by the ground, either because the soil cover is thin or non-existent, or because permafrost prevents absorption.

1. (a) **What advantages do lakes provide?**
   (b) **Explain how the distribution of large lakes is not to the advantage of most Canadians.**
2. **How were many of the large lakes in the Precambrian Shield formed?**
3. (a) **Why does coastal B.C. have the highest runoff rate in Canada?**
   (b) **Which problem might occur in coastal B.C. if a river experienced a period of exceptionally high runoff?**
   (c) **Why is the proportion of runoff to precipitation quite high in the Arctic?**
4. (a) **Locate and name three river basins shared between Canada and the U.S.A.**
   (b) **Which problems could arise from sharing a river basin?**
5. (a) **Name two Canadian cities or towns located at a river mouth.**
   (b) **Which advantages would these cities have because of their location?**
   (c) **Why would the mouth of the Mackenzie be a less desirable location for a large urban settlement than the mouth of the Fraser? Give at least three reasons. Use an atlas when answering the question.**
6. **Why is the source of a river a less likely choice for the location of settlement or industry, than its mouth?**

238

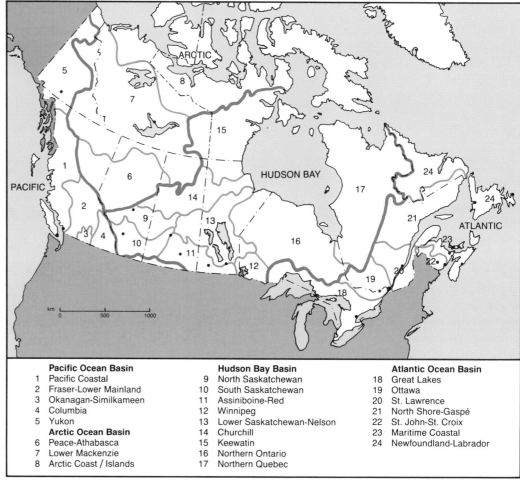

4-68 Drainage basins of Canada.

Pacific Ocean Basin		Hudson Bay Basin		Atlantic Ocean Basin	
1	Pacific Coastal	9	North Saskatchewan	18	Great Lakes
2	Fraser-Lower Mainland	10	South Saskatchewan	19	Ottawa
3	Okanagan-Similkameen	11	Assiniboine-Red	20	St. Lawrence
4	Columbia	12	Winnipeg	21	North Shore-Gaspé
5	Yukon	13	Lower Saskatchewan-Nelson	22	St. John-St. Croix
Arctic Ocean Basin		14	Churchill	23	Maritime Coastal
6	Peace-Athabasca	15	Keewatin	24	Newfoundland-Labrador
7	Lower Mackenzie	16	Northern Ontario		
8	Arctic Coast / Islands	17	Northern Quebec		

A drainage basin includes all the land drained by a river, or several rivers. The sources of the tributaries which flow into the main rivers can be seen. The tributaries join the major rivers at confluences, or junctions.

1. **Which ocean basins include rivers which flow to the north?**
2. **What is the source of many of the river basins in the prairies?**
3. **Explain why the Pacific basin has a relatively large water flow despite its small size.**
4. **Why has the large Arctic basin a relatively low water flow?**

New Words to Know
Runoff
Slough
Hydrologist
Drainage basin

Water Demand

In the past, Canada's small and largely agrarian population placed few demands on the water supply. Today, as population, urban centres, industry and agriculture expand, so does the demand for water. Still, there is little fear of actually running out as yet. Some parts of Canada, such as the semiarid prairies, do at times suffer water shortages. This is a distribution problem. Fresh water is not always available where people want it, when they want it. The Great Lakes basin contains nearly 70 percent of Canada's people and about three-quarters of its industry. The Great Lakes supply drinking water to more than 35 percent of all Canadians. In the very dry summer of 1988, the Toronto region experienced water shortages because the filtering system was overworked. This was not a water *shortage*: the problem was the lack of facilities to provide good-quality water.

Withdrawal water is water which has been taken from its natural source, which may be a lake, a river or an underground source, depending on the region. Usually this water is returned to its source, although not necessarily in its original pure condition. It may return warmer, or containing chemicals, wood fibres or untreated sewage. Sometimes part of the withdrawal water is lost. For example, when a house in Toronto uses withdrawal water from Lake Ontario for washing, cleaning, etcetera, most of the water is returned to the lake through the sewage system. However, if the water is used to sprinkle a lawn, some is "lost" into the atmosphere through evaporation and transpiration. This is classified as **consumed water**.

4-69 **Nanticoke is the site of a large Ontario Hydro thermal-electric plant in southwestern Ontario, on the shore of Lake Erie.**

Thermal-electric plants are the single greatest users of water in Canada. They require enormous amounts of water for steam production, and particularly for cooling. Most of these plants are in the heavily developed Great Lakes basin.

The Pacific Coastal and St. Lawrence river basins are also population and manufacturing centres, but use much less withdrawal water. Both have an abundant supply of hydro-electricity, which does not consume water.

Manufacturing also withdraws great quantities of water — for processing materials, for cleansing, for steam generation, for cooling equipment and for diluting waste materials. Two of the largest users of water are the pulp-and-paper and steel industries. To produce a tonne of newsprint requires 230 000 L of water. To produce a tonne of steel requires 180 kg of steam to power the equipment alone. Other industries use water as an ingredient — for chemicals or soft drinks, to cite two examples. Most of the water withdrawn by indus-

239

try is returned to its source. Unfortunately, this returned water often contains pollutants.

Agriculture withdraws much less water than the thermal-electricity and manufacturing sectors. However, in the semi-arid South Saskatchewan and Okanagan-Similkameen River basins, a great deal of water must be withdrawn to irrigate fruit orchards and vegetable fields. Though agriculture uses relatively little water, it *consumes* a higher proportion of what it uses than any other sector. Over two-thirds is lost to evaporation and transpiration. During dry years, irrigation in Alberta accounts for over half the consumed water in Canada!

The three sectors already mentioned, along with the mining industry, account for the bulk of water withdrawal in this country. Of course, Canadians use water for many other purposes, which are classified as non-withdrawal uses. Hydro generation, transportation and recreation all use water, but they do not remove it from its source.

1. **Explain the difference between withdrawal water and consumed water.**
2. **Why is water withdrawal highest in the Great Lakes basin?**
3. **Name five or more ways manufacturing uses withdrawal water.**
4. **"The amount of available water varies depending on the geographic regions of Canada and the nature of the lakes and rivers themselves." Explain this statement in your own words.**

New Words to Know
Withdrawal water
Consumed water

Ground water

Where does fresh water come from? The average Canadian would probably talk about such features as lakes, ponds, rivers and streams, since these are obvious in everyday life. Yet buried beneath the earth's surface is a very large but hidden supply of fresh water — **ground water**. Scientists estimate there is an astonishing 37 times more water underground than on the surface!

The importance of ground water to Canada can be seen in the following:

- Twenty-six percent of all Canadians (6.2 million people) rely on ground water for their domestic water supply.
- Eighty-two percent of rural Canadians (nearly 4 million people) rely on ground water for domestic or household purposes.
- Prince Edward Island depends 100 percent on ground water.
- Nova Scotia, Saskatchewan, New Brunswick and the Yukon depend on ground water for 50 percent of their water supply.
- Thirty-eight percent of all municipalities rely partly or totally on ground water.
- Nearly all the water used in the raising of livestock comes from ground water.

Water is retained underground in much the same way as it is in the small holes of a sponge. When precipitation falls to earth some of it is absorbed into the soil, or seeps into cracks in the rock. Gravity keeps drawing it down. As the pores at lower levels become filled, the level of the underground water rises. The uppermost level of the ground water is called the **water table**. Occasionally the water table rises high enough to reach the surface. This is one way springs form.

People extract ground water by drilling or digging wells. Usually a well is drilled many metres below the existing water table, to ensure an adequate water supply in case the water table should vary from one year to the next. In Canada, there are nearly one million recorded wells, supplying ten percent of this country's water needs.

1. **Which is the most common use of ground water and who are the largest users?**
2. **Why does Prince Edward Island rely so heavily on ground water?**
3. **Explain how precipitation ends up as ground water.**
4. **(a) Why might a well go dry?**
   (b) In which season might this well again contain water? Explain your answer.

New Words to Know
Ground water   Water table

Save It Or Sell It?

With such an enormous supply of fresh water, it seems strange that anyone would think to transfer water from one part of Canada to another. Yet in Canada there are several large projects involving the diversion of water from one drainage basin to another. The James Bay Project in Quebec, the Churchill River Diversion in Manitoba, and the Churchill Falls Project in Labrador are the largest of these. There are more than 50 smaller ones. Most have one purpose — to increase the flow of a river in order to generate more hydroelectric power.

Very few interbasin transfers provide water for municipalities, industries or irrigation. A number of schemes of this type have been proposed, however. Most of these would take "surplus" water from the Arctic or the Hudson Bay ocean basins and divert it southward. It would then be used in water-deficient areas of Canada or the U.S.A.

These schemes used to be dismissed as futuristic, too costly or just harebrained. However, the frequent dry periods of the 1980s and the concern over water during the free-trade debates, have renewed interest in them.

In the west, a plan called the North American Water and Power Alliance (NAWAPA) has been proposed. Water from rivers in the Northwest Territories, Yukon, Alaska and British Columbia would be diverted through a series of dams, reservoirs and canals to water-deficient areas on the prairies and in southwestern and midwestern U.S.A. In the east, another massive proposal called the Grand Canal Plan would build a dam across the mouth of James Bay, allowing the bay to fill with fresh water from the rivers flowing into it. This would be diverted by huge canals to the Great Lakes, then distributed to dry areas on the Canadian prairies and in the U.S. southwest. This plan would provide water to nearly 200 million Canadians and Americans, but at a cost of at least $100 billion!

1. (a) **Which ocean basins do most water diversion plans involve?**
   (b) **Why would you expect this to be the case?**
2. **Write a letter to the Minister of the Environment opposing or supporting the Grand Canal project.**

4-70 **Exporting our water** *(printed with permission from the Toronto Star)*

Exporting our water

No alarm bells went off. That's the most disturbing thing about a Sault Ste.Marie entrepreneur's plan to export water to Asia.

It didn't occur to the two provincial bureaucrats in Thunder Bay who granted the permit that there might be far-reaching implications. They treated it as a routine application to draw water from Lake Superior.

It didn't occur to their boss, Environment Minister Norm Sterling, to slam on the brakes. Water exports aren't his responsibility.

It wasn't until the controversy landed in Ottawa that officials appreciated the seriousness of the case. Foreign Affairs Minister Lloyd Axworthy pledged to find a way to block the sale.

In the end, government action wasn't necessary. John Febbaro, the businessman who triggered the controversy, offered to surrender his permit if Ottawa, Ontario and the United States all agreed to ban Great Lakes water sales.

The three governments are working on it. Canada asked the U.S. to jointly refer the question of Great Lakes water exports to the International Joint Commission, which oversees boundary waters. It will report back in the fall.

This is not the first time commercial interests have tried to turn Canadian water into a profit-making commodity.

In the early '90s, a similar water diversion project was halted, at the last moment, in British Columbia. A local company won a permit to ship water to drought-plagued California. Soon, dozens of others were lining up for permits. Worried, the B.C. government banned the export of water in anything larger than a 20-litre container.

These incidents fit the scenario that trade critics warned of, a decade ago, when Canada and the United States negotiated their historic free trade agreement.

They feared that private companies would start selling our fresh water to the world's parched regions; and once begun, such exports would be impossible to stop.

Federal officials dismissed their concerns. Canada would never allow water to become a tradeable commodity, they insisted. There was no need for a specific law banning bulk water sales.

It is clear they were wrong. All it took was a couple of bureaucrats with tunnel vision and a provincial minister who didn't look beyond his own jurisdiction to prove that we do need reliable safeguards.

Fortunately, this case seems headed toward a benign conclusion. But there's a lesson here: A worrisome precedent was set - almost by accident.

We cannot afford to be so careless with a precious inheritance.

Concern for Water Quality

Although Canada has an abundance of fresh water, the quality of this water has been threatened by human activities. Pollution is most serious in eastern Canada, but no part of the country is immune. Researchers have found pesticides in rivers across the country and even in the Arctic basin.

Pollution is caused by people. Municipalities, manufacturers, farmers and mining and forestry companies have for years used lakes and rivers to dispose of their wastes. Some waters have become so contaminated that they pose a health hazard for humans and aquatic life. In some cases, pollution is actually killing or severely damaging wildlife. No one human sector or type of pollution can be pinpointed as "the main polluter" of Canadian waters. Instead, the cause varies from region to region. Two of the worst pollutants are **toxic chemicals** and **acid rain**.

Toxic Chemicals   Toxic (poisonous) chemicals are found throughout Canada's river basins. Nowhere are they more prevalent than in the Great Lakes basin, where over 800 different chemical compounds have been discovered in the water! Many of these have found their way into water as a result of years of neglect and ignorance. Only recently have these invisible and odourless poisons been detected. As monitoring equipment improves, more and more are found each year.

Toxins such as PCBs, Mirex, Dioxin TCDD, mercury and arsenic are extremely dangerous, even in very small concentrations. They enter the body through drinking water, contaminated food or by absorption through the skin. The safe level of Dioxin TCDD contamination in fish and shellfish is 20 parts per *trillion*, yet shellfish in several parts of Canada have been found that far

4-71   Problem areas of toxic substances in the Great Lakes.

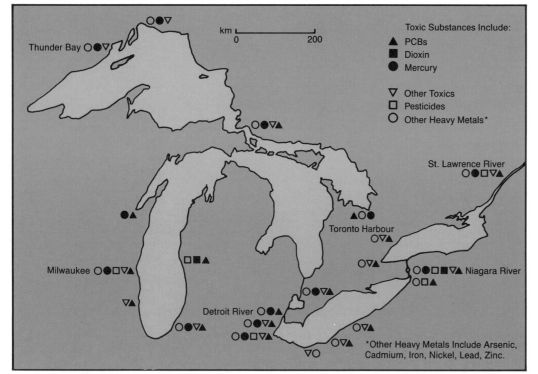

The improper storage and handling of toxic chemicals has contributed to the contamination of the Great Lakes. In 1998, the International Joint Commission on the Great Lakes noted the lakes' environment had improved significantly (due to massive cleanup efforts and government bans on DDT and PCB release) but critical pollutants still remained in the Great Lakes Ecosystem.

4-72   *The introduction of exotic, non-native species into an ecosystem causes changes to habitat. These are not always positive. This shopping cart was left in zebra mussel-infested waters for a few months. The mussels have colonized every available surface.*

exceed this limit. Unfortunately, the long-term effect of exposure to many of these toxins is not fully known; nor is the exact concentration that poses a threat to human health. However, just the sheer number of toxins present is cause for concern. Health officials do know that the build-up of mercury in the body causes irreversible nerve damage, or **Minamata disease**, and that Mirex, dioxins, lead, and PCBs contribute to cancers, birth defects, infertility and liver disorders.

The concentrations of these chemicals vary throughout the Great Lakes, but are present in every part of the system. Areas near industrial and urban centres show the greatest concentration and variety, particularly the St. Clair and Niagara rivers. Most toxins are artificial chemicals used in industrial processes. They enter the lakes in waste water, through accidental spills, and by careless handling and storage at factories and waste-disposal sites. Each year tonnes of heavy metals (from smelting industries and car exhaust) and dioxins (from city incinerators burning plastics) descend on the lakes, often travelling hundreds of kilometres from their sources. Regulations now limit further emissions of the most toxic of these substances, but getting rid of existing toxins will be very difficult and costly.

The problems of PCB storage and disposal were dramatically brought to the attention of Canadians after a fire at St-Basile-le-Grand, Quebec in 1988. PCBs were once used as a cooling agent in electrical transformers, in dyes and printer's ink, and in the pulp-and-paper industry. They are now banned, but there are still 25 million tonnes of them in storage across the country. In the past, disposing of PCBs meant burying them in waste-disposal dumps. This allowed them to escape into the soil and ground water, and from there into the lakes. Burning, a common method of disposing of wastes, was also unsuccessful: they did not burn completely, and smoke from the incinerators carried PCBs and other toxins many kilometres downwind. This is a problem with many toxins — they are almost indestructible. Since nature didn't create them, it has difficulty breaking them down into safe substances. They are long-term poisons.

Removing PCBs and other toxins from lake water is not economically feasible, and often technically impossible — at least so far. It is hoped that, over time, the natural flushing action of the lakes, along with strict regulations limiting the use of toxins, will reduce the hazards.

1. **What are the sources of most water pollution in Canada?**
2. **Why is society only now becoming aware of toxins in water?**
3. **How can toxins enter the human body?**
4. **Which health problems have been associated with toxins?**
5. **How do toxic chemicals get into the water system?**
6. **Why is it difficult to dispose of toxic chemicals such as PCBs?**

New Words to Know

Toxic chemical   Toxin
Acid rain   Minamata disease

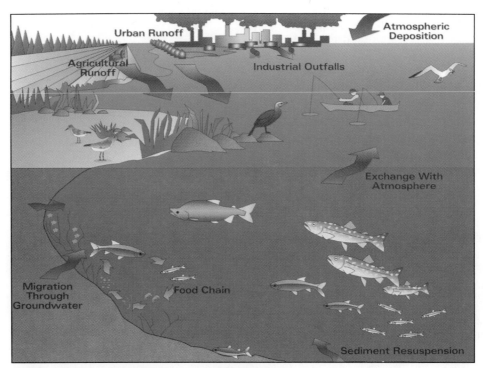

4-73   Sources and Pathways of Pollution.

Acid Rain Of all the forms of water pollution, acid rain receives the most publicity. It has become a routine item on the agenda whenever politicians from Canada and the U.S.A. meet to discuss problems.

Acid rain is created when emissions of sulphur dioxide and nitrogen oxides are released into the atmosphere. These gases react with water in the atmosphere to produce weak solutions of sulfuric and nitric acid. The prevailing winds — generally west to east in Canada — carry these acids through the atmosphere until they return to earth as a component of rain — or snow, or hail, or fog. Some people fear that acid precipitation (a more accurate term for acid rain) will burn their skin. Of course, it does not — at least for the time being. However, even in this weak state, the acidity can have a harmful effect on trees, organisms in lakes and streams, and on buildings.

The major sources of sulphur dioxide are coal-burning thermal-electric plants and smelters. Among the largest Canadian contributors are the smelters at Sudbury, Ontario, but they are not the only Canadian source.

The main source of nitrogen oxides is burning fuel. Automobile exhausts account for 40 percent, and thermal-electric plants for 33 percent; the remaining 27 percent comes from homes and industries. In North America, some 50 million t of sulphur dioxide and nitrogen oxides enter the atmosphere every year from these sources.

The effects of acid rain are greatest in eastern Canada. There they pose a threat to fish and aquatic life in many lakes and rivers. These life forms are at the base of the food chain; when they are eaten by other wildlife nearer the top of the chain, such as loons, those higher forms may come into jeopardy. Young fish are particularly sensitive to acidic conditions and often die. Because of the local rock and soil composition, many water bodies have only a limited ability to neutralize increases in acidity. In Canada, more than 300 000 lakes are vulnerable to acid rain; 14 000 have already been acidified, and another 10 000 to 40 000 are at serious risk.

Acid rain affects more than water. It also is killing sugar maples in Quebec, eroding concrete bridges and buildings, corroding automobiles, and aggravating lung problems in humans.

Removing acid rain from the environment is a costly and complicated process. In 1985, Canada imposed standards that would cut sulphur-dioxide emissions

4-74 Acid deposition in eastern Canada.

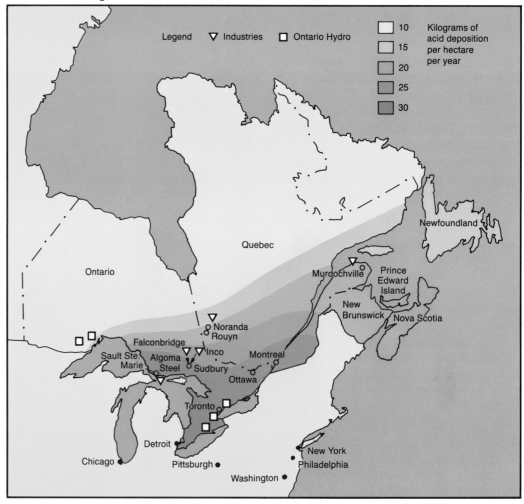

244

from thermal-electric plants and industry to half their present levels by 1994. These restrictions, along with voluntary ones by large polluters, have already led to improvements. Many companies, such as INCO and Ontario Hydro, have installed costly smokestack scrubbers, which remove much of the sulphur dioxide before it reaches the atmosphere. These companies are testing new devices to reduce emissions even further. Ontario Hydro has switched to low-sulphur coal, and washes it to further reduce the sulphur content. Tougher automobile-emission standards came into effect in 1987.

Acid rain is also an international issue. More than half the acid rain falling on eastern Canada comes from the U.S.A.; in some areas, this figure is 75 percent. Unfortunately, the United States has been slow to tackle the problem. Automobile-emission standards are more lax than those in Canada, and the allowable level of sulphur-dioxide emissions is higher. A similar situation exists in Europe, where acid emissions from one country often fall as precipitation on a neighbouring country. This problem cannot be solved until international standards are agreed to by all nations of the world.

1. **In your own words, describe how acid precipitation is formed.**
2. **What negative effects does acid rain have on our society and environment?**
3. **Explain some of the steps taken by industries and governments to reduce airborne emissions.**
4. **Why must acid rain be considered an international problem?**

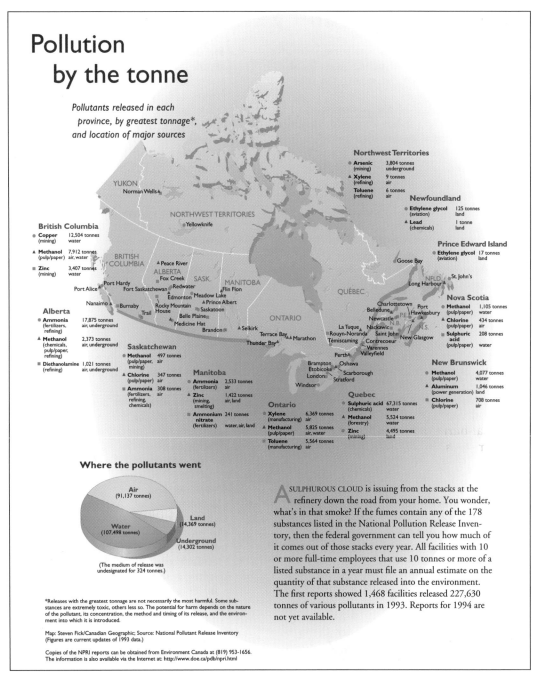

Pollution by the tonne

Pollutants released in each province, by greatest tonnage*, and location of major sources

Northwest Territories
- Arsenic (mining) 3,804 tonnes underground
- Xylene (refining) 9 tonnes air
- Toluene (refining) 6 tonnes air

Newfoundland
- Ethylene glycol (aviation) 125 tonnes land
- Lead (chemicals) 1 tonne land

Prince Edward Island
- Ethylene glycol (aviation) 17 tonnes land

British Columbia
- Copper (mining) 12,504 tonnes water
- Methanol (pulp/paper) 7,912 tonnes air, water
- Zinc (mining) 3,407 tonnes water

Alberta
- Ammonia (fertilizers, refining) 17,875 tonnes air, underground
- Methanol (chemicals, pulp/paper, refining) 2,373 tonnes air, underground
- Diethanolamine (refining) 1,021 tonnes air, underground

Saskatchewan
- Methanol (pulp/paper, mining) 497 tonnes air
- Chlorine (pulp/paper) 347 tonnes air
- Ammonia (fertilizers, refining, chemicals) 308 tonnes air

Manitoba
- Ammonia (fertilizers) 2,533 tonnes air
- Zinc (mining, smelting) 1,422 tonnes air, land
- Ammonium nitrate (fertilizers) 241 tonnes water, air, land

Ontario
- Xylene (manufacturing) 6,369 tonnes air
- Methanol (pulp/paper) 5,825 tonnes air, water
- Toluene (manufacturing) 5,564 tonnes air

Quebec
- Sulphuric acid (chemicals) 67,315 tonnes water
- Methanol (forestry) 5,524 tonnes water
- Zinc (mining) 4,495 tonnes land

Nova Scotia
- Methanol (pulp/paper) 1,105 tonnes water
- Chlorine (pulp/paper) 434 tonnes air
- Sulphuric acid (pulp/paper) 208 tonnes water

New Brunswick
- Methanol (pulp/paper) 4,077 tonnes water
- Aluminum (power generation) 1,046 tonnes land
- Chlorine (pulp/paper) 708 tonnes air

Map labels: YUKON, Norman Wells, NORTHWEST TERRITORIES, Yellowknife, BRITISH COLUMBIA, Port Hardy, Port Alice, Nanaimo, Burnaby, Trail, Peace River, Fox Creek, ALBERTA, Redwater, Fort Saskatchewan, Edmonton, Meadow Lake, Rocky Mountain House, Belle Plaine, Medicine Hat, Brandon, SASK., Flin Flon, Prince Albert, Saskatoon, Selkirk, MANITOBA, Thunder Bay, Terrace Bay, Marathon, ONTARIO, La Tuque, Nackawic, Rouyn-Noranda, Saint John, Témiscaming, Contrecoeur, Varennes, Valleyfield, Perth, QUÉBEC, Brampton, Etobicoke, Oshawa, London, Scarborough, Stratford, Windsor, Goose Bay, Long Harbour, NFLD., St. John's, Charlottetown, Belledune, Newcastle, Port Hawkesbury, New Glasgow, P.E.I., N.B., N.S.

Where the pollutants went

Air (91,137 tonnes)
Water (107,498 tonnes)
Land (14,369 tonnes)
Underground (14,302 tonnes)

(The medium of release was undesignated for 324 tonnes.)

A SULPHUROUS CLOUD is issuing from the stacks at the refinery down the road from your home. You wonder, what's in that smoke? If the fumes contain any of the 178 substances listed in the National Pollution Release Inventory, then the federal government can tell you how much of it comes out of those stacks every year. All facilities with 10 or more full-time employees that use 10 tonnes or more of a listed substance in a year must file an annual estimate on the quantity of that substance released into the environment. The first reports showed 1,468 facilities released 227,630 tonnes of various pollutants in 1993. Reports for 1994 are not yet available.

*Releases with the greatest tonnage are not necessarily the most harmful. Some substances are extremely toxic, others less so. The potential for harm depends on the nature of the pollutant, its concentration, the method and timing of its release, and the environment into which it is introduced.

Map: Steven Fick/Canadian Geographic; Source: National Pollutant Release Inventory (Figures are current updates of 1993 data.)

Copies of the NPRI reports can be obtained from Environment Canada at (819) 953-1656. The information is also available via the Internet at: http://www.doe.ca/pdb/npri.html

4-75 Pollution by the tonne.

Mind Benders and Extenders

1. Prepare a report on one form of water or air pollution affecting your local area. Include in the report the following:
   (a) the source of the pollution.
   (b) the effect it has or could have on humans and the environment.
   (c) any present or future plans by the polluter to control or eliminate it.

2. The location of new storage and disposal facilities for toxic wastes has become a serious problem throughout Canada. Some communities are against having these facilities near them, while others welcome them. Compile a list of points for and against having such a facility built near your community.

3. Research a mineral of your choice and find out:
   (a) location of the main producing areas in Canada.
   (b) method(s) of extracting the mineral.
   (c) location of the main processing centres.
   (d) major uses of refined product.
   (e) markets, domestic and foreign, for the refined product.
   (f) future demands for the refined product.
   (g) problems (environmental, social, political or economic) associated with the mining and processing of the mineral and sale of the final product.

4. In small groups debate the issue: "Should a prairie wheat farmer receive government subsidies and other forms of financial aid in order to stay in business?"

5. It has been suggested that Canada, which has a surplus of wheat, should give it to countries in need of food. In small groups debate advantages and disadvantages of such a proposal.

6. The federal and provincial governments provide farmers with aid and assistance through a number of programs and marketing boards. Investigate and prepare a report on how one of these serves farmers and consumers.

7. "The first forest was a gift of nature. The next one is up to us." Write a one-page explanation of this quotation.

8. Canada has enormous reserves of fossil fuels locked in difficult-to-develop frontier areas. Research and write a report, accompanied by maps and diagrams where possible, on one frontier area. Note in particular:
   (a) the location and potential size of the resource.
   (b) the problems in searching for, and transporting the resource to markets.
   (c) environmental concerns related to extracting and transporting it.
   (d) problems, if any, between government and aboriginal people over ownership of the resource.
   (e) the future impact and concerns the development of the resource will have on the people and economies of the areas involved.

9. Construct a chart to illustrate the characteristics of offshore and inshore fishing. Consider such items as: size of the boat, crew and catch, where and how the fish are processed, etcetera.

10. Conduct a survey to list all fish, beef, poultry and pork products eaten by your family for at least two weeks. Also, make a list of the fish products sold in your local supermarket and, where possible, note the origin of the product.
    (a) Note the amount of seafood you eat, compared to beef, poultry and pork.
    (b) How does your consumption compare to that of other members of the class?
    (c) Which reasons can you give for the relatively small quantities of fish eaten by Canadians?
    (d) How have the processing companies attempted to diversify their marketing of fish products?

11. Your provincial government has decided to grant a lumber company the right to cut thousands of hectares of virgin forest. The company claims it needs the trees in order to stay in business. Other groups, however, wish to preserve the forest in its wild state as a provincial park. Also, access to this forest requires building a road through land claimed by local aboriginal people.

    In a group, role-play one of the following, stating their support or opposition to the plan.
    (a) the owner of the company
    (b) the president of the loggers' union
    (c) a spokesperson for the aboriginal people
    (d) a nearby tourist outfitter
    (e) a resident of a nearby town
    (f) an American tourist

12. Research one form of alternative energy such as wind, solar, tidal, geothermal, biomass conversion or hydrogen. For your choice note:
    (a) the advantages of this energy source.
    (b) the areas where this source of energy is most abundant.
    (c) the methods of converting the energy for everyday use.
    (d) any functioning or experimental applications which are presently in use, as well as future projects.
    (e) the drawbacks to your choice.

CANADA
EXPLORING NEW DIRECTIONS

Commerce & Industry

*"The shift from conventional manufacturing
leads economists to conclude that prosperity lies in the quality of the work force,
in its skills, inventiveness, creativity and entrepreneurship."*
Walter Pitman
Chairman, Ontario Institute for Studies in Education

Commerce, Industry and Jobs

Every Canadian cares about commerce and industry because every Canadian cares about jobs. In simple terms, the workers in **industry** make the goods Canadians need, while those in **commerce** buy and sell them. When Canada's economy is healthy, employment is high and the country is said to be "in good times" or in a "boom." Businesses increase their sales and must hire workers to expand production or make more services available. However, when sales in the marketplace decrease, businesses begin to reduce production and services. That means there will be idle machinery, idle factories and, most important of all, idle, unemployed workers. When this occurs, the country is suffering "bad times," or is in a "bust." Economists call these bad times a **depression**. If conditions are not too severe or long-lasting, they are called a **recession** instead.

During the 1930s Canada, along with the world, suffered through the Great Depression. Millions of Canadians were out of work in that decade, leading to poverty, hardship, and labour unrest across the country. Since then the country's economy has experienced several periods of boom and bust. Fortunately, the busts have been considered recessions rather than depressions. Neither governments nor economists who study the **business cycle** — the ups and downs of the economy — have been able to find ways of guaranteeing that all times will be good.

One reason the Canadian economy is hard to understand, let alone control, is its huge size and complexity.

In August 1997, the Canadian labour force totalled almost 16.7 million workers. Of that total, about 15.2 million were employed. The **unemployment rate** was about 9.7 percent. This rate is determined by the following formula:

$$\frac{\text{Total number of jobless}}{\text{Total labour force}} \times 100 =$$

5-1   The sectors of industry.
(a) Primary industry: a tractor hauling logs out of the forest.
(b) Secondary industry: a garment factory producing goods for consumers.
(c) Tertiary industry: a nurse providing services in a hospital.

Full employment in good times is difficult to guarantee. Pitman's comment, which began this chapter, is important to understand: he is saying that *quality* is what really matters in this country's work force. This in turn means that the more highly skilled and educated its workers are, the better their chances of being employed. A skilled labour force also ensures that Canada's economy will be able to compete internationally and hold onto the industries it now has — along with the jobs they provide.

The unemployment figure for Canada can be very misleading. To some people it suggests that the picture for Canada's jobless is the same across the country. Statistical chart 5-2 clearly shows, however, that the employment situation varies greatly from province to province. Chart 5-3 shows that there is also a great variation between cities. Note that Regina's rate is less than half that of two other cities.

One other way of looking at the economy and employment is in terms of the different parts or **sectors of industry**. Most economists place industries in one of three sectors: primary, secondary and tertiary.

Primary industries are those that work directly with securing natural resources. They extract the raw materials for processing by the secondary industries. Thus farming, fishing, forestry, and mining are all primary or **extractive industries**. In the nineteenth century, the vast majority of working Canadians were involved in these industries, particularly agriculture. Today this sector employs the fewest Canadians.

Secondary industries are involved in manufacturing and construction. The manufacturing sector takes raw materials and processes them; the steel industry, for example, smelts coal, iron ore and limestone to produce iron and steel. Other manufacturers — auto makers are

one example — take semi-processed materials or manufactured parts and assemble them into final products. The construction industry, which raises buildings, is considered a secondary industry. The number of Canadians employed by secondary industry is large, but declining.

Tertiary industries are harder to define because they are involved in services to people, governments or businesses. Rather than making products, workers in this sector do things for others. As a result, this group is often known as the **service industry**. It includes a huge variety of workers: truck drivers, retail merchants, wholesale merchants, sales people, bankers, government employees, doctors, teachers, repair technicians, hairdressers, accountants, hotel-keepers, entertainers, professional athletes —

5-2 Unemployment in Canada.

PROVINCE		CITY	
Newfoundland	20.4 unchanged	St. John's	13.7
Prince Edward Island	16.2 +	Charlottetown	n/a
Nova Scotia	13.0 +	Halifax	9.0
New Brunswick	13.0 −	Saint John	12.4
Quebec	12.2 +	Quebec	10.6
		Montreal	11.9
Ontario	9.0 −	Ottawa-Hull	9.0
		Hamilton	8.4
		Toronto	7.8
		London	7.9
		Windsor	8.9
		Sudbury	8.7
Manitoba	7.1 −	Winnipeg	7.4
Saskatchewan	6.3 −	Regina	5.8
Alberta	6.9 +	Edmonton	6.8
		Calgary	6.1
British Columbia	8.0 −	Vancouver	8.7
		Victoria	7.8
Canada	9.7 unchanged		

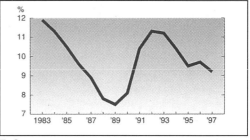

5-3a *Unemployment Rate 1983 – 1997 (Annual Average).*

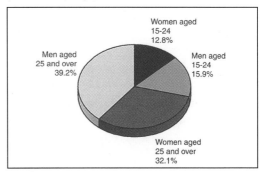

5-3b *Who Are the Unemployed? – 1997 (Annual Average).*

249

5-4 Summary of Canadian Labour Market Conditions, 1996, projected to 2001.

			Labour Market Ratings			
	M	P	T	I	L	All
Business Finance/Admin	G/G	G/G	F/F	F/F	–	F/F
Natural/Applied Sciences	G/G	F/G	F/F	–	–	F/G
Health	F/F	G/G	F/F	F/F	–	G/G
Social Science/Edu/Govern/ Services & Religion	G/G	F/F	F/F	–	–	F/F
Art, Culture, Rec & Sports	F/F	P/P	F/P	–	–	P/P
Sales & Services	F/F	–	F/F	P/P	F/P	F/F
Trades, Transp & Equip Operators	F/F	–	F/F	P/F	P/P	P/P
Primary Industry	G/G	–	F/F	P/P	P/P	F/F
Processing, Manuftg & Utilities	F/F	–	F/F	F/F	F/F	F/F
All	G/G	F/F	F/F	P/F	P/P	F/F

G indicates good; F indicates fair; and P indicates poor. The first letter refers to labour market conditions in 1996. The second letter is the projection for 2001.

there are hundreds of other groups. These workers form the majority in the labour forces of modern industrial nations such as Canada. In Canada, the service sector provides employment for more than twice as many people as the other two sectors combined.
Between 1991 and 1997 small businesses (those with 20 employees or less) created over 85 percent of all new jobs in Canada.

Some economic analysts define a fourth or **quaternary** (information) **sector** along with the primary, secondary and tertiary sectors. The quaternary sector includes the host of new occupations associated with **information technology** – the production, storage, retrieval, and transfer of information. Researchers, librarians, computer operators, program-

mers, journalists, and others are included within this sector which is the most rapidly developing business area in the country. Because Statistics Canada has not yet broken out specific figures for the quaternary sector, however, it remains within the tertiary sector for purposes of this text.

In 1996, the Department of Human Resources analyzed the Canadian job market and the availability of jobs within specific skill levels and market segments. These figures were then **extrapolated** to present a picture of the Canadian job market as we might expect to see it in 2001. The results are presented in Fig 5-4. Young people entering the job market might do well to study these projections.

1. (a) What was the unemployment rate in 1997? What is it today? Using chart 5-2, list in order, those provinces which had a unemployment rate lower than the national average. Then, list those whose rate was higher.
   (b) Why do some provinces have lower unemployment rates than other provinces?

2. In your local area, name:
   (a) two primary industries.
   (b) three secondary industries.
   (c) ten tertiary industries.

3. Take a class survey to determine the percentage of parents working in the three industrial sectors. Using this data, prepare a neat, labelled circle graph, and compare it to Fig 5-39b.

4. What conclusions can you draw from Fig 5-4? What are those areas in which the job market was weakest in 1996? What areas are projected to be weak in 2001? What areas are predicted to be strongest in 2001? Why? How might some of these projections be misleading?

New Words to Know
Industry
Commerce
Depression
Recession
Business cycle
Unemployment rate
Sector of industry
Primary industry
Secondary industry
Tertiary industry
Extractive industry
Service industry
Quaternary sector
Extrapolate
Information technology

Canada's Economic Development

Canada — The Richest Underdeveloped Country in the World

To understand Canada's modern economy will require some knowledge of how it evolved. The present economy is vastly different from the one found here 200 years ago. It remains true, however, that some industries have been important here virtually from the beginning.

Canada possesses many advantages for industrial development. Some were mentioned in Chapter One. It also possesses some disadvantages. One of the latter is that this country's population has always been considered too small for the task of developing the vast resources found here. As a result, people from other countries sometimes call Canada "the richest underdeveloped country in the world." Canadians have barely scratched their huge mineral, forest, water and agricultural resources, in the view of many experts. Also, winters here are long, cold and costly. Even in the 1990s they increase unemployment in the country. Another problem is that the distances across the country east to west are much greater than the ones to the large population centres of the U.S.A. just over the border. Trading goods across a long east-to-west axis continues to be less logical economically than trading across the border north to south. However, political development and a growing sense of unity have kept the country together, and have proven more powerful than the economic arguments.

Though only four percent of its land area is suitable for crops, that small figure still amounts to more actual cropland than is available in most countries. On that four percent, Canada's farmers have produced enough to export massive quantities of food to nations that are unable to feed themselves. Canadians have never had to face famine.

Of course, there are some limitations to Canada's resource base. Even with their country's vast resources, Canadians are learning that "wise management" must be their motto.

The Beginning of Commerce and Industry

In many ways, commerce and industry led to the discovery and exploration of Canada. In the sixteenth century, European fishermen came to haul in rich harvests from its Atlantic waters. Farming attracted French settlers to the valley of the St. Lawrence River in the seventeenth century. Soon the fur trade had also begun. Mining had its beginnings there, as iron ore from near Trois-Rivières was smelted to make simple

5-5 Factory scene, early twentieth century.
Here, shells are being produced for World War I. How would a modern factory differ from this one?

implements. Forestry too had its start at that time, as trees were harvested to make tools, furniture, houses and other goods.

The fur trade spurred the French, and later the English with their Hudson's Bay Company, to explore more of the continent's interior. Extensive transportation lines had to be established so that furs could reach Montreal or the trading posts along Hudson Bay. The problem of how to move bulky staple products across thousands of kilometres of wilderness began to be solved over 300 years ago. Those raw materials were heading for foreign markets, still a major aspect of Canada's commerce.

Events in Europe had a considerable influence on Canadian development. When beaver hats were fashionable in Europe, the fur trade couldn't help prospering. During the Napoleonic Wars, the British navy needed Canadian timber to build ships. This gave a great boost to Canada's young forest industry. When imperial preferences in **tariffs** gave Canadian wheat easy access to the British market, this country's farmers prospered.

Manufacturing Gets Its Start Canada's economy was dominated in its early years by primary industries, particularly agriculture. Almost all manufactured goods were imported, most of them from Europe and, to a lesser extent, from the U.S.A. Canada's population, a mere 3.5 million at the time of Confederation in 1867, was too small and scattered to encourage the development here of much manufacturing. Confederation, however, eventually led to the building of railways from the Atlantic to the Pacific. As a result, the building of railway rolling stock became one of Canada's first major secondary industries. This industry required iron and steel, so another manufacturing industry developed. Canada's population grew quickly around the turn of the century. Waves of European immigrants began to arrive, particularly on Canada's prairies. This encouraged still more industry, as there was now a larger domestic market for manufactured goods. With the opening of the West, the manufacture of farm implements became a major industry in central Canada. This industry expanded to answer the domestic demand and then began to export its goods beyond Canada.

By the twentieth century, the demand for newsprint in the U.S.A. had led to the building and expansion of pulp-and-paper mills in Canada. The expansion of the railway system led to the discovery of mineral deposits on the shield, most notably near Sudbury. Smelters were built to smelt the nickel, copper, silver and gold ores. Electricity began to replace steam and falling water as the main energy source for industry. Waterfalls on the shield and at Niagara were harnessed to provide it.

The First World War gave a particular boost to the manufacture of metals in Canada. In 1917, late in that war, even the manufacture of aircraft began in Canada. The expanding population and rising living standards of the 1920s led to the manufacture of stoves, refrig-

5-6   Time chart of major highlights in Canada's economic development.

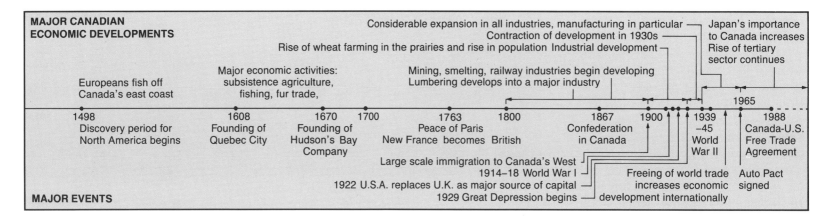

252

erators, radios, washing machines and countless other consumer goods. The automotive industry expanded during this period and began to play a major role in economic development.

A very important change took place in 1922: for the first time, the major source of money for development was the U.S.A. Until then, Great Britain had been the main source of capital. It was U.S. money that made possible the tremendous growth of Canadian industry in the middle decades of the twentieth century. Of course, this also led to Canada's having a **branch-plant economy**, as American **parent companies** built branch plants here, controlled in the U.S.A. Industrial growth came to a sudden halt in the early 1930s as a result of the stock-market crash of 1929, which resulted in the terrible hardships of the Great Depression.

Manufacturing Matures   The value of manufactured goods simply "took off" during the Second World War. The war created a great demand for aircraft, trucks, ammunition, metals, and electrical and electronic goods. All of these were provided in great quantity by Canada, whose factories were safe from bombing raids. The economy boomed even more strongly during the 1950s, as factories retooled to provide peacetime consumer goods. In addition, new mineral discoveries, improvements in **technology**, and a growing market all spurred production.

The postwar period saw an increase in **foreign investment** in Canada. This came from many countries but particularly from the U.S.A. Foreign domination of Canada's industry concerned many

Canadians during this period. The tremendous growth in world trade, and the reconstruction following the Second World War and the Korean War, all encouraged the growth of Canada's industries. Within Canada, transportation networks improved with the building of the St. Lawrence Seaway and the expansion of the national highway system. Technology grew more in scope and usefulness. Throughout this time of change, the domestic market continued to expand. All of these things aided commercial and industrial growth.

In 1965 the already well-developed Canadian automobile industry was given renewed growth when Ottawa signed the Auto Pact with the U.S.A. This greatly reduced the tariffs on both automobiles and parts. Existing car-assembly and parts plants expanded and new ones were started. This had a major effect on many other industries, for the automobile requires many products — steel, base metals, plastics, rubber, chemicals and others.

In the 1970's and into the 90's a completely new factor began to influence Canada's industrial development: the industrializing of the Asian Pacific Rim, particularly of Japan. For several centuries the United Kingdom and the U.S.A. had been Canada's major trading partners. They had also supplied most of its development money. Japan's rise to world prominence as the second-largest indus-

Manufacturing in Canada had a very slow start but saw strong increases during the two world wars. Since 1970, the increase has been incredible. It must be remembered that part of the increase is due to inflation. **Value added by manufacturing** *is that portion of a*

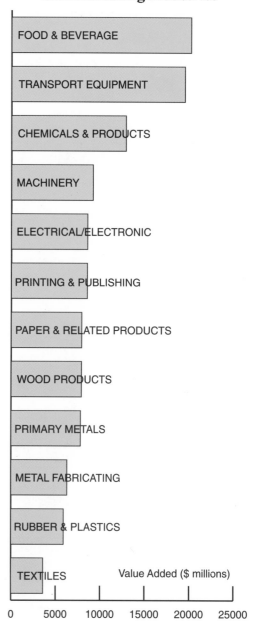

5-7 Value added in Canada's top manufacturing industries

FOOD & BEVERAGE

TRANSPORT EQUIPMENT

CHEMICALS & PRODUCTS

MACHINERY

ELECTRICAL/ELECTRONIC

PRINTING & PUBLISHING

PAPER & RELATED PRODUCTS

WOOD PRODUCTS

PRIMARY METALS

METAL FABRICATING

RUBBER & PLASTICS

TEXTILES

Value Added ($ millions)

0   5000   10000   15000   20000   25000

product's value that is created during the manufacturing process; it excludes the cost of the raw materials.

trial power in the free world has had global effects that Canada has not escaped. Japan replaced the U.K. as this country's second-largest trading partner in the early 1980s. In fact, this country now trades more with Japan and the other Pacific Rim countries than it does with European countries, including the U.K. Japan's great corporations — Sanyo, Toyota, Honda and many others — have established plants and dealer organizations in Canada. Less obvious, but equally important, are Japan's powerful banks, which have been making investments in everything from real estate to Canadian government bonds. Canadians, once so concerned about American control of their economy, have become concerned about Japanese influence.

The Canadian economy is small compared to that of the United States, Japan, or the **European Union (EU)**. For most of our history, Canadian industries have been protected by tariffs. These made imported goods so expensive that Canadians were forced to "buy Canadian." Since the Second World War, there has been a lowering of tariffs all around the world, and international trade has expanded. In order to secure an even larger market for Canadian industrial goods, and to help them expand, the Canadian government proposed a free-trade agreement with the U.S.A. in the late 1980s. This touched off a national debate, for some Canadians felt that Canada's industries would be unable to compete when placed on an equal basis with American industries. Others were concerned about the possible loss of sovereignty and cultural identity.

There is one other recent change in the Canadian economy which deserves attention, and that is the rise of the tertiary or service sector. Though the value of this country's manufactured products has continued to increase, improved technology has greatly reduced the proportion of the work force required to produce those goods. By the early 1990s the number of workers actually making goods in Canada may amount to only 15 percent of the labour force.

1. **Name three major events and explain how each one affected Canada's economic development.**
2. **Place the following products in order, from earliest to latest, in terms of when they were of most importance in Canada's economic development: wheat, cars, lumber, Japanese electronic goods, fish, furs.**
3. **Explain four ways in which Canada's economic development has *not* changed much over the last 200 years.**

New Words to Know
Tariffs
Branch-plant economy
Parent company
Technology
Foreign investment
Value added by manufacturing
European Union (EU)

5-8 European Union (EU)

The European Union, an association of sovereign states, is a centrally regulated market of 374 million people. Institutions within the Union include the Council of Ministers, European Commission, European Parliament (626 members) and the European Court of Justice. With capitals in Brussels, Strasbourg, and Luxembourg, the Union has 15 members: Belgium; Germany; France; Italy; Luxembourg; Netherlands; Denmark; Ireland; United Kingdom; Greece; Spain; Portugal; Austria; Finland; and Sweden. EU's basic currency unit (introduced in 1999) is the Euro (1997 exchange rate, $1.00 CDN = .64 ECU). The EU national holiday is May 9 (Schuman Day). The GDP is $8 billion US, and the unemployment rate is 10.6 percent. Canada exported $14.4 b.illion Canadian to the EU in 1997. Canadian imports from the EU for the same period were $26.8 billion

Location Factors of Industry

One of the most important questions for a geographer studying the Canadian economy is this: "Why are industries located where they are?" The location factors of industry are those which explain why companies place their factories or businesses where they do.

Most of these factors operate on very simple principles, or just common sense. For example, even a child with a lemonade stand knows that it should be in a busy place. The more people walking by, the better — it should mean more sales. Though the child doesn't even know the words "**market**" or "customer," their importance is still understood. If the stand becomes extremely busy, this young **entrepreneur** will also learn some of the problems of transportation — of how to get the ice, water and lemon concentrate to the stand. **Capital** will become important too, for without profit (or a loan from a parent) how can more concentrate be bought for tomorrow?

There are seven factors that decide industrial location:

(1) Capital: Money Used to Make Money
(2) Raw Materials: Something to Process
(3) Energy: Power to Make Things Happen
(4) Transportation Facilities: Moving the Goods
(5) Labour: People Power for the Process
(6) Markets: Industries Need Customers
(7) Other Factors: Government or History May Decide

Determining the location for any factory always requires a consideration of the first six factors. However, sometimes the seventh or *special* factor or need can be the most important of all.

Capital: Money Used to Make Money

Perhaps capital is the best place to start this discussion, for nothing can happen unless some surplus money is available for investment. That money — which is what capital is — is needed to buy the land, build the factory, buy the machinery and the raw materials, and pay the workers. The economic system of Canada is called capitalistic because individual citizens may use capital to build or invest in the industries that produce the goods that people want. In a **communist economic system** only the state, or government, is allowed to own factories and industries, or to provide the capital needed to build and operate them. The state, in such a system, also decides what is manufactured. In a capitalist system individual citizens and the marketplace do this.

Capital, in the huge quantities needed for industry, has four major sources. A company may rely on one or more or all of them. They are:

5-9 The widget factory and its locative factors.

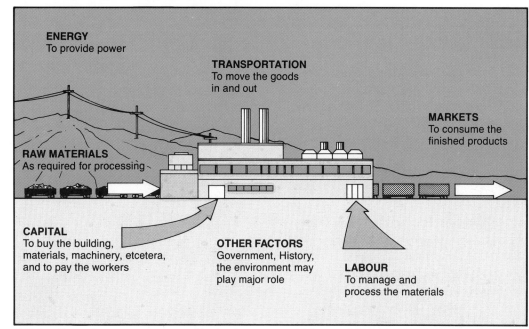

ENERGY
To provide power

TRANSPORTATION
To move the goods in and out

MARKETS
To consume the finished products

RAW MATERIALS
As required for processing

CAPITAL
To buy the building, materials, machinery, etcetera, and to pay the workers

OTHER FACTORS
Government, History, the environment may play major role

LABOUR
To manage and process the materials

"Widget" is a humorous term used by business to stand for anything that is manufactured.

255

The company itself An existing company or corporation wants to expand production. It may buy the new machinery and materials — or even a new factory — from its own profits.

A loan A company may secure a loan from a bank or other financial institution.

Stocks and bonds A company may sell **stock** on a **stock exchange**. People who hold stock have a share in the company's ownership. A company could also sell **company bonds**, which don't involve ownership, but do require the company to repay the money, with interest added, at a specific time.

Governments Governments at all levels — federal, provincial and municipal — are involved in helping industries to increase production. Sometimes they will provide a grant, which is money that doesn't have to be repaid. Sometimes they offer loans instead, with little or no interest. In some cases municipalities and provinces will provide free land, or relief from taxes for a certain number of years. This means companies don't have to raise as much capital from elsewhere. This of course attracts them to the regions where the assistance is offered.

Securing capital to establish a new factory will be almost impossible unless the plant is located in the right place. No one will want to assist a company or entrepreneur unless the choice of site makes economic sense. Canada has attracted a great deal of capital from other countries because it has been a law-abiding stable democracy, a safe place in which to invest. No one wants to lose millions by investing in a country where a dictator may decide to keep the profits or take over the company.

When governments provide capital or assistance of any kind, they naturally will have much to say about where the factory is located. After all, when a government offers assistance, it is so employment will go up in its own area. Also, a new factory brings in more taxes, so governments most often get their money back eventually.

Chart 5-10 gives some idea of how much money can be involved in capital-investment schemes. It also lists other data that will be of interest in other sections in this chapter. The figures under "company assets" include the values of all the factories, equipment, inventories and so on owned by the companies. These assets have been built up over the years through capital investment. Much of the money for that investment came from the companies' past profits. Note the difference between the capital assets of the last three companies listed. Canada Life has $31 750 million, Canada Safeway $1 257 million, and Amoco Canada Petroleum $3 622 million. Of course, these companies differ widely in nature. Canada Life, as an insurance company, has huge capital holdings, with deposits, bonds, and capital reserves as assets. Canada Safeway, as a food distributor, has stores, vehicles, and warehouses as

5-10 Canada's top ten corporations by revenue, 1996

CORPORATION (headquarters)	REVENUE ($ million)	ASSETS ($ million)	PROFITS ($ million)
1. GENERAL MOTORS OF CANADA *automotive/Oshawa, Ontario*	27 300	N/A	679
2. FORD MOTOR CO. OF CANADA *automotive/Oakville, Ontario*	25 500	N/A	423
3. CHRYSLER CANADA *automotive/Windsor, Ontario*	17 060	5 854	152
4. SUN LIFE ASSURANCE OF CANADA *insurance/Toronto, Ontario*	10 720	71 817	485
5. MANULIFE FINANCIAL *insurance/Waterloo, Ontario*	9 893	47 243	503
6. IBM CANADA *technology/Markham, Ontario*	9 500	N/A	N/A
7. McCAIN CAPITAL CORP *management/Florenceville, New Brunswick*	7 346	2 467	N/A
8. CANADA LIFE ASSURANCE *insurance/Toronto, Ontario*	5 152	31 750	229
9. CANADA SAFEWAY *food distribution/Calgary, Alberta*	4 795	1 257	71
10. AMOCO CANADA PETROLEUM *oil and gas production/Calgary, Alberta*	4 512	3 622	378

The highest revenue firm was also the highest in profit. The ranking of the other nine firms, however, changes when they are evaluated by profit. Suggest reasons why rankings by revenue, profit, and asset are really quite different.

assets, with a total value of about one twenty-sixth that of Canada Life! Amoco Petroleum owns assets of machinery, land rights and storage facilities totalling $3 622 million, or about three times that of Canada Safeway. Canada Life and Amoco Petroleum are more **capital-intensive** than Canada Safeway.

One very important aspect of economic development involving capital is **research and development**, or **R and D** as it is called in the business world. The purpose of R and D is to find new and better technology, for example, new and better ways of making goods or providing services. New technology might mean that the same product can be made faster, or that a higher-quality item can be made for the same cost. Sometimes entirely new products are created. Many experts claim that Canada doesn't invest enough in R and D. Most industrial nations spend twice as much as Canada on R and D, as a proportion of gross national product.

Two reasons are generally given for Canada's weakness in R-and-D spending. One is that it has a branch-plant economy. R and D is usually carried out where the company has its main plants. The other is that Canada's federal and provincial governments do not provide as much capital for research as do governments in most other industrial nations. When Canadian scientists and engineers are involved in R and D, their work has often resulted in new discoveries welcomed around the world. These have brought profits to Canada and provided more jobs.

Canada's aerospace sector is a leader in R&D. Aerospace Industry Association

5-11 Research and Development at Canadian General Electric. *A researcher is conducting tests on silicone products with the aid of a computer.*

President, Peter Smith, notes: "This sector undertakes 15 per cent of all industrial R&D conducted in Canada. Every year, Canada's aerospace firms re-invest more than 12 per cent of revenue in the development of new and innovative products and services."

In 1996, the Canadian government launched its **Technology Partnerships Canada (TPC)** initiative to support R&D in Canadian knowledge-based

industries. The program provides for repayable investments in advanced manufacturing, and biotechnology, information technology, and evironmental technology firms. TPC announced 23 projects in 1997 to result in an investment of $470 million over several years, with an additional $2.5 billion worth of investment by the private sector.

257

Other governmental programs include the **Foundation for Innovation** with some $2 billion invested over five years for research at universities and hospitals, and the Industrial Research Assistance Program to help smaller businesses with technology support.

Total Canadian expenditures on research and development in 1997 were $13 billion.

1. **Give two reasons why the loonie you may have in your pocket is not really classed as capital funds.**
2. **How is an industry's location important in attracting capital?**
3. **How are entrepreneurs who wish to come to Canada generally welcomed? (refer to Chapter 2, Immigration)**
4. **Give reasons why some Canadians are for, and some against, government grants to industries.**
5. **Why does limited R & D occur in Canada?**

New Words to Know
Market
Entrepreneur
Capital
Company stock
Communist economic system
Company bond
Stock exchange
Capital-intensive
Research and development (R&D)
Consortium

Raw Materials: Something to Process

Canada is a major source of raw materials. Many of Canada's primary processing plants are located near the resources themselves. This is because the cost of transporting the refined material — whether it is metal, oil, wood or apples — may be much less than the cost of shipping the raw material. Food, being perishable, is usually processed at canneries and other plants close to the supply. It generally makes more economic sense to do it this way, and this means higher profits. If large, heavy materials must be shipped, companies will try to use water transportation, for it is the most inexpensive way to send them.

Because of this, metal smelters and refineries are located near mines throughout Canada, from Bathurst, New Brunswick, to Trail, British Columbia. Similarly, sawmills and pulp-and-paper mills are scattered throughout the forest regions. The fish canneries and other seafood-processing plants are along the coasts, where fresh fish can be delivered directly to the plants. These and other food products are often processed in the source region, particularly if this is located far from urban markets. However, much food processing is done in urban areas. Canada's larger cities dominate the food and beverage industries, which suggests that market location can be the most important factor in deciding where a processing plant is built.

It is important to remember that "raw material," for most factories, is not *natural* or *unprocessed* material, but

5-12 The manufacture of roof trusses, Ajax, Ontario. *This factory uses lumber, already processed in a sawmill, and produces a pre-fab product for the construction industry.*

consists of parts or **semi-processed material**, such as metal, rubber, lumber and the like. Many factories assemble parts that were made in other factories. This helps to explain how the large industrial cities grew: to keep costs down, the parts factories generally needed to be close to the plants they supplied.

Energy: Power to Make Things Happen

The manufacture of products requires some source of power. Various energy sources have been used to turn the wheels of industry. For most of recorded history, these included animal power, slaves and the wind. Before the Industrial Revolution, waterfalls were harnessed to power factories. This meant these early factories had to be located near rivers that could provide a constant supply of water. The invention of the steam engine meant that factories could be built away from river banks. When that happened the new problem became coal supply: a factory had to be near a coal mine that could supply the fuel to generate the steam.

It was the development of electric power that finally freed industry from the restrictions placed on it by the location of the energy supply. Electricity can be transmitted economically for thousands of kilometres at high voltages to where it is needed.

It is still true, however, that in some instances power can determine where an industry is located. The best example of this is the aluminum-smelting industry, which requires huge amounts of energy. Even though Canada is a major producer of aluminum, none of the metal is actually mined here. Instead, **alumina**,

5-13 Saw mill in New Brunswick. *In pioneer Canada saw mills had to be located along the rivers to use the power of water directly. Today that power comes as electricity in wires, and mills and factories need not be located along rivers.*

the partly processed ore, is brought thousands of kilometres from Jamaica and other source regions to smelters in places such as Arvida, Quebec, or Kitimat, B.C., where there are huge supplies of low-cost hydro power. Other industries that require a great deal of electricity — some chemical producers and steel-makers — achieve benefits by locating near a large, secure and inexpensive energy supply.

Several regions of Canada are especially rich in hydro-electric power, the cheapest major power source in the world. About 65 percent of all the electricity used in this country is generated by hydro plants. In most countries the figure is much lower: in the U.K. it is 2.3 percent, in West Germany 5 percent, in both Japan and the U.S.A. about 14 percent. Canada still has huge, undeveloped hydro potential, particularly in the north. Inexpensive hydro power may yet attract certain industries to Canada. Quebec has already developed some of its hydro potential for sale to American markets.

1. Why is the term "raw material" not the best term to use to describe what many factories actually process?
2. (a) Name all the various sources of power noted in this section.
   (b) Name some other sources of power not mentioned in this section.
   (c) Which source of power do you think has the greatest potential in the 21st Century? Why
3. How did the invention of the steam engine make the problem of locating industries easier to solve, yet create a new type of restriction?
4. Why is Canada the free world's leading producer of aluminum, though no aluminum ore is mined in this country?

Transportation Facilities: Moving the Goods

Transportation facilities are essential both before and after the manufacturing process. The raw materials must be brought to the plant, and the finished products sent to market.

All manufacturing companies try to build their plants as close as possible to existing transportation lines. The major highways into Canada's cities are often lined with factories. This saves these companies time, and that generally means lower costs as well. Older factories in Canada's city centres often relocate to the suburbs: the high cost of moving is soon balanced by the reduced cost of transportation. The land is usually far cheaper as well.

Some industries have especially high transportation costs. The steel industry is a good example: the raw materials for steel-making are heavy and bulky. Steel mills must be located on a waterway, because the cheapest way to move iron ore and coal is by water.

5-14 Factories along the Queen Elizabeth Way, Mississauga, Ontario.
A major locative consideration is easy access to transportation, so factories often line the highways leading into major cities.

Shipping over land by pipeline is also inexpensive, but it is only suitable for some products such as oil and gas. Oil refineries and chemical plants will locate near pipelines — sometimes near the oil fields, sometimes near the markets. Railways are still the most economical method of land transportation for bulky, heavy products for which pipelines cannot be used. Trucks may haul the wheat from a Saskatchewan farm to the nearest grain elevator, but that elevator is always beside a railway line. It is trains that transport the wheat to a market or a port.

In recent decades, trucks have replaced trains and even ships for the movement of many products. A truck can take the materials quickly and directly to the factory where they are needed. Since industries tend to cluster in manufacturing regions to be near both their suppliers and their markets, the trip can often be accomplished overnight.

Aircraft are costly but they have the great advantage of speed. Perishable products such as flowers and specialty foods, and small but valuable products such as diamonds or "information," are often sent by air.

1. (a) Explain why it is often an advantage for an industry to move to the suburbs from a city's core.
   (b) Why would some industries located in central urban areas not relocate to the suburbs?

260

2. Why are trains not used to pick up the wheat from prairie farms?
3. (a) If a factory had to spend five times as much to ship the finished product as to send the raw material the same distance, how would the location of the factory be affected?
   (b) Name four industries which would be similar to the one described above.

Labour: People Power for the Process

Some manufacturing firms, such as the clothing and furniture industries, are labour-intensive. They require a lot of workers in order to produce their products. Some industries require very little labour; for example, ten skilled workers can operate a modern chemical plant to make ethylene. Of course, that plant may have cost $500 million to build.

More and more industries now require skilled workers; to make it easier to find and hire these people, most firms try to locate in a large urban centre. Although some people are prepared to move to find work, the majority resist doing so. This is particularly true with skilled, experienced workers, who probably have good jobs already.

One of the best examples of labour as a locative factor concerns the so-called **information industries**. These industries require a pool of highly specialized and skilled personnel, and will locate their operations in order to be near it. The computer industry is a good example: IBM chose Don Mills as the site of its North American software laboratory. It did so because the Toronto area has seven universities and because a number of other computer companies are located

5-15 International Business Machines complex, Don Mills, Ontario.
IBM has a huge complex of factories and offices in this Toronto location for several reasons. What would some of the major locative factors be?

there. Also, Toronto is only a few hours away from IBM headquarters in Armonk, near New York City. Today the Toronto lab employs about 1500 computer experts. This helps make Toronto one of the most important high-tech centres in the world. Ottawa and Kitchener-Waterloo, both in Ontario, are also important centres for the computer industry.

A large urban centre will often grow at the expense of smaller ones, since it has a huge and varied labour pool. Of course, smaller centres do have some advantages that may make up for their lack of skilled labour. Sometimes wages in smaller cities are lower than those in larger ones — perhaps because work is

harder to find. Also, many people enjoy living in smaller communities. Certainly, land costs will be lower outside the major centres. Every company has to weigh all the different factors when deciding where to locate. The ability to attract skilled workers is often a major factor.

Unemployment in Canada's labour force has been a serious problem for many years. Three major reasons are foreign competition, automation, and the decline in the fishing industry.

Canada's major foreign industrial competitors have generally enjoyed fuller employment for their workers. Between 1970 and 1997, the unemployment rate in Canada averaged 9.1 percent.

France's unemployment was 11 percent, Japan 3 percent, the United States 5.6 percent, and the United Kingdom 8 percent.

High wage rates have forced some companies to relocate outside Canada. The weekly average industrial earning in Canada in 1996 was $586 — well above that of most other countries. For example: it was estimated that Alcan, Canada's major aluminum producer, would have to spend $51 million annually to employ a thousand Canadians. To employ a thousand Brazilians would take only $10 million. Obviously there is great financial pressure on Canadian companies to locate in countries where wage costs are lower.

Automation is another factor sometimes blamed for high unemployment in Canada. Industries today are still finding faster and less expensive ways to produce goods. In the eighteenth century crude machines replaced workers; now robots, computers and complex automated machines are replacing simpler machines. But at the same time, automation opens up other jobs: when robots reduce the number of workers in a car-parts factory, they also increase the number of workers in the robot-making factory. The same principle applies to computers and complex machines: as many workers may well be needed, but they will be different *kinds* of workers. Governments, workers and business managers all hope that automation will create more jobs than it destroys, just as the original Industrial Revolution did.

Automation will be essential if Canada is to compete with other nations

5-16 *A welder at work. Arc and spot welding are two popular application areas for robots.*

in the global marketplace. The Robotics Industry Association (based in the U.S.) reported that unit shipments of robotic equipment in Canada and the United States have increased 132 percent over the past five years. The dollar increase is even more dramatic as close to $1.1 billion U.S. dollars worth of equipment was shipped in 1997 – an increase of 136 percent over 1996. 1997 saw orders placed for 12 149 robots in North America.

During the 1990s, use of robots increased in the materials handling industries, in addition to their more traditional uses in the automotive field. Canadian robotics manufacturers have designed and built equipment for use in food analysis, laboratory automation, packaging, genomics automation, palletizers and drug discovery.

Automation puts some people out of work, but one of the areas in which Canada has a serious shortage of workers is the blue collar trades. Welders, tool-and-die-makers, and machinists are essential, even in the automated age.

Markets: Industries Need Customers

The "market" refers to the people who are available to buy the goods or services a company has to offer. It could be considered the first locative factor: no company or intelligent entrepreneur would invest millions of dollars unless there was a demand for the product. But the market can also be considered the final factor: selling products to customers is the final stage in the business process.

There are basically two kinds of markets. One is well-known, for it includes products for people to use — everything from orange juice to shoes to cars. These are called **consumer goods**. However, many companies make products that are used by companies or individuals to manufacture other goods, or to provide services. Such products are called **capital goods**, and they include 20-t metal presses, five-gram industrial needles for a clothing manufacturer, automotive transmissions, dentists' drills and large trucks.

The market's location often decides a factory's location. Being close to the market obviously saves on transportation costs. Cities act as magnets for industries that make consumer goods. Urban centres predominate in the food-and-beverage industry because the final products — baked goods, for example — are more perishable than the flour and other materials needed to make them. In the case of capital goods, map

5-17 West Edmonton Mall, Edmonton, Alberta. *A large shopping mall must attract large numbers of customers. Its location in or near large metropolitan centres is essential. Note two other factors in this photo which will attract customers.*

5-18 shows how the car-parts manufacturers locate reasonably close to their market, which is the auto-assembly plant.

Automobiles are assembled near the markets because parts are easier and cheaper to ship than the vehicles themselves. In order to manufacture automobiles at a reasonably low cost, the factories must be huge — able to make hundreds of thousands of cars every year. As a result, the assembly plants cluster near the huge markets in Ontario and Quebec. It makes more economic sense to ship finished vehicles to smaller markets later — and these "smaller" markets, for auto-makers, include some

very big cities such as Vancouver and Winnipeg.

When a manufacturer sells a "made in Canada" product in Canada, it is being sold on the **domestic market**. If the product is being sold throughout the country, it is being sold on the **national market**. If it is sold only near where it is made, it is considered to be for the **local market**. Many Canadian companies have sold, or hope to sell, goods in the U.S.A. and other countries. Any foreign country is part of the **foreign market**. Finally, more and more companies are assembling goods in one country from parts made in several other countries and selling them in many countries

263

around the world. This is the global market, and it is growing quickly, especially in the electronics and automotive industries.

1. (a) Why is it less costly to ship one tonne of lumber than one tonne of furniture the same distance?
   (b) How does the answer to (a) affect the location of furniture factories?
2. (a) Name ten or more consumer goods which are within ten or more metres of where you are now.
   (b) Name the items in your answer which might be sold on the global market.
   (c) Name five capital goods located in your community or school.
   (d) How might one car be considered a consumer good, but another car be considered a capital good?

New Words to Know

Consumer goods	National market
Capital goods	Local market
Domestic market	Foreign market

Other Factors: Government or History May Decide

In addition to the six major factors determining the location of an industry, there is a seventh that can sometimes be more important than the other six combined. This seventh is the influence of government, history or a special factor.

Canada's federal government, through a number of different programs and agencies, often provides grants to industry. Such grants may require that the

5-18 Automobile parts factories cluster near their market—the assembly plant.

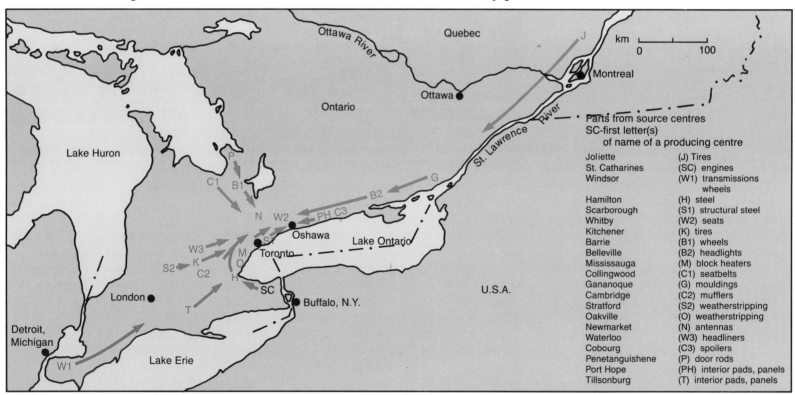

This map shows only a few of the locations and parts involved in the manufacture of just one model, the GM Lumina, in Oshawa. Most of them stretch along the main street of Canada's industrial heartland, Highway 401 in Ontario. GM attempted to find suppliers within a four-hour trucking radius of Oshawa. Some plants are not shown and some are beyond the map in the U.S.A. Oshawa itself has many plants.

1. Using your atlas, prepare a map similar to 5-18 and name all the parts-producing centres.
2. Refer to map 5-50. Why would many of the plants on map 5-18 be closely related to the centres that appear on map 5-50?

264

company locate in a depressed area of the country where unemployment is high. A provincial or municipal government may also offer a company temporary relief from taxes, grants, free land or interest-free loans. A company's decision on where to locate may well be based on which government is ready to provide the largest grant. Grants often total millions. It was estimated that in 1997, over $8 billion was provided to businesses in this country by governments at all levels. Some Canadians object to this assistance, especially when it is given to huge, profitable corporations. Others feel that the grants create jobs, strengthen the local economy and generate tax revenue.

To mention only one example, the 1998-9 federal government funding of the Atlantic Canada Opportunities Agency was budgeted at $260 million. This agency attracts small businesses to the Atlantic Canada region which needs new industry badly. The unemployment rate for the area is considerably above the national average, and the decline in the fishing industry makes the problem worse.

Another way the federal government affects the location of industry is through the granting of industrial contracts. The government is a huge purchaser of goods, particularly for defence. It tries to spread the contracts across the country so all regions benefit. Competition among companies, cities and regions is usually very keen.

Historical factors can also affect the location of an industry. Some firms established themselves long ago in older, central areas of cities. The reasons at the time were likely good ones — perhaps the location was near a railway line or, as was the case in Montreal, near the Lachine Canal. But the conditions may have changed over the years. If they have, with increasing costs resulting, such industries may still have to stay where they are because the costs of moving are even more prohibitive than the costs of staying.

Safety and the environment can also be crucial concerns when it comes to locating plants involved in certain industries, particularly chemical and nuclear plants, or those with hazardous wastes or emissions. Their location will be restricted by zoning and environmental regulations. Even earthquake risk becomes a factor when the storing of nuclear material is being considered.

On a smaller scale, a factory may require a specific type of site. Generally speaking, industrial plants require large, flat lots on which the services have already been installed. These lots have to be accessible to public transportation. Once a decision is made to locate in Calgary, Belleville or St-Jean, then a suitable site must still be found in the chosen city. Sometimes the suitability of the available sites is a key factor in the company's decision.

5-19a Retail trade in Canada 1997.

	VALUE ($ MILLION) 1997
Canada	**237 277.7**
Food	55 949.5
Supermarkets and grocery stores	51 655.4
All other food stores	4 294.1
Drug and patent medicine stores	12 297.8
Clothing	13 384.9
Shoe stores	1 649.9
Men's clothing stores	1 569.5
Women's clothing stores	4 335.2
Other clothing stores	5 830.3
Furniture	11 605.2
Household furniture and appliance stores	9 305.3
Household furnishings stores	2 299.9
Automotive	92 765.0
Motor vehicle and recreational vehicle dealers	62 830.1
Gasoline service stations	16 306.7
Automotive parts, accessories and services	13 628.2
General merchandise stores	26 182.6
Other retail stores	25 092.7
Other semi-durable goods stores	8 187.8
Other durable goods stores	6 008.2
All other retail stores	10 896.7

Source: Statistics Canada.

5-19b Wholesale Sales 1997.

	VALUE ($ THOUSANDS) 1997
Canada	**325 512 007**
Food Products	50 248 734
Beverage, drug, tobacco products	21 252 635
Apparel and dry goods	6 199 272
Household goods	9 124 729
Motor vehicles, parts, accessories	57 468 580
Metals, hardware, plumbing and heating equipment and supplies	21 537 401
Lumber and building materials	25 576 565
Farm machinery, equipment and supplies	8 770 589
Industrial and other machinery, equipment and supplies	48 398 808
Computers, packaged software and other electronic machinery	29 228 764
Other products	47 705 932

Source: Statistics Canada

1. Staple goods are defined as those products which are absolutely necessary for life. What are the staple goods in Chart 5-19a.
2. How does chart 5-19a differ from 5-19b? Explain.

3. A third method of sale tracked by Statistics Canada is called a direct sale. What do you think this term means? Cite some examples of direct sales.
4. What are the various ways in which governments might assist companies to attract them to certain areas.

Secondary Industries

The secondary industries consist of both manufacturing and construction, but far and away the true bread-and-butter industry of Canada is manufacturing. In 1997 it employed 2 191 000 people, almost three times the number in the primary sector. If the 748 000 workers employeed in construction are added, then 2 939 000 workers in all were employed in secondary industry. Service or tertiary industries employed three times as many people as secondary ones, but the former depend on the goods and wealth created by the latter. Manufac-

turers provide the goods that retailers sell, governments tax, truckers ship, advertisers promote and bankers finance. Much of the money that pays for services is generated by manufacturing activity.

Manufacturing is usually broken into two types. **Primary manufacturing** processes the natural resources that the primary industries produce. Steel mills, flour mills, fish canneries and pulp-and-paper plants are all primary manufacturers. **Secondary manufacturing** takes the products made by the primary

manufacturers and turns them into finished products: they turn textiles into clothing, plastic into toys, paper into books. The secondary class is much larger than the primary. In all, there are over 47 000 factories in Canada involved in primary and secondary manufacturing.

As a nation, Canada has many advantages which encourage the growth and development of manufacturing. Canada has rich — though limited — resources. The country has, in many areas, a huge potential for hydro-electric power, so

5-20 Secondary industry. *Secondary industry includes both manufacturing and construction. The photographs show a smelter in* *Sudbury, where ore is turned into metal, and a hotel under construction north of Toronto.*

much in fact, that Quebec can sell its excess hydro to the U.S.A. Canada's labour force is reasonably large, skilled and well-educated. This country also has considerable capital for developing its resources. It is also a stable democracy, which helps attract additional capital from other countries. Our industries are varied and have been in place for many years. They are supported in a variety of ways by all levels of government.

These advantages are not found to the same degree in every province; they are found most particularly in southern Ontario and Quebec, Canada's industrial heartland. In terms of total value, the two produce about 77 percent of Canada's manufactured goods. Well over half of Canada's market is found in those areas as well. They have one distinct advantage over the rest of the country: they are very close to the huge markets and manufacturing centres of the northeastern U.S.A.

Manufacturing in Canada is often more costly than in the U.S.A., for several reasons. This country's large size makes transportation expensive, especially for industries selling on the national market. Canada's domestic market is small: its population is only about one-fifth that of Japan, one-tenth that of the U.S.A. and one-thirteenth that of the EEC. For Canadian industry to be competitive, it *must* sell to foreign markets. The benefits of **mass production** — and therefore of lower production costs — are only possible in a large market. This is a major reason for Canada's industries trying more than ever to export their products and encourage free trade.

5-21 Manufacturing. *There are two kinds of manufacturing, primary and secondary. The processing of potatoes at McCain's in New Brunswick is an example of primary manufacturing; the fabrication of metal file cabinets at Steelcase in Markham, Ontario, illustrates secondary manufacturing.*

Manufacturing in Canada is portrayed in graphs 5-22a & 5-22b. Fig 5-22a shows what percentage of the Canadian economy is devoted to manufacturing, while Fig 5-22b shows how Canada's manufacturing is distributed by province. Notice how widely the provinces differ in the value of shipments – the value of manufactured goods they produce.

Chart 5-23 shows Canada's industries by type and relative importance. There are several very large industrial groups: Transportation Equipment; Primary Metals; Food Products, and Wood Products, to mention a few. Note that the total value of the eleven smallest industries does not even come close to the value of the largest industry.

5-23 Manufacturing shipments in Canada by industry.

Industry Group	Value in Billions	
	1997	1988
1. Transportation Equipment	99.0	47.3
2. Food Products	49.9	37.9
3. Chemical Products	30.3	22.5
4. Paper Products	29.6	22.3
5. Electrical/Electronic Products	29.1	17.7
6. Primary Metals	27.9	21.6
7. Wood Products	27.0	14.9
8. Fabricated Metal Products	22.7	17.2
9. Petroleum Refining	22.5	13.8
10. Machinery	17.5	9.8
11. Printing & Publishing	15.3	12.4
12. Plastic Products	9.2	5.3
13. Non-metallic Minerals	7.7	7.8
14. Beverages	7.1	6.0
15. Clothing	6.6	6.5
16. Furniture	6.1	4.2
17. Rubber Products	4.3	2.3
18. Primary Textiles	3.8	3.2
19. Textile Products	3.5	3.2
20. Tobacco Products	2.8	1.8
21. Leather Products	1.0	1.3
22. Other Manufacturing	7.6	5.4

5-22a The structure of manufacturing in Canada.

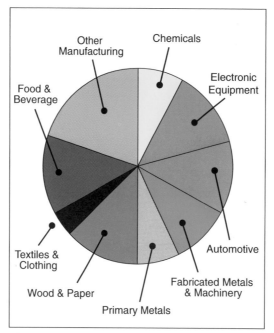

5-22b Manufacturing shipments in Canada by province.

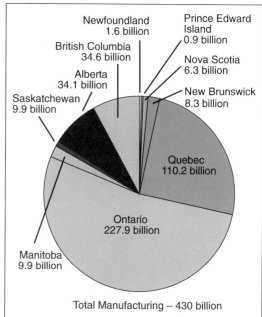

5-23b *The Canadian Polystyrene Recycling Association (CPRA) was formed in 1989 to "recycle products made from polystyrene plastic in an environmentally acceptable manner, and to increase public awareness to encourage polystyrene recycling." The association operates a 82 000 km² recycling plant in Mississauga, Ontario, where it receives product from over 500 industrial, commercial and institutional companies in the Greater Toronto Area. The photograph shows CPRA's post-consumer processing line.*

1. **Using chart 5-23, determine:**
   (a) what percentage of the total value of shipments for Canada is produced by the top ten industries.
   (b) at least five pairs (or trios) of industries that are closely related to each other. (There are scores of relationships among the industries listed; for example, many foods require steel or aluminum as well as chemicals for canning.)
   (c) those industries which are directly related to the manufacture of automobiles by the goods they produce, and which are primary-manufacturing industries.

New Words to Know
Primary manufacturing   Mass production
Secondary manufacturing

Manufacturing in Canada's Regions

The Atlantic Region   The Atlantic provinces are weak in manufacturing compared with Canada's other regions. Newfoundland, Nova Scotia, New Brunswick and Prince Edward Island are all among the five lowest provinces in terms of value of manufactured shipments. For many years the federal government has been attempting to improve the Atlantic region's economy and decrease unemployment there. Both problems are closely related to the low level of manufacturing there.

Equalization payments have been given to the Atlantic provinces, and **incentive grants** have been made to the industries there, in an effort to strengthen the economy. These efforts have only been partly successful.

Though economically weak compared with the other regions, the Atlantic region does have several advantages for manufacturing. It has a big and varied resource base. Mining, forestry, farming and, of course, fishing are all important in the region, though their distribution is generally scattered. As well, the region has considerable energy resources. There are good reserves of coal and huge hydro-electric potential. Most of the latter is in Labrador. Finally, the region has a well-educated population which could be trained for industrial jobs.

But the region is also saddled with some severe handicaps to industrial development. The first and foremost of these is its small market. The total population of Atlantic Canada is smaller than that of Alberta, Canada's fourth-largest province by population. And that population is scattered: along the coasts, along the river valleys and in isolated forestry or mining towns. The largest city and only true metropolis is Greater Halifax. It is the region's financial and commercial centre with 300 000 people.

The sea is the main geographic factor in this region; it both helps and hinders the economy. Goods shipped by sea must generally pass through one or more break-of-bulk points (where goods must

5-24   Saint John River Valley, New Brunswick. *This landscape is suggestive of the scattered settlement pattern in much of the Atlantic Provinces. Huge urban areas are the exception.*

move from one carrier, say a ship, to another, perhaps a train) to reach the inland consumer. This adds to their cost. However, ocean-borne products such as crude oil need refining, naval stations must be supported and fish-processing plants must be supplied. All of these economic activities are best located along a seacoast, as are repair and ship-building facilities.

Because the region's natural-resource base is large and varied, primary manufacturing — to process these materials — dominates the economy. Metals are smelted at Belledune, New Brunswick (lead), and Sydney, Nova Scotia (iron ore); pulp and paper is produced at centres such as Grand Falls, Newfoundland, Port Hawkesbury, N.S., and Bathurst, N.B.; fish is processed in many centres, such as Lunenburg, N.S., and St. John's, Nfld.

Food processing from farm products is also important, with potatoes being prepared in plants in New Brunswick and Prince Edward Island, particularly for the fast-food market. Apples from the Annapolis Valley are also processed in a variety of ways, for shipment to other parts of Canada and abroad. This survey barely begins to outline the considerable resource-based primary manufacturing in the Atlantic region.

Manufacturing in the Atlantic region is concentrated in the larger cities.

In addition to major oil refineries in Halifax and Saint John, each city has shipbuilding and repair industries. Offshore oil development and the increasing size and role of the Canadian navy have also brought growth to that city. Halifax has long been Canada's chief naval base.

5-25 Irving refinery, Saint John, New Brunswick. *This is just part of Canada's largest oil refinery. It is dependent on foreign sources for its crude oil, which is delivered by tanker.*

The Navy has also had a major effect on industrial development in St. John, New Brunswick, which has long been one of the most important ports on the eastern North American coast (see Fig 7-8). St. John has five dry docks, including the largest one in Canada. Often considered the industrial hub of Atlantic Canada, St. John has Canada's largest oil refinery, two large pulp and paper mills, and the Lantic Sugar refinery – the country's largest. Potash and petroleum are the two major cargoes of the port. The New Brunswick Telephone Company, Moosehead Breweries, and New Brunswick Power, Irving Oil, Irving Paper and St. John Shipbuilding are major employers.

There are important secondary-manufacturing activities in such urban centres as Moncton in New Brunswick and Trenton, Lunenburg and Bridgewater in Nova Scotia, but again the industries are generally related to transportation.

1. Draw a sketch map of the Maritimes, making New Brunswick a square, Nova Scotia a rectangle with a hook, and so on. Label the names of the provinces, every urban centre named above, and the major water bodies. Add two useful latitude lines, two longitude lines, a north arrow and a scale. Then mark economic data on the map, according to where this text says industries are found, by devising small symbols for pulp and paper, shipbuilding and so on. Remember that the text only mentions examples of the industries; it does not provide all the centres where any one industry is found.
2. How could the distribution of each of the following in Atlantic Canada be briefly described?
   (a) fishing villages
   (b) farming areas
   (c) pulp-and-paper mills
3. Explain briefly the major locative factor which has resulted in there being less manufacturing in the Atlantic region than in Canada's other regions.
4. The capitals of New Brunswick and Prince Edward Island were not mentioned in this section. Why were they not mentioned? How would you describe these cities in terms of their function?
5. How do the following terms or phrases relate to manufacturing in Atlantic Canada?
   (a) mass production   (d) Saint John
   (b) food industries   (e) Volvo
   (c) farming products

New Words to Know

Equalization payments   Frigate
Incentive grants   Dry dock

The Central Provinces Ontario and Quebec are Canada's largest provinces in both area and population. They are also the largest in terms of the value of

5-26 Highway 401, *shown here in the city of North York, is Canada's busiest transportation link. The trucks indicate how important it is in linking industry to industry and industry to consumer.*

manufactured goods produced. In 1997, $327 900 000 worth of goods was shipped from plants in those two provinces. This was 76 percent of Canada's total. These two provinces also have 62 percent of Canada's population, which makes it clear why Canadian manufacturing is concentrated where it is. The concentration of the work force is even more obvious; about seven out of ten manufacturing jobs in the country are in this region.

The vast majority of manufacturing takes place in the south, in the Great Lakes/St. Lawrence Lowlands. It is often called "Canada's Industrial Heartland." It has earned this nickname especially because of its manufacturing power. The rest of Canada, beyond this heartland, is referred to by some geographers as the **hinterland**. It is the

hinterland which provides many of the raw materials processed in the heartland, and some of the market for finished goods.

The industrial-heartland region is also referred to sometimes as the Windsor-to-Quebec axis or corridor. It makes a lot of sense to refer to a manufacturing region by its urban centres, for it is in those centres that the manufacturing is concentrated. Cities dominate the industry and commerce of the region. This is particularly true of Toronto and Montreal, each of which has over three million people. With their huge and varied production, and their large numbers of corporate headquarters (Toronto alone has 40 percent of the nation's total), they strongly influence the economy of the nation as a whole. Economic decisions made in Toronto and Montreal

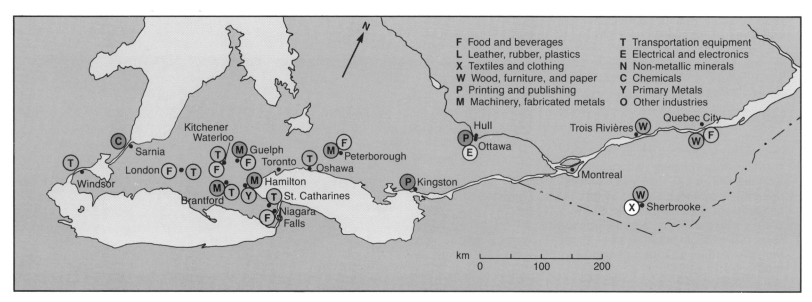

5-27 Manufacturing in major urban centres in the industrial heartland.

Map legend:

- **F** Food and beverages
- **L** Leather, rubber, plastics
- **X** Textiles and clothing
- **W** Wood, furniture, and paper
- **P** Printing and publishing
- **M** Machinery, fabricated metals
- **T** Transportation equipment
- **E** Electrical and electronics
- **N** Non-metallic minerals
- **C** Chemicals
- **Y** Primary Metals
- **O** Other industries

5-28 *Royal Mat, based in Quebec's Beauce region, is a pioneer in the manufacture of products from recycled tires. Royal Mat's reclaimed rubber products include soundproofing panels, splash guards for trucks and trailers, solid tires for wheeled bins and trolleys, and honeycomb mats for equestrian centres, fitness clubs, and daycare centres. The company recycles over one million tires a year.*

5-29 Ontario and Quebec share of manufacturing

INDUSTRY GROUP	PERCENT PRODUCED BY	
	ONTARIO AND QUEBEC	ONTARIO ALONE
1. TRANSPORTATION EQUIPMENT	95	82
2. FOOD	65	42
3. PAPER & ALLIED PRODUCTS	59	28
4. CHEMICALS & CHEMICAL PRODUCTS	74	51
5. PRIMARY METAL	86	53
6. ELECTRIC, ELECTRONICS	90	60
7. FABRICATED METAL	79	58
8. WOOD	38	14
9. PETROLEUM & COAL PRODUCTS	54	36
10. PRINTING, PUBLISHING	78	50
11. MACHINERY (NON-ELECTRICAL)	71	53
12. NON-METALLIC MINERAL PRODUCTS	69	45

1. Which types of industries have the highest percentages?

2. If an industry has a low figure in the first column, what does that likely reveal about its distribution in Canada?

3. How do the percentages in the first column compare to the percentages of the population in Ontario and Quebec? Why is this comparison essential to understand what the percentages actually mean?

4. Which industry (or industries) has a higher percentage located in Quebec than Ontario?

affect Canada from coast to coast.

The region's largest urban centres cluster along or close to the Macdonald-Cartier Freeway (Ontario Highway 401) and along the St. Lawrence River and Trans-Canada Highway (Quebec Highway 20). These transportation links, the busiest in Canada, are sometimes referred to as "Canada's Main Street."

There is a great variation in how industries are distributed, not only across the country, but across the heartland as well.

Every major industrial category is represented in southern Ontario and Quebec.

Ontario's importance is obvious in chart 5-29. Ontario is the leading province in 16 of the industries on chart 5-23, by value of shipments. Quebec is number one for paper and paper-related products, clothing, primary textiles, textile products and tobacco products. British Columbia, as might be expected, leads in the value of wood produced; in every other industry, however, Ontario and Quebec are the two leading manufacturing provinces.

Ontario and Quebec almost overwhelm the rest of the country in some manufacturing sectors. Other industries, often involving primary manufacturing — food, paper, wood and the like — are more evenly spread through the country.

5-30

Warehousing, materials management and distribution, or logistics – getting raw materials or finished goods to the right place at the right time – is important to any manufacturing operation. This warehouse is operated by Ryder Integrated Logistics, a leading provider of logistics services in Canada. Bar coding, radio frequency communication to electronically controlled inventory, pallet flow or pick to light racking, and remote operation lift trucks are some of today's modern logistics tools.

5-31 Technology levels in industry.

HIGH-TECH
MOTOR VEHICLES
COMMUNICATIONS
ELECTRICAL PRODUCTS
RUBBER AND PLASTICS
METAL MINES
TRANSPORTATION EQUIPMENT
MOTOR VEHICLE PARTS AND ACCESSORIES
MACHINERY
CHEMICALS
TRADE
CONSTRUCTION
COMMERCIAL SERVICES
FINANCE, INSURANCE, AND REAL ESTATE
PETROLEUM AND NATURAL GAS

MID-TECH
AGRICULTURE
MISCELLANEOUS MANUFACTURING
TEXTILES
NON-METAL MINES (EXCLUDING COAL)
COAL MINES
PAPER
IRON AND STEEL

LOW-TECH
NON-METALLIC MINERAL PRODUCTS
PRINTING AND PUBLISHING
FISHING AND TRAPPING
LEATHER
FURNITURE AND FIXTURES
WOOD
FABRICATED METALS
KNITTING AND CLOTHING
TOBACCO
NON-FERROUS METALS
FOOD AND BEVERAGES
FORESTRY
PETROLEUM AND COAL PRODUCTS

The greater the amount or complexity of technology required to make a product, the higher the technology level. Thus to produce a computer requires much more complex materials, more skilled labour, and more complicated processes, than to produce a piece of lumber. In the near future, the high-tech industries are expected to see the greatest growth, much of it in Ontario and Quebec.

Jobs for the Future, Today

Consumer confidence rose for the ninth straight quarter in early 1998, and so far this year, retail sales have grown by 10 per cent.

Housing starts increased by 25.6 per cent in 1997. Our Land Transfer Tax refund has helped more than 28,000 families buy their first new homes. The value of business building permits issued in 1997 rose by 32.6 per cent.

Ontario's international exports are up by 8.3 percent so far this year. The Export Marketing Task Force is working to increase Ontario's share of the global export market.

Women entrepreneurs are playing an ever-increasing role in the expanding Ontario economy. The Joseph L. Rotman School of Management at the University of Toronto is developing new business leaders with its highly successful management program. In partnership with the private sector, it is recognizing outstanding successful women entrepreneurs through its Women Entrepreneur of the Year Award.

The 1997-98 Ontario deficit will be $5.2 billion. This represents a reduction of almost $1.4 billion from the $6.6 billion target for 1997-98 set out in the 1997 Budget, and includes $725 million for the cost of an agreement with teachers providing for an early retirement opportunity and other benefit enhancements.

The deficit for 1998-99 will be $4.2 billion, $0.6 billion lower than the deficit target for this year set in the Balanced Budget Plan.

The Balanced Budget Plan will ensure that the deficit is eliminated by the year 2000-01.

The Ontario economy created 265,000 net new private sector jobs between February 1997 and February 1998. This was the largest number of jobs created in a 12-month period in the province's history. In the first quarter of this year, the Ontario economy created jobs at a rate unequalled in the past 15 years.

Ontario's unemployment rate declined again in March to 7.4%.

Overall, 1998 promises to be one of the best years in Ontario's history for job creation....

Excerpted from the 1998 Ontario Budget Speech presented by The Honourable Ernie Eves, Q.C.

5-32 *Ontario Minister of Finance Ernie Eves delivered the 1998 Ontario Budget Speech –* **Jobs for the Future Today** *on May 5, 1998. The excerpts above give an upbeat outlook for the province as it heads into the new millenium. What is a strong indicator of economic growth? What is a budget deficit? How does Finance Minister Eves intend to address Ontario's deficit? What do you think areas of potential concern for the province might be?*

The Locative Factors Favour the Heartland Tens of thousands of manufacturing plants are located in the heartland of southern Ontario and Quebec. Clearly, the heartland's locative factors must be very favourable.

Capital The heartland region is not only the manufacturing centre of Canada, it is also the financial centre. The stock exchanges in Toronto and Montreal are the biggest in the country; the Toronto Stock Exchange alone handles 80 percent of the transactions in Canada, by value. Most of the major **securities**, trust and insurance firms have their headquarters in Toronto or Montreal. The great financial markets of New York City are less than two hours away by plane — though only split seconds away by telephone and computer link. Telecommunications also link the region to the great financial centres of London and Tokyo. The region's many advantages for manufacturing make it relatively easy for companies to secure capital for expansion or for new plants.

Raw materials Quebec and Ontario have huge and varied resources available in their forests, farms, mines, rivers and lakes. Many are found in the northern parts of these provinces. The primary-manufacturing industries — the paper plants, sawmills and smelters — tend to be in the north, near the raw materials. In the industrial heartland, secondary manufacturing dominates: materials, often partly processed, are brought down from the north to be finished and assembled in the south.

Energy Both Ontario and Quebec are leading producers and users of energy in Canada. Quebec, in particular, has so much hydro power, both developed and potential, that it has made multibillion-dollar agreements far into the future to export power to the U.S.A. Neither province mines coal, however. Small quantities of oil are produced in southern Ontario, but the total is not significant; both provinces import almost all their oil and natural gas requirements. Energy will not be a restriction on the heartland's future growth; in fact, inexpensive hydro power is a major reason why manufacturing has developed there.

Transportation The St. Lawrence River and the Great Lakes provide the heartland region with easy, though seasonal, access to the ocean. The largest cities in the region are all either on the Great Lakes, or on the St. Lawrence or one of its major tributaries. Of more importance to most of the region's manufacturing plants, however, is the well-developed networks of railways and highways which link the various parts of the region. The importance of the major highways shows up very clearly on map 5-50, which shows the location of the auto-assembly plants. They appear like beads on a string, the string being the major expressways in Canada and the **interstate highways** in the U.S.A. Ships bring crude oil to the region from the east, and pipelines bring both oil and gas from the west.

Labour The labour pool in the heartland is the largest, best educated and most varied in Canada. Many companies place advertisements for skilled workers in this region even when the jobs themselves are located outside of it. The region has the highest concentration in

Canada of universities and colleges. These train the skilled personnel manufacturers need.

Market Over 18 million people live in Ontario and Quebec, most of them in the southern heartland. They form the largest, most concentrated market in Canada. Manufacturing industries pay high wages, which means that "purchasing power" is higher in this region than in the rest of the country. It is generally

more economical for a company to produce goods near this large market and ship them to the smaller, scattered markets in the hinterland according to the demand in those areas. Finally, the heartland is close to the huge market in the northeastern U.S.A. Most geographers look on this Canadian region as an extension of the American northeastern industrial belt.

5-33 Employment by detailed industry and sex, 1997.

	Number employed (thousands)		
	Both sexes	**Men**	**Women**
All industries	13 941	7 649	6 292
Goods-producing industries	3 769	2 863	906
Agriculture	423	285	138
Other primary industries	292	251	41
Fishing and trapping	36	32	4
Logging and forestry	79	69	10
Mining, quarrying and oil wells	177	151	26
Utilities	140	106	34
Manufacturing	2 167	1 556	611
Construction	747	665	82
Service-producing industries	10 172	4 786	5 386
Transportation, storage and communications	897	662	235
Transportation and storage	582	473	109
Communications	315	189	126
Trade	2 386	1 339	1 047
Wholesale trade	655	468	187
Retail trade	1 731	872	860
Finance, insurance and real estate	795	311	484
Finance and insurance	544	173	371
Real estate and insurance agencies	252	138	113
Services	5 303	2 027	3 275
Business services	1 005	569	435
Educational services	962	362	600
Health and social services	1 425	306	1 120
Accommodation, food and beverage industries	898	368	530
Other service industries	1 013	423	591
Public administration	791	446	345

Source: Statistics Canada

1. **What percentage of the total Canadian population was employed in 1997?**
2. **What fields do women employees dominate? Why?**
3. **What industry employs the largest number of people? Why do you think this is so?**

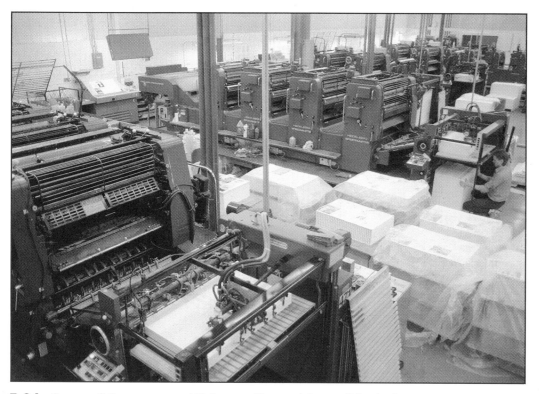

5-34 Some of the presses at Friesens Corp., Altona, Manitoba.
This book manufacturing plant is the largest one in Canada. Altona is just south of Winnipeg.

Other factors The heartland region has the historical advantage of having been settled first in Canada. Industries were already established in southern Ontario and Quebec when the West was still being explored and opened for development. As well, the rich agricultural areas of the heartland attracted their own share of farmer-settlers, who provided a good market for industrial goods.

This "early start" is a distinct advantage in manufacturing, as it attracts workers, which leads to the establishment of towns and cities. These urban areas then become markets for even more industrial growth.

1. Why are southern Ontario and Quebec often referred to as "Canada's Industrial Heartland"?
2. Four industries are identified in 5-29 as having extremely high concentrations in the heartland. Explain briefly why each of these industries is concentrated in that region.
3. Ontario is not the leading producer for six of the industries listed in 5-23. Name the six and give the leading province for each. Explain why these industries are concentrated where they are.

New Words to Know
Hinterland
Securities firm   Interstate highway

The Prairie Provinces In many ways, the manufacturing sector in the Prairie provinces resembles that of the Atlantic region. The emphasis is on primary manufacturing — on the processing of raw materials such as crude oil, wood, minerals and foodstuffs. Secondly, the market in the Prairie region is relatively small, totalling about 4.5 million people. This is about 1.7 million more than the total in the Atlantic region. Lastly, in both regions the population is spread out over a large area. However, there are more metropolis-sized urban areas on the prairies: Edmonton, Winnipeg and Calgary have over 600 000 people each. Each of these cities forms a small, concentrated market.

The population distribution is more uniform than that of Atlantic Canada. The pattern basically follows the region's rich soil and extensive flat areas, which are highly suitable for extensive farming of grains. This is a far different pattern from that in Atlantic Canada, where the landscape is stony and hilly, with only a few scattered pockets of good farmland. The northern parts of the Prairie provinces are sparsely inhabited, with occasional isolated urban areas founded on an extractive industry such as mining. This is the pattern found in much of Atlantic Canada.

Energy is reasonably abundant in the Prairie region. Oil and gas dominate there, just as hydro does in Atlantic Canada. Coal is found in both regions; both export a great deal of energy.

The labour force in both regions is well-educated. Because of its larger population and more diversified economy, there is a greater variety of skills available in the Prairie region, mainly in the oil-and-gas industry. In Atlantic Canada, oil is a developing industry.

In both the Prairie and Atlantic regions, transportation costs add considerably to the price of moving goods to outside markets. The major Canadian market for both is the central provinces. The prairies are even farther from major U.S. and European markets, so it is little wonder that cost of shipping goods by rail has been an important concern ever since the region was settled. If railway freight rates were not subsidized to some degree, most manufactured goods would not be able to compete in either the Canadian or the global markets. The laws of economics do not encourage secondary manufacturing in the region.

The dominance of the oil industry should be noted. In terms of sales, six of the top ten manufacturing companies on the prairies are oil companies. This is in contrast to Ontario, where only three oil giants are in the top ten by sales, and where three of the top five are automotive companies. Ontario also has two retail firms in its top ten; Quebec has three.

Alberta comes fourth in manufacturing among the provinces, after Ontario, Quebec and British Columbia. Manitoba is a distant fifth. It has about one-half of Alberta's population but only one-third the manufacturing shipments. Saskatchewan ranks eighth, ahead of only Newfoundland and Prince Edward Island.

Winnipeg is the major city on the prairies when it comes to secondary manufacturing, with the transportation and other machinery industries being the largest represented there. That city is also the most productive in the region

5-35 *Edmonton is a major transportation and supply centre to the North, and one of only two international cargo centres in Canada. It was cited by the Organization of Economic Cooperation and Development as one of the top eight cities in the world for continuing education. Edmonton is home to over 20 major chemical producers and the largest petroleum refineries in Canada. Edmonton's manufacturing sector employs close to 10 percent of the labour force, and contributes over $2 billion to its economy. The city is the centre for research and design in heavy oil production. It has 25 major shopping centres and produces in excess of $7.8 billion worth of retail sales.*

for leather and textile products. The Edmonton CMA had a population of 862 597 in 1996. It is interesting to compare Edmonton with Hamilton, which had a 1996 population of 624 360. These two cities present quite different pictures as manufacturing centres: one is from the industrial heartland, the other from a region where primary manufacturing dominates.

1. Explain three ways in which manufacturing in the Prairie region is different from that of the Atlantic region.
2. Why are railway freight rates of such importance to manufacturers on the prairies?
3. In the provincial ranking in manufacturing, Alberta is the most important province in the Prairie region, followed by Manitoba and Saskatchewan. Give three or more reasons why this is the case.

The Pacific Province A summary of manufacturing will not be presented here for this region — at least, not in words. Instead, prepare your own summary after studying the graphs, maps and photographs supplied in this text. Try using, in addition to a good atlas, the following sources in particular:

(a) graph 5-22, showing the value of shipments by province

(b) chart 5-23, showing the value of shipments by industry

The following facts should be included in your account:

(a) British Columbia's population was 3 724 500 in 1996.

(b) British Columbia is third- or fourth-ranked among the other provinces for most of the industries on chart 5-23.

(c) British Columbia leads all other provinces in wood industries with $14.6 billion worth of shipments in 1997, well ahead of Quebec, the second-ranked province.

(d) The CMA of Vancouver is Canada's third-largest with a 1996 population of 1 831 665.

(e) The CMA of Victoria had a population of 304 287 the same year.

5-36 Base map of British Columbia.

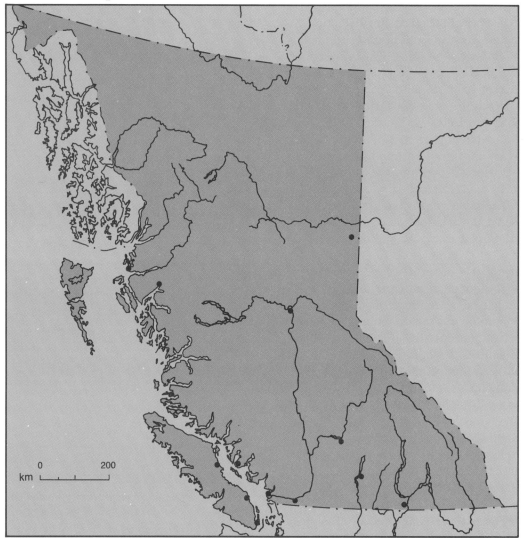

Copy this map and use it as part of your report on manufacturing in British Columbia. Major rivers, lakes, and urban centres, bordering lands, and seas should be named. Data should be added from other references, such as urban populations, mining and forestry towns, etcetera.

(f) The same year, three other cities in British Columbia had over 50 000 people: Prince George, Kamloops, and Kelowna.

First decide on a plan of organization.

You could compare British Columbia to another region. Another approach is to list the industries in British Columbia, describe their features, and discuss the reasons for those features.

278

Tertiary or Service Industries

The tertiary industries work not at making products, but at providing services. The tertiary or service sector is dominant in Canadian industry, and continues to grow. People like lawyers, tv repair personnel, truck drivers, insurance agents, medical doctors, secretaries, bankers, internet service providers, and civil servants are all members of this tertiary sector. They do not make things, they do things for other people, companies and governments.

The tertiary sector is divided into the following categories: business, community, and personal services; trade and commerce; transportation, communications, and utilities; public administration; and insurance, finance and real estate.

Community, business, and personal services

Canadians employed 5 415 000

This group comprises over one-half of all Canadian workers in the service fields. It is also a rapidly growing employment area. Occupations include nurse, teacher, barber, mechanic, advertising executive, architect, engineer, employment counsellor, management consultant, doctor, dentist veterinarian, coach, announcer, and psychologist. Over 923 000 people were employed in education and related services in 1997 compared to 1 207 700 in health and social services, and 827 900 in the accommodation food and beverage services. Many persons in the community, business, and personal services fields are self-employed, a factor which supports the rapid growth in this area.

Trade and commerce industries

Canadians employed 2 561 000

Occupations in this group include retail merchant, importer, salesperson, clerk, cashier, stockperson, telemarketer, direct mail consultant, and wholesaler. The term commerce refers to the buying and selling of goods. People working in the tens of thousands of stores across this country are known as **retailers** – they sell directly to the public from commercial outlets. 1997 retail sales in Canada were $237 billion, compared to $194 billion in 1993. **Wholesalers** supply retailers. Wholesalers buy products in bulk from producers and manufacturers for redistribution in smaller quantities to retail outlets. Wholesalers operate from a warehouse. A recent phenomenon is the rise of wholesale or warehouse stores. These no-frill outlets stock items in bulk, and are self-service operations with few customer amenities. Sometimes these stores require that customers have a store membership and proper identification prior to being able to shop within. Warehouse stores tend to discount goods to offer a price lower than the regular retail. Thus warehouse stores may operate on less profit margin than retail stores.

5-37 *Food retailing is a major service-oriented tertiary industry.*

Transportation, communications and utilities

Canadians employed: 1 056 000

This sector of the economy includes people in the air, railway, water, truck, pipeline, and public passenger transport industries. It takes in occupations in the fields of storage and warehousing, telecommunications – broadcasting and carriers, courier services, electric power, gas or water distribution systems, and other utilities. It is the area of greatest potential future growth for Canada as this country has a level of communications interconnectedness well beyond that of most other countries in the world today.

Public administration

Canadians employed: 938 000

This group consists of all levels of government employee – federal, municipal, provincial. The variety of fields in this level is large because our governments are involved across the board in most aspects of Canadian life. Occupations classed under public administration include M.P., M.P.P., food inspector, postoffice worker, urban planner, statistician, meteorologist, firefighter, busdriver, transit worker, member of council, mayor, law enforcement officer, customs agent, and crown prosecutor.

Finance, insurance and real estate

Canadians employed: 835 000

This section includes people employed as bankers, insurance adjusters, investment counsellors, real estate brokers, accountants, bookkeepers, collection agents, mortgagors, and stockbrokers.

Virtually every Canadian, at some time or other, has dealt with members of this sector. It is also important to remember that this sector is very much involved with

5-38(a) Canada's labour force by economic sector, 1968-1997.

	PERCENTAGE OF WORKERS IN THE LABOUR FORCE				NO. OF WORKERS, AUG. 1997
	1968	1978	1988	1997	
PRIMARY SECTOR					
AGRICULTURE	6.8	4.8	3.6	3.0	422 000
FORESTRY, FISHING, TRAPPING,& MINING	2.9	2.6	2.6	2.1	289 000
SECTOR TOTAL	9.7	7.4	6.2	5.1	711 000
SECONDARY SECTOR					
MANUFACTURING	23.3	19.5	17.2	15.6	2 191 000
CONSTRUCTION	6.5	7.0	5.9	5.3	748 000
SECTOR TOTAL	29.8	26.5	23.1	21.0	2 939 000
TERTIARY SECTOR					
COMMUNITY, BUSINESS, & PERSONAL SERVICES	24.6	28.1	32.9	38.1	5 335 000
TRADE & COMMERCE	17.0	17.4	18.0	17.1	2 397 000
TRANSPORTATION, COMMUNICATIONS, & UTILITIES	8.3	8.4	7.4	7.5	1 049 000
PUBLIC ADMINISTRATION	6.2	7.0	6.6	5.7	794 000
FINANCE, INSURANCE, REAL ESTATE	4.4	5.2	5.8	5.7	793 000
SECTOR TOTAL	60.5	66.1	70.7	74.0	10 377 000
ACTUAL WORKERS (IN MILLIONS) =	(7.7)	(10.8)	(13.1)	(14.0)	14 017 000

5-38(b) Changes in economic sectors by percentage of workers employed, 1968-1997.

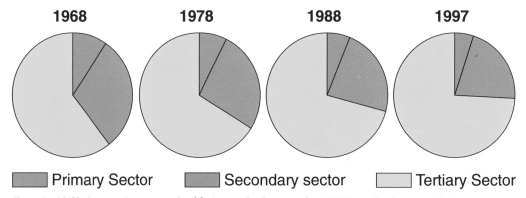

1968 **1978** **1988** **1997**

■ Primary Sector   ■ Secondary sector   □ Tertiary Sector

Even in 1968 the service sector had become the largest: by 1997 it truly dominated the economy.

manufacturing and with virtually all other industries as banks, trust companies, insurance and real estate firms are all sources of capital required for the growth and development of the economy.

Note how the tertiary sector has come to dominate Canada's economy since 1968

(Fig 5-38b). Note too, how the number of people employed in the business, community, and personal services alone, is greater than the total number of people employed in both the primary and secondary sectors. Fig 5-39c examines the changes in employment by class of worker since

1995. Note the significant rise in private self-employed persons. Changing economic climates have curtailed the permanent, long-term employment patterns which used to characterize the Canadian workforce. Downsizing of traditional industries, casual or part time labour, plus shifting consumer patterns in Canada and abroad have radically altered the work-force profile. Most self-employed persons are service providers to the rest of the population.

Tertiary Industries and Location Factors

A person or company providing a service from an office is hard to compare with a factory when it comes to deciding upon a particular location. Service industry location factors are not dependent upon local materials, energy supply, capital or transportation. A service industry's location is determined primarily by market, and most major service industries employ expert advice to determine where the best site might be. What, for example, if you opened a ice cream stand on the north side of what appeared to be a busy downtown street, only to discover that most pedestrian traffic travelled on the street's sunny south side, and traffic was too busy to allow an easy crossing. It is impossible to provide a service if the people or companies requiring it can't get to it. Why do you think fast food outlets are often close to shopping malls? Where are gas stations usually found? Newspaper kiosks? Trucking companies? Resorts? Government offices? Stockbrokers?

Some cities have a very high percentage of service workers. These cities are usually political, financial, regional, or educational centres.

5-39(a)   Canada's labour force by province, 1996 annual average (000's).

	NFLD	PEI	N.S.	N.B.	QUE	ONT	MAN	SASK	ALTA	B.C.	CANADA
PRIMARY SECTOR											
AGRICULTURE	-	5	9	6	80	127	38	71	84	34	454
OTHER PRIMARY INDUSTRIES	21	4	19	17	54	42	8	12	79	61	317
SECTOR TOTAL	**21**	**9**	**28**	**23**	**134**	**169**	**46**	**83**	**163**	**95**	**771**
SECONDARY SECTOR											
MANUFACTURING	20	6	48	39	649	992	64	30	113	214	2175
CONSTRUCTION	19	5	27	24	186	352	31	24	106	166	940
SECTOR TOTAL	**39**	**11**	**75**	**63**	**835**	**1344**	**95**	**54**	**209**	**380**	**3051**
TERTIARY SECTOR											
COMMUNITY, BUSINESS, & PERSONAL SERVICES	88	23	160	124	1300	2082	199	171	534	733	5414
TRADE & COMMERCE	44	10	80	63	613	976	95	87	261	332	2561
TRANSPORTATION, COMMUNICATIONS, & UTILITIES	19	4	31	30	247	385	50	37	113	140	1056
PUBLIC ADMINISTRATION	21	7	39	29	232	341	43	31	90	105	938
FINANCE, INSURANCE, REAL ESTATE	7	-	20	15	189	360	30	24	74	116	835
SECTOR TOTAL	**179**	**44**	**330**	**261**	**2581**	**4144**	**417**	**350**	**1072**	**1426**	**10804**
TOTALS	**239**	**64**	**438**	**347**	**3596**	**5708**	**563**	**487**	**1462**	**1914**	**14818**

5-39(b)   Comparison of labour force in Ontario and British Columbia.

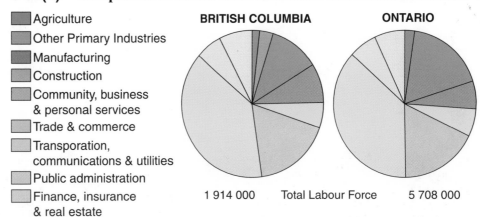

Agriculture
Other Primary Industries
Manufacturing
Construction
Community, business & personal services
Trade & commerce
Transporation, communications & utilities
Public administration
Finance, insurance & real estate

BRITISH COLUMBIA   ONTARIO

1 914 000   Total Labour Force   5 708 000

The graphs of Ontario and British Columbia display some striking differences. In Ontario only .7 percent of the labour force is involved in Other Primary Industries; in B.C. the percentage is four times greater, because of the importance of fishing, mining, and especially forestry Note that the number of workers in those industries in Ontario, is actually two thirds of the number in B.C. Percentages of workers in some sectors may vary between provinces; is this true in the tertiary group?

Another great difference is in the number of workers in manufacturing, 17.4 percent in Ontario, and 11.2 percent in B.C.

1. **Choose two other provinces, draw similar graphs, and compare them to each other, or to B.C. or Ontario.**
2. **Graph the percentage of workers in the service sector across the country. Explain the results.**

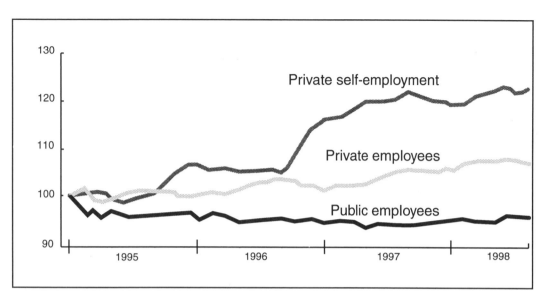

5-40(a) Index of employment growth by class of worker, January 1996 – 1998.

New Words to Know
Civil servant
Retailer
Wholesaler

5-40(b) Fast-food restaurants are a perfect example of a service industry that must choose its locations according to its market.

1. (a) Conduct a class survey of wage earners in your families. The proportions in each sector could then be compared to the national or Ontario figures.

   (b) Conduct the same survey with different groups, or with other grade levels. The differences should prove interesting. Explaining them would help you understand both your school and your community.

   (c) Conduct a similar survey in your neighbourhood. Everyone in your class should do the same. For example, each student could determine in which sector the wage earners in his or her home and the homes on either side should be classified. Collect the results: they may well indicate much about the community and its demographic patterns.

   (d) The issue of job change or transition could also arise. A survey of different jobs held by parents in their working years might reveal how much change has taken place in the sectors involved, and the advantages of possessing certain marketable skills and attitudes.

2. Name ten other occupational groups not named in the text, and place them within the five sections of the service industry. At least one occupation should be named for each section.

3. (a) Which occupational groups in services did you need in the last month?

   (b) Which occupational groups in services are likely required by a local motel owner or a local dentist?

4. McDonald's wishes to locate a restaurant in a particular community. The community has energy resources, and nearby urban centres to provide the food, and Mac's sales have provided ample capital. What will be some of the factors involved, however, when locating the exact site within the community?

The Iron-and-Steel Industry

The Basic Manufacturing Industry

In 1995 the greatest producers of steel in the world, in order, were Japan, the U.S.A., China, Russia and Germany. Most experts would also say that the world's greatest economic powers at that time were the same five nations. Canada in 1995 was the thirteenth steel producer in the world, producing just over 15 million tonnes that year. Japan produced about 102 million tonnes the same year. Whether Canada is also thirteenth as an economic power is hard to determine, but it is clear that all of the great industrial powers produce steel. It is an industry that signifies both economic power and military might. It is the basis of industry; most of the **less developed countries (LDCs)** of the world wish they had steel mills. One of the first steps on the path to industrialization for such nations is often the building of an iron-and-steel industry.

Steel is by far the most important metal in the world today. In 1995 the world produced 753 million tonnes of crude steel. Gold production is measured, sold and priced in ounces; steel is measured, sold and priced in tonnes. During 1997 gold was valued at about $446 per ounce, copper at about $1.30 a pound, and steel at $652 per tonne.

The Steel Industry Feeds the Auto Industry

The story of the automobile industry, which follows this section, actually began in Chapter Four, in which the mining of iron ore and other minerals was described. Iron ore, along

Steel is one of the most recycled materials in the world

- In North America, over 78 million tons of steel scrap is used each year for the production of steel
- Electric Arc Furnaces, like the one at Dofasco in Hamilton, use up to 100% scrap as a raw material to make steel
- Our steelmaking furnaces can recycle up to 2 million tons of scrap steel per year
- More than 60% of the steel made in North America is collected and recycled
- Steel can be recycled infinitely without losing any of its strength or other qualities

Source: Dofasco's 1997 – Environment & Energy Report

5-41   Huge continuous caster in the Stelco plant, Hamilton, Ontario.
Continuous casting is a technological improvement which is much more efficient, less costly and produces a high quality steel product. The molten steel goes directly from the oxygen furnace into the caster which produces finished products continuously.

with coal and limestone, are the major minerals needed to produce steel. After the steel is produced, a large percentage of it ends up in the auto industry's parts and assembly plants. The two major sources of steel for Canada's auto industry are the steel giants in Hamilton, Ontario — Stelco and Dofasco. They send about one-third of their iron and steel to the automotive industry and its suppliers.

The average car contains about $700 worth of steel. Auto makers have been lowering the weight of their products by replacing steel with plastics and aluminum, but steel is still by far the most important metal in every car. As an indicator of how concerned the steelmakers are about the auto-industry, consider the recent technological and production spending of Dofasco, which, in 1996, installed an Electric Arc Furnace (E.A.F.) and slab caster facility for $400 million (see diagram on page 286). This equipment produces **flat rolled steel** mostly used by the auto industry. In addition, Dofasco announced late in 1997, the building of a $180 million hot-dip galvanizing line mostly for the production of exposed auto body panels. A tube mill began production in 1997, again for use mainly in auto manufacturing. In the early 90s, Dofasco, initiated a 50-50 partnership to form Gallatin Steel in Kentucky. Gallatin has a minimill to produce flat rolled steel for cars using the E.A.F./slab caster technology. Stelco and Dofasco are both involved in a consortium (U.L.S.A.B.) of 35 steel corporations from 15 countries to develop lighter and stronger steel to be used in automobiles.

Federal U.S.A. laws require better and better gas mileage from automobile manufacturers. One of the best ways to meet this requirement is to make cars lighter. The Ultra Light Steel Auto Bodies organization, lobbies to maintain the use of steel in automobiles, rather than have it replaced by plastics or aluminum.

A steel mill is classed as a primary manufacturing plant, for it takes materials in a raw or almost raw form to produce a product. The industry is capital intensive: huge amounts of investment money are needed to build the massive blast furnaces, coking ovens, basic oxygen or electric arc furnaces, rolling mills and other facilities. Between 1991 and 1997, Dofasco alone spent one billion dollars improving and adding to its facilities in order to maintain its competitive edge and provide special steels for its major market, the auto industry. Steel often leaves the Hamilton mills of Stelco and Dofasco going directly to auto parts manufacturers and then on to auto assembly plants. There are hundreds of plants processing the steel into car parts. The Hamilton Wire Co. receives 80 percent of its materials from Stelco.

Though construction is the largest market for steel in Canada, the automotive industry is the second largest user. In 1997, when auto sales were high, Canada's steel production hit a record high. As noted above, steel producers are constantly seeking better products – lighter steels, more corrosion-resistant steels – for the auto industry. Though the steel industry "feeds" the automobile industry, in an important way the reverse is true! The recycling rate for steel from automobiles in 1998 was an amazing 98% – for steel cans it is 58%

(only 15% in 1988), and for appliances it is up from 23% to 75% over the same period. Thus the auto industry certainly "feeds" the steel industry, and steel, known widely as the recyclable metal, could truly be called the "Environmetal" of the future.

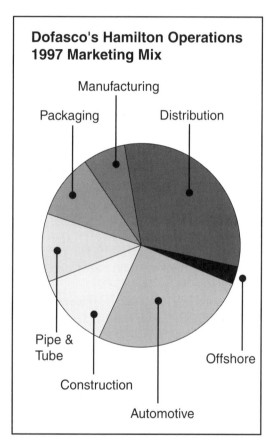

5-42   Dofasco's 1997 Market Mix.
This circle graph provides another indicator of the importance of the auto industry to just one steel corporation. Approximately what percentage of the steel went into automobiles? What would "offshore" signify?
(Source 1997 Annual Report Dofasco Steel Corporation)

The World Can Make More Steel Than It Needs

The World Can Make More Steel Than It Needs Overcapacity has been a worldwide problem in the steel industry ever since the mid-1970s. Demand has levelled off, and at the same time new technology has greatly increased production while cutting back on workers required. Canadian steel producers are managing to remain competitive, but they have been forced to spend millions of dollars on plant upgrades and automated equipment. Canadian steel production rose to 16 million tonnes in 1984, and levelled off to 15 million tonnes in 1995, but 1997 was a banner year, and Canadian steel consumption reached record highs. Larger Canadian steel companies bought out smaller companies, but the levelling off of the steel market did not hit Canadians as hard as it did members of the European Economic Community, or our American neighbours. In fact, there have been so many plant closures in the areas of Pittsburgh and Buffalo that that erstwhile **steel belt** region was dubbed the **rust belt**.

Canadian steel companies have had to be innovative both in their use of technology, and in maximizing their employee skills in order to remain profitable in the changing business climate. Collectively, they have invested over $4 billion into the industry as technological upgrades ($700 million in 1997 alone). Productivity is up over 50 percent from 1988 to 1997, and expenditures in employee training have doubled since 1994.

They are focussing on new products (over 50 percent of the automotive sheet steels have been newly developed since 1989) including new steels for use in construction, and in pipelines, as well as moving into new markets like electrical transmissions and housing.

New Words to Know

Less developed country (LDC)
Steel Belt
Rust Belt

1. Why does a strong steel industry signify "economic power and military might"?
2. Explain "Steel mills feed the auto industry, and vice versa."
3. (a) What is a major disadvantage of more efficient technology in making steel?
   (b) Must new technology be purchased?

5-43   The Canadian iron and steel industry and its sources of raw materials.

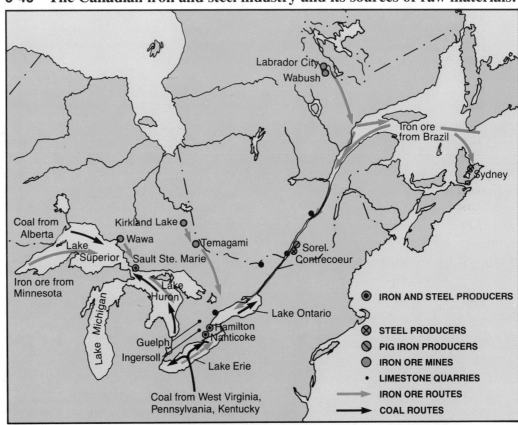

The industry is centred in the industrial heartland, especially Ontario, as this map shows. Raw material sources are mostly Canadian, except for coal suitable for making coke, which is imported from nearby American states. The importance of water transportation, and the availability of water for the processing plants is obvious.

285

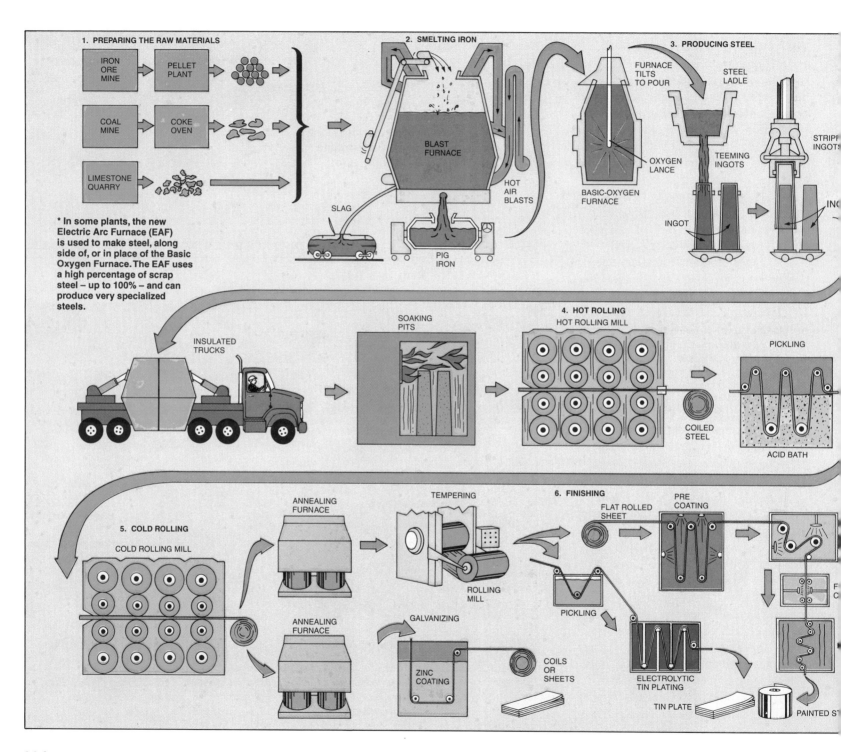

1. PREPARING THE RAW MATERIALS

IRON ORE MINE → PELLET PLANT →

COAL MINE → COKE OVEN →

LIMESTONE QUARRY →

* In some plants, the new Electric Arc Furnace (EAF) is used to make steel, along side of, or in place of the Basic Oxygen Furnace. The EAF uses a high percentage of scrap steel – up to 100% – and can produce very specialized steels.

2. SMELTING IRON

BLAST FURNACE

HOT AIR BLASTS

SLAG

PIG IRON

3. PRODUCING STEEL

FURNACE TILTS TO POUR

STEEL LADLE

OXYGEN LANCE

TEEMING INGOTS

BASIC-OXYGEN FURNACE

INGOT

STRIPPING INGOTS*

INGOT

INSULATED TRUCKS

SOAKING PITS

4. HOT ROLLING

HOT ROLLING MILL

COILED STEEL

PICKLING

ACID BATH

5. COLD ROLLING

COLD ROLLING MILL

ANNEALING FURNACE

ANNEALING FURNACE

TEMPERING

ROLLING MILL

6. FINISHING

FLAT ROLLED SHEET

PRE COATING

GALVANIZING

PICKLING

ZINC COATING

COILS OR SHEETS

ELECTROLYTIC TIN PLATING

TIN PLATE

PAINTED ST

286

5-44 Making steel.

Iron is important because it is both very strong and very common—i.e. cheap. But though it is strong, iron is relatively brittle. Most iron is used in the form of **steel**, which is iron that has been refined, **alloyed** with small amounts of other substances, and treated in any of a number of ways. Depending on how it is made, steel may have great toughness, resistance to corrosion, flexibility, or other valuable properties.

This diagram shows the major steps usually followed in making steel, from iron ore through to the flat sheets commonly used in the automobile industry. Although modified from a diagram developed by Dofasco, the basic steps are similar to those used in modern mills throughout the world.

1. Preparing the raw materials: It takes one-and-a-half tonnes of iron ore, half a tonne of **coke**, 200 kg of limestone, and 200 kg of steam to produce one tonne of iron. Of these four raw materials, two must be pre-processed before steel-making can begin: the iron ore is concentrated into high-grade pellets; and coke must be prepared by heating coal to drive off impurities that would contaminate the metal.

2. Smelting iron: When the raw materials are ready, they are loaded into a **blast furnace**—a tall, 20-storey tower that is perhaps the most striking part of a steel mill. The coke burns because of a blast of superheated air forced upward through the mixture. This raises the temperature to about 1650°C, melting the iron. The molten metal flows downward, and is drained off at the bottom as **pig iron**. The limestone combines with impurities to form **slag**, which rises, and is drawn off further up the furnace.

3. Producing steel: Pig iron is fed into the **basic-oxygen furnace** and heated. Pure oxygen is forced into the molten metal, burning off further impurities such as carbon and silicon. When ready, the steel is emptied into huge ladles, which pour it into moulds, where it hardens into **ingots**, metal slabs of up to 15t.

4. Hot rolling: While still hot, the ingot may be passed under a series of rollers at high speed and great pressure until it is reduced to a long, thin strip a few millimetres thick. It may be sold in this form as hot-rolled steel, or it may be **pickled** by passing it through a bath of acid, which cleans off surface scale.

5. Cold rolling: When it is cold, the steel may be further reduced in thickness by being rolled again. This allows the thickness to be controlled very precisely, but increases the brittleness of the steel on the inside.

6. Finishing: Finally, various finishing processes are used to prepare the steel for the needs of specific customers. **Annealing** softens the steel by heating it in an oxygen-free atmosphere. **Tempering** restores the steel to whatever grade of hardness is required, and also improves the finish. As cold-rolled steel it is then shipped for use in automobiles.

Other processes may involve coating the steel. **Galvanizing** covers the steel with a layer of zinc to protect against rust. **Tinplating** uses **electrolysis** to add a minute layer of tin (0.00038 cm) over the steel to avoid rust—a ''tin'' can is almost entirely made of steel. There are also special paints that are baked onto the steel.

1. **What is iron? What is steel, and how does it differ from iron?**
2. **What is coke and how is it produced?**
3. **Why is limestone added to the blast furnace, and what by-product results?**
4. **Name eight products that might be sold by a steel mill.**

New Words to Know

Steel	Ingot
Alloy	Pickle
Coke	Anneal
Blast furnace	Temper
Pig iron	Galvanize
Slag	Tinplate
Basic-oxygen furnace	Electrolysis

Canada's Steel Industry

There are 17 Canadian steel plants, located in six provinces (Alberta, Saskatchewan, Manitoba, Quebec, Nova Scotia, and Ontario), employing a total of 33 400 people. 1997 sales in the Canadian steel industry totalled in excess of $11 billion, of which exports accounted for $3.6 billion. Canadian steel manufacturers supply the transportation, appliance, oil and gas, packaging, and construction industries.

Ontario produces 70 percent of Canada's steel, and production is centred in Hamilton, a city with a large protected harbour, well located to receive raw materials and ship out finished product.

Hamilton is in the centre of the **Golden Horseshoe,** the highly populated, industrialized area which lies at the western end of Lake Ontario.

When steel-making began in Hamilton, in the late nineteenth century, there was no golden horseshoe of industry. The city's location on Lake Ontario gave the mills ready access to the coal of West Virginia and Pennsylvania, and to the iron ore of the Mesabi Range in Minnesota. The other main ingredient, limestone, was quarried locally.

Today, most of the coal continues to come from the United States; some is shipped all the way from the prairie provinces. Iron ore comes from a variety of sources in Ontario, Quebec, and Newfoundland, because the Mesabi Range is no longer able to supply major amounts. Limestone comes from local Ontario quarries such as those in Beachville, Guelph, and Ingersoll.

Canada's two largest steel producers, Stelco and Dofasco, are both in Hamilton.

Every Canadian steel centre except one is located somewhere along the Great Lakes/St. Lawrence system. The availability of huge quantities of water, as well as access to low-cost water transportation, is essential to the industry.

Near several of the iron ore mines are pellet plants that produce **iron ore pellets**. These are round, marble-sized balls of concentrated ore. By concentrating the iron in the ore into pellets at or near the mines, considerable savings in transportation are realized: most of the **non-ferrous minerals** are left behind at the pellet plant. The making of these partly refined pellets reduces the smelting time required at the mill, as well as the amount of limestone needed by the blast furnace. The ores may be only 35 percent iron; the pellets typically have an iron content of about 65 percent.

Steel Mills and the Environment

Iron-and-steel plants, by their very nature, have been notorious from an environmental standpoint. Solid wastes, fumes and polluted water were for many decades considered one of the costs of the benefits steel mills provided. In the early part of this century a city or town was often proud of the fact that its industries belched smoke and gases — it meant that companies were thriving there and that the town was prosperous. In the latter part of this century, however, the environmental damage is being recognized. Society has come to realize that part of the cost of the products it consumes must go toward environmental safeguards and controls.

All of the steel companies in Canada are fully involved with the government agencies in matters affecting the environment. Hundreds of millions of dollars have been spent by the steel companies to reduce and sometimes completely eliminate some of the land, water and air pollutants that the production of iron and steel caused in the past. The Canadian steel industry is committed to increasing energy efficiency by 1% a year between 1990 and 2000 by reducing carbon dioxide emissions. It is on target for doing so. Carbon dioxide (CO_2) is a natural by-product whenever fossil fuels like coal, oil and gas are burned, but less CO_2 is released when natural gas rather than oil is used as the fuel. Another key environmental iniative is waste reduction which is facilitated by selling and recycling increasing amounts of steel by-products, sending less and less material to landfill or final treatment facilities.

Pollution abatement is not cheap: environmental controls for one hot-mill alone cost over $30 million. Stelco spent over $64 million to meet its environmental goals and government standards in 1997, *in addition to $11 million in capital costs.*

1. "Some steel plants in Canada are where they are in part because of industrial inertia." Explain.
2. In one sentence, explain the significance of the following places in the Canadian steel industry: Guelph, Sydney, Nanticoke, auto-assembly plants, Sault Ste. Marie, Newfoundland.
3. Why was a factory smokestack belching smoke once considered a "good thing"?

New Words to Know
Golden Horseshoe
Iron ore pellets
Non-ferrous minerals

5-45   The "Rougher" in Dofasco's #2 Hot Strip Mill. *The operator rolls a red-hot slab of steel into a long, flat strip, which can be further processed into a broad range of specialized grades of steel for different end uses.*

The Automobile Industry

A love affair exists between North Americans and their automobiles. The automobile was invented in Europe, but much of its development has taken place here, and its influence has been greatest on this continent. Perhaps the long distances and wide-open spaces of North America have helped make cars and trucks so important here.

As an industry, the manufacture of cars and trucks dominates Canada's economy; as a vehicle, the automobile often dominates the lives of Canadians. Cars have even influenced the English language: when Canadians speak of "putting on the brakes" or "stepping on it," cars are not usually on their minds, yet that is where both phrases come from. If someone asks, "Have you got wheels?", everyone knows the wheels are car wheels. Almost everyone knows what these expressions mean: "Model T," "Edsel," "Cadillac of the industry," "the Bug," "lemon." These terms, and others you may be able to add, are part of the automobile culture.

It is difficult to grasp the importance of the automobile to Canadians' lives. Cars have affected the development of this country's cities and towns and the location of its industries, and are a big factor in everyone's lifestyle. Parking garages, drive-in movies, freeways, heritage auto clubs, safe-driving courses and drunk-driving campaigns — all have been spawned by the automobile.

The automobile is the most important single product manufactured in Canada. Canadian manufacturing accounts for 18% or $430 000 000 of Canada's total economic output, and the automotive

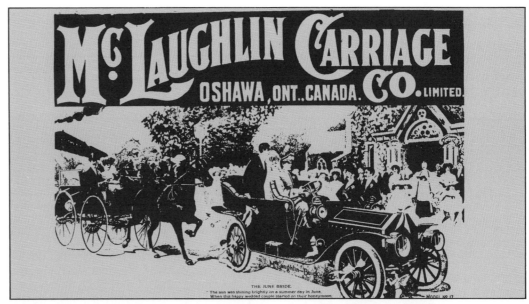

5-46 Advertisement for McLaughlin's cars and carriages, 1910.
The McLaughlin Carriage Company grew to become the General Motors plant in Oshawa.

5-47 GM concept vehicle, Buick Signia.

sector comprises 12% of that amount. One out of every six manufacturing jobs relates to the automotive industry. Four out of every five Canadian households owns an automobile. In 1997 alone:

- 739 669 cars and 650 527 light trucks were sold in this country
- 1 373 561 cars and 1 161 194 light trucks were assembled here.
- The Canadian automotive industry sold over $64 000 000 000 worth of vehicles and parts – compared with $9 000 000 000 for the furniture and appliance industry
- The average passenger car cost $22 797 as opposed to $17 157 in 1992.
- Every car in Canada pumps out an average of over four tons of pollutants annually.
- Ontario accounts for over 15% of the cars, trucks, and minivans built in North America.
- Canada's auto industry is the world's sixth largest, producing more cars than Italy.
- In 1997 there were 445 vehicles on the road for every 1 000 people, compared with 46 in 1950.
- Ford Canada, one of the "Big Three" automobile manufacturers, invested $4 00 000 000 into its operations between 1992 and 1997.

Canadian Production Centres

Charts 5-48 and 5-50 clearly show that Canada's production of cars and trucks is centred in Ontario. They also show that three centres and three companies dominated production here. All the major **assembly plants** in Canada are shown in Fig 5-50. The same map also shows many of the major assembly plants in the U.S.A. Many automotive experts see the Canadian

5-48 Canadian Car and Light Truck Production 1997.

	Car Production	Light Truck Production*	Total Share
Big Three			
Chrysler	203 988	423 169	24.7%
Ford	228 616	402 213	24.9%
General Motors	590 712	304 052	35.3%
Big Three Total	**1 023 316**	**1 129 434**	**84.9%**
New North American Manufacturers (NNAM)**			
CAMI***	69 940	31 760	4%
Honda	165 040	–	6.5%
Toyota	108 717	–	4.3%
Volvo	6 548	–	.3%
NNAM total	**350 245**	**31 760**	**15.1%**
Canadian total	**1 373 561**	**1 161 194**	**100%**

* Light trucks includes vans and minivans, and trucks under 16 000 pounds

** NNAM is used to designate new vehicle assembly plants built in Canada over the last decade or so by foreign manufacturers

*** CAMI stands for the joint GM/Suzuki car assembly plant in Ingersoll, Ontario

The vast majority of cars assembled in Canada are produced in just three cities – Oshawa, Windsor and Oakville. In recent years, Canada has produced more cars than were sold here. The additional cars were exported, mostly to the U.S.A., giving Canada a very favourable balance of trade in automobiles. The car companies are able to produce cars more efficiently and more economically by having each model produced in one country, even though the product is sold in both Canada and the United States. The interrelationship between production and assembly between the two countries remains very close however, and a shortage of automotive parts manufactured in the United States (either through lack of raw material or labour walkout) can disrupt or even close down assembly plants in Canada, and vice versa.

Figures provided by Des Rosiers Automotive Consultants

and American plants as part of one long Can-Am Auto Alley or axis. Because of the **Auto Pact**, the so-called **Big Three** companies (General Motors, Ford and Chrysler) can virtually ignore the international border, shipping parts and models across it without serious restrictions.

In the U.S.A., there are many other assembly plants outside the Auto Alley, near major population areas. These are not shown. In Canada, there are only three assembly plants not part of the axis, and they are small: Halifax (Volvo), Kelowna (Western Star Trucks), and Burnaby, British Columbia (Freightliner Trucks).

Fig 5-50 only shows assembly plants. It would be impossible to mark the

thousands of factories producing automotive parts on a map of this size. These factories are found everywhere, but are clustered most densely in the Alley. This cluster is especially dense in southwestern Ontario, Michigan, Ohio, and Indiana.

Fig 5-50 shows three newer Ontario assembly plants (not in full production until 1990 or later): the Toyota plant in Cambridge, the GM/Suzuki plant in Ingersoll, and the Honda plant in Alliston. All of these plants required huge start up capital investments. The Honda plant, alone, cost over $800 million, but by 1998, it had doubled its annual capacity to over 270 cars.

It is not only new plants that require a huge capital investment. The Big Three have made even larger investments in new technology and larger plants. Between 1993 and 1996, G.M. alone made capital investments in Canada totalling $3.7 billion.

New Words to Know

Assembly plant   Big Three
Auto Pact

1. In 30 – 50 words, explain how the automobile is important to your life.
2. How can Canada export over 80 percent of the cars and trucks it produces, and still sell as many as it makes?
3. Why are labour unions happy to see companies making large plant investments? Why are company shareholders happy to see the same thing?
4. Find the locations mentioned above on an Ontario map.
5. (a) What companies mentioned above are related to Japanese interests?
   (b) What is the total number of jobs beings created by Japanese investment in the above report?
   (c) How many jobs are being created by a German company, based in North America?
6. Using the Can-Am Auto Alley map, name:
   (a) three urban centres in Canada where GM has an assembly plant
   (b) three plants which did not begin production until 1990 or later.
   (c) three Japanese companies which have assembly plants in the U.S., but not in Canada
   (d) six automobile-producing centres in Ontario located along Highway 401
   (e) Why have automotive plants concentrated in this small geographic area?

Recent Auto Industry Investment

Chrysler Canada announced plans to add a third shift at its Bramalea assembly plant and hire 1000 workers.

Meritor Automotive Canada Inc in Bracebridge will add 100 new jobs by 2001, for a total of 523 employees.

The village of Arthur welcomed the Musashi Seimitsu Industry Co Ltd of Japan. The company will provide automotive suspension components for Alliston's Honda of Canada manufacturing plant. The $3 million facility employs 30 workers.

Shelburne became home to KTH Shelburne, a new plant that will make automotive frame components for the Honda mini-van plant in Alliston. The $26.6 million investment means an estimated 250 new jobs by 2001.

In Elora, Jefferson Elora Corporation broke ground for a $23 million plant that will manufacture automotive body components for Honda in Alliston. The plant will create 70 new jobs within two years.

In Bradford, West Gwillimbury, Kumi Canada Corporation broke ground for a $14 million plant that will supply parts to Honda of Canada. It is expected to employ 70 workers by 2001.

Denso Corp., one of the world's largest parts companies, broke ground on a plant in Guelph, Ontario. Denso, which is affliliated with Toyota Motor Corp., will make air conditioners for that auto maker and other Japan-based companies that assemble vehicles in Canada. It will create 150 jobs at its 104,000 square foot, 6.1 hectare facility.

Moriroku Co. of Japan, and its subsidiary Greenville Technology of Greenville, Ohio invested $16 million in a new plant in Listowel that will create 150 jobs within five years.

Showa Corporation of Japan announced that it will invest $10 million to build an auto parts plant in Schomberg, Ontario. The new company, Showa Canada Inc., will be a subsidiary of Showa Corporation and will employ 800 people by the year 2000.

(as reported in the April 1998 Ontario Report to Taxpayers)

5-49   *Recent Auto Industry Investment*

7. (a) Sketch the Canadian portion of map 5-50
   (b) Mark on your sketch the names of major centres where auto-assembly plants are located
   8. Approximately what percentage in both cars and light trucks is the NNAM total of the "Big Three" total?

8. In August 1998, Volvo Canada announced it would close its assembly plant in Halifax, Nova Scotia. What factors do you think influenced this decision? How will the closing affect the pricing of Volvo cars in Canada? Why? Volvo employed 200 people in Halifax. Explain how this closing will affect them. What other areas of the Halifax economy will it affect?

5-50   The Can-Am auto alley.

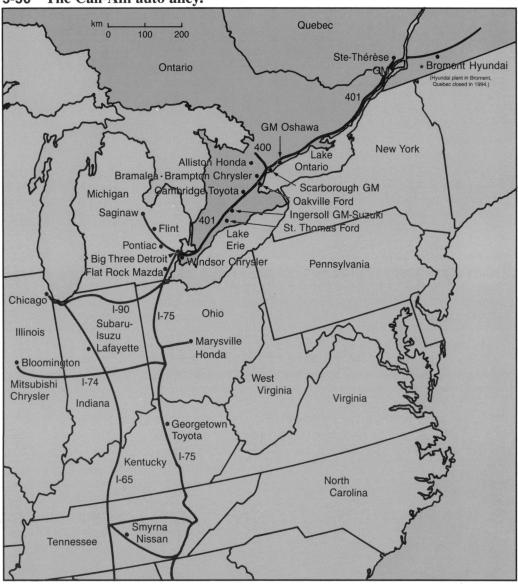

All the automotive assembly plants in Canada located in the area are indicated, but only some of those in the U.S.A. U.S. plants are concentrated in the south Michigan and Ohio area centred on Detroit, and along the interstate highway system. The pattern shows the importance of the expressways. New Japanese plants in both countries are shown, to make clear the importance of this trend. This is an incomplete map as far as the total industry is concerned. Hundreds of other parts plants could not be shown, along with the scores of towns and cities where they are located. Without them, the assembly plants would grind to a halt within days. Many parts plants are located well outside the area shown.

From Benz, Ford and McLaughlin to Honda and Toyota

There remains some question as to who invented the automobile, but it is generally acknowledged that it was Carl Benz of Mannheim, Germany. Just over 100 years ago, in 1885, he built his Velociped, a three-wheeled open carriage that used a coal-and-gas-fired, water-cooled **internal combustion engine**. The engine was similar to the one that two other Germans, Otto and Daimler, had been developing. It is fitting that the name of one of the most famous companies in the automotive world is Daimler-Benz AG of Stuttgart, Germany. These two inventors did considerable pioneering work.

It took many years, and a man called Henry Ford, to make the automobile not just a rich man's toy, but something that everyone could own and enjoy. He founded his Ford Motor Company in 1903. He adapted the production techniques of many other inventors, European and American, and added his own to produce a simple, low-priced car. In 1908, Ford began to mass-produce just one model, using a moving assembly line. Conveyers brought the car to each worker, who would carry out only one or two simple tasks in the assembly process. This process, along with other techniques Ford developed, enabled him to produce his famous **Model T** or ''Tin Lizzie'' in great numbers and at a low cost. In 1909 he startled the auto industry by producing 10 000 Model Ts, which sold at $950 each. By 1927 he had sold 15 million in all, and reduced the price to under $300.

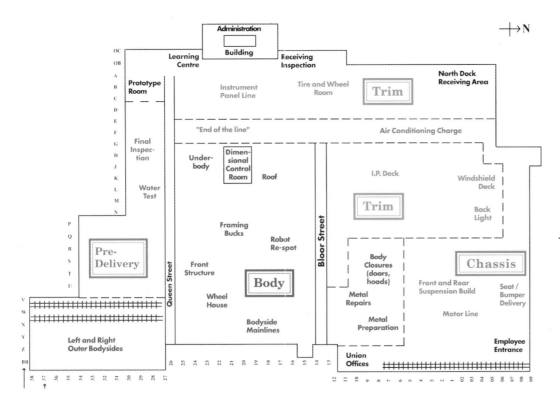

5-51 Layout of Ford Canada's Oakville Assembly Plant

Ford's Oakville Assembly plant, located on a 1 203 ha. site, produced its first car on May 11, 1953. The assembly line is just under 39 km long, and today, the plant produces Ford Windstar minivans. The one-millionth Ford Windstar produced in Oakville rolled off the assembly line in February 1998. Note how the various manufacturing processes are positioned to maximize assembly efficiency. Ford's high tech paint shop is missing from this chart because it is in a separate building, connected to the main plant by an above ground tunnel. This Oakville plant is one of the largest manufacturing plants anywhere in Canada.

5-52 McLaughlin Oshawa Carriage Works, built in 1867.
This was the beginning of the GM story in Oshawa. Robert McLaughlin is the third from the left.

Ford's techniques would be useful later in many industries. By concentrating on producing a simple vehicle that was easy to assemble, had interchangeable parts and could be sold in the millions, Ford pioneered mass production. This production method saved costs, for the same machine which had to be purchased to make hundreds or thousands of parts could now be used to make hundreds of thousands of parts. His moving assembly line that brought the work to the worker greatly increased productivity — that is, the same number of workers could now produce more cars. Many people criticized the assembly line as dehumanizing, but it is still used today in almost all industries.

No one person "invented" the auto-

5-53 An old-style assembly line once used at GM. *The body is being "dropped" onto the chassis, the frame on which the car is constructed.*

5-54 Robots at work welding, GM Oshawa. *The latest type of assembly line has the body of the car being spot welded by robots as it moves along the line.*

mobile, but Henry Ford definitely popularized it. He fulfilled his dream, for he had said, "I will build a motor car for the great multitude. It will be so low in price that no man making a good salary will be unable to own one — and enjoy with his family the blessing of hours of pleasure in God's great and open spaces." Cars still provide pleasure and employment, and a great deal of wealth for people throughout the world.

Developments in Canada

In Canada, the most famous name in the early history of the automotive industry was R.S. "Sam" McLaughlin.

He was born in 1871 and raised in the town of Enniskillen north of Oshawa, Ontario, where his father owned a carriage works. The family moved to Oshawa in 1876, because, McLaughlin said, it had two things Enniskillen did not have — a railway and a bank. The McLaughlin Carriage Works was a very successful business, but in 1907 R.S. McLaughlin realized the carriage would soon be replaced by an even better vehicle — the automobile, or "horseless carriage." In that year the McLaughlin Motor Car Company of Oshawa was formed, using Buick engines and capital to make a new line of cars, the

McLaughlin-Buick. This eventually developed into Canada's largest automaker, GM of Canada.

The other two of Canada's Big Three auto-makers also became established in Canada fairly early in the industry's history. Ford came to Windsor in 1904. Chrysler followed, though much later, in 1925, when it purchased the Maxwell-Chalmers auto company. The buying out of Maxwell-Chalmers was only one of scores of auto-company mergers, buy-outs and bankruptcies. Early in this century there were over 1000 automobile companies in North America. Now there are only a handful.

Since those early years, the car-making industry has grown tremendously. It is now Canada's biggest manufacturing industry. In 1926, for example, Chrysler's Windsor plant produced 7857 cars and employed 243 workers. In 1996, Chrysler Canada produced 705 000 vehicles and directly employed 14 000 workers.

The automotive industries of Canada and the U.S.A. have developed in similar ways since the early years of this century. The automobile has undergone constant change as better technologies have improved and refined it. The car-making industry has become number one in the economy of both nations. Because of this, economists and governments watch it carefully. Sales figures are monitored weekly in newspapers, for if the auto industry is strong, so is the economy. Charles Wilson, a former

president of GM summed it up with this: "What's good for [America] is good for General Motors and vice versa."

In 1965 an important agreement was signed between Canada and the U.S.A. in regard to the trade in automobiles. This agreement, known as the Auto Pact, permitted duty-free trade in new motor vehicles and original equipment parts between the two countries. This duty-free trade has two restrictions: First, in each company of the Big Three, for all vehicles sold in Canada, an equivalent value of vehicles has to be manufactured in Canada. Second, for every American-built car sold in Canada, 60 percent of the value added to its manufacture has to be Canadian. (This Canadian component generally takes the form of Canadian-made parts.)

The Auto Pact led to an expansion of both factories and production in Canada. The auto-makers could now build a particular model in a plant in one country for sale to customers in either country. The Pact meant that Canadian production plants would produce fewer models but far more cars of each model. The other models Canadians wished to purchase could be imported from the U.S.A. without any tariff.

The Auto Pact balance in recent years has been in Canada's favour; that is, Canada has shipped a higher value in cars and parts than it has imported. In 1994 Canada sold $3.1 billion more in automotive products than it purchased.

5-55 Ford's Quality Controlled Paint Shop.
*The bodypanel on a new Ford **Windstar** is checked by computer-guided laser scanner for fit, and to ensure that its surface is ready for final finish application. Note the computer on the right which controls the measuring machine.*

Thus Canada has a surplus with the U.S.A. in automotive exports. For many years, however, this country bought so many cars, trucks and parts from countries of the world other than the United States, that we ran a deficit in the overall auto trade.

1. **Why has the sale of automobiles increased so significantly since Henry Ford's first cars?**
2. **What new techniques were introduced by Henry Ford which made his name the best-known in the auto industry.**
3. **Describe briefly, in your own words, the founding company that produced the McLaughlin-Buick car. Be sure to include the reasons for that car's development.**
4. **How many more vehicles were made in 1997 than in 1926 by Chrysler Canada? How many more workers were employed by that company over the same time?**
5. **Why is what's good for General Motors also good for the U.S.A.?**
6. **When was the Canada-U.S. Auto Pact signed, and what benefits has it brought to Canada?**

New Words to Know
Internal combustion engine
Model T

The Automotive Industry Undergoes Major Changes

Beginning in the late 1960s, major changes began to occur in the auto industry. Many experts believe the major cause of the change was the tremendous growth in sales of Japanese products, all over the world but particularly in Canada and the U.S.A. In fact, in 1980 the Japanese auto industry overtook that

5-56 Sales of vehicles in Canada, by manufacturer, 1997.

Manufacturer	Cars	Light Trucks	Market Share
Chrysler	87 292	168 870	18.4%
Ford	109 008	202 792	22.4%
GM	244 915	205 824	32.4%
"Big Three"	**441 215**	**577 486**	**73.2%**
Honda	90 303	16 782	7.7%
Hyundai	19 285		1.4%
Mazda	22 195	6 000	2.0%
Nissan	20 570	16 541	2.7%
Subaru	7 944	1 374	.7%
Suzuki	4 883	4 167	.7%
Toyota	80 128	26 169	7.6%
Asian	**245 308**	**71 033**	**22.8%**
BMW	7 117		.5%
Jaguar	1 020		.1%
Mercedes Benz	5 703	459	.4%
Lada	646	703	.1%
Land Rover		1 115	.1%
Volkswagen	26 541	846	2.0%
Porsche	759		.1%
Volvo	10 245		.7%
European	**5 2031**	**3 123**	**4%**
Total	**739 669**	**650 527**	

figures courtesy Des Rosiers Automotive Consultants, Inc

Sales fluctuate from year to year, as do the most popular models. Note the high number of lights trucks being produced compared to the number of cars. Sales of European made cars have declined dramatically, with the exception of Volvo, who make their cars here in Canada. Why might this be the case? Where is the Lada made? What might account for the drop in Hyundai numbers?

of the U.S.A., making Japan the world's number one car manufacturer.

In this "Japanese Invasion," as it has been called, the Japanese were selling high-quality cars at prices lower than those for comparable North American cars. The invasion also led to the building of plants in North America (map 5-50). For a time the reason for the invasion was considered to be wage

Sales of vehicles in Canada, by manufacturer, 1988.

	CARS	TRUCKS
DOMESTIC VEHICLES		
GENERAL MOTORS	379 589	179 146
FORD	207 539	148 336
CHRYSLER	158 660	113 500
TOTAL	**745 788**	**440 982**
FOREIGN VEHICLES		
HONDA	75 045	—
TOYOTA	70 400	15 736
VOLKSWAGEN	35 553	934
HYUNDAI	31 013	—
NISSAN	29 812	13 118
MAZDA	27 656	11 597
SUBARU	7 759	—
SUZUKI	6 588	5 438
VOLVO	6 176	—
BMW	5 016	—
MERCEDES-BENZ	4 221	—
JAGUAR	2 154	—
SAAB	1 369	—
LADA	970	158
SCOCAR	378	—
TOTAL	**304 110**	**46 981**

rates: the Japanese could build cars more cheaply using Japanese labour than using high-wage North American workers. However, in the 1960s, Japanese wages began to approach those of American and Canadian workers. At that point people began to realize that in many ways the Japanese had better production and quality-control techniques than the North Americans.

This has led to major changes in North America's auto industry. The Japanese car-makers were so successful in the North American market that the Canadian and the American governments both came under pressure to slow down the flow of Japanese car imports to protect the jobs of North American auto-workers. Both countries established quotas restricting the number of Japanese cars that could be imported. To get around the quota, the Japanese decided to start making some of their cars in North America. Map 5-50 shows most of these new assembly plants. Almost all are located in the Auto Alley between Ste Therese, Quebec and Smyrna, Tennessee.

There have been other recent changes in the North American auto industry. The general lowering of global trade barriers has made cars and parts much more important in international trade. Mergers across international borders are becoming more common. Daimler-Benz and Chrysler announced a merger-partnership in May 1998, Volkswagen and Volvo were rumoured to be discussing a similar arrangement in June of the same year. The price of fuel skyrocketed in the 1970s (as a result of price-setting by the OPEC oil cartel) and has remained high ever since. And not only fuel, but the cost of cars themselves increased substantially in the 1990s. Technology or "know-how" has improved enormously, and, finally, the public's demand for safer, environmentally responsible vehicles has become a dominant factor in the industry.

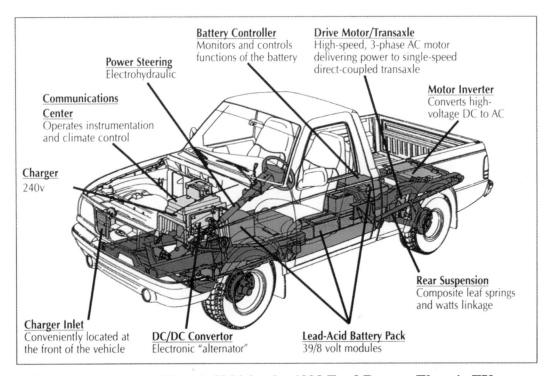

5-57 An Innovative Electric Vehicle, the 1998 Ford Ranger Electric EV Pickup Truck *Based on Ford's best-selling compact truck, the Ford Ranger EV is powered by lead acid batteries (39/8 volt), and equipped with third-generation, electric vehicle-specific, powertrain components and electronics. 0-80 km/h acceleration is 12 seconds, top speed is 121km/h and range is 80 km. Gross vehicle weight is 2450 kg. The truck is equipped with an on-board charger, re-programmable electronics and regenerative breaking.*

The "world car" concept With the general decline in trade barriers over the past decade, international trade in cars and car parts has greatly increased. Car parts are now produced in many countries, shipped to another for assembly and shipped again to many countries for sale. Because the workers in less developed countries have lower wage rates, companies often use them to make parts at lower cost. In the auto industry this is called **global sourcing**. A number of Pacific Rim countries such as South Korea, Taiwan, Singapore and, of course, Japan have become particularly active in the auto industry. But so have many others, from Brazil to Belgium.

The just-in-time system (JIT) This term is also relatively new to the industry. It refers to the arrival of parts at the assembly plant "just in time" for them to be placed on the vehicle as it moves along the assembly line. This system creates big savings in both space and storage costs, and is being adopted by many other industries as well. The Japanese corporations are generally credited for introducing this technique, which has had a considerable effect on where parts industries are located. Supply is more secure if the parts plants are within a few hours' drive of the assembly plant.

Smaller, lighter automobiles The automobile has changed considerably along with the industry. Cars are now much more complex, with built in computers, anti-lock breaking systems, safety air-bags, and traction assistance.

All wheel drives (AWDs) are common, and minivans one of the most popular configurations available. When the price of gasoline rose sharply in the 1970s, due to demands for higher revenues by the producing nations, a barrel of crude oil that cost $2 USD early in that decade rose as high as $30 USD a few years later. During the 1990s, crude oil prices have stabilized at just under $20 USD. Smaller, lighter vehicles are much more economical than large ones, because the latter burn more fuel. It is a truism that the era of cheap oil is long gone.

Car manufacturers have worked hard to make their cars more fuel efficient. They have improved engines. They have also lightened the cars. A 1990s car has much more aluminium, plastics, and lighter steels than one built in the 1980s.

"Smarter" cars Automobiles now have computers aboard. These are useful in monitoring the engine, and making the driver aware of any required maintenance. The auto industry has become one of the major markets for the computer industry.

Improved technology and productivity Car-makers have led most other industries when it comes to improving the technology of production. Plants are now much more automated and computerized; many now use robots. All of this cuts costs and increases production. Sometimes an entire plant is computer-regulated. The more computerization, of course, the fewer workers required. The Japanese in particular have developed much of this advanced technology. Only a decade or so ago, North American assembly lines produced 65 cars annually per worker. Today that figure is out the window as the Ford Windstar assembly line alone produced 1200 Windstars a day in 1994.

Quality control has become much more important as well. Japanese companies have led the way in this field.

In some factories the time-honoured assembly lines are being removed. In GM's new **Autoplex** plant in Oshawa, workers work in teams and perform many operations together. The car is brought to them on an **automated guided vehicle**, or AGV. The team carries out its assigned tasks before the

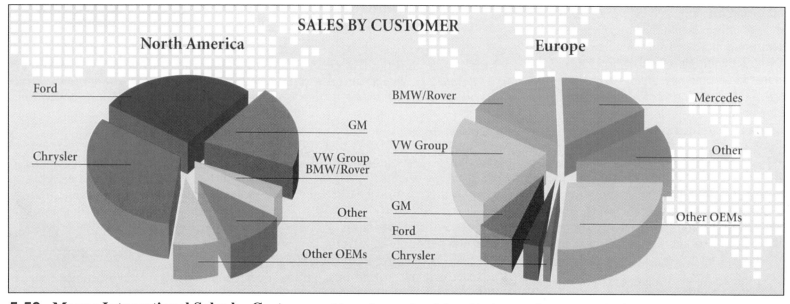

5-59 Magna International Sales by Customer. *Magna International, based in Aurora, Ontario, is a leading global supplier of interior and exterior body and chassis systems to every major automobile manufacturer in North American and Europe. Magna products range from front and rear bumper systems to body panels, to turnkey assemby lines, instrument panels, seating, mirrors, air vents, grills, brake pedal assemblies, thermal electric systems, heating and cooling systems and much, much more. Magna employs over 36 000 people in 128 manufacturing facilities around the world (50 in Ontario alone). It has 22 sales offices and 26 product development centres. 1997 sales were $7.7 billion.*

AGV moves on. This assembly method, which was developed in Sweden, relieves the worker of some of the pressure of keeping up with an ever-moving assembly line. It also gives workers more varied tasks, which cuts down on boredom and thus improves quality.

Automobile combinations There is an old saying in politics and business that "If you can't beat them, join them." This approach has become common in the automotive world. The Big Three auto-makers in North America have all made financial and technical arrangements of various kinds with other auto-makers. In particular, they have forged many links with Japanese producers.

Map 5-50 shows a number of auto plants that are American-Japanese joint ventures. Chrysler has made connections with Mitsubishi (Japan), Hyundai (South Korea) and China Motor (Taiwan). Ford is involved with Mazda (Japan), Kia (South Korea) and Ford Lio Ho (Taiwan). GM, the biggest of the Big Three, is tied in with Suzuki, Toyota and Nummi (all of Japan) as well as Daiwoo (South Korea) and Kuo Zui (Taiwan). The financial and manufacturing arrangements are too complex for detailed explanation here. However, the point is that car-making is turning into a highly international industry. In the future, many cars will be "crossbreeds" developed and perhaps manufactured by several auto companies. Even the People's Republic of China is starting to get into the game, planning to manufacture cars in Changchun in co-operation with Chrysler.

Overcapacity a Serious Concern
There are many car plants now, and

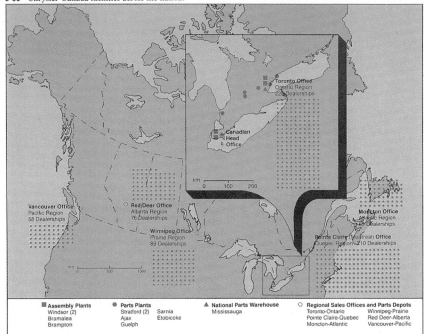

5-60 Chrysler Canada facilities across the nation.

Assembly Plants	Parts Plants		National Parts Warehouse	Regional Sales Offices and Parts Depots	
Windsor (2)	Stratford (2)	Sarnia	Mississauga	Toronto-Ontario	Winnipeg-Prairie
Bramalea	Ajax	Etobicoke		Pointe Claire-Quebec	Red Deer-Alberta
Brampton	Guelph			Moncton-Atlantic	Vancouver-Pacific

The distribution of the Chrysler Canada facilities across Canada is similar to that of the other major automobile manufacturers. The concentration of manufacturing is in Ontario, for having plants close together reduces costs. The sales force, however, must be organized to care for a national market. Though the sales offices are located in major cities, is size the only factor determining their location? Use an atlas to determine the names of the various cities where the facilities are located.

many in the process of being built, yet world-wide sales have not been rising for the past few years. One result is **overcapacity** in the auto industry. In 1978, North American auto plants produced over ten million vehicles. That figure hasn't been equalled since. It means that the industry in North America — and in the world as a whole — can produce far more cars than it can sell. Several auto plants were closed in the U.S.A. in the 1980s, and there were threats of closure in Canada. As the different companies fight for their market shares, some Canadian auto-assembly plants will likely have to close. This would cause unemployment and displacement for thousands of workers, including many who work in other industries.

1. (a) What was the Japanese Invasion, and why did it take place? What have been some positive results in the industry in general?
   (b) **Which other factors led to the many changes in the auto industry?**
2. **Why is the JIT system being introduced in the auto and many other industries?**
3. (a) **How did OPEC have an effect on how cars were built?**
   (b) **How were cars affected? Name five ways.**
4. (a) **Explain the Swedish approach to assembling autos. What are its merits?**
   (b) **Where is this approach being used in the Canadian auto industry?**
5. **Research links North American auto companies have with companies in Europe.**
6. **Explain why overcapacity is a threat to the auto industry, or any industry.**

New Words to Know
Global sourcing
Just-in-time system
Autoplex
Automated guided vehicles
Overcapacity

Driven to Merge. *The following article is an excerpt of one written by Ross Laven in MacLean's Magazine, May 18, 1998. It is an excellent example of economic globalization – how world events affect Canadians.*

No, Levi-Strauss is not in talks to merge with Italy's Armani. Nor, as far as anyone knows, is McDonald's planning to team up with a chain of snooty French restaurants. But in the category of seemingly improbable corporate pairings, few other deals can compare with last week's odd-couple marriage of Detroit-based Chrysler Corp. and Germany's Daimler-Benz AG, maker of Mercedes-Benz cars. The former is a mass-market American automaker known for its snazzy vehicle line-up and sometimes less-than-stellar quality. The latter, a polished if plodding symbol of German engineering excellence, churns out solid and conservatively styled luxury cars for wealthy buyers the world over. Talk about a culture clash.

Yet when North America's third-largest carmaker announced a $57-billion merger with Germany's biggest industrial firm, almost no one in the global auto sector questioned the logic of the deal. Union leaders, rival manufacturers and financial analysts all said the partnership makes perfect sense, giving Chrysler the international presence it has always lacked and establishing Daimler-Benz as a major player in North America, where it is weak. What remains unclear is whether the deal—the world's biggest merger between manufacturing companies – will trigger a wave of similar alliances in the auto industry, which is struggling with chronic overcapacity, skyrocketing research-and-development costs and cutthroat international competition.

The answer to that will depend in part on how successful Daimler-Benz and Chrysler are in uniting their operations under one banner. DaimlerChrysler, as the new entity will be called, will have 420,000 employees and 24 car and light-truck assembly plants in six

countries. With $189 billion in annual revenues, it will rank as the world's third-largest auto company, behind General Motors and Ford but well ahead of Toyota.

The numbers are formidable but so are the challenges – not only in reconciling two dramatically different corporate cultures, but in making sure that the whole is greater than the sum of the parts. The effective takeover of Chrysler by a German-controlled company – under the deal, Daimler-Benz shareholders will own 57 percent of the new company – also calls into question the future of the Canada-U.S. Auto Pact. Under that 33-year-old treaty, North American automakers can import vehicles and parts duty-free from anywhere in the world, while foreign manufacturers, including Daimler-Benz, are required to pay a 6.7-percent tariff.

The irony is that, for years, Chrysler has been the most strongly nationalistic of Detroit's Big Three automakers. After a brush with bankruptcy in 1980 – during which it was rescued with U.S. and Canadian government loan guarantees – the company fought back with a lineup of new cars and minivans and repeated appeals to consumers to "buy American." More recently, its Canadian subsidiary has been at the forefront of a campaign to preserve Canada's 6.7-percent duty on imported cars, including Mercedes-Benz.

That Chrysler has now decided to gamble its future on German ownership is a measure of how much the auto industry has changed during the past two decades – and is likely to continue changing. During the late 1980s, a wave of consolidation saw Chrysler take over American Motors, including its Jeep division, while Ford acquired Britain's

Jaguar and a minority interest in Mazda of Japan. Meanwhile, GM took control of the Swedish automaker Saab and bought a chunk of Japan's Isuzu. Later, BMW took over Britain's Rover Group, and Sweden's Volvo came close to joining forces with Renault.

All of these deals have one thing in common: a desire to wring further efficiencies out of an industry that faces intense competition, declining profit margins and stagnant markets in industrialized countries. Of the world's major automakers, Chrysler has been among the most successful in adapting to the harsh new realities of the auto business. Its near-death experience in 1980, coupled with a similar financial crisis at the beginning of the 1990s, forced the company to streamline operations from top to bottom, eliminating layers of management and implementing Japanese-style product-development teams. Those changes have made Chrysler one of the industry's most innovative and nimble manufacturers, capable of designing and introducing a new vehicle from scratch in less than 36 months – roughly a year faster than its North American rivals. Thanks to the popularity of its minivans and four-wheel-drive sport utility vehicles, it is also the world leader in per-vehicles profitability.

In an era of global brands and converging consumer tastes, the auto industry may have no other choice but to follow the example of Daimler-Benz. "This industry is consolidating." Eaton said. "We believe this is just the beginning." Gentlemen, start your engines.

With Ruth Abramson in Toronto and Regina Wosnitza in Berlin

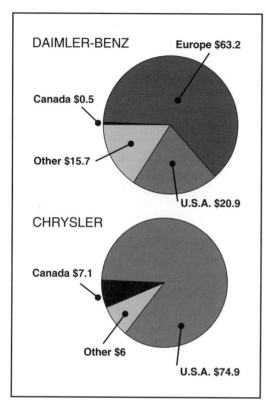

5-61 Worlds Apart, Revenue by region, in billions.

These graphs show the differences between Chrysler and Daimler-Benz, at the time of their announced May 1998 merger with regard to revenue sources.

1. **(a) Just how big will the new company be?**
   (b) List in order, the world's four largest auto companies.
2. **What are three of the major problems automakers faced?**
3. **Why does the Canada-U.S. Auto Pact create some possible problems for the new company?**
4. **In what way, other than the Auto Pact policies, has government greatly affected Chrysler?**
5. **Why were the beneficial effects of Chrysler's two "brushes" with financial disaster?**

The Auto Dealer: Providing a Service

There were 15 724 638 vehicles on the road in 1995, 33.2% were one to five years old, 40% were six to ten years old, and 26.8% were ten years old and older. 30.5% of these vehicles were light trucks, the rest were passenger cars.

The final stage in the auto industry is the selling of the vehicles. Across Canada automobiles account for one-third of all retail sales. Except for a house, a car is the most expensive item most Canadians ever buy. Almost every student who reads this section will some day be buying a car (if you have not done so already). You will discover the role of the automobile dealer in the auto-industry chain.

There were 3 608 car, light truck and minivan dealers in Canada as of April 1997. Ontario and Quebec have the most dealerships at 1 234 and 960, respectively. British Columbia has 412, Alberta 354, Saskatchewan 175, Manitoba 149, Nova Scotia 144, New Brunswick 99, Newfoundland 58 and Nova Scotia 23. Sales per dealer ranged from an average of 356 vehicles per annum in Alberta to 170 per annum in PEI.
Ontario's average sales per dealer was 354. The annual average is 286, down from 318 in 1994. The most popular cars in Canada, ranked in order are the *Cavalier, Sunfire/Sunbird, Civic, Escort,* and *Neon.* Bestselling light trucks are the *C/K* and *Sierra,* the *F-Series, Caravan, Windstar,* and *Voyager.*

There are actually more people involved in selling the cars and servicing them than there are in either manufacturing or assembling them. Dealers need to employ tens of thousands of mechanics, receptionists, salespeople, accountants, secretaries, clerks, computer operators, and cleaners, either full or part-time. In addition, automobile dealers require banking services, newspapers, radio and TV stations, advertising agencies, legal firms, trucking companies and so on. The auto-retail business, like the auto-assembly business, is a complex one.

Auto dealers are the largest retail dealers in Canada, selling about $50 billion in new and used vehicles, and about $15 billion in parts every year. Almost one-third of provincial sales tax revenues are collected by auto dealers; in addition, the local community benefits from the municipal taxes on land and buildings.

Auto dealers usually have a **franchise** arrangement with one car manufacturer. Chrysler has close to 1000 franchised car dealers in this country. Manufacturers want dealers who can not only sell their products, but who are able to operate a successful business as well. Dealers must be good at advertising the products, and be able to secure a reasonable share of the community's automotive sales. Dealers who provide poor service backup not only ruin their own reputations, but also those of the manufacturers they represent.

The dealers, for their part, want to be sure that the manufacturers will deliver the cars when they need them. They also expect the car-makers to provide ample national advertising, build quality vehicles and make technical and sales assistance available. The car-maker and the dealer depend on each other, and are constantly improving the telecommunication and computer systems which link them together. The manufacturers provide a great deal of technical training for their dealerships' mechanics — thus the term "factory-trained mechanic," which signifies quality.

5-63 Canadian sales of imported vehicles by manufacturer, 1997.

	Automobiles	Light Trucks
Honda	47 152	16 782
Toyota	46 009	26 169
Volkswagen	26 541	846
Mazda	22 195	6 000
Nissan	20 570	16 541
Hyundai	19 285	–
Subaru	7 944	1 374
BMW	7 117	–
Mercedes-Benz	5 703	–
Suzuki	2 311	–
Jaguar	1 020	–
Porsche	768	–
Lada	646	703
Range Rover	–	1 115

These statistics indicate the total number of vehicles imported, but not their source. More detailed data would indicate that 80 187 cars came from the USA, 24 550 from Mexico, and the rest from "off-shore" or non-North American assembly plants. It is interesting to note that the companies ranked highest in regard to cars, are not in the same order with regard to light trucks.

1. **How many cars were produced from "off-shore" manufacturers? Compare your answer to the number of cars produced in Canada (found in Fig 5-48).**
2. **(a) List the light truck manufacturers in order of the quantity of vehicles imported. (b) Why is the ranking of truck import numbers so different from the ranking of car import numbers?**
3. **In what countries would you find the headquarters for the car companies listed above? Name as many as you can.**
4. **Why would car companies based around the world chose to manufacture cars in the USA or Mexico and then import them into Canada?**

5-62 Jim Davidson Motors is a typical full-service auto dealership.

302

The franchise agreement makes the business arrangement between the auto-maker and the dealer very clear. Dealers are not employed by the auto company; they operate their own businesses, and must buy the cars from the manufacturer as soon as they arrive on the transporter or **car float**. Generally, bank loans are required at times during the year when the inventory must be high. Dealers must be very skillful to ensure that they aren't saddled with unsold cars. Many dealers branch out beyond the selling of new and used cars, into leasing and renting as well.

Some dealers are known as **megadealers**. They run huge dealerships which sell more than one make of car. We have well over 200 such dealerships in Canada now, and more are set up every year. Mega-dealerships use real estate, personnel and equipment to maximum advantage. The more cars a dealer sells, the lower the capital cost. In this way, the retail business is like mass production, more efficient than a small single-franchise dealership. Such an arrangement also makes it possible to concentrate on whichever car is most popular at any given time.

1. (a) Name two local automobile dealers in your community, and the company whose cars they sell.
   (b) Briefly describe the location of one of the dealers named and explain why you believe it to be a good (or poor) location for that business.
   (c) Is either of the two a ''megadealer''?
2. Why would you like, or not like, to be a car dealer in your local community?

New Words to Know
Inventory
Franchised dealer
Car float
Megadealer

5-64a and b *Chrysler's CATIA computer design system makes the design of new models more efficient and faster as cars can be created in virtual reality on the computer screen, rather physically assembled. Designers can explore all aspects of design and its implications prior to commiting to concept car production. This Plymouth Prowler concept car is slated for production in 1999.*

An Interview with Senior Officials at Honda Canada

The following interview was conducted with Vaughn Hibbits, Vice President, and Lisa Timpf, Senior Administrator, on May 21, 1998 – twelve full years after the very first car assembled at Honda's then new manufacturing plant just south of Alliston, Ontario, rolled off the line.

1) Why did the company locate here in Alliston, a small town in rural Ontario?
Many factors encouraged Honda to locate here – plenty of space, 180 ha. at a good price; close to major urban centres, but not in them; ample local transportation and a nearby railway line; plenty of labour to draw on – those were some of the major advantages we felt this Alliston site had.

2) I understand you now have two plants here.
Yes, one completed in 1986 costing $800 million, and another, which will start production this August, costing $300 million. The first plant produces 170 000 vehicles a year. The second plant at capacity, can produce 120 000. Right now, the former produces Civics and Acuras, and the new one will produce full-size Odyssey minivans. By June, 1997, we had built one million cars in Canada.

3) With a new plant coming on stream, you have obviously been hiring new workers.
Are we ever. The new plant made it necessary to hire 1 200 new **associates** – that's what we call our workers here. All won't be in the new plant, for about 30% of its workforce will be experienced personnel from our original plant. When hiring is complete, there will be 3 200 associates here.

5-65a   Honda plant. Alliston, Ontario.

4) When hiring, what kinds of associates are you looking for?
We consider hiring a new associate a very important matter, for we consider it a long term commitment. Hiring for 1 200 positions this year has flooded us with 20 600 applicants, so we have been very busy! If a candidate passes our two initial interviews, he or she is given a five and three quarter hour series of tests. We have determined five skills and characteristics to look for based on our needs: the abilities to read, write, listen, observe, and to work in a team. Other aspects of importance include education level, manufacturing experience, flexibility, dexterity, physical stamina, the ability to focus on a task, and the desire to do a good job. We want people we can count on for stable attendance, and who are motivated to want promotion.

5) What is the source of your workforce?
Over 90% live within a driving distance of about 45 minutes from the plant. Some specially trained people, engineers or accountants, of course, may come from hundreds of miles away – moving closer when hired. What's amazing is how few have automobile manufacturing experience.

6) Then you must train them here?
Certainly do. Our orientation and specific job training is very intensive. All work on "the line," and may need more training for specific tasks in a special department. We will also pay an associate up to $1 000 a year to further his/her education. Many of them take auto-related courses at Georgian College in Barrie.

7) What is the gender mix here?
At present, about 18% of our associates are women – many of them working on the production lines. Recent hirings have run about 26% women, a figure that reflects the percentage of applicants. We simply hire the best candidates, but we've ended up with a higher ratio of women than that normally found in the auto industry.

8) Have you been successful in your recruiting process?
We think so. According to surveys we've conducted, our associates enjoy working here, and are proud of being Honda associates. We have awards for safety, teamwork, quality production, and for those who suggest useful improvements, among other incentives. We believe in the Honda philosophy of respect for the individual. That builds teamwork and gives our associates a common purpose. They truly are associates, you know; we are one team. There are no special parking areas, lunch rooms, uniforms (we all wear white) – and only the nurse has a private office. Our absenteeism is well under comparable auto plants – as is our annual turnover of 2 1/2% – a quarter to a fifth of the industry average. One result is a mature, experienced workforce here. The average age of our associates is 33 years and rising.

5-65b Workers inside the facility at the Alliston Ontario Honda complex.

9) Well, you certainly are expanding. What's the reason for such a large expansion?
Our vehicles are being well received in the Canadian market, and Honda believes in building cars where it sells cars. This belief benefits Ontario, Canada, and Honda.

10) What are the sources for the parts you need and don't manufacture here?
Our North American content in parts is similar to that of the "Big Three." We draw on suppliers in Canada, the U.S., and Japan. Purchases will double with the new plant in place. We draw the majority of our needs from sources only about an eight hour drive from Alliston. The parts all come in by truck, but we ship 90% of our finished product out by rail. Ontario-based plants are our biggest suppliers – and Ontario is our largest single market. We do ship in our engines from our Anna, Ohio plant.

11) What is the future for Honda here?
Obviously we believe it is very secure or we wouldn't be opening a huge new plant this year. No one can predict the future, but we hope to be a good corporate citizen, here for the long term. That has always been our philosophy.

1) Which of the responses above surprised you the most, and why?
2) Which of the major locative factors was not mentioned?
3) a: What is your opinion of what Honda looks for in a new associate?
   b: "Honda hires associates who have good personal qualities." Discuss the truth of this statement.
4) Briefly explain the hiring policy at Honda as it involves women associates.
5) Prepare at least one question you would like to ask these Honda officials if you had the chance.

New Words to Know
Associate

Mind Benders and Extenders

1. Briefly discuss all seven of the locative factors as they would relate to the location of a child's lemonade stand in a residential area.

2. There are four major sources of investment capital. What are they, and what are the advantages and disadvantages of each to a small business that is trying to raise capital?

3. (a) Using statistical chart 5-10, prepare two lists. In one list, rank the top ten companies in order of assets; in the other, rank the same companies in order of profits. Study the two lists, and chart 5-10, and explain the possible reasons for the distribution you see.
   (b) Comment on the locations of the headquarters of these top ten companies.

4. Prepare a research report on a manufacturing industry in or near your community. Which locative factors help to explain its location?

5. "Neglecting R and D is a short-sighted industrial policy." Discuss this statement, using facts to back your argument.

6. Canada ships a great deal of electric power to the U.S.A. Does this involve any danger for Canada? For the U.S.A.? Whether or not we should export electricity to the U.S.A. would make an excellent debate topic.

7. In the section of this chapter on the central provinces, only a brief summary was given of the locative factors. Prepare a research paper of 400 to 500 words, that gives a fuller description of *two* of the following locative factors in that region: raw materials, market, labour force, transportation. An atlas, and other chapters of this text, would come in handy.

8. "In the foreseeable future, Canada's Industrial Heartland will continue to be the nation's manufacturing region."
   (a) Give reasons based on facts to support this statement.
   (b) Use this topic as the basis for small-group discussion. Each group must be prepared to present its major findings to the class.

9. Using the chart of labour-force statistics 5-39a, draw circle graphs for the construction and trade-and-commerce industries by province. Then prepare a brief analysis of 500 to 600 words, describing and explaining the graphs.

10. Why is recycling an important element of manufacturing today? What examples of recycling can you find in the industries in your community?

11. Visit a local automobile dealership and write a brief report on how it compares with the average dealer as described in this chapter.

12. Choose an occupation in the service industry which you might be interested in following as a career. Which section of the five described in the text would include your choice? Discuss that section in a brief report. What changes may occur in your chosen occupation in the next 20 years?

13. "The Auto Pact was one of the most significant trade deals ever made in Canada's history."
    (a) Prove this statement using the facts — including statistical evidence — given in this chapter.
    (b) Show how this agreement revolutionized the way cars are manufactured and sold in the U.S.A. and Canada.

14. (a) What seven major changes, as noted in this chapter, have occurred in the auto industry in recent years?
    (b) Prepare two reports of 100 to 200 words each. Each report should describe one of the changes and your personal reaction to it.

15. "The Canadian steel industry is similar to, yet different from, the steel industries of Japan and the countries of the E.U." Research this topic and discuss it under appropriate organizer headings.

16. Prepare a map showing the countries, companies and quantities indicated on statistical charts 5-57 and 5-63. Briefly explain how you feel about Canada importing so many cars and why you hold those opinions.

17. Choose any five photos from this chapter and for each one explain the geographical ideas it illustrates. Sketch any two of the photos selected and label the features you have listed.

CANADA
EXPLORING NEW DIRECTIONS

Transportation & Communication

*"Transportation is intricately woven into the fabric
of Canadian history. Its genesis lies in the will to conquer a vast new land.
It evolved to a network of arteries carrying the nation's lifeblood."*
Science Council of Canada

Transportation and Communication

Transportation and communication both involve movement. Transportation is the movement of goods and people. Communication is the movement of information and ideas. Both are trademarks of modern civilization. Almost every thing we use — food, clothing, the materials of our homes, factories and offices has been moved from somewhere else. People, too, move about, sometimes great distances, every day. And information is moved between near and distant places even while it is happening, so that people as far apart as Windsor and Whitehorse can watch a hockey game while it is actually being played in Winnipeg.

Although Canadians take this modern movement of goods, people and information for granted, they may sometimes forget that creating these systems required special efforts in this country, and that they are vital to the very existence of Canada as a nation.

Canada is the second largest country on earth in physical area, but it is only about 28th in population size. Most of the population is spread along a rather narrow belt near the southern border; smaller pockets of people are scattered throughout the vast distances of the country. All of these people need to be linked by transportation and communication. As a result, Canada has more kilometres of railways and roads per capita than its wealthy neighbour to the south. But linking widely separated and remote communities is costly, and a small population means there are relatively few people to share the cost.

In addition to great distances, both landforms and climate help to make linking Canadians difficult and costly. The mountains of the Cordillera and the rugged terrain of the Canadian Shield, with its rock and areas of muskeg, have provided builders of roads and railways with enormous challenges both past and present. So have the extremely cold temperatures and permafrost. In southern Canada, temperature changes from freezing to thawing break up road surfaces. Even such a basic job as removing snow from roads, railways, parking lots and airport runways is yet another expense made necessary by our climate.

In spite of such difficulties, Canada's transportation and communication systems are among the best in the world. Without them there would be no Canada.

6-1 A freight train winds along the Columbia River in British Columbia.
What routes will road and railway builders take whenever possible in mountainous country?

Water Transportation

"Canada in the first half of the 19th century was the child of her waterways." So said Arthur Lower, one of Canada's most respected historians. Indeed, it was by water that the earliest explorers came from Europe to our coasts. It was by water that much of the interior was explored. Canada's magnificent network of rivers and lakes formed the "highways" of those days.

The St. Lawrence River provided the earliest gateway into the interior. Cartier explored this majestic river as far as the site of present-day Montreal. Here rapids stopped him from going any further. Tales of great wealth beyond these rapids led some to believe that China would be found to the west, and so the rapids were named "Lachine," the French word for China. Champlain explored further up the Ottawa into the Great Lakes as far as Georgian Bay. Many others entered through the St. Lawrence and risked their lives in penetrating the interior of the continent, eventually even crossing it. Other explorers looked for a northwest passage through the waters of the Arctic. Still others approached the continent from the Pacific and explored what is now Canada's western coast. Names like Hudson, Mackenzie, Fraser, Thompson, which have been given to various rivers and bays, are a permanent reminder of those brave and adventurous men.

These explorers were really the first geographers of our land. Their rough maps and notes were the first "geography books" of Canada. It was their description of the landscape and the location of its physical features that

6-2 Birchbark canoes like this used Canada's water highways to explore the land and transport furs.

6-3 The route of the voyageurs.

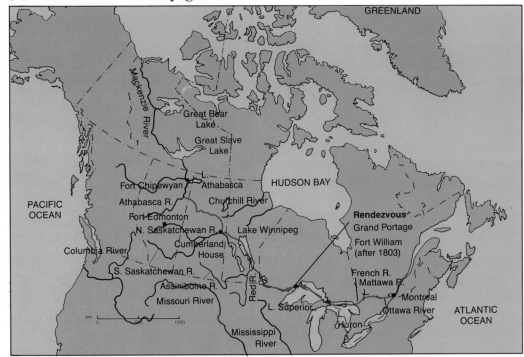

Canada's waterways made it possible to export furs from as far inland as Lake Athabasca. From across the north and west, furs were brought to Grand Portage on Lake Superior, where they were picked up by the voyageurs from Montreal.

paved the way for the fur trade and later the lumber and wheat trade.

These journeys would have been even more difficult had it not been for the help of the native people, who guided explorers and fur traders over water routes familiar to them. In the eastern woodlands they had also developed a craft that was especially well-adapted to the conditions of that wild land — the birchbark canoe. It was the canoe that made possible the transportation of furs from the interior of the continent to Montreal. They were amazing vessels, sometimes as long as twelve metres, weighing less than 150 kg, yet capable of carrying four tonnes of crew and freight. The tough and often colourful men who paddled these frail canoes through rapids and gales, often for more than 15 hours a day, were known as **voyageurs**.

Canada's earliest exports depended on water transportation. Fish from Atlantic coastal waters were the first natural resources to be sent to Europe. Furs became important later on. A third export, lumber, also depended greatly on water transportation. The easiest way to get lumber from the forests of Canada to the shipbuilders of England in the early 1800s was by rafting it down the Ottawa and St. Lawrence river systems to Quebec City. From there the squared timber was shipped across the Atlantic to wood-hungry England. Just when the timber trade weakened, new processes were developed to use wood for making paper. Many rivers of eastern Canada were used to float logs to the new pulp and paper mills established downstream. Rivers still carry logs in some parts of Canada.

Water transportation was also of very practical importance to early settlers. In the days before highways, railways and airplanes, it was for the most part the best way to move both goods and people. Canada's earliest settlements were all on navigable waters. The difficulties of travelling by land even by 1800 are suggested by this item written by Mrs. Simcoe, the wife of Upper Canada's first Lieutenant Governor: "I have lately had the misfortune to ride on the roads, I saw distressed families in wagons breaking down, falling into deep gullies and bridgeless creeks."

Today, many of the "water highways" of Canada have been replaced in importance by highways of concrete and asphalt, and by railways of steel. But water transportation continues to be extremely important in certain parts of Canada, especially in the Great Lakes-St. Lawrence region, in island communities along our coasts, and in the ports which deal with our trading partners across the oceans.

1. **What makes it more difficult and expensive to have a modern transportation network in Canada than in many other countries?**
2. **Choose a brief statement from the text that best emphasizes the importance of transportation and communication.**
3. **Why was water transportation so important in Canada's early history?**
4. **List the main waterways used by the**

6-4 Logs on the Gatineau River. *These are on their way to a paper plant in Hull, Quebec. Of what larger river is the Gatineau a tributary?*

voyageurs to move furs from the Athabasca region to Montreal.

5. Explain how waterways were vital to Canada's first three major export commodities.

New Words to Know
Voyageur

The Great Lakes-St. Lawrence Seaway

Reaching some 3769 km from the Atlantic into the heart of North America is the greatest inland waterway in the world — the Great Lakes-St. Lawrence system. No other continent has a waterway like it. Since the St. Lawrence flows into the Atlantic, it provided a natural corridor into the continent for explorers, traders and settlers from the other side of the Atlantic. (Just think how different the history of Canada might have been if the St. Lawrence had flowed into the Arctic or Pacific Ocean). Until the mid-1800s, most goods and passengers travelled by water. Since then, competition from railways, automobiles, trucks and pipelines has gradually narrowed the role of Great Lakes-St. Lawrence water transportation until it has evolved, for the most part, into a specialized handler of bulk cargos.

Overcoming Natural Obstacles

The native people who paddled these waters called the St. Lawrence "river without end." But while the water seemed to flow on forever, vessels had to end their journeys or at least interrupt them in several places because of natural obstacles. The oldest and simplest way of overcoming these

6-5 Major natural obstacles to navigation on the Great Lakes-St. Lawrence.

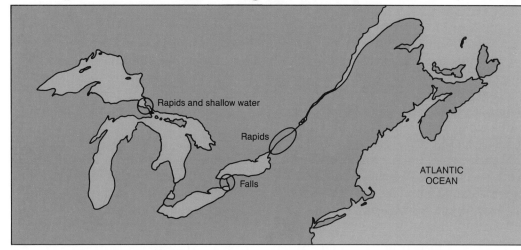

For a view of one of these obstacles, see photo 1-30, page 29.

obstacles was for travellers to get out of their vessel and carry it around the falls or rapids. **Portaging**, as this is called, was taken for granted by the early explorers and fur traders, but when vessels became larger and heavier than canoes, portaging was not possible.

To overcome these natural barriers, canals and **locks** were built. Locks are necessary whenever there is a major change in elevation. The earliest canals and locks were built in the 1780s on the St. Lawrence west of Montreal by Royal Army engineers to carry small vessels from Lake St. Louis to Lake St. Francis. These canals had a depth of only 0.76 m.

In the late 1790s, the Northwest Company constructed a canal and lock along the St. Mary's River between Lake Huron and Lake Superior. Here trappers and traders had been forced by the "sault," or falls (they were actually rapids), to portage their ever-increasing cargoes of furs and merchandise. This

canal was used by canoes and other small boats until 1814 when it was destroyed by American troops during the War of 1812. Not until 1855 was another, much larger, canal built here, this time by the Americans.

In 1824 the first successful Lachine Canal was opened. It cut a 13.5-km waterway through the island of Montreal with seven locks and a depth of 1.5 m. Vessels could now bypass the rapids that had stopped Jacques Cartier some three centuries earlier.

But there were still rapids in the upper St. Lawrence and on the Niagara River which prevented water passage from Lake Erie to Montreal. This became of great concern to Canadians when, in 1825, the State of New York opened the Erie Canal, which connected Buffalo on Lake Erie with Albany on the Hudson River, and thus with the port of New York. This offered an uninterrupted water link between the industrial heartland of North America and the

Atlantic Ocean. It posed a very serious threat to Canadian shipping and to the development of Montreal as a major port.

Canadians met this challenge with a burst of canal building. The first Welland Canal was opened in 1829; canals were also built along the St. Lawrence at Cornwall and Beauharnois. In 1848, an improved Lachine Canal was also completed. A Canadian water route was now open from Lake Huron to the Atlantic with locks twice as large as those on the Erie Canal.

But the Great Lakes-St. Lawrence waterway did not immediately play as big a role as the canal builders had hoped. Just as the early canals were being completed, railways began to make their appearance. By the late 1850s, railways had been built more or less following the same routes as the canals, and were moving people and goods not only faster, but year round, while the water route was often frozen for up to five months. If inland water transportation was to compete with railways it had to give better service.

What water transportation does best is to move bulky goods very cheaply between places located on navigable water routes. **Bulk commodities** like wheat or coal, for which speed is not important, are shipped long distances most cheaply by water. However, to ship bulk goods efficiently requires large ships, and the vessels wanting to use the Great Lakes and St. Lawrence had been steadily getting larger. What was needed was a water system with fewer delays, deeper and wider canals, and deeper, bigger, fewer and more efficient locks.

6-6   How a lock works.

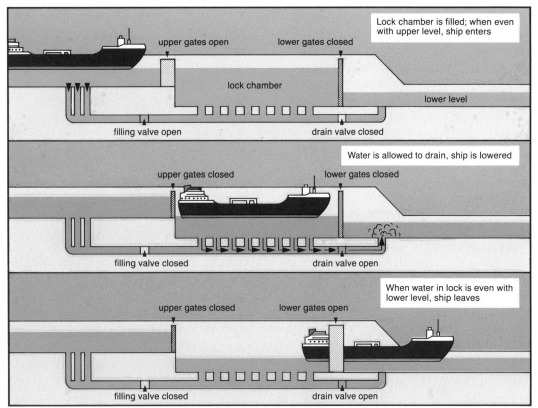

Lock chamber is filled; when even with upper level, ship enters

upper gates open   lower gates closed

lock chamber

lower level

filling valve open   drain valve closed

Water is allowed to drain, ship is lowered

upper gates closed   lower gates closed

filling valve closed   drain valve open

When water in lock is even with lower level, ship leaves

upper gates closed   lower gates open

filling valve closed   drain valve open

Describe the steps that would be followed to enable a ship to go upstream.

6-7   Locks 4, 5 and 6 on the Welland Canal at Thorold, Ontario.
These twinned locks allow two-way traffic. The average time taken for vessels to go through the entire 42-km canal, is about twelve hours. Find the location of these locks on the topographic map on page 336.

312

Beginning around the 1850s, a continuing series of improvements began to be made, which over the years resulted in the Great Lakes-St. Lawrence system as it is today. The two biggest projects were the improved Welland Canal between Lake Erie and Lake Ontario, and the St. Lawrence River canals between Lake Ontario and Montreal. The chart, maps and diagrams show the main features of these projects.

Continuing Challenges

When the present Seaway system was completed in 1959, a fairly large volume of **general cargo** began to reach Great Lakes ports, especially from overseas. Much was made of the fact that 80 percent of the world's ocean ships could now sail to the heart of the continent, and it was expected that general cargo would continue to be a major factor in Great Lakes-Seaway traffic.

However, only a small percentage of the total tonnage moving through the Welland Canal is accounted for by general cargo. This is the result mostly of containerization, and the demand for more speedy delivery of goods. A ship moving through locks and canals takes much longer than a truck on a highway. A container full of manufactured goods from Europe, for example, will get to Detroit far more quickly if it is unloaded in Montreal and shipped by truck or rail to its final destination. Competition from speedier road and rail transportation is one of the Seaway's problems. It is aggravated when traffic is heavy and ships sometimes find themselves lined up waiting to get through a particular lock in the Welland Canal. For a ship, waiting is costly. A

6-8 The Great Lakes-St. Lawrence waterway.

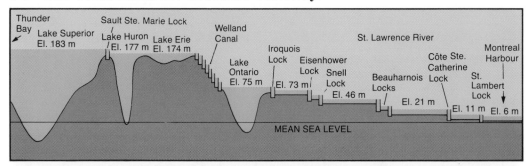

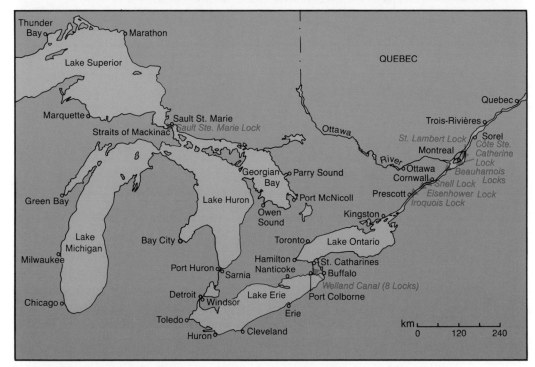

1. Name the lakes through which a ship would pass going from Montreal to Thunder Bay.
2. What is the difference in elevation between Lake Superior and the St. Lawrence at Montreal? More than half of this difference in elevation occurs between what two lakes? What canal connects these two lakes? What makes this canal necessary?
3. What is the drop in elevation between Lake Ontario and Montreal? How many locks are required to enable ships to overcome this difference in elevation?
4. How many locks would a ship have to go through to get from:
   (a) the Atlantic to Toronto (b) the Atlantic to Cleveland (c) Thunder Bay to Nanticoke
   (d) Thunder Bay to Montreal (e) Chicago to Buffalo?
5. What limitations do users of the St. Lawrence Seaway face?

6-9   Welland Canal improvements.

	LOCKS	LENGTH BETWEEN GATES	WIDTH	DEPTH OF WATER OVER SILLS
FIRST WELLAND: Started 1824—Completed 1829	40	33.5 m	6.7 m	2.4 m
SECOND WELLAND: Started 1842—Completed 1845	27	45.7 m	8.1 m	2.7 m
THIRD WELLAND: Started 1873—Completed 1887	26	82.3 m	13.7 m	4.3 m
FOURTH WELLAND: Started 1913—Completed 1932	8	261.7 m	24.3 m	9.1 m
WELLAND BYPASS: Completed 1973, eliminating the need for six lift bridges in the city; city road traffic was improved; ship traffic through canal was speeded up.				

1. In total, how many Welland Canals have been built? Why has more than one been necessary?
2. In what ways was each new canal an improvement over the previous one?

6-10   Commodities shipped on the Seaway, 1996.

COMMODITY	TONNAGE	PERCENTAGE OF TOTAL (MILLION TONNES)
IRON ORE	13 374	26.8
GRAINS	12 789	25.6
COAL	4 504	9.0
OTHER BULK*	13 374	26.8
GENERAL CARGO	5 895	11.8

* Iron and steel products represent 10% of the total Seaway tonnage. Also included in this Other Bulk category is salt, coke and stone.

1. Grains and iron ore are the two most important commodities. Where would these products originate, and what would be their major destinations?
2. Would it be possible for the General Cargo to be of a higher value than the grains shipped? Explain your answer.

related problem is that for three to four months of the year the Seaway is frozen. Rather than earning money all year long, lake ships must sit in frozen harbours waiting for warmer weather.

Of concern in recent years has been the decline in wheat, coal and iron ore shipments on the Seaway. Less demand for Canadian wheat from Europe and greater demand from Pacific Rim countries has sent more wheat to British Columbia ports. At the same time less steel is being made by mills around the Great Lakes. When changes like this occur, there is a loss of income for shipping companies and their workers, as well as for workers at grain- and iron-ore-handling ports like Thunder Bay, Ontario, and Sept-Îles, Quebec.

Seaway authorities are constantly trying to find ways to enable ships to pass through the locks more quickly. Over the years the system has been continually enlarged and improved by widening canal sections, improving lighting for night navigation, extending lock approach walls, and adding a highly sophisticated, computerized traffic-control centre.

In spite of its problems, the Great Lakes-Seaway remains a vital transportation route. Not only has it played a major role in the development of the Great Lakes basin in the past, but it continues to provide the most economical transportation for bulk commodities in Canada's industrial heartland. The well-being of Canada's steel industry, electric-power generation, and wheat production continues to be closely involved with this unique waterway.

1. (a) You are taking a boat trip in 1750 from Montreal to the present site of Thunder Bay. Name the rivers and lakes through which you would pass and the places where you would have to portage.
   (b) Why would portaging not be necessary if you were taking that same trip today?
2. How long ago was the present Seaway system of locks and canals completed?
3. (a) If you wished to get from the Atlantic to the Great Lakes in 1830, which water route, other than the St. Lawrence, could you take?
   (b) Why were Canadians concerned about this route, and what did they do about it?
4. (a) What advantage does water transportation have over other forms of transportation?
   (b) What major disadvantages does it have?
5. In order, which three commodities make up the greatest tonnage carried on the Great Lakes-St. Lawrence waterway?
6. The Seaway earns income by charging fees or "tolls" from the ships that use it. The greater the number of ships and the tonnage carried through the Seaway, the better off it is financially. Why have the Seaway authorities had to worry about finances?

New Words to Know

Portage	Bulk commodity
Lock	General cargo

Ports

Table 6-24 lists Canada's top 10 ports ranked according to total tonnage. Tonnage, however, does not necessarily give a true picture of a port's importance. Port Cartier, for example, handles more tonnage than Montreal. However, most of this is iron ore being shipped in bulk to steel mills. In Montreal the cargos are much more varied, and approximately equal amounts are unloaded inbound and loaded outbound. Many more people and a much greater variety of equipment are needed. In terms of employment and general economic impact, Montreal is far more important than Port Cartier.

Montreal Stretching about 25 km along the St. Lawrence River, Montreal is, in most respects, Canada's greatest eastern port, even though it is located some 1600 km from the Atlantic Ocean! In fact, after New York, it ranks as one of the most important ports in all of eastern North America.

Montreal's location has had much to do with its development into a major port. It has already been mentioned that Jacques Cartier was stopped here by the Lachine Rapids in his journey up the St. Lawrence. Such a point is called the **head of navigation**. Here ocean-going vessels had to stop; passengers and goods wanting to go further upstream had to be transferred to smaller vessels. Increasing numbers of people wanted to go upstream to the Great Lakes and beyond. In other words, there was a need for a port. Soon Montreal became the jumping-off point for western exploration and a centre for the export of furs to Europe.

6-11 *Port of Montreal in winter. What is the river?*

By 1850, when the early Welland and St. Lawrence canals had been built, Montreal became the great "middleman" of the St. Lawrence system. It served as the transfer point for goods on their long journey between Europe and the interior of the continent. This position was strengthened by the development of steamships, which could more easily navigate the narrower waters and currents of the St. Lawrence and thus bypass Quebec City.

In addition to the St. Lawrence, other rivers and valleys focussed transportation routes on Montreal. When the building of railways and roads began around the mid-1800s they tended to follow these same valley routes, reinforcing Montreal's role as a **transportation node**. Large industries were established to take advantage of Montreal's location, particularly along the harbourfront and the Lachine Canal. All of this attracted increasing numbers of people who provided a growing labour force. As a result of the multiplier effect, further business and industry were attracted to the city.

Today road, rail, air and water transportation all come together to make possible the interchange of goods and people at Montreal. Water transportation started it all, and continues to be a prime generator of wealth. In fact, the

315

6-12   Containers being unloaded in Montreal.
Note the size of the crane. What two methods of transporation will most likely be used in the further distribution of these containers?

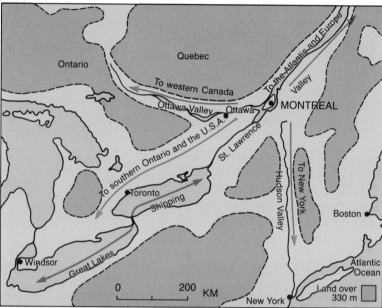

6-13   The location of Montreal.

A number of river valleys focus transportation routes naturally on Montreal. Name the river valleys and the important places they lead to.

port of Montreal is directly and indirectly responsible for the employment of over 17 000 people. These people do a great variety of jobs including operating, maintaining and repairing equipment used for loading and unloading; ship repair and maintenace; cargo checking; trucking; administration; security; and many other functions.

Table 6-15 shows the main commodities handled by the port of Montreal. By far the largest category is "miscellaneous cargo." Close to 5 million tonnes of this category is made up of container cargo. Containers are large metal boxes of standard size (2.4 m by 2.4 m by 4.9 m or 9.8 m) used by shippers to carry all kinds of valuable products. Containers have become especially popular for overseas shipments. They can be loaded with goods at a manufacturer's plant, transported to a port by truck or train, and loaded onto a ship by crane. Most of the ships and equipment used are especially designed to handle containers. Before containers were developed, the goods would have had to be unloaded piece by piece onto the dock, reloaded onto, and later unloaded off, the ship. Containerization speeds up handling, reduces costs, especially of labour, and cuts down on the possibility of damage or theft.

Montreal is Canada's foremost container port. It ranks third in container handling on the North American east coast after New York and Baltimore. Six modern terminals move over half of all the container traffic in Canada every year. Montreal's success as a container port is based partly on the fact that it is still, in a way, the head of navigation. Many large ocean freighters prefer to unload and load their entire cargos at Montreal and quickly head back across the ocean, rather than spend the time and money to go through the Seaway.

Goods unloaded at the port can easily be shipped on in a variety of ways: by water along the Great Lakes-Seaway; by rail (both major railways have terminals at the port); by air (26 domestic and international airlines use Dorval and Mirabel airports); and by truck, with expressways only minutes away from the port. Montreal can still boast that it is on the most-direct shipping route between the great markets of Western Europe and those of Central Canada and the American Midwest.

6-14   The Montreal area.

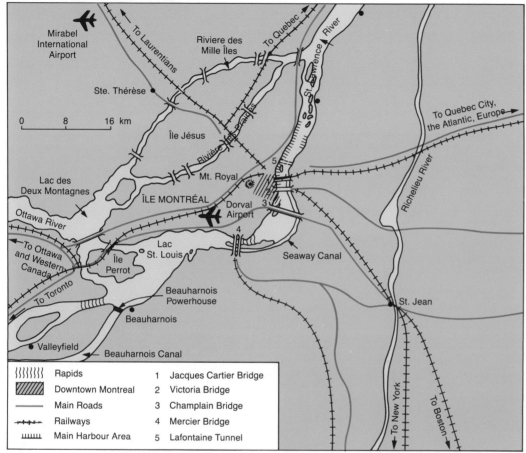

Montreal is located on an island, which is almost completely built up by now. The City of Montreal is the largest of many municipalities on the island.

Map legend:

	Rapids	1	Jacques Cartier Bridge
	Downtown Montreal	2	Victoria Bridge
	Main Roads	3	Champlain Bridge
	Railways	4	Mercier Bridge
	Main Harbour Area	5	Lafontaine Tunnel

Map labels:
Mirabel International Airport · To Laurentians · Riviere des Mille Îles · To Quebec · St. Lawrence River · Ste. Thérèse · Île Jésus · To Quebec City, the Atlantic, Europe · Lac des Deux Montagnes · Mt. Royal · ÎLE MONTRÉAL · Rivière des Prairies · Ottawa River · Dorval Airport · Richelieu River · To Ottawa and Western Canada · Île Perrot · Lac St. Louis · Seaway Canal · To Toronto · Beauharnois Powerhouse · St. Jean · Beauharnois · Valleyfield · Beauharnois Canal · To New York · To Boston

6-15 Principal commodities handled by the port of Montreal.

COMMODITY	TONNAGE LOADED	UNLOADED	TOTAL
MISCELLANEOUS CHEMICALS	866 285	1 109 716	1 976 001
FUEL OIL	286 405	1 599 633	1 886 038
IRON ORE	783	1 702 428	1 703 211
GASOLINE	129 198	1 565 254	1 694 452
WHEAT	636 841	904 865	1 541 706
MACHINERY/EQUIP. & MISC. CARGO	1 154 151	1 110 662	2 264 816

1. **Where does the commodity with the greatest tonnage come from, and where, in general, is it going?**
2. **Explain how at least two of the commodities emphasize that Montreal is an important transportation centre.**

6-16 Canada's top four container ports, 1996

PORT	CONTAINER TONNAGE
MONTREAL	7 875 227
VANCOUVER	5 141 179
HALIFAX	4 076 101
SAINT JOHN	246 898

1. **How does Montreal's container tonnage compare with the total tonnage of the other three ports?**

With its combination of natural and man-made advantages, Montreal can expect to continue to flourish as one of the world's busiest inland ports. For the world it is the "Gateway to the Great Lakes;" for much of Canada it is the "Gateway to the Atlantic."

1. **How does Montreal rank as a port (a) in Canada (b) in North America?**
2. **Summarize the factors, both natural and human, that have made Montreal Canada's major eastern port.**
3. **Explain the advantages of containerization.**

New Words to Know
Head of navigation
Transportation node

Vancouver   Vancouver is by far Canada's number one port based on tonnage handled. In fact, in all of North America, only the port of New York handles a larger amount. The port has become Greater Vancouver's most important generator of employment; it is estimated that over 50 000 jobs are directly or indirectly connected to the port.

317

Most of the port of Vancouver has been built along the shores of Burrard Inlet. With a large protected water area and a depth of 15 m at low tide, Burrard Inlet is recognized as one of the world's finest natural harbours. It is also connected to the best break through the Coast Mountains — the Fraser Valley. This makes it by far the best Pacific outlet for the goods of southern B.C. and the prairies.

In spite of the size of Burrard Inlet, the port has expanded its facilities outside of the main harbour to provide even better service for bulk cargo. The expense of handling and transporting

6-17 Vancouver and Roberts Bank.

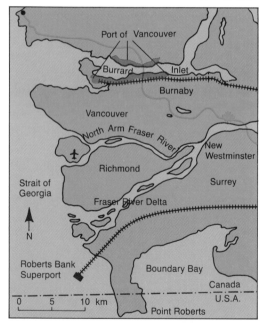

1. How far is Roberts Bank from Vancouver's main harbour?
2. How can you tell that Roberts Bank is a better place to load huge bulk carriers than the harbour in Burrard Inlet?
3. What makes Point Roberts unusual?

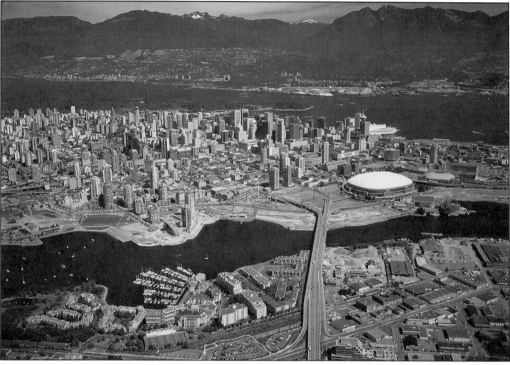

6-18 The port of Vancouver.

6-19 Loading coal at Roberts Bank.

bulk goods like coal, sand, wheat, oil and others makes up a large part of their final cost. The less they are handled and the more efficiently they can be delivered, the more cheaply they can be sold. For this reason enormous bulk carriers, able to hold more than 100 000 t, have been built. Such huge vessels need more room to manoeuvre and less traffic than Burrard Inlet can offer. So a brand new artificial island was created about 40 km south of the main harbour. Built from sand and silt dredged from the ocean floor, the 80-ha island is connected to the mainland by a 5 km **causeway**. By 1970 a "superport" called Roberts Bank had been built on the island. Its main purpose is to handle the huge quantities of B.C.'s coal, which are exported chiefly to Japan. Here, in

water that is 20 m deep, with little interference from other ships, bulk carriers can easily dock, have 100 000 t of coal or more poured into their holds by conveyor belts, and be ready to head back across the Pacific in less than one day.

The coal comes from the Rocky Mountain area of eastern British Columbia in special **unit trains**, which can be as much as three kilometres in length. They are made up of about 100 specially designed cars, each carrying roughly 100 t of coal. When the train reaches the unloading point, each car is mechanically turned over, thus dumping out its coal, and then set upright again without being detached from the rest of the train. It takes about a minute and a half to unload each car, so that the whole trainload of 10 000 t is dumped in

under two hours. The coal moves along a conveyor belt either to huge stockpiles or directly to the ship loaders. Meanwhile, the empty train begins its journey back to the coal mines for another load. It makes the 2400-km round trip in about three days.

1. In which respect does Vancouver rank as Canada's number one port?
2. Which type of commodities make up most of its tonnage?
3. How important is the port to the city and people of Vancouver?
4. Name the body of water that forms Vancouver's harbour. Which natural features make it such a good harbour?
5. Explain the function of Roberts Bank and why it was built.

New Words to Know

Causeway   Unit train

6-20 Principal commodities handled by the port of Vancouver 1996

COMMODITY	TONNAGE LOADED	UNLOADED	TOTAL
COAL	27 445 397		27 445 397
WHEAT	7 798 710	71	7 798 781
SULPHUR	5 193 778	19 600	5 213 378
POTASH	3 645 789	22 059	3 667 848
MISCELLANEOUS CHEMICALS	2 897 232	65 505	2 962 737
LUMBER AND SAWN TIMBER	2 748 438	127 354	2 875 792
WOODPULP	2 567 152	108 430	2 675 582
RAPESEED	2 352 931		2 352 931
OTHER ORES and BASE METALS	827 387	509 020	1 336 408
FODDER and FEED	1 189 180	39 384	1 228 564
FUEL OIL	540 028	743 400	1 283 438
MACHINERY/EQUIP. & MISC.CARGO	523 578	814 900	1 338 478
BARLEY	1 199 644		1 199 644
PULPWOOD	985 063	7 579	992 642
LOGS, BOLTS, etc.	652 754	284 991	937 745
OTHER COMMODITIES	4 687 969	3 407 942	8 095 900
TOTAL	**65 255 030**	**6 150 235**	**71 405 265**

1. Is most of Vancouver's tonnage exported or imported? How can you tell this from the table?
2. What percentage of total tonnage is accounted for by: (a) coal (b) wheat (c) raw materials?
3. List the items that originate principally in the prairie provinces. Approximately what percentage of the total do they account for?

6-21 *Container ship in Vancouver's Burrard Inlet*

Halifax To the native Micmacs, Halifax was *Chebooktook* — "the greatest harbour." To the British governor of Nova Scotia in 1749, Edward Cornwallis, it was "the finest harbour on the east coast of North America." Most people would still agree with this.

Halifax's harbour is spacious, deep enough to handle the largest ships afloat, practically ice-free, and has no narrow channels or dangerous shallows. Furthermore, Halifax is 800 km closer to Europe than any other major port on the North American mainland.

But if Halifax harbour has such outstanding natural advantages, why is Halifax not Canada's foremost eastern port? Chiefly to blame are the St. Lawrence River, the location of Canada's main industrial region, and the steam engine.

Before about 1850, Halifax was indeed British North America's most thriving port. But from this period on, when steamships began to replace sailing ships, Halifax steadily lost its lead. Steamships were more powerful than sailing ships and so they could get up the St. Lawrence River more easily. They could reach Montreal, which was more than 1500 km closer to the rapidly developing industrial areas of southern Ontario and Quebec. Thus Montreal replaced Halifax as Canada's leading Atlantic port. Until the 1960s, activity would increase in the port of Halifax during the winter because the St. Lawrence at Montreal would freeze over for a few months. Since 1964, however, icebreakers have kept the port of Montreal open year round.

Nevertheless, Halifax is Canada's leading Atlantic coast port. Its modern

6-22 Autoport Limited, a part of Halifax Harbour.
A subsidiary of Canadian National Railways, Autoport is one of North America's leading auto terminals. (Find it on the map.) Ships load and unload at a 200-m floating dock.

facilities handle both raw materials and other cargo, including containers and automobiles. In fact, Halifax has an automotive terminal that ranks with the best in North America. Called Autoport Limited, this 36-ha open-storage area for imported cars handles over 1000 vehicles. When one of the specially designed ships, carrying as many as 3000 cars, docks at Autoport, vans full of drivers are brought into the ship to move the vehicles onshore to the storage area. They can unload the ship in about

eight hours. Autoport also has facilities to clean the vehicles, repair any damage, install radios and air conditioning, and provide any other services that the auto companies might require.

Halifax has two well-equipped container terminals. It is also a regular port of call for several **Ro-Ro** (short for "roll on and roll off") **ships**. These are so designed that trucks can drive right on board, thereby reducing the time taken to get containers on and off the ship.

6-23 Halifax Harbour.

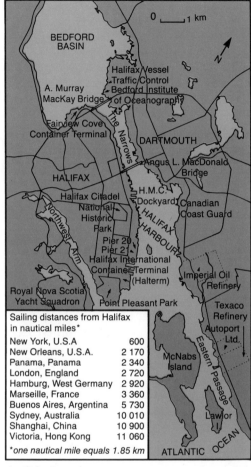

Sailing distances from Halifax
in nautical miles*

New York, U.S.A	600
New Orleans, U.S.A.	2 170
Panama, Panama	2 340
London, England	2 720
Hamburg, West Germany	2 920
Marseille, France	3 360
Buenos Aires, Argentina	5 730
Sydney, Australia	10 010
Shanghai, China	10 900
Victoria, Hong Kong	11 060

*one nautical mile equals 1.85 km

1. Calculate:
   (a) the width of the harbour at the Narrows and at Halterm
   (b) the width of Bedford Basin
   (c) the distance from Autoport to the MacKay Bridge
   (d) the length of the ferry route from Halifax to Dartmouth.
2. Approximately how long would it take to walk across the MacDonald Bridge?
3. On an outline map of the world, mark the places in the table and draw routes to them from Halifax. Show the distances of these routes in kilometres.

6-24 Cargo tonnage at leading Canadian ports, 1996.

	LOADED	TONNAGE UNLOADED	TOTAL
VANCOUVER	65 255 030	6 150 235	71 405 265
SEPT-ILES/PT.NOIRE	20 607 150	1 976 781	22 583 930
PORT CARTIER	18 038 202	3 691 165	21 729 367
SAINT JOHN	10 253 258	10 321 574	20 574 832
MONTREAL/CONTRECOEUR	7 085 611	12 122 261	19 207 872
QUEBEC/LEVIS	6 470 541	10 516 429	16 986 970
HALIFAX	7 064 864	6 522 142	13 587 006
HAMILTON	784 561	11 972 318	12 756 879
THUNDER BAY	9 723 055	377 044	10 100 099
PRINCE RUPERT	9 381 201	69 940	9 451 141
TOTAL	**154 663 473**	**63 719 887**	**218 383 360**
ALL OTHER PORTS	70 471 848	70 900 818	139 372 665
TOTAL OF ALL PORTS	**223 135 321**	**134 620 705**	**357 756 025**

1. On an outline map of Canada, locate the ports listed in the table.
2. Group the ports into three columns under the headings: Pacific, Great Lakes/St. Lawrence, and Atlantic. Show the total tonnage for each port and each group. Which group has the greatest tonnage? Why?
3. List three ports where far more tonnage is unloaded than loaded. What is the main commodity in each?

Halifax was founded in 1749 as a British naval base. Today it has become Canada's largest naval and Coast Guard base. Thus the harbour has always been especially busy in times of war. During World War II, huge fleets of cargo ships and tankers were assembled in the inner harbour formed by Bedford Basin. Here they were loaded with food, fuel, ammunition and other supplies. When ready they steamed out of the harbour in **convoys** taking their vital cargos to the Allied forces in Europe.

Although it seems unusual, a good deal of Canada's trade with Australia passes through Halifax because it is closer to the large markets of southern Ontario and Quebec than Vancouver is.

So exporters from Australia and New Zealand send their products via the Panama Canal to the big population centres of the U.S. northeast and to the port of Halifax, Canada.

1. Summarize the natural features that make Halifax "the finest harbour on the east coast of North America."
2. Explain why Halifax has not become Canada's foremost eastern port.
3. Explain why a lot of Australia's trade with Canada goes through the port of Halifax. What makes this unusual?

New Words to Know

Ro-Ro ship   Convoy

6-25 Major commodities handled in Canada's Atlantic Region and top Ports (1995 Tonnage)

COMMODITIES	LOADED	PORTS
GYPSUM	5 833 667	HALIFAX
CRUDE PETROLEUM	5 095 360	PORT HAWKESBURY
FUEL OIL	4 521 138	ST. JOHN
GASOLINE	3 077 838	ST. JOHN
POTASSIUM CHLORIDE	1 535 911	ST. JOHN
NEWSPRINT	1 311 130	CORNER BROOK
COMMODITIES	UNLOADED	PORTS
CRUDE PETROLEUM	19 077 803	ST. JOHN
FUEL OIL	1 467 308	ST. JOHN
COAL	1 096 828	BELLEDUNE
MACHINERY/EQUIPMENT & MISC. CARGO	1 082 533	HALIFAX

1. **What accounts for the largest tonnage handled by Halifax?**
2. **Why is "Machinery/Equipment" more important to a port than the tonnage statistic indicates?**

6-26 *The Confederation Bridge's main or deep water section consists of 89 segments (44 box girder spans and 60 metre drop in spans). The typical elevation is 40m above the water, with a distance between piers of 250 m. The navigation span is 60 m above the water's surface. 800 000 tons of aggregate, 200 000 tons of cement, and 53 000 tons of reinforcing steel (rebar) were required to build the bridge.*

Ferry Services

When most Canadians decide to travel to another province they think simply of taking the most convenient highway. Some Canadians, however, must include water transportation in their travel plans (unless they go by air, of course). These are the people who live on islands, which include two provinces — Prince Edward Island and Newfoundland — and our largest island off the Pacific coast, Vancouver Island. Surface links between such places and the mainland are provided by ferry boats. Virtually everything islanders use that is not produced locally must be shipped by water.

Communities that depend on ferries sometimes find themselves isolated from the mainland. This can happen when storms or ice conditions make crossing difficult, or when mechanical problems or strikes by ferry workers keep the boats from operating. When P.E.I. entered Confederation in 1873, it insisted the Federal Government maintain a year-round ferry service between the island and the mainland. This connection cost the government over $25 million annually in recent years. The service has been slow and is regularly delayed by ice jams in the winter months, and tourist line-ups in the summer. Several plans for a **fixed link** across the Northumberland Strait via causeway, tunnel, bridge, or combination of same have been considered over the years. Such a link would have several advantages. Island fish and farm products would be more competitive in mainland Canada and the United States. Goods imported onto the island would be less expensive. It would be easier for tourists to visit the island. The island would be more attractive to light industry.

6-27 Marine Atlantic's **M.V. Caribou** enters the harbour at Port aux Basques, Newfoundland. Built in 1986, the ferry can carry 1 200 passengers and 350 cars, or 77 tractor trailers. The **Caribou**, and her sister ship **M.V. Joseph and Clara Smallwood** maintain the link between Newfoundland and mainland Canada (North Sydney, N.S.) on a daily basis.

6-28 BC Ferrys has 40 vessels serving 46 ports of call along 1 000 kilometres of British Columbia coastline. The most advanced vessels, **Spirit of British Columbia** and **Spirit of Victoria** handle the busy Vancouver-Victoria crossing, but other ferries call at ports on the Queen Charlotte Islands, the Northern and Southern Gulf Islands, the Discovery and Sunshine coasts, and the Inside Passage.

Opponents of the fixed link felt that the increased traffic on the island would destroy its unique character and way of life. Others worried about the effects of a causeway and other structures on the currents, temperatures and marine life in the Northumberland Strait (and hence the fishing industry).

In 1993, following a public plebiscite, the Canadian government contracted with Strait Crossing Development Inc, a private consortium, to built a 2-lane highway bridge connecting New Brunswick and Prince Edward Island across the narrowest point of the Northumberland Strait. Costs of the bridge design, construction, maintenance and operation were paid for by the private sector. The 12.9 kilometer bridge developed as three separate sections – the P.E.I. approach (600 m), the New Brunswick approach (1 300 m) and the main bridge (11 000 m). Construction

began in 1993 with the determination of pier locations in the Strait, and testing of surface and bedrock. Fabrication yards were built to allow for manufacture of bridge components on land. During 1995-96, bridge components were assembled in the Strait itself.

Opened for business on May 31, 1997, the Confederation Bridge cost $1 billion dollars Canadian to build. Between 5-6 000 persons (96% from Atlantic Canada) were employed on the project. 1998 fare for the bridge (one automobile) is $35.50. By comparison, ferry service between Wood Island P.E.I. and Caribou, N.S. (across a wider section of the Strait) is $46.00. Operation of the bridge will remain the responsibility of Strait Crossing Development until the year 2032, at which time ownership of the structure will be transferred to the Canadian government.

1. (a) Check an atlas to see which straits you would have to cross to reach the mainland from each of the islands mentioned.
   (b) Give the approximate width of each strait at its narrowest point.
2. At a speed of 20 km/h, and adding a half hour each for boarding and disembarking, about how long would it take to get across each of the above straits?
3. Name three of Canada's busiest and most important ferry services in terms of people and goods handled.
4. Discuss the advantages and disadvantages of a fixed link with the mainland from a Prince Edward Islander's point of view.

New Words to Know
Fixed link

Railways

Canada's first railway began operating in Quebec in 1836. It was a very small railway stretching only 23 km between La Prairie (a ferry boat ride across the St. Lawrence from Montreal) and St. Jean on the Richelieu River. It was basically a "portage" railway, bridging the gap between the St. Lawrence at Montreal and the water routes that led to New York. It didn't take long, however, for people to realize that railways could do more than just supplement waterways. Travel in those days was uncomfortable and inconvenient, and undertaken only when absolutely necessary. Roads between communities hardly existed: carriage rides were bumpy and bruising; waterways couldn't be used during the winter because they were frozen. The railway, then, was hailed as a means of making travel more comfortable and much speedier, and it was able to handle both passengers and freight all year round. In 1850 there were only 110 km of railway lines in what is now Canada; by 1860 there were over 3500. Canada was experiencing a railway boom.

However, the boom was not without its problems, especially with regard to finances. Some leaders in the British colonies began to think that transportation problems could be better handled by a larger government of a united British North America, which would have more capital resources. Transportation between the colonies became even more urgent in the 1860s when the Civil War was raging in the United States and there was talk of the victorious Union Army marching north to absorb the British colonies into the U.S.A. A railway link between the Maritime colonies and Upper and Lower Canada was vital. Also, a railway across the vast uninhabited stretches of the West to the Pacific might prevent that area from falling into the American orbit.

By 1867, the year of Confederation, the Intercolonial Railway between southern Ontario and Nova Scotia was under construction. Four years later British Columbia became part of Canada on the understanding that a railway link with the rest of the Dominion would be built. The line would have to span an incredible 4000 km through the rugged rock and railway-swallowing muskeg of the Canadian Shield and the even more difficult passes of the unexplored western mountains. These were days before bulldozers and power shovels: most of the work was

6-29 An artist's view of the official opening day of Canada's first railway.

done by hand labour and by real horsepower. Against unbelievable odds, the last spike in the Canadian Pacific Railway was driven home on November 7, 1885. At the time it was the longest single railway system in the world!

In the next 30 years, other rail lines were built to compete with the CPR. By 1914 Canada had more railway lines per capita than any other country. But there were not enough people nor enough business to support all of these railways. Some went bankrupt. After many studies and much debate, the federal government took over several financially shaky lines to form the Canadian National Railway, in 1923. The Canadian National and the Canadian Pacific are still Canada's two principal railway companies today.

Promoting Development

When the first "iron horses" of the railways chugged across the mostly unsettled lands of the west they brought in a new age. The prairies, which to Canadians in the East had once seemed as remote as another continent, were now accessible. People began to move west in ever larger numbers. The land was found to be fertile and good for growing a basic human food — wheat. Between 1886 and 1914 a flood of people moved into the West in one of the greatest migrations in Canadian history. The railways not only carried the people and their supplies to the region, but they exported the new grain crops and brought back the manufactured products needed to operate their farms. By connecting the West with the newly formed Dominion in the East they helped create a country out of the northern half of the continent, and kept it politically separate from the U.S.A.

It was not only on the prairies that railways had such a great influence. In areas already settled they also changed peoples' lives. Before there were trains, rural people had a hard time earning money. But railways paid local men to lay tracks, and when the railway was built farmers could send their surplus eggs, chickens, milk, butter, fruit and grain to city markets. Furthermore, farmers could sell wood to the railways to fire the boilers of the early steam locomotives. Railways also brought newspapers and letters from distant places in a few days rather than in the weeks or months required by horses and

6-30　Over 10 000 Chinese men came to Canada to work on the railway.

6-31　Building the railway. *Enormous amounts of human labour were needed to build Canada's early railways. The difficulties in the mountains and shield were far greater than here on the prairie.*

stage coaches. Telegraph lines, built by the railway company to send messages from one station to another, were soon providing services for local communities as well. Thus the coming of the railway brought prosperity, growth, and a feeling of being closer to the rest of the world. No wonder that most people looked on railways with enthusiasm and wanted a line to come through their community.

Railways were also vital in the development of Canada's resources. It was in 1883, during the construction of the CPR, that the great bed of copper-nickel ore near Sudbury was found. Twenty years later, the rich silver deposits near Cobalt were discovered while the Temiskaming and Northern Ontario Railway was being built. The early growth of northern Ontario was based on minerals such as these. The

6-32 Trains, like this one carrying wheat in Saskatchewan, are vital in bringing the resources of the prairies to world markets.

railways also made it possible for the lumber and pulp-and-paper industries to move their commodities to markets in the South.

In other provinces, too, the development of resources was made possible by railways. In Manitoba the Hudson Bay Railway to Churchill, the province's only port, was completed in 1930. As a major export route for prairie grain the railway has been a disappointment. But in more recent years it served to open up the mineral, power, and forest resources of northern Manitoba, such as the nickel mines of Lynn Lake and Thompson, and the hydro-power developments on the Nelson.

In Alberta the 370-km Alberta Resources Railway was completed as recently as 1969 to link the resources of the Peace River country with the CN near Jasper. In British Columbia the second Canadian Pacific transcontinental line, through the Crowsnest Pass, helped to tap the forest and mineral wealth of the Kootenays. The Pacific Great Eastern Railway, now the British Columbia Railway, between Vancouver and Prince George, opened up the forest industries of the northern interior. In Quebec during the 1950s, the Quebec North Shore and Labrador, and the Cartier Railway, both connected the iron mines of the interior to the St. Lawrence.

Adapting to Change

For well over half a century railways were the chief means of moving both passengers and freight between Canadian cities. However, by 1950 they were faced with ever tougher competition. Good highways had been built; more and more people were driving better cars and taking buses;

trucks offered door-to-door freight service. People in a hurry were traveling by airplane. The railway companies have attempted to cope with this competition in two ways: by getting rid of services which do not pay, and by improving those that do.

By the late 1940s, railway passenger service was losing a lot of money. In 1940, Canadian railways still carried about 38 percent of total **intercity** travelers; by 1970, this had dwindled to 5 percent and the railways were losing millions. The simplest remedy was to cancel the services to communities where there were too few passengers to make it profitable. This often brought cries of protest, but, with grudging government permission, many lines were closed. Still there was dissatisfaction on all sides, so in 1977 the federal government formed VIA Rail. This is a **crown corporation**, which is responsible for all intercity passenger trains that were previously operated by CN and CP. VIA Rail, according to its stated purpose, was supposed to "revitalize passenger rail services in Canada and manage and market them in an efficient commercial business."

VIA Rail

By 1997, Via Rail could boast: "We have successfully transformed VIA Rail into a low-cost, high quality passenger service;" train ridership was on the increase. This was achieved by cutting costs and at the same time improving trains and service. An example of cost cutting was the closing of the Halifax and Toronto maintenance centres and consolidating maintenance activities in Montreal, Winnipeg, and Vancouver. On the improvement side: refurbished stainless steel equipment was introduced on train services in northern Manitoba, northern Quebec, and southwestern Ontario. This equipment provided greater reliability and improved comfort. More trains running more frequently and with improved service were added in southwestern and eastern Ontario to meet the needs of the highly competitive medium-to-long distance commuter market. Major improvement in northern services were expected to attract people interested in adventure and eco-tourism since railways allow travelers to reach wilderness destinations in comfort.

VIA is also involved in joint promotions with tourism authorities and private operators across Canada and the U.S. advertising the many winter carnivals, summer festivals, and countless other Canadian attractions to which rail travel gives easy access. Increased co-operation with the great AM Track rail network in the U.S. is also seen as a way of attracting more passengers.

With such policies and efforts VIA hopes to demonstrate that when managed effectively, passenger rail can survive and thrive as a vital part of Canada's transportation mix in the 21st century.

6-33   A GO train picks up passengers destined for Toronto.

For certain distances, trains actually seem to be more suitable than planes. Consider two cities about 500 km apart. At a speed of over 300 km/h – a speed at which some trains in the world can travel – a train can actually move people from one city centre to another faster than a plane. This is because air travelers have to spend a lot of time getting to and from the airport, which is usually some distance outside the city.

For shorter distances, passenger trains have made a remarkable comeback. One of the best Canadian examples is the Ontario-government-sponsored GO Transit system, which daily carries thousands of commuters from cities and towns around Toronto into the city centre. Even busier are the subway trains and light rail vehicles which move hundreds of millions of people annually within Canada's largest cities.

1. **Why did transportation between the British North American colonies become a specially urgent topic in the 1860s and 1870s?**
2. **(a) Give two reasons why the government wanted the Canadian Pacific Railway built.**
   (b) When was it completed?
3. **When and why was Canadian National Railways formed?**
4. **How did railways help open the West?**
5. **On an outline map of Canada show the location of those railways mentioned as playing "a vital role in the development of Canada's resources."**
6. **(a) Why were the railways in trouble by the 1950s?**
   (b) How have they attempted to cope with their problems?
   (c) Which types of "trains" have no trouble attracting passengers?

New Words to Know

Intercity   Crown corporation

6-34   Rogers Pass.

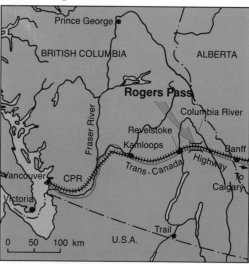

In 1989 one of the biggest improvements ever made by the CPR was completed — a new 34-km railway line through Rogers Pass in British Columbia's towering Selkirk Mountains. It includes a 14-km tunnel, the longest rail tunnel in the western hemisphere. Before this tunnel was built, freight trains, already being pulled by five engines, were forced to stop at the foot of the pass and add six more locomotives in order to make it through the pass. Since the new railway has reduced the grade from 2.2 percent to 1 percent, this is no longer necessary. Westbound trains can now run from the interior to the coast two hours faster than before. (Eastbound trains, often hauling empty cars, will continue to use the old line.) With this saving in time, CP can increase the number of trains travelling through the pass from a maximum of 15 to 24. This is of great importance in handling the increasing demands of countries on the Pacific Rim for Canada's coal, potash, grain and sulphur.

327

Hauling Freight by Rail

In transporting freight the railways have coped with their competition more successfully than in transporting passengers. However, here too they point out that they lose money by having to maintain too many old branch lines that serve small communities. Nevertheless, new technology and equipment and operating methods have made rail freight service faster, less expensive and more suited to customers' needs.

Although the railways made many changes and improvements in their first hundred years, it is in the last half century that the greatest changes and adjustments have been made. Following World War II, old worn-out railway cars were replaced by new ones that were easier to load and unload and could carry more freight. Steam locomotives were replaced by new diesel-electric locomotives, which are more powerful, yet less expensive to operate and maintain.

Costs have been cut by reducing the number of people needed to operate the trains. For example on long trains going through the Rockies, diesel locomotives are placed not only at the head of the train but also in the middle and at the end. Yet only one engine crew is needed to run the entire train — the crew in the front engine controls all locomotives with the help of a computer and radio signals.

Computers perform a great variety of functions for the railways. CP's operations centre, tracks every one of their 55 000 freight cars, 1 250 locomotives and 6 400 repair cars by computer. The location of 20 000

6-35 *Churchill, Manitoba, Canada's most northerly seaport, is the terminus for the Hudson Bay Railway (which runs north from Flin Flon, connecting, through CN, down to Winnipeg). Grain exports through Churchill were 400,000 tonnes in 1996. The HBR is also a popular means of transport for ecotourists on their way to view polar bears.*

foreign railway cars, scattered across Canada and the United States is available on the computers.

Computers also assist in managing the **marshalling yards**, generally on the outskirts of major cities, where freight cars are assembled into trains. Information on incoming trains is sent to the yard by computer before the train arrives. As the train approaches, it is videotaped and monitored on closed-circuit television, to ensure that the cars match the computer's description. The trains then move over a small hill called a **hump**. The cars are pushed to the top of the hump and uncoupled, then roll by gravity downhill through a series of switches controlled by computers to link

up with the correct outgoing cars.

Computerized control panels allow operators to regulate the traffic on the main lines. The computers warn of breakdowns or traffic jams, so that the trains can be rerouted. Computers also are used to schedule equipment maintenance.

The job that railways do best is to move bulk goods long distances overland. To do this most efficiently and cheaply, the railways developed the unit train. A unit train carries large amounts of only one product to only one destination in specially designed cars. Once the locomotive and the cars are linked, they remain as a unit, to be separated only for maintenance. The unit trains carrying coal from the Rockies to Roberts Bank have already been mentioned. Another 104-car unit train carries coal from Corbin, B.C. to Thunder Bay to be shipped on to Ontario Hydro's thermal-electric plant at Nanticoke. Other unit trains carry iron ore from the interior of Quebec and Labrador to ports on the St. Lawrence, and wheat from the prairies to Vancouver and Thunder Bay.

In carrying freight, other than bulk goods, the railway's most serious competition comes from trucks. Until recently, trucks could do something railways generally could not — provide door-to-door pick-up and delivery. So the railways decided to combine the advantages of shipping by rail with the benefits of trucking. The result was the **piggyback** system. The railway sends a truck and trailer to load a customer's freight; this is hauled to a railyard, where the trailer is lifted onto a **flatcar**.

The train gives the trailer a piggyback ride to a railyard in the designated city. There it is unloaded, and a truck delivers it to its exact destination. For overseas shipments, railways provide truck-to-rail-to-port service (or vice versa) using special containers and truck trailers. Since different kinds or modes of transportation are involved in such operations, they are called **intermodal**. Both CP and CN have acquired railways in the United States in order to be more competitive in their freight rates. In 1994, CP enlarged its Detroit River tunnel to accommodate larger freight cars (previously these had been shipped by barge between Windsor and Detroit, a process that took 12 hours). CN opened a tunnel under the St. Claire River in 1995 to carry trains between Ontario and Michigan, slicing twenty-four hours off the old travel time. As much as 25 percent of the goods shipped between Canada and the U.S. crosses at this southern border area.

6-38 CN combines rail, water and highway transportation to link Newfoundland (which has no railway) to the mainland.

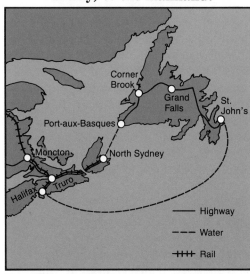

Describe two different routes by which goods might be sent from Halifax to St. John's, and the types of transportation involved.

6-36 Two trains meet at Rogers Pass tunnel in B.C.

6-37 Intermodal transportation. *This train in the mountains near Jasper includes a number of flatcars ''piggybacking'' trailers, which will be hauled from the train to their final destination by trucks.*

329

Moving Wheat to Market

Wheat has long been one of Canada's most important exports. The wheat that ends up as part of a meal in China, Russia, Japan, or in several other parts of the world, starts its journey in a truck on a prairie farm. The truck is loaded right in the field by a combine which has separated the grain from the straw. The truck hauls the wheat either to bins on the farm for temporary storage, or directly to the local **grain elevator**. Grain elevators, once found beside the railway tracks in almost every prairie town, have become a symbol of the West, "the sentinels of the plains."

The farmer typically brings a load of about seven tonnes of wheat at a time to the elevator. Commercial truckers with much larger vehicles may haul over 20 t at a time. At the elevators, the wheat is weighed, graded, and then dumped into a hopper from which it is carried to the top of the elevator by large metal cups on a vertical belt called an **elevating leg**. (You can see how these buildings get their name.) At the top, the wheat is directed to any one of some twenty storage bins. A seven-tonne load can be moved from hopper to storage in about three minutes. The total amount a country elevator can hold varies. In an old one it could be around 550 t, while newer models might hold over 4200 t (150 000 bushels).

In addition to wheat, elevators handle other grains such as oats, barley, flax and rye. "Handling" may include not only receiving and storing the grain, but also cleaning, drying and mixing.

When the grain is scheduled to leave the elevator, the railways take over. The wheat pours from the elevator by gravity

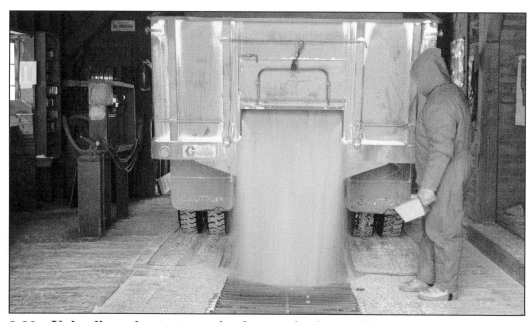

6-39   Unloading wheat at a grain elevator in the prairies.

into special 100-t hopper cars. The most up-to-date elevators can fill about three such cars an hour. When a hundred cars or so are loaded, perhaps from several elevators, the powerful diesel engines begin their haul to one of several ports. Millions of tonnes of grain move along the rail lines every year from prairie elevators to the giant concrete **terminal elevators** through which Canadian wheat is loaded into ships for overseas markets. The operations at the terminal elevators are much the same as at a country elevator, only on a much larger scale. One terminal in Vancouver, for example, has a capacity of 199 000 t.

As noted earlier, the largest tonnages of wheat are shipped from Vancouver. It has the advantages of a year-round shipping season, and lies across the ocean from some of Canada's largest customers. Prince Rupert, to the north

6-40 *Thunder Bay has nine grain elevators with a capacity for storing 1.2 million tons of milling wheat, durham wheat (used for pasta), mustard and canary seeds, lentils, flax, peas, barley, and canola. 1997 was a good year, and the elevators handled 10 million tons of grain.*

along the B.C. coast, shares these advantages. In sharp contrast is Churchill, Manitoba, where the season lasts only about three months. Thunder Bay's shipping, on the other hand, is halted for about three months by ice. But even in winter some grain is moved from this lake port by rail to eastern ports. Some ocean-going ships do load at the Lakehead and sail directly to overseas ports. However, most wheat moves from Thunder Bay by lake boats, called **lakers**, to ports like Montreal, Port Cartier, or Baie Comeau, where it is stored for loading onto ocean-going vessels. Large lakers carry as much as 27 000 t of wheat. This is enough to make almost three loaves of bread for every man, woman and child in Canada. Some ocean-going vessels may carry twice that amount, or more.

In view of what is involved in moving wheat from farms to market — trucks, trains, elevators, ships, ports, the people who run and service all of these transportation facilities, the factories and people who build them, the bankers who handle the money to operate them, the insurance companies that insure them, and the many other organizations who supply transportation companies with an endless variety of goods and services, it is clear that wheat is important to the entire country.

1. **Which technological improvements have railways made to improve service?**
2. **Which job do railways do best?**
3. **Summarize the steps involved in getting wheat from a prairie farm to (a) the Atlantic (b) the Pacific.**

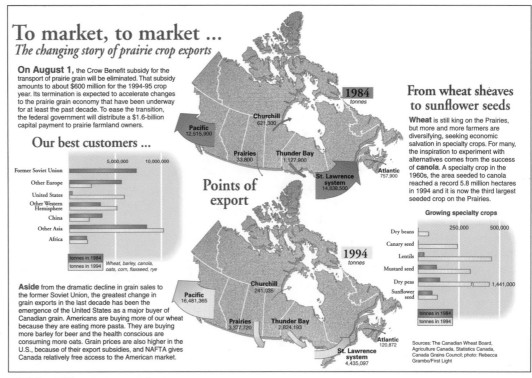

To market, to market ...
The changing story of prairie crop exports

On August 1, the Crow Benefit subsidy for the transport of prairie grain will be eliminated. That subsidy amounts to about $600 million for the 1994-95 crop year. Its termination is expected to accelerate changes to the prairie grain economy that have been underway for at least the past decade. To ease the transition, the federal government will distribute a $1.6-billion capital payment to prairie farmland owners.

Our best customers ...

Former Soviet Union
Other Europe
United States
Other Western Hemisphere
China
Other Asia
Africa

tonnes in 1984
tonnes in 1994

Wheat, barley, canola, oats, corn, flaxseed, rye

Aside from the dramatic decline in grain sales to the former Soviet Union, the greatest change in grain exports in the last decade has been the emergence of the United States as a major buyer of Canadian grain. Americans are buying more of our wheat because they are eating more pasta. They are buying more barley for beer and the health conscious are consuming more oats. Grain prices are also higher in the U.S., because of their export subsidies, and NAFTA gives Canada relatively free access to the American market.

Points of export

1984 *tonnes*
Pacific 12,515,900
Churchill 621,300
Prairies 33,800
Thunder Bay 1,127,900
St. Lawrence system 14,838,500
Atlantic 757,900

1994 *tonnes*
Pacific 16,481,365
Churchill 241,035
Prairies 3,377,720
Thunder Bay 2,824,193
St. Lawrence system 4,435,097
Atlantic 120,872

From wheat sheaves to sunflower seeds

Wheat is still king on the Prairies, but more and more farmers are diversifying, seeking economic salvation in specialty crops. For many, the inspiration to experiment with alternatives comes from the success of **canola**. A specialty crop in the 1960s, the area seeded to canola reached a record 5.8 million hectares in 1994 and it is now the third largest seeded crop on the Prairies.

Growing specialty crops

Dry beans
Canary seed
Lentils
Mustard seed
Dry peas — 1,441,000
Sunflower seed

tonnes in 1984
tonnes in 1994

Sources: The Canadian Wheat Board, Agriculture Canada, Statistics Canada, Canada Grains Council; photo: Rebecca Grambo/First Light

6-41 Wheat to market.

Canada's major wheat growing region is in the interior of the continent; its major customers are overseas. The wheat must therefore be moved to ports for overseas delivery.

1. **How many ports compete for this business (a) on the Pacific (b) on the Atlantic?**
2. **What is the Port of Churchill's main disadvantage?**

6-42 Grain being loaded from a terminal elevator into the hold of a large laker.

New Words to Know
Marshalling yard
Hump
Piggyback
Flatcar
Intermodal
Grain elevator
Elevating leg
Terminal elevator
Laker

Road Transportation

Road transportation is the most common method of moving goods and people in North America. Most Canadians live in heavily populated areas and are used to having a dense network of roads all around. It may shock them, therefore, to see that on a map most of Canada has no roads at all. This, however, is misleading: in fact, Canada's settled areas are all extremely well provided with a dense network of top-quality roads. Over 80 percent of all travel in the country is by road. So widespread has the use of auto-mobile, trucks and buses become, that we simply take them for granted. It is interest-ing to note, therefore, that in 1922 there were only about 80,000 km of roads in all of Canada; by 1990 – just one lifetime later – this had grown to an amazing 900,000 km. Traveling on these roads are over 12 million cars and more than 4 million trucks and buses.

Automobiles

The reasons for the popularity of the automobile are easy to understand. It allows people to travel comfortably, quickly, and in privacy directly from their homes to practically anywhere on the continent. Furthermore, automobiles can travel great distances as easily as they can to the nearest shopping centre. Indeed, the automobile has transformed the lives of Canadians.

The motor vehicle has had an enormous influence on how we use the earth's surface. Consider, for example, the amount of space used for streets, expressways, interchanges, and parking lots. Because of

6-43 Canada's major road and rail networks.

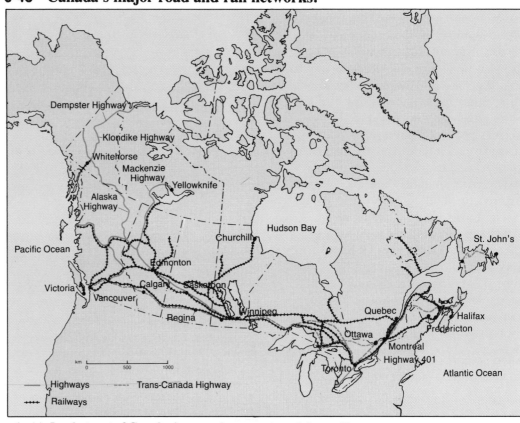

1. (a) In what part of Canada, in general, are most roads located?
   (b) With what other patterns of Canada's geography do the road patterns coincide?
2. Name the main highways that connect the following places:
   (a) Victoria, B.C., and St. John's, Newfoundland
   (b) Windsor, Ontario, and Montreal, Quebec
   (c) Dawson Creek, B.C., and Whitehorse, Yukon
   (d Dawson, Yukon, and Inuvik, NWT
   (e) Yellowknife, NWT, and Edmonton, Alberta.
3. The northern halves of which provinces have no roads? (Note, however, that within the small communities in these areas there are local streets.)
4. Name the most northerly community that can be reached by road from the south in:
   (a) Quebec
   (b) Ontario
   (c) Manitoba
   (d) Saskatchewan.

6-44 *Ontario's Electronic Toll Road, Highway 407, provides a four-lane, and in some cases, a six-lane expressway, 69 kilometres long, between highway 403 in Oakville, and Highway 48 in Markham. Its construction involved the excavation of 17 million cubic metres of earth, and the building of 120 bridges, 29 interchanges, and 15 grade separations. The 407 is one of the world's first electronic toll roads, and requires no stopping, and no toll booths. A photo-based license recognition system levies tolls. Vehicle owners can also purchase a transponder which allows a radio frequency-based reader to identify vehicles as they enter and exit the highway. Owners are billed based on distance travelled.*

cars, people are able to live in one community and work in another. This has contributed to the growth of sprawling suburbs. Shopping centres with their huge parking lots are popular, because more people can get to them easily by car. Gas stations, motels and restaurants all cater to a motoring public. Giant industries, like automobile manufacturing, fuel produc-

tion, the tourist business, and many others, exist because of our love of mobility and the vehicles that provide us with it.

Unfortunately, the popularity of motor vehicles also brings serious disadvantages, like traffic congestion – which creates demand for more highway lanes. Difficult decisions have to be made about how much space should be devoted to private motor vehicles. Heavy traffic contributes to accidents, although other factors are also responsible for the large number of deaths and injuries on our highways. Then, too, cars are powered by burning gasoline, and as with all combustion, gases are created that pollute the atmosphere. There is increasing concern about just how much pollution the earth's atmosphere can absorb. There is concern, too, about what happens to our economy and way of life when gasoline, a non-renewable resource, begins to run out.

Urban planners are particularly concerned about how to persuade people to leave their cars at home. Many Canadian

6-45 *Workers put the finishing touches on the scenic Cobequid Pass Route in Nova Scotia*

cities simply have no more room to build new roads or to add more lanes at any kind of reasonable cost. Many planners agree that more people would take public transit or form car pools if the cost of driving cars into the city was high enough. This could mean charging tolls, a much more simple matter nowadays because we can use electronic means without interrupting traffic. Higher car costs plus convenient public transit may be the way to bring some relief to ever increasing car congestion and the unpleasant results it brings.

Intercity Buses

Although the greatest users of the road network are private car owners, well over 20 million passengers use buses every year to get from one Canadian city to another. This is more than three times the number who used trains. Buses have, in fact, replaced trains for short journeys between many cities and in rural areas; in high-density traffic areas like the Windsor-Quebec corridor, they compete with VIA Rail. Buses also provide charter, tour, and sightseeing services.

Trucks

Most of the things we use in our daily lives – including the food we eat, the clothes we wear, and the building materials used to construct our homes – are brought to us by truck, at least part of the way. In fact, about the only commodities we use daily that are not brought by truck are electricity and telephones, which come to us by wire, and water and natural gas, which come to us by pipeline. Over half of Canada's exports (by dollar value) to the United States, and 80% of Canada's imports from our southern neighbour, are moved by truck.

333

Facts like these make it easy to appreciate that trucking plays a major part in our every day life and in the Canadian economy, accounting for a very important share of income and jobs across the country.

Trucks are the most versatile of the freight-carrying vehicles. They have become the most important movers of consumer goods in Canada. Unlike trains or ships, they are not restricted to tracks or waterways. They can deliver goods door to door, or haul them across the continent. The trucking industry has also developed a great variety of specialized trucks and trailers to handle specific commodities – liquids like gasoline or milk; solid bulk goods like sand, cement or flour; perishable goods like meat, fruit and vegetables.

The manufacturing industries are responsible for the greatest amount of truck activity. Bringing together the raw materials and parts manufacturing plants need, and distributing the products they produce to consumers throughout Canada and the U.S.A. keeps a major part of the trucking industry in business. Many of these trucks, especially those on Ontario highways, are involved in bringing together the thousands of parts that go into assembling automobiles and transporting the finished vehicles to dealers. And there are countless other manufacturers besides those involved in the automotive industry which rely on trucks for most of their transportation needs.

1. **How important is road travel in Canada's total transportation picture?**
2. **Give two good reasons why the automobile is so popular.**
3. **How has the automobile influenced our use of space on the earth's surface?**
4. **What problems are caused by the increasing use of motor vehicles?**

6-46 *Greyhound Canada Transportation Corporation operates one of the largest intercity bus companies in the country, as well as the very popular Gray Line Sightseeing.*

5. **Write a brief report on the importance of busses - intercity or intracity.**
6. **What is it that keeps the largest part of the trucking industry busy? Why has the trucking industry become important?**

A Large Trucking Company

One of Canada's largest trucking companies is Trimac Transportation System with head offices in Calgary, Alberta. Together with its subsidiary truck leasing company – Rentway – Trimac is a major North American transportation company.

Trimac specializes in hauling bulk commodities. It owns and operates trucks and trailers; it services, repairs and cleans them, and dispatches them from its many terminals; it hires the drivers; it is responsible for moving cargos from source to destination. By the 1990's Trimac had become the largest bulk commodity carrier in North America. What some of these trucks carry and where can be seen in the accompanying pictures.

Rentway, on the other hand, does not operate the trucks that it owns; it rents or leases them to companies who prefer not to spend the large sums needed to buy their own trucks. Rentway also services and repairs the trucks that it leases, freeing its customers from much of the concern about transportation issues, and allowing them to concentrate on whatever business they are in.

Rentway has customers involved in a wide variety of activities. Some of them produce plastic products which they sell to various retailers in the U.S.A.; some of them are in the business of delivering parcels and packages; others transport a large variety of consumer goods. But the largest number of their trucks, in the Greater Toronto area, for example, are busy transporting two categories of products. One is food, such as dairy products, poultry, ketchup, fresh and frozen meat, etc., which is transported from manufacturers to supermarkets in many parts of Ontario and beyond. The other category is transporting automobile components which are gathered from manufacturing plants in various locations in Ontario and nearby states of the U.S., and brought to automobile plants in Alliston, Brampton, Oakville, Oshawa, and Windsor.

(Asphalt unit; Edmonton, Alberta)
ASPHALT Type of products hauled: asphalt. 338 units, North America

(Woodchip unit; Bulk Systems, Canada)
WOODCHIPS Type of products hauled: woodchips. 311 units, North America.

(Lumber unit; Calgary, Alberta)
FLAT DECK AND VANS Type of products hauled; package goods, machinery, lumber products. 195 units, Canada.

(Acid units: Vallelyfield, Quebec)
CORROSIVES Type of products hauled: Acids. 147 units, North America.

6-47 a *(top)*
A cross-section of trucking equipment and products handled by Trimac Transportation which has 96 locations in North American, and 4 015 trailers.

6-47 b *(bottom)*
Trimac has been managing CN North America's CargoFlo bulk intermodal operation since 1984. The pneumatic unit pictured here is loading plastic (polyethylene) pellets at CN's largest CargoFlo site in Concord, Ontario. This terminal handles over 330 000 tonnes of product, including plastic, grains, hydrochloric acid, glycols, and de-icing fluids.

Niagara Map Study

1. The area shown on this map is part of what peninsula? Name the bodies of water that make it a peninsula.
2. Name the bodies of water that extend along the eastern and western edges of the map.
3. What two cities dominate the map area? (N.B. Only the edge of one is shown.)
4. The broken line running along the Niagara River indicates an international boundary. What countries, and what province and state, does it separate? Find another boundary on the map and tell what political units it separates.
5. Name four methods of transportation shown on the map that are discussed in this chapter. Which one dominates this area?
6. Name three multi-lane highways shown on the map. For each, indicate what urban centres it connects. What major urban centres are linked to St. Catharines by the Queen Elizabeth Way?
7. Why is there such a concentration of highway traffic in the map area?
8. In what direction does the Niagara River flow? How can you tell this from the map?
9. What two bodies of water are connected by the Niagara River? Why is the river not used for transportation between them? What does provide a transportation link?
10. According to the evidence on the map, what two advantages does the new Welland Canal have over the old?
11. A fifth kind of transportation is shown on the map. The word "transmission" is sometimes used to refer to this, rather than "transportation." Find this form of transportation, tell what is being transported, and draw a sketch map to show the routes involved.
12. Find and briefly describe the location of the largest industrial plant on the map. Here General Motors manufactures transmissions, which are used in cars assembled in Oshawa. Why is this a good location for this plant?
13. For what two specific purposes is most of the farm land used in the northern part of the map? Is this land above or below the escarpment? Explain the advantages of this area for this kind of farming.
14. What feature shown on the map do at least eight motels have in common?
15. Give the shortest, direct distance between:
    (a) Allanburg and the Garden City Skyway
    (b) Horseshoe Falls and The Whirlpool
    (c) the cement plant in Thorold South and the golf course east of St. Davids.
16. Give the actual distance by road between:
    (a) the auto plant and Horseshoe Falls
    (b) the senior citizens' home at 538715 and the Rainbow Bridge
    (c) the cement plant in Thorold South and the golf course east of St. Davids.
17. What is located at the following points?
    (a) 483795 (b) 539787 (c) 465723.

6-49 Part of Niagara topographic map 30 M/6. *Contour interval is 10 m in Canada, 25 feet in the U.S.A. For scale and key, see page 405.*

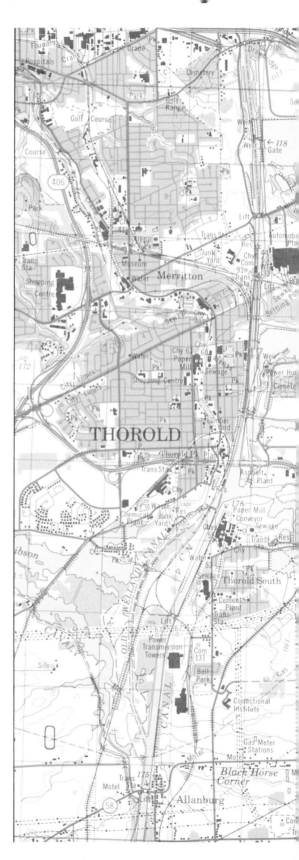

6-50 Two views, looking north and northeast, from the Skylon tower opposite the American Falls. *Using clues in the photos, find this tower on the map and work out the areas shown in the pictures. How many features can you identify on the map?*

336

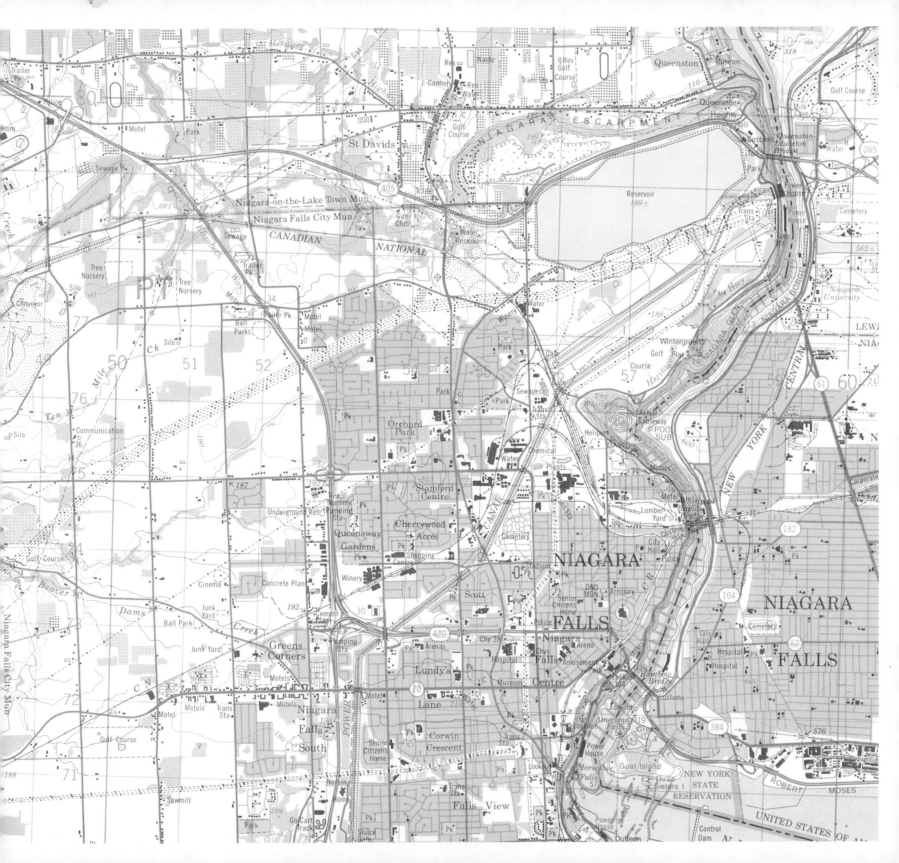

Pipelines

Most of Canada's oil and natural gas resources are found in the more remote areas of this country, far away from the Canadian and international cities and markets they serve. Canada's vast network of pipelines, a network of more than 540 000 kilometres in length brings 95 percent of these resources to domestic and export markets.

In 1997, Canadian pipelines transported over $32 billion of natural gas and crude oil products – including 5.3 trillion cubic feet of natural gas and 745 million barrels of liquid petroleum.

Canada is a leader in pipeline development, and a world leader in safe, reliable, long-distance energy transportation, particularly through difficult terrain and climate.

Movement through a pipeline requires pressure. The government has established the maximum operating pressure for each pipeline system. Some natural gas systems operate at almost 100 times atmospheric pressure. Compressors, powered by turbines, piston engines and electric motors are the most common sources of pipeline pressure. Pumping stations are located at 50 – 100 km intervals to boost pressure and maintain the correct flow.

The first major Canadian pipeline was built in 1912 to carry natural gas from Bow Island, Alberta, to Calgary, a distance of 270 km. Pipeline development expanded dramatically after the discovery of the gas and oil fields near Leduc, Alberta in 1947.

Pipelines which service the oil and gas industry's producing or upstream sector (the oil or gas wells), are called flowlines or gathering systems. These pipes range from 50 – 600 millilitres in diameter. They

6-51   Constructing the Norman Wells to Zama pipeline   *Large sections of steel pipe are welded together. After the welds are tested, the pipe is wrapped in a protective coating and buried in a deep trench.*

bring the resource from individual wells to major collection points. Canada has over 200 000 km of flowlines.

Pipelines which service the refineries, and the marketing and gas distribution areas of the industry (the downstream sector) are called distribution systems (natural gas), or product pipelines (refined petroleum products). These pipes are usually 25 – 150 millilitres wide, although

they have been as large as 900 millimetres. Today, over 50 percent of these pipelines are made of plastic.

Transmission pipelines, which can be over a metre in diameter, link the downstream and upstream sectors. Canada has over 90 000 kilometres of transmission pipeline, operating 24 hours a day, 365 days a year.

Crude oil pipelines carry a variety of

338

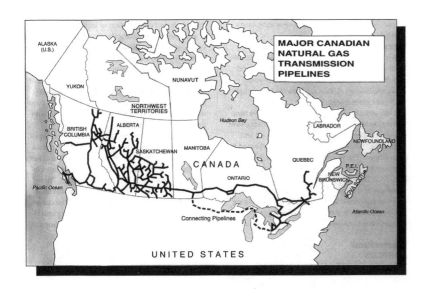

6-52 a & b Canada's major pipeline networks

Pipelines are necessary because most of Canada's oil and gas is found in the west, while the biggest markets are in Ontario, Quebec, and the U.S.A.

petroleum products: several crude oil grades, natural gas liquids, and petrochemicals. Each item is shipped separately, in a batch, and travels about five kilometres an hour. Tank farms along the route act as holding areas, like railway switching yards, allowing operators to control the amount and flow of transmission. Cost of transportation ranges between six and ten percent of the cost of the resource. Oil was $27 a barrel throughout much of 1997, with a transmission cost of $2.70 per barrel.

At the refinery, crude oil is processed into lubricants, fuel oils, petrochemicals, and gasoline. These products are then distributed to retailers by railway, by truck, or product pipeline.

Interprovincial Pipe Line (IPL), with its 11 700 of pipe, operates the world's longest oil transmission system, and transports close to 75 percent of Canada's

oil production. IPL extends from the Northwest Territories into five provinces and seven states.

Natural gas transmission pipes link up directly with distribution systems to carry gas into residences, institutions, and businesses. Natural gas pipelines transport natural gas only, at a rate of 20 kilometres per hour. 1997 transportation costs averaged $1.90 per 1 000 cubic feet. These costs run between 25 and 55 percent of the cost of the resource. Almost half of our natural gas is exported to the U.S.

Pipelines transmit almost two-thirds of Canada's energy. Natural gas deliveries increased 55 percent from 1990 – 1996, and crude oil increased 32 percent. The National Energy Board projects crude oil production to rise 7 percent and natural gas to rise 5 percent by the year 2000. The Canadian pipeline industry has proposed $17 billion in pipeline expansions to meet

these new requirements and to reach new U.S. markets.

In 1996, Canadian pipelines produced 2.7 percent of Canada's greenhouse gas emissions (or 16.9 million tonnes). These consisted of carbon dioxide, nitrous oxide, and methane from the pumps and compressors driving the pipeline flow. Canada's agreement to the Kyoto Protocol in 1997, means that we must reduce our greenhouse emissions below 1990 levels by the years 2008-2012. The pipeline industry is working to help achieve that goal.

1. **Which are the most important Canadian products moved by pipelines?**
2. **How does a pipeline work?**
3. **What are the different types of pipelines?**
4. **What are greenhouse gas emissions? Why is a reduction in emissions important?**

Air Transportation

In few countries of the world is air transportation as essential as it is in Canada. The reasons for this stem from Canada's great size, widely separated cities, and isolated northern settlements. Airlines form yet another transportation network that is vital in keeping this country together and economically healthy.

The Growth of Air Transportation

Canada's first successful airplane was the Silver Dart, which flew for half a mile near Baddeck, N.S. on February 23, 1909. Designed by Alexander Graham Bell and his associates, it was similar to the very first aircraft flown six years earlier by the Wright brothers in North Carolina.

By the 1920s, airplanes had demonstrated that they could go where other forms of transportation could not. They were especially useful in flying prospectors into the underdeveloped wilderness of the North to search for minerals.

Flying into the northern bush continued to increase in the 1930s in spite of the Great Depression. Between 1931 and 1937, for example, freight carried in northern operations increased tenfold to over 11 million kilograms. In no other country was as much freight being carried by air. The daring exploits of Canada's bush pilots became known throughout the world. As early as 1926, one bush pilot and businessman by the name of James Richardson formed Western Canada Airways. This company united with several others in 1942 to become Canadian Pacific Airlines.

Today it is Canada's second largest airline, and is known as Canadian Airlines International.

Canada's largest airline, Air Canada, was organized by the Canadian government and began service in 1937 as Trans-Canada Airlines. This marked the beginning of Canadian airline services in a modern sense. By the end of the 1930s, TCA had begun regular passenger services between Montreal, Ottawa, Toronto and Vancouver.

There is no question that travel has been revolutionized by the airplane, at least as much as by the railway. In 1850, people wanting to get from the east coast to the west might have travelled by ship around South America

6-53 **Air transport has become a major method of moving goods and people in Canada's north.**

— a dangerous journey that could last more than five months. Thirty-five years later, when the CPR was completed, the same journey could be completed in comfort and safety in about seven days — an incredible improvement! Today, regularly scheduled jets take passengers from Halifax to Vancouver in less than seven hours, an equally remarkable accomplishment.

Air travel is popular because it is fast and easy. Other reasons why air travel is popular include higher personal incomes, increased leisure time, longer vacations, lower air fares, and more efficient, comfortable, and safe aircraft. By 1965, Canada's airlines were already carrying some 7 000 000 passengers. By 1997, that figure had tripled to over 22 000 000. Canada has 1 138 airports across the country, 816 of which have paved runways.

Lester B. Pearson International Airport

Canada's busiest and biggest airport is Lester B. Pearson International Airport, servicing Toronto, and the Greater Toronto Area. Pearson handled over 26 million passengers and over 375 000 aircraft in 1997 to rank it as the 25th top airport in the world.

Pearson is currently operated by the Greater Toronto Airports Authority (GTAA), a not-for-profit organization that took over control from the government in December 1996. The Authority's goal and mandate is to create an airport system for the GTA that supports and advances the economic vitality of the region.

The airport is currently undergoing a massive, $2.8 billion redevelopment program which will upgrade airport facilities for the 21st century. One of the major projects is the construction of a new terminal building to replace the existing Terminals 1 and 2. Two additional runways and additional taxiways, an infield cargo area, improved de-icing facilities, two new firehalls, and an improved system of road access are also in the works. Scheduled for completion in the year 2010, the dramatically re-designed airport will have the capacity to handle up to 50 million passengers per year.

Pearson is a fundamental economic asset to the Greater Toronto Area and in 1997, it generated an estimated $11.5 billion in business revenue, $3.2 billion in wages, $2.3 billion in tax revenue, and nearly 112 000 jobs.

6-54 Air passenger traffic.

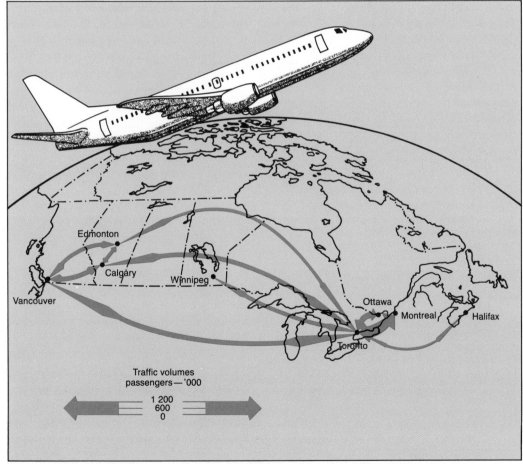

Traffic volumes
passengers — '000

1 200
600
0

List, from heaviest traffic volume to lightest, the ten pairs of cities that are linked by air.

341

As the air traffic pressure on Pearson International Airport continues to rise, other alternatives are being explored. One proposal is to force smaller airlines to use other airports like that in Hamilton, although this displeases many people because of the extended distance from downtown Toronto.

In 1998, the government revisited its plans to build a new, international airport north of Pickering in Durham County, to service the Greater Toronto Area. Plans for an airport in the Pickering area were originally drawn up in the 1970s, and large amounts of private farm land were expropriated by the government in order to develop an airport site. Public outcry and opposition, however, forced the government to postpone its new airport idea. By 1998, opposition seemed more muted, and it is possible that Toronto may indeed have a new airport to cater to its new millennium needs.

6-55 Top 20 airports in Canada, 1997.

	TOTAL AIRCRAFT MOVEMENT AT TOWERED AIRPORTS
TORONTO/LB PEARSON INTL ON	31 859
VANCOUVER INTL BC	27 065
MONTREAL/ST HUBERT QC	21 902
CALGARY INTL AB	20 342
BOUNDARY BAY BC	20 070
MONTREAL/DORVAL INTL QC	16 352
VICTORIA INTL BC	15 492
OTTAWA/MACDONALD-CARTIER INTL ON	15 297
HALIFAX INTL NS	14 390
ABBOTSFORD BC	14 237
WINNIPEG INTL MB	13 860
QUEBEC/JEAN LESAGE INTL QC	13 764
TORONTO/BUTTONVILLE ON	13 265
TORONTO CITY CENTRE ON	10 917
CALGARY/SPRINGBANK AB	10 268
KITCHENER-WATERLOO-GUELPH ON	9 766
LONDON ON	9 685
HAMILTON ON	9 611
EDMONTON INTL AB	8 989
SASKATOON/JG DIEFENBAKER SK	8 908

Locate and label the cities on an outline map of Canada. Show the comparative importance of each airport by means of a graphic symbol to represent numbers of passengers.

Moving Goods by Air

The amount of freight carried by air has increased dramatically over the years, although in total tonnage it is small by comparison with truck and train freight. The important role of bush pilots in opening up the North has already been mentioned. In the 1950s, planes again made a special contribution in developing the resources of the North. One dramatic example was the construction of a hydro-electric plant built in advance of the rest of the project to provide power for the new iron ore mines and the town of Schefferville, Quebec. With no roads or railways leading into the area, everything — people, food, machinery, fuel, building materials, construction equipment, even 190 000 bags of cement — was brought in by air. On the other side of the country, airplanes were providing equally spectacular service in the development of the hydro-electric plant and the aluminum smelting town of Kitimat, B.C.

Today the range of goods carried by air, and the speed with which they are delivered, is greater than ever before. Both of Canada's major airlines emphasize that "if it fits through our aircraft doors, we'll take it." And so an endless list of goods is shipped in, out of, and around the country. Regular flights from London bring goods for a chain of British department stores in Canada. Three times a week an Air Canada plane brings passengers and 10 to 15 t of clothing, brassware, vegetables and others goods from Bombay, India. From Pacific Rim countries come electronic

6-55b A Profile of Canadian air carriers, 1992-1996.

	1992	1993	$ THOUSANDS 1994	1995	1996
Operating revenues	7 548 389	7 535 257	8 406 365	9 319 702	10 038 104
Scheduled services	5 884 911	5 792 261	6 330 483	7 030 379	7 693 554
Passenger	5 280 370	5 180 855	5 700 484	6 347 320	6 996 286
Goods	604 540	611 406	629 998	683 059	697 268
Charter services	1 395 349	1 439 459	1 734 719	1 851 761	1 906 124
Passenger	1 239 915	1 237 154	1 505 640	1 549 200	1 643 987
Goods	155 434	202 305	229 080	302 561	262 137
Other flying services	33 389	55 796	66 980	71 766	68 424
Subsidies	6 185	2 481	1 237	2 014	538
Net incidental air transport related revenue	228 556	245 260	272 947	363 781	369 464

Source: Statistics Canada.

goods and automotive parts. From Florida and California come fruits and vegetables in winter. Furs may be shipped to London, England, and oil-drilling equipment to Aberdeen, Scotland. Fresh seafood is flown from the Maritimes to Montreal, Toronto and Winnipeg. Flowers come from Colombia and the Netherlands to Toronto, and then are reshipped to centres across the country. Huge quantities of envelopes and packages are guaranteed overnight door-to-door delivery between Canada's major cities. As Air Canada boasts,

"We have 2000 employees dedicated solely to serving the cargo shipper."

In general, goods shipped by air have one or more of the following characteristics: they are relatively light in weight, generally of high value, perishable, or required quickly. The "Just in Time" policy of some manufacturers, which saves much of the cost of warehousing and storage, has also brought a greater demand for air freight. Because of trends like this, and because more and more people are demanding faster delivery of mail and cargo, continued growth is predicted for air freight. To meet this demand, both major Canadian air lines were expanding their operations at the end of the 1980s and had over a billion dollars worth of new aircraft on order.

6-56 Canadian air cargo shipment being loaded at Schiphol airport in Amsterdam. *While some planes carry freight only, many passenger planes carry cargo as well.*

1. When and where was Canada's first successful airplane flown?
2. Who were the "bush pilots" and what was their major contribution?
3. Name Canada's two largest airlines. Why does a country with such a small population have two major airlines?
4. (a) Which route had to be taken to get from the east to the west coast in 1850?
   (b) How long might such a journey take at that time?
   (c) Which route would a ship take to-day to get from one coast to the other?
   (d) Check with an airline to see how long it would take to get from Halifax to Vancouver by air.
5. Why has air travel become so much more popular?
6. Give one example of how airplanes helped develop northern resources.
7. In general, which kinds of goods are shipped by air?
8. What are the disadvantages of air travel?

Communication

Human progress can be attributed to our ability to communicate — the movement of information and ideas from one person to another. The spoken and written word have been our primary forms of communication for hundreds of years. Today, because of technological advances such as computers and satellites, our communication has become very sophisticated.

In ancient times, communicaton methods were limited to the scope of the human voice. Some societies used (and may still use) smoke signals or the beat of drums to communicate over extended distances. These have limited application.

The invention of the printing press during the fifteenth century dramatically improved communication. It gave it a world view. The printing press made possible newspapers, books and encyclopedias. Circulation of this material increased in direct proportion to improvement in transportation methods. The popularity of the train and of steam boats, for example, made national and international communication faster and more efficient.

Mechanical/Electronic Communication

In 1835, Samuel Morse, fascinated by the fact that an electric current could flow a great distance through wire in a fraction of a second, developed the idea of the **telegraph**. His system of dots and dashes became the famous Morse code. For the first time, news could travel as quickly as electricity. By the 1860s, most North American cities were linked by networks of telegraph lines. A submarine cable even connected the North American continent with Europe.

It was Alexander Graham Bell who, in 1875, succeeded in finding a practical way of transmitting the human voice over wire. Using his invention, the **telephone**, we can now talk to other people anywhere on earth. The company that carries his name — Bell Canada Enterprises (BCE Inc.) controls a group of companies whose accomplishments in the design, development, manufacturing, marketing and operation of electronic telecommunications' products and services give Canada a respected worldwide reputation.

Both the telegraph and the telephone require wire to carry messages. In 1895, an Italian inventor, Guglielmo Marconi, put together the ideas of several other scientists to send signals through the air. In 1901, he built a transmitter in southwest England and sent signals to a receiving station in St. John's, Newfoundland. Marconi called his invention the wireless telegraph. It became better known as the **radio**.

By the 1920s, improvement to broadcasting technology made the radio affordable and widely used. It revolutionized communication. The 'wireless' was the first communication media to provide entertainment, sports, and news. Information flowed to people everywhere through the radio receiver. The Canadian government realized the potential of radio as a communication tool and established the Canadian Broadcasting Corporation (CBC) in the early 1930s. The CBC eventually would establish radio and later, television networks throughout the country.

By the 1950s, television replaced radio as the most popular means of communication. The invention of the picture tube provided TV with a visual image to

6-57 Alexander Graham Bell, inventor of the telephone, at work in his study in 1913.

accompany the auditory message. Television has changed considerably over the last fifty years. Originally, TV relied on small light-bulb-like "tubes" to conduct its signals. These tubes had a limited life as they often became too hot and burned out. This resulted in frequent repairs, and a low quality black and white picture. These internal tubes were ultimately replaced by transistors which were more efficient in transmitting signals to the picture tube. This made possible larger, colour picture tubes which required less repairs. Today, because of electronic refinements and the use of satellites, our modern televisions are able to provide live broadcast of world news, sports and entertainment.

Time Zones

Vast distances complicate communication within a country the size of Canada, and between countries around the world. Suppose you were having an early lunch in St. John's, Newfoundland and wished to phone a friend in Vancouver. It would be important for you to know that your west-coast friend would probably still be in bed. When it is 12 noon in St. John's, it is only 7:30 a.m. in Vancouver. Why do we know this?

The concept of world time zones was designed by Sandford Fleming, a Canadian engineer, in 1884. Fleming proposed that the world be divided into **24 time zones**, starting from Greenwich, England. He

proposed twelve (12) zones extending east from Greenwich and 12 west. Each zone would represent one hour within the 24 hour period that it takes the earth to make one complete rotation on its axis. Fleming centred each zone on a longitude line.

Longitude lines or **meridians** are always a multiple of 15°, such as 30°, 45°, 60°, etc. The 12 longitude zones would total 180° east of Greenwhich and 180° west. Therefore, one revolution of the earth would equal 360°. The determination of time in each zone would be done by comparing that zone with the time in Greenwich, England. Thus time in the first zone to the east of the central meridian (Greenwich) would be one hour later. The time in the first zone to the west of Greenwich would be one hour earlier. Halfway around the globe from Greenwich (180°) was designated as the **International Date Line** and the calendar date changes once that line is crossed.

The Postal Code

Canada's Post office has its own way of dividing the country into regions or units so that mail can be sorted and delivered more efficiently. It introduced Canada's unique postal code in 1971. The code is given as six characters; ANA – NAN: wherein "A" represents a letter of the alphabet and "N" a number from 0 to 9. A typical example is L5A – IXI. The first three characters can be referred to as the **"area code,"** and the last three as the **"local code."**

The first letter of the area code represents an entire province, or a heavily populated area within a province. For example the letter "A" represents the province of

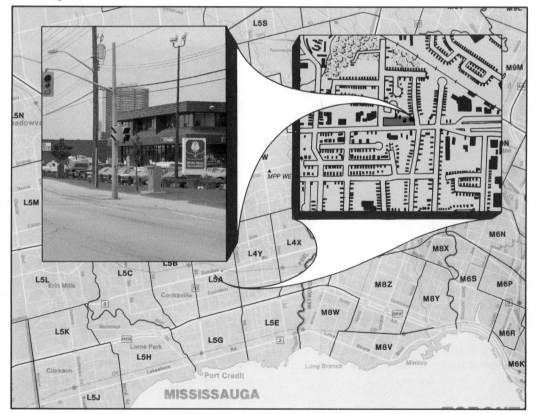

6-58   L5A 1X1. *This is the small section of Dundas Street in Mississauga which is designated by Canada Post as L5A 1X1.*

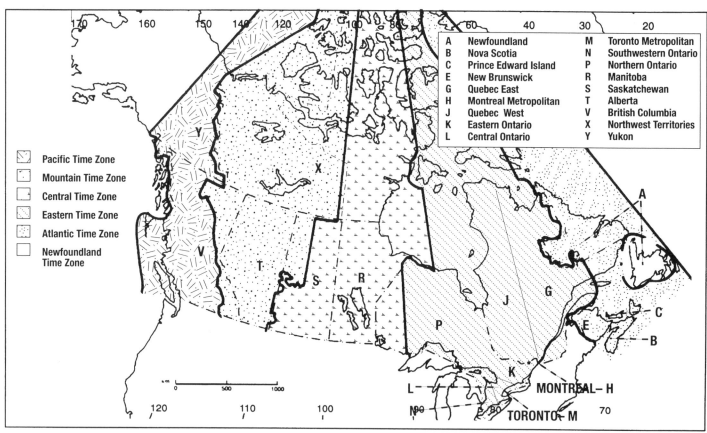

A	Newfoundland	M	Toronto Metropolitan
B	Nova Scotia	N	Southwestern Ontario
C	Prince Edward Island	P	Northern Ontario
E	New Brunswick	R	Manitoba
G	Quebec East	S	Saskatchewan
H	Montreal Metropolitan	T	Alberta
J	Quebec West	V	British Columbia
K	Eastern Ontario	X	Northwest Territories
L	Central Ontario	Y	Yukon

Pacific Time Zone

Mountain Time Zone

Central Time Zone

Eastern Time Zone

Atlantic Time Zone

Newfoundland
Time Zone

MONTREAL– H

TORONTO– M

6-59 Canadian Standard Times Zones and Postal Codes.

Newfoundland; the letter "L" represents south-central Ontario. The letter "V" represents British Columbia; the letter "H" the city of Montreal. Thus, L5A is an urban area in south-central Ontario.

The local code, 1X1, pinpoints the address further. The local code can designate a part of a street, an apartment building, or office building or even a local post office. The code, L5A 1X1, denotes a small section of busy Dundas Street East, in the city of Mississauga, Ontario. The Post Office holds new codes in reserve to allow for population growth. To eliminate possible confusion, the letters "I" and "0" (which look like the numbers one and zero, are never used at all).

1. **How many times zones are there across Canada?**
2. **In what time zone do you live?**
3. **When it is 9 a.m. where you are, what time is it in the capitals of each province? of each territory?**
4. **What postal codes will be used for the new Nunavut territory?**

New Words to Know
Time Zone
Postal Code

Modern Telecommunications

The speed and effectiveness of telecommunication increases almost daily as the result of improvements and inventions. Telecommunication or "distant communication," systems can send and receive sound, printed materials and visual images by means of electrical impulses. These signals reach the receiving stations by

346

6-60 FOTS: Fibre Optic Transmission system – *an 8 000 km fibre optic line was built from Halifax to Vancouver. Engineering on the $240 million project began in 1986 and construction in 1987. At its peak, about 600 people were employed on the project – surveyors, equipment operators, lawyers, suppliers and even archeologists. The environment – rivers, swamps and mountains – posed many challenges. Great care was taken not to disturb wildlife migration patterns or the sacred sites of ancient Aboriginal peoples.*

6-61 Microwave and satellite systems make communications possible across Canada's vast distances and rugged terrain, even in remote areas.

travelling through wires, cables, and by electromagnetic waves through space.

Microwave radio and television broadcasts are usually sent through the air by radio waves. Television signals that must be sent over long distances use very short radio waves called microwaves. Microwaves are also used for many long distance telephone calls. Microwaves travel in straight lines, and because the earth is curved, a cross country system of microwave stations spaced about 50 km apart is necessary to relay the signals and to amplify them. There are three microwave systems in Canada, each of which spans the country: one is owned by CNCP Telecommunications, the other two by Telecom Canada (a company formed by a group of ten telecommunications companies, including Bell Canada and B.C. Tel.)

Fibre Optics Optical fibres are thin strands of glass, about the thickness of a human hair. These fibres are so thin that hundreds of them are needed to make up a single cable. Fibre optic cables are replacing copper wires as conductors of information. The sounds, data or pictures

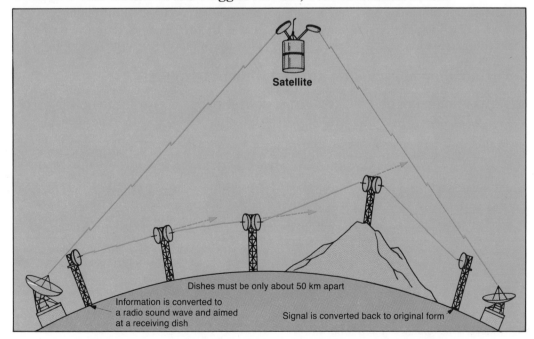

Satellite

Dishes must be only about 50 km apart

Information is converted to a radio sound wave and aimed at a receiving dish

Signal is converted back to original form

347

to be transmitted are converted into code and sent through the optical fibre by a pulsing beam of light which flashes 45 million times a second. A receiver decodes these light pulses back into their original information form. Two optical fibres can carry over 32 000 telephone conversations at the same time. Fibre optics have revolutionized the sending of messages by "wire." Two trans Canada fibre optic trunk lines were completed in the late 1980s, and now form a vital part of Canada's expanding telecommunications network.

Satellites Canada was the first country to launch a non-military satellite for communication within its national boundaries. Anik, an Inuit word meaning "brother," was the first satellite launched in 1972. Several other satellites have been launched since Anik 1. Today, Canadian T.V. radio and telephone companies use a satellite network to facilitate national and international communication.

Satellite communication is especially important to people in remote areas who are not served by wire or microwave networks. These people can receive a variety of television and radio programs in both English and French. Canada's Aboriginal Peoples develop and broadcast programs in their own languages via satellite.

Satellites are important in the transportation, oil and gas industries, but they are essential to Canadians living in sparsely populated northern parts of this country.

Computers and communication

The first computer was built in the 1940s. It was a mechanical device which performed many calculations at the same time. It was extremely inefficient, far too awkward, expensive and, by today's standards, extremely large — and consequently well out of the public's reach. By the 1970s, government (particularly the military) and large corporations were using high-level computers designed for intensive calculations. These computers were called **mainframes**. It was not until the early 1980s that the **personal computer** or **PC** became affordable.

Technological improvements in the CPU (central processing unit), mother boards and video monitors made computers affordable. This affordability paralleled an improvement in user-friendly applications. Operating systems, like DOS, Windows NT, Unix, Linux and OS2 allowed companies to develop a wide range of communications software. Soon, **hardware** (computers and their equipment) and **software** (computer programs and instruction) manufacturers became some of the fastest growing and most successful businesses in the world. Word processing, data base and spreadsheet programs have revolutionized personal, educational, and business communications.

New areas of research have produced creative communication solutions. Some software developers are already marketing 'voice' operated software. An individual speaks or dictates into a microphone and the software translates the voice into a typed screen message. Keyboard typing is eliminated. There is no question that such programs will be widely used within a few years.

The integration of business computers, fibre optics and satellites have reduced the cost of local, national long distance and international phone calls. Moreover, these inventions have created new markets for the cell phone, pagers and voice mail.

6-61 Canadian households with and without telephones.

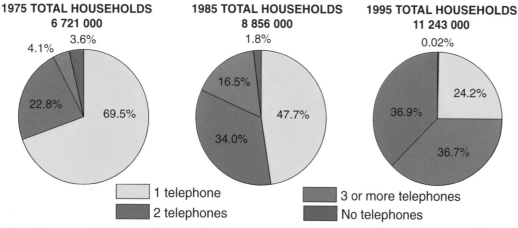

| 1975 TOTAL HOUSEHOLDS 6 721 000 | 1985 TOTAL HOUSEHOLDS 8 856 000 | 1995 TOTAL HOUSEHOLDS 11 243 000 |

1 telephone
2 telephones
3 or more telephones
No telephones

What changes took place in the period shown?

Paperless data storage

New hardware like CD-ROM readers and writers have changed the nature of data storage and communication. Encyclopedias are now stored on CD-ROMs, reducing production costs and creating greater affordability. Scanners can scan or "photocopy text." OCR (Optical Character Recognition) software, which recognizes characters (letters and numbers), allows one to save content immediately as text in word processing format. This process is making keyboard typing of text obsolete. Modern communication and publishing have become electronic industries.

Communication by FAX

Today, personal and business computers transmit digitalized information by way of computer **facsimile** or **fax**. This process is essentially paperless. But it does require a **modem** — a piece of computer hardware which translates the desired message. The modem transmits messages by telephone line, fibre optic cable, or satellite to its destination. The receiving fax machine or computer, can print a hard copy of the message if one is required. National and international fax transmission have become an integral component of today's business world. They are made possible because of the personal computer, modem hardware and communications software. The modern computer has redefined **"mass media."**

The World Wide Web

The **Internet** is a community of computers linked together around the world.

6-63 *The internet cafe combines Canadians' love of coffee, socializing, and international, on line, internet chat.*

The world wide web, an international collection of interlinked documents or web pages, is part of the Internet, as is usenet, a bulletin board system composed of thousands of special interest newsgroups. The internet provides an electronic meeting place so that people all over the planet can exchange information on any conceivable topic — either through **e-mail** (**electronic mail**), by visiting an individual **web site**, by joining a **newsgroup**, or through **real-time chat**. Internet access is through personal computers linked by the world's telephone systems, or by cable, to a series of interlinked **Internet service providers** (ISPs). Most people visiting the Internet have a unique address to facilitate the receipt of e-mail. Usually this address is provided by the ISP. The Internet address of the Government of Canada is http://canada.gc.ca.

People exploring the resources of the Internet are said to be surfing the net. Millions of people around the world use the Internet daily to communicate with each other. Internet users can send and receive text and/or visual files, photographs, videos, sounds, and links to specific web sites. Web sites are not restricted to businesses, or corporations, or institutions. Private, special interest web sites created by enthusiasts make the **web** a truly unique landscape.

E-Commerce

People are now able to transact business electronically on the Internet. Companies can reach markets throughout the world (without leaving their offices) by setting up electronic store fronts. These **electronic commerce suites** can be accessed by anyone using the internet. A company can display, market, sell, invoice and arrange for delivery of its products from its electronic store front. Every function therein is an electronic one. Thus overhead is low, and potential profits are high. Many Canadian banks and investment firms and airline companies are heavily involved in e-commerce.

Electronic commerce suites can be rented from suite providers. INTERNET FRONTIER (IFront – a division of MICROFORUM Inc.) is a suite provider based in Toronto. With the "IFront suite" program, vendors can: set up the "store front" displays; market the products through presentations and descriptions; transact all purchases including invoices, packing slips, sales reports, process credit card: electronic payment, fraud prevention, verification; arrange for delivery and shipping

Many well known companies (Business Depot, Bell Canada, McCain, and CIBC – to name a few) already have business establishments on the internet. The general public is not far behind. Many consumers, however, are reluctant to do business over the "net" because they are concerned about potential leaks in confidential information (credit card number, bank account number, or the like). Some potential customers are worried about phony retailers who might "take their money and run." The personal, face to face contact is still very important to many consumers. In time, most of us will probably become comfortable with E-Commerce, because it provides us with choices that span the globe.

An Interview with L.R.Wilson

L.R. Wilson retired as CEO of BCE on May 6, 1998, although he remains with the company as Chairman of the Board. Mr. Wilson reflects on the leadership role Canada and BCE must play in the emerging medium of cyberspace, and in telecommunications in general.

Q. *Last year the federal government committed to making Canada the most connected nation in the world by the year 2000. What does this mean for Canada?*

6-64 *BCE is Canada's global communications company, with revenues of $33 billion, and 119 000 employees (of which 70 000 work in Canada). Note the wide range of services BCE provides – from wireless communication* with Bell Mobility; the design, building, and integration of communications networks with Northern Telecom; the supply of telecommunications services to emerging markets with Bell Canada International; telephone directory publications with Tele-Direct; and, of course, Bell Canada, this nation's largest supplier of telecommunications services, which carried 9.9 billion conversation minutes in 1997. (Chart and figures as of December 1997)

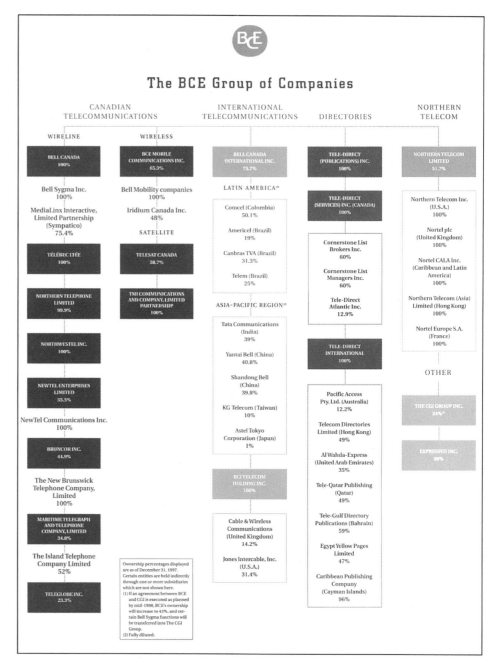

A. This suggests two things: first a recognition that access to information is vital to the emerging knowledge-based economy; and second, that Canada has a historic opportunity to space the future of "cyperspace," which refers to the medium of interaction by which millions of people around the world are linked through computers, phone lines and satellites.

Q. *Becoming the most connected nation in the world seems quite ambitious. Is it possible?*
A. Actually, we are already among the best connected. Our phone network is nearly 100 percent digital – a proportion unequalled even by the United States. We also have the best combination of telephone, cable TV and personal computer penetration among G7 countries.

What we need beyond this is leadership. The private sector has the responsibility to marshall its skills, innovation and investment to create infrastructure and applications. Governments can lead by fostering a supportive regulatory and policy environment. They can also stimulate new services by exercising their enormous purchasing power in health care, education and information.

Q. *What's around the corner in cyberspace?*
A. Undreamed of technologies and applications are in the wings which will allow seamless and intuitive entry to the information highway. We must develop and perfect practical applications in health care, education and electronic commerce to spur continued technological growth. This will not only make us leading exporters of technology, but will also attract even more investment in knowledge-based businesses for Canada.

Q. *Can you give some examples of that you mean by practical applications?*
A. In the field of tele-medicine, for example, we're now transmitting diagnostic X-rays and mammograms over high-bandwidth lines and doctors at major hospitals are examining patients in outlying areas via interactive video links. In the field of training and education, Queen's University in Kingston, Ontario, now offers MBA courses from St. John's to Whitehorse via the telecommunications network. Federal and provincial governments, working with the Stentor group of companies, Telsat Canada and others, are nearing the completion of Schoolnet, which will connect all of Canada's 16 000 schools and libraries to the Internet.

Q. *What are the technical or market-related considerations that will drive growth on the Internet and consequently hasten introduction of new applications?*
A. We need a much faster Internet connection for homes and small businesses – one capable of supporting a reasonably high quality video image on a PC screen. Second, we need to make sure that access to the information highway can become as habitual and intuitive as consulting a telephone directory. This means a connection that is always on and ready to accessed with no need to turn on a PC and wait. The connection should be as simple as a dialtone, but shouldn't tie up your telephone lines or Bell's switches.

Finally, and most importantly, this faster, simpler connection must be cheap and widely available so as to stimulate rapid take-up. In short, we need the cyber-space equivalent of Henry Ford's original strategy for the automobile industry. Once the possibilities of connectedness begin to be understood by everyone, new applications will multiply and the investment needed to further upgrade the infrastructure will be readily forthcoming.

Q. *You have expressed great confidence that a man-made information highway is part of our future. What is BCE doing to make this a reality?*
A. I view this as a nation-building exercise, and every BCE company is getting involved in the initiative.

For example, homes and businesses need faster and simpler Internet connections that are "always on" and don't interfere with voice calls. Bell's asymmetric digital subscriber lines (ADSL) fit the bill perfectly and service roll-out has already begun. In a similar vein, Nortel recently unveiled the 1-Meg modem, an Internet access technology that's considerably faster than the fastest analog modem.

Nationally, Bell and its fellow Stentor telephone companies are using equipment from Nortel, Newbridge and others to build the next cross-country multi-media network, based on a new transmission method called asynchronous transfer mode (ATM). Through MediaLinx, we're investing in Internet content development in English and French for our Sympatico Internet service. Finally, Bell Canada's new Bell Emergis division is working with entrepreneurs and researchers to develop network software to power new applications.

But remember, cyberspace is not a world waiting to be discovered. It's a world we have to create together through ingenuity and talent.

Mind Benders and Extenders

1. "Almost everything we use has been moved from somewhere else." To support this statement, find the source of some of the items you use daily—the district, city, province, state or country where the items originated. Choose at least six of the following: milk, apples, oranges, cereal, sugar, shoes, shirts, jeans, bicycle, TV set, car, bricks, windows.

2. What are the advantages and disadvantages of travelling by ferry?

3. Considering that less energy is required to transport goods by water than by any other method, under what circumstances might the Great Lakes/ St. Lawrence waterway be used more than it is at present?

4. Draw a sketch map to illustrate how Montreal is the "gateway to the Great Lakes."

5. Using Canadian ports and statistics to support your answer, explain why total tonnage does not necessarily indicate the general importance of a port.

6. Using the following categories as organizers, compare the ports of Montreal and Vancouver: location; physical setting; cargo handled; area served (hinterland); major destinations and origins of cargo.

7. Explain why Halifax might be expected to be Canada's foremost eastern port, and why it is not.

8. "In the second half of the nineteenth century Canada was the child of her railways." Explain what this statement means and how true it is.

9. Choose a place in Canada that you would like to visit. Then complete a chart similar to the one below to determine the best way to get there. You can get information from airline, railway and bus companies, travel agencies, maps, people who have taken such a trip, etcetera.

A TRIP TO _____

MEANS	DISTANCE	TIME	COST
Car			
Train			
Bus			
Plane			

What other factors should you consider in choosing which type of transportation you would take?

10. Prepare arguments for and against the following topic: "The Windsor-Quebec City corridor needs and could support a modern, rapid passenger rail system." Two groups could be involved, each group preparing one side of the issue.

11. Explain why Canada needs such vast and expensive oil and gas pipeline networks.

12. Find the price of wheat and calculate the value of the load of wheat in (a) a railway car (b) a laker.

13. What types of transportation—truck, train, ship, plane, pipeline—would you choose as the best way to move the following goods? Give reasons for your choices.
(a) 30 000 t of potash from Saskatchewan to Rotterdam, Netherlands.
(b) Six automobiles from Oshawa, Ontario to one dealership in Moncton, N.B.
(c) One thousand textbooks from Toronto to Ottawa.
(d) 50 000 t of sheet steel from Hamilton to Calgary.
(e) Fresh flowers from Amsterdam to Montreal.
(f) Natural gas from the Beaufort Sea to Edmonton.
(g) Ten ship containers from Halifax to Windsor.
(h) Fresh tomatoes from Leamington, Ontario, to North Bay; to Churchill, Manitoba.

14. Write an essay on the topic "Modern telecommunications help keep Canada united."

15. Explain why satellite communication is especially important for Canadians living in the far north.

16. Find a map in the telephone directory showing area codes. Which provinces have only one area code? How many area codes do Ontario and Quebec have? Explain the variation in the numbers of area codes found in different provinces.

17. The following exercise will illustrate the confusion that would exist if communities were to operate on sun time rather than standard time.

Draw a horizontal line to represent the distance between Windsor and Quebec City and label these cities at each end of the line. On the line mark with dots the approximate locations of London, Toronto, Kingston and Montreal. Using an atlas, determine the approximate difference in degrees of longitude between each of these cities. Based on this information calculate the approximate sun time at each of the cities when it is 12:00 noon at Quebec City.

The key to this problem is the time difference for each degree of longitude. When you remember that the earth makes one rotation of 360° in 24 hours, you can easily calculate how long it takes to move 1° of longitude.

CANADA
EXPLORING NEW DIRECTIONS

Global Connections

"One lesson I have learned is that international policy, far from being foreign, lies at the heart of everyday interests of Canadians. In this world of instant communication and nuclear weapons, in this modern trading country, foreign policy is domestic policy. Isolation is not an option for Canada. We could not draw back from the world, even if we wanted to. And Canadians don't want to."
Joe Clark

Internationalism: A Two-Way Street

Since the end of the Second World War, the world has entered an era of rapid technological change, increasing economic dependence, and has witnessed the growth of large transnational corporations. Canada is far from self-sufficient with regard to economic prosperity and is deeply involved internationally. Increasingly, the well-being and the prosperity of Canadian citizens depends on how they collectively respond to the constraints and opportunities which flow from the international market.

As cartoon 7-2 shows, Canadians would have a very different culture and environment if they were to isolate themselves from the world. Roy Peterson, a well-known Canadian cartoonist, aimed his joke at Canadians, but his point could apply to many countries. Canada is only one of the countries that are being "internationalized," and Canadians are not the only people who sometimes become upset about it. Citizens in Moscow and Beijing no doubt complain about Pepsi stands in their cities; in Paris and Bonn, people complain that English words are invading their language; in Tehran or Cairo, many people object to Western dress, believing it violates religious codes; American popular music is heard, and often condemned from Vietnam to Afghanistan. Popular culture is crossing borders to an extent unmatched in history. The world is in Canada, but it is also in virtually every other nation.

Internationalism is not a one-way street. The world is in Canada, but Canada is also in the world. Foreign trade is vital to Canadian interests. One-third of Canadian Gross Domestic Product (GDP) depends on exports. While the majority of these exports have come from the primary industries based on natural resources, there has been a growth in the export of Canadian technology and knowledge-based consulting services. Canadian engineers have designed dams in China and telephone systems in Saudi Arabia; Canadian real-estate firms own and have developed land and buildings worth billions of dollars in London and New York City, the world's two great financial centres; Canadian nickel, aluminum, wood, salmon, wheat, and computer software are found in countless countries; hundreds of thousands of Canadians have investments abroad in stocks, bonds, and real estate. One Canadian export, ice hockey, is known the world over. The Canadian government maintains almost 200 offices in 164 countries to serve Canada's interests and its citizens abroad.

Official relations outside our border are conducted by two basic methods. One involves maintaining active multilateral relations particularly through international institutions. The other is accomplished by cultivating selective bilateral relations. In turn, hundreds of foreign countries, companies, and individuals have embassies, consulates, offices, investments and other involvements in Canada.

Canada's role in world affairs continues to develop. Among nation states Canada is one of the most powerful both economically and politically. This relative position makes it mandatory that Canada participate fully in international organizations like the United Nations and its specialized agencies. Canada's international importance has also increased because of its economic integration with the rest of the world. Canada is involved in the world; the world in involved in Canada.

7-1 The Canadair Regional Jet, *an important Canadian export. The Montreal-based transportation giant Bombardier has received 363 orders for the very successful Canadair Regional Jet. Sales of Bombardier's aircraft have taken off so rapidly that the company's share of the world market for regional jets has increased substantially.* Photo credit: Bombardier Regional Aircraft.

7-2 "Mr. and Mrs. Average Canadian."

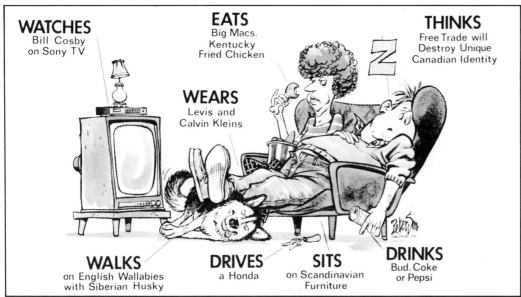

WATCHES
Bill Cosby
on Sony TV

EATS
Big Macs.
Kentucky
Fried Chicken

THINKS
Free Trade will
Destroy Unique
Canadian Identity

WEARS
Levis and
Calvin Kleins

WALKS
on English Wallabies
with Siberian Husky

DRIVES
a Honda

SITS
on Scandinavian
Furniture

DRINKS
Bud. Coke
or Pepsi

Do you think these people are close to average? Perhaps they work for the Swiss firm Nestlé, begin the day with Brazilian orange juice (imported by an American company), and just love kiwi fruit. In what other ways might they be undergoing "internationalization"? What makes this cartoon humorous?

Canada and Canadians: Strongly International

A very strong argument can be made that Canada and Canadians are more involved internationally than most of the world's countries and people. In a period of 18 months between 1996-1997, Canada was the host for three major international events:

- an international conference on AIDS held in Vancouver, July 7-11, 1996
- the Asia-Pacific Economic Cooperation (APEC) summit in Vancouver, November 21-25, 1997
- the signing by 121 countries of the treaty to ban land mines in Ottawa on December 3, 1997

Hosting just one of these events would have placed Canada on the front pages of newspapers throughout the world. Few other countries with a population as small as Canada's could host so many international events. No other country is a member of the G-7(8) Group, the Commonwealth, La Francophonie, and the Security Council.

In addition to these three major meetings, Canada plays a strong role in the United Nations, international peacekeeping efforts, and the worldwide environmental movement. Canada's reputation as a supporter of peace and keeper of the environment has given it a strong international voice.

Another simple way to show that Canada and Canadians are strongly international is to look at a major Canadian newspaper. The following headlines appeared on the front page of *The Globe & Mail*, September 30, 1997.

- *France To Back Quebec's Choice*
- *Mideast Peace Talks To Restart*
- *Scientists Keen To Catch Flue Virus*
- *Screening What's On The Screen*
- *Link Found In Mad Cow Disease*
- *Cartoon Feature*

Each headline suggested, directly or indirectly, links between Canada and other countries:

France As Canada wrestles with the possibility of Quebec's separation, the question of whether other nations will recognize Quebec becomes a critical one for both Quebecers and other Canadians. Premier Bouchard visited France and found some, but by no means complete, support for Quebec independence.

Mideast Canadians are very concerned about the never-ending conflict, involving terrorism and bloodshed, between Israel and various Arab organizations. Hundreds of thousands of Jewish and Arab Canadians are especially interested.

Scientist Flu viruses often spread from country to country and various strains may directly affect Canadians and even bear names indicating the source, such as "Asian flu". This article indicated that Canadian scientists are searching in the Norwegian tundra for genetic "footprints" of the 1918 Spanish flu which killed hundreds of thousands of Canadians just after World War I.

Screening In this article a national rating scheme was described regarding TV programs. Needless to say, virtually all Canadians see American-produced TV in particular, and others from scores of countries such as Britain, Germany, France, Brazil, etc.

Link One of the most serious results of internationalization has been the spread of human disease from country to country, for disease has few boundaries today. The mad-cow disease which affects cattle causing their death was found to be linked to human beings. This disease had a serious outbreak in the U.K. in 1996.

Cartoon The cartoon feature on the front page on September 30, 1997, showed the Pope in a possible and humorous reaction to a visit to him in Rome, the day before by Bob Dylan, the famous American pop hero, singer and composer. There are over twelve million Roman Catholics in Canada, as well as millions of Christians, very interested in Pope Paul's activities.

1. Study cartoon 7-2. Name other things that are not in the cartoon but that reflect the world's influence on Canadians.
2. "Internationalization" is not a one-way street. Give five examples, involving Canada, of each side of the two-way street.
3. Mark on a world map all the continents and countries which are named or referred to indirectly in the newspaper headlines on the front page of the *Globe & Mail* of September 30, 1997.

4. (a) Scan the front page of a major newspaper for today, and note the countries mentioned.
   (b) Other than international stories, which other kinds of news seem important enough for page one?
5. How many international references appear on the first pages of the other sections of the paper such as business, sports and entertainment?
6. Which of the headline topics from September 30, 1997 still concern or interest Canadians?

New Words to Know

La Francophonie   Group of Seven (G-7)
Commonwealth

There are many reasons why Canada cannot avoid the world, and the world cannot avoid Canada. This chapter will explore five of the main ones.

The trade link Canada is one of the greatest trading nations in the world. It is more dependent on export trade than

7-3 Canadian tourists in China. *At any time of year, particularly summer, Canadians can be seen throughout the world. For what reasons other than tourism might they be in foreign countries?*

any other major industrial nation except Germany. This country must maintain a huge international trade or the standard of living of every Canadian will drop considerably.

The political or strategic link
Canada's geographic position and economic power make it important in world politics. Canada is in a unique position because of its strong ties to the U.S.A., and its special links with Great Britain and France. Canada is also a member of the United Nations' Security Council, the North Atlantic Treaty Organization, the Group of 7 (8) economic powers, the Commonwealth, La Francophonie, and many other international agencies. All of which give Canada a far greater influence on world affairs than its population might suggest.

The aid link Most Canadians feel they are fortunate to live in this country, and wish to share their resources with the less fortunate throughout the world. The proof lies in the billions of dollars in aid they have given to other countries for aid, economic development and help in emergencies.

The tourist link Canadians enjoy visiting other countries; in turn, this country attracts many visitors from abroad. Tourists cannot help becoming interested in the lands and peoples they visit.

The family link Canada is a nation of immigrants, who have come from all parts of the world. Cultural and family ties with their countries of origin give most Canadians a special interest and concern about events in many homelands around the world.

The Trade Link

"Trade made my day!" Very few Canadians would ever think of making a statement like this, but for almost every person in Canada it is completely true. A Canadian may wake up to the sound of a clock-radio from Hong Kong, have a breakfast that includes coffee from Brazil or orange juice from the U.S.A., then drive to work in a Korean-built car. At school, students may use an atlas from Britain, a TV from Japan, an encyclopedia from the United States. A lunchtime chocolate bar may be made of cacao from Ivory Coast and peanuts from Senegal. After class, they may practise with the school band using instruments made in France, or drop by the camera club to try out equipment made in Germany. For dinner they may have Mexican lettuce, New Zealand lamb, Indian tea and Chilean grapes. And, before bed, they may watch an Italian movie on their VCR, which may have been made in Taiwan.

Trade has even more influence than this. No mention was made of *marmalade* or *raisins* or *perfume* or *shoes* or *designer clothes* or *home furnishings* or *jewellery* or *rain coats* or *skis* or *bananas* or *hair dryers* or, or, or — the list is endless and these are only the consumer goods. Perhaps you can name the countries where these imports were made, and name other foreign-made products which become exports to Canada. Can you imagine a lifestyle without these products, many of which could be produced in Canada only at great expense, if at all? Would you prefer a world in which there was no international trade? What are its advantages?

Canada trades with many countries, but the vast majority of our trade, over 70 percent, is with the U.S.A. Japan is our second major trading partner, and the United Kingdom our third. It is a reasonably simple matter, if you have the money, to walk into a store and purchase an imported shirt, package of peanuts or opal brooch. However, trade is in reality a very complex matter and a great many skilled people and rules and regulations are involved in your simple purchase.

1. **On a world map, label the products, and their source countries, which are mentioned in the paragraph beginning *"Trade made my day."***

A geographer studying Canada's international trade must answer three basic questions:
- Why does Canada trade with other countries?
- Who are Canada's trading partners?
- What makes up Canada's trade?

7-4 Trade show in Toronto's Convention Centre. *A scene like this is repeated many times a year in Canada's large cities, as buyers and sellers from many nations meet to trade.*

357

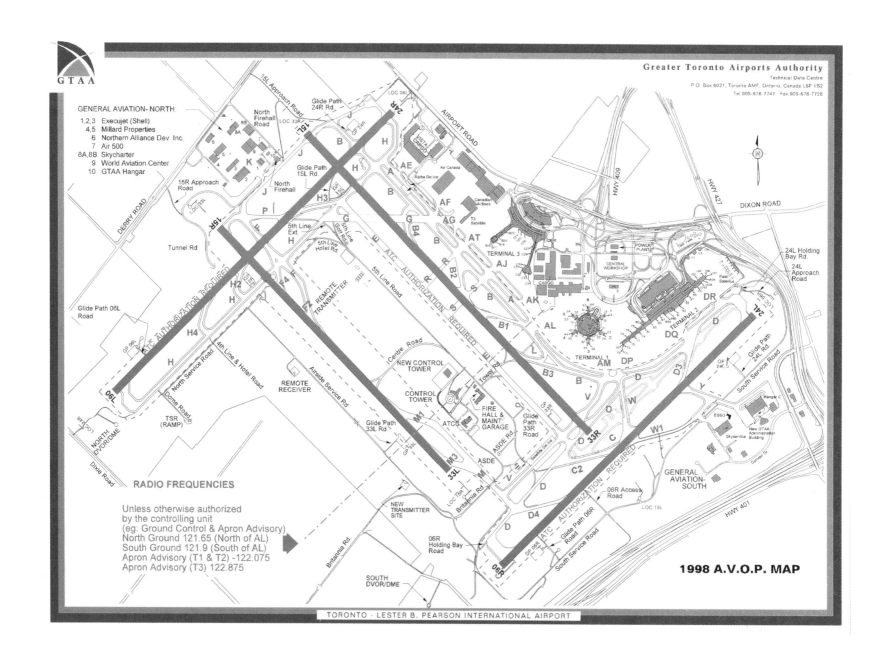

7-5 Pearson International Airport Aviation Operations Map.

7-6 Trade statistics for the Group of Seven (G-7) – 1997

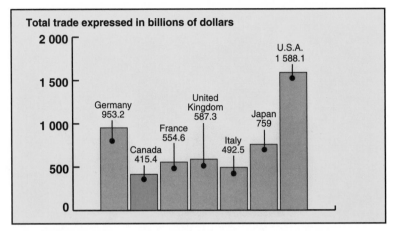

Total trade expressed in billions of dollars

Ten years ago, in 1987, trade figures for the G-7 countries were as follows (expressed in billions of dollars): Germany (west) 693; Canada 252; France 407; United Kingdom 379; Italy 315; Japan 506; USA 894. Compare the 1987 and 1997 figures. Which country had the greatest trade growth? Which had the least? Why?

7-7 Canada—busy exporting and importing.

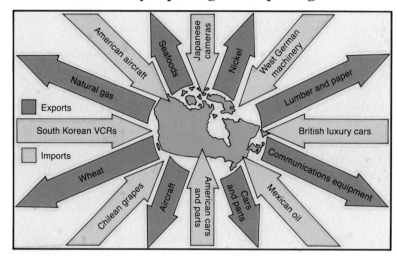

This diagram shows only a small part of Canada's trade. To name all goods and their destinations or sources would need a room-sized page. Name one country receiving each of the exports shown.

Why Trade?

Different commodities The most obvious reason for two countries to trade is that each has something the other wants and doesn't have. Thus Canada exports products such as pulp and paper, aluminum, asbestos, nickel and maple syrup, to countries that cannot produce those things. For exactly the same reason Canada imports things it could not possibly produce. A good example is foodstuffs — bananas, coffee and off-season grapes and lettuce cannot be grown here. (Or they could be, but only at enormous cost.) In some instances, as with lumber, wheat and apples, other countries may be able to produce the goods, but not in sufficient quantities. Similarly, Canada imports oil for use in the eastern part of the country even though it is both a producer and an exporter of oil in the West.

An odd fact about trade is that industrial nations carry out their most extensive trade with other industrial nations. This seems a **paradox**: it would seem more likely that each industrial country would try to produce everything it needed. If each did so, there would be no need to trade at all. The reality is that industrial countries tend to specialize in areas where they have a particular advantage, or special knowledge and experience. Japan, for example, is well-known for producing electronic goods and cameras. It has a large supply of experienced workers, technicians and engineers, and experience with designing, marketing and selling these types of products. Japanese companies have developed vast sales and distribution networks around the world. Jaguar, Ferrari and Mercedes are brand names that are instantly recognized by millions

the world over; each has a long-established reputation for quality craftsmanship. Do you know the source countries for these cars? Customers know and like these products, and have confidence that they are of high quality.

Location For obvious reasons, countries tend to trade with countries that are nearby. It is often easiest and quickest to establish business contacts in a neighbouring country, and it reduces transportation costs. Thus companies in the U.S.A. can save money by using nickel from Canada rather than from New Caledonia, which is near Australia. If the price of New Caledonian nickel were greatly reduced by lower labour costs, it might become cheap enough to make up for the higher transportation costs. Canadian companies buy their bananas from Central America rather

359

than from Malaysia or Indonesia. This greatly reduces transportation costs and also prevents the shipments from spoiling.

Some countries have locative advantages in producing certain goods. Canada can easily produce aluminum because it has huge and inexpensive supplies of hydro-electric power. The U.S.A. and several other large countries have large enough domestic markets to produce commercial aircraft, as well as the technology and skilled workers to make them.

Friendship Trade and political relations are closely connected. Countries naturally tend to trade with countries that are friendly to them. Friendship encourages trade, and trade encourages friendship. Nations in The Commonwealth and La Francophonie have much contact with Canada, and a higher level of trade is one result. For many decades the **Cold War** between the capitalist and communist worlds greatly reduced the trade between them. The break-up of the Communist bloc and the establishment of the Russian Republics has opened up new trading possibilities.

Wealth A nation's wealth will determine to a great degree whether it can purchase goods from other countries. The great trading nations of the world are naturally those which can afford to buy imports from other countries. Generally these imports are paid for by selling exports. India has a huge population of over 950 million people, but the U.K., with only 59 million people, is a

7-8 *The Port of St. John, New Brunswick, at the mouth of the St. John River as it flows into the Bay of Fundy, covers 109 ha. and has 27 berths, with a ferry terminal, a container terminal, and a shipbuilding yard. The Navy Island Forest Products Terminal ships out over one millon tons of forest products, including paper, all over the world. The port is fully computerized, and a regular cruise ship stop.*

much better market for Canada's goods. This is mainly because India doesn't have the money to purchase all the goods it would like to buy from Canada. The *wealthy* nations, not necessarily the largest ones, are the major markets for Canadian exports.

1. **The fact that Canada both produces and imports oil is a paradox. Explain why this apparent contradiction is true.**
2. **Name two major exports from France,** or two from Australia.
3. **In your own words, give several reasons why countries that produce manufactured goods are generally the world's greatest trading partners.**
4. **How could nickel from New Caledonia undersell Canadian nickel in the U.S.A., despite the higher transportation costs?**

360

Canada's Trading Partners

Canada trades with many countries, but the vast majority of its trade — over 70 percent — is with the U.S.A. Japan is Canada's second-biggest trading partner, the U.K. the third-biggest.

Canada's major trading partners are shown in charts 7-10 and 7-11 and on map 7-9. Though Canada traded with scores of countries (38 in Africa alone!), only the top 12 are listed for both imports and exports. Note that in these lists, only 15 countries are represented. Nine countries were involved "both ways" in trade with Canada. The six others appear on only one list: they were important in either importing from or exporting to Canada, but not both. Thus Norway was a major supplier of imports, but was not one of this country's top 12 export markets.

Mexico's trade was even more unbalanced: it received about $1251 million worth of exports from Canada, but exported $6033 million worth of imports to Canada. The value of the peso and low labour costs allows Canadians to purchase inexpensive products from Mexico, yet many Mexican people cannot afford products made in Canada.

Some Canadians might be surprised to

7-9   Canada's major trading partners, 1997.

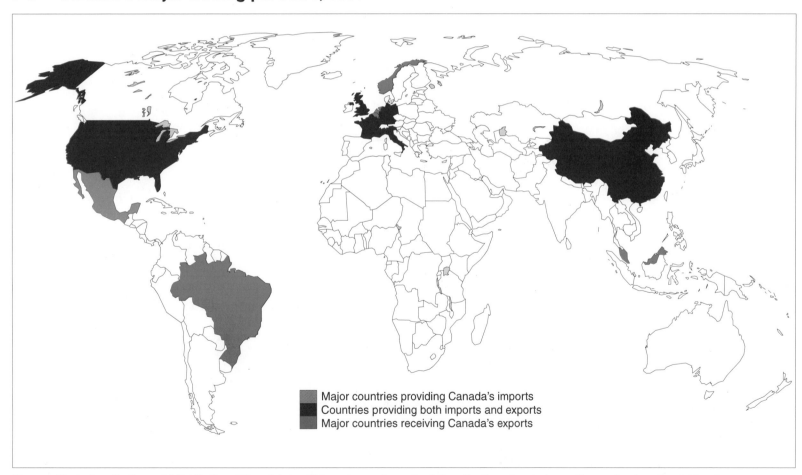

Major countries providing Canada's imports
Countries providing both imports and exports
Major countries receiving Canada's exports

How would you describe the world distribution of Canada's top 12 importers and exporters? Describe the pattern in terms of general and continental locations.

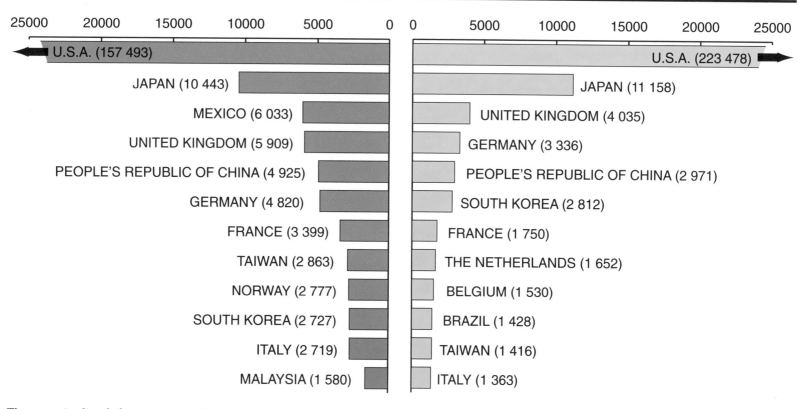

7-10 Top 12 origins of imports into Canada, 1996.

U.S.A. (157 493)	
JAPAN (10 443)	
MEXICO (6 033)	
UNITED KINGDOM (5 909)	
PEOPLE'S REPUBLIC OF CHINA (4 925)	
GERMANY (4 820)	
FRANCE (3 399)	
TAIWAN (2 863)	
NORWAY (2 777)	
SOUTH KOREA (2 727)	
ITALY (2 719)	
MALAYSIA (1 580)	

7-11 Top 12 destinations for Canada's exports, 1996.

U.S.A. (223 478)	
JAPAN (11 158)	
UNITED KINGDOM (4 035)	
GERMANY (3 336)	
PEOPLE'S REPUBLIC OF CHINA (2 971)	
SOUTH KOREA (2 812)	
FRANCE (1 750)	
THE NETHERLANDS (1 652)	
BELGIUM (1 530)	
BRAZIL (1 428)	
TAIWAN (1 416)	
ITALY (1 363)	

The countries listed above are only a fraction of the scores of countries that trade with Canada. International trade is one of the most vital and extensive of many ties between Canada and the world.

1. **What fraction of the total value traded involves the U.S.A., on each list?**

The External Affairs Department spends about one-third of its budget on managing trade connections. Note that these countries were our major partners in 1996.

2. **What fraction of that total involves the top 12 on each list?**

see markets such as Hong Kong, Taiwan, South Korea and Brazil so high on the trade lists. These countries belong to a group known as the **newly industrialized countries** (NICs). Most are on the Pacific Rim. They have become major trading and financial powers only since the mid-1970s. One of their biggest trade advantages has been low labour costs. Generally, they export small manufactured products such as computers, radios, VCRs, tools and hardware, and small motors. This isn't always true — some also build cars and supertankers.

The rise of Hong Kong, South Korea, Taiwan and Singapore as economic powers has earned these Asian NICs the nickname **the Four Tigers**. Along with Japan, China and some smaller countries, these nations form the Asian Pacific Rim countries which have surged forward in trade with Canada.

Statistics show that Canada, which once generally faced east to Europe, is now increasingly facing west to the Pacific. In 1996 Asia took about eight percent of Canada's export; that same year, Europe took only seven percent. Since 1982 Canada has traded more with the Asian Pacific Rim than with its traditional markets in Europe.

1. Why would you expect to find most of the same countries on both chart 7-10 and chart 7-11?
2. Why would African countries have little trade with Canada, even though they obviously export very different products?
3. Certain countries are on only one of the lists in charts 7-10 and 7-11. Organize them under the headings "Major Importers Only" and "Major Exporters Only."
4. Canada's trade for decades has mainly been with the U.S.A., Britain and Western Europe. Recently its trade with Asian Pacific Rim countries has grown tremendously.
   (a) What are the reasons for this sudden growth?
   (b) Compare the total value of goods Canada trades with Pacific Rim countries with that for European countries from the top-12 lists in charts 7-10 and 7-11.
5. How do countries which buy more from Canada than they sell pay for the Canadian goods?

New Words to Know
Newly industrialized countries
The Asian Tigers

7-12 Automobiles shipped by rail.
Automobiles are by far the most important trade item between Canada and the U.S.A.

Canada-U.S. Trade: The World's Largest

The two dominant statistics in charts 7-10 and 7-11 concern the huge volume of trade between Canada and the United States. In 1997, the U.S. received $229 billion worth of Canada's exports, 74 percent of the total exports. In turn, Canada purchased $183 billion worth of American goods, approximately 68 percent of Canada's imports. These **bilateral** trade figures are the largest between any two nations in the world. They dwarf the rest of Canada's trade. Canada's second-largest trading partner, Japan, accounts for less than five percent of our total trade.

The strong trade relationship between Canada and the U.S. is based on factors discussed earlier. The two countries are friendly neighbours sharing a long, undefended boundary. There are countless ties between the two countries including everything from family ties to mutual-defence policies. Their cultures are similar, and most Americans and Canadians speak the same language. Both are democracies with capitalistic economies. Both are among the richest in the world, which makes them excellent markets. To some degree the economies are complementary: Canada has no subtropical region, so American oranges, grapefruits,and off-season vegetables find a ready market here. The U.S. has large coniferous forests, but many Americans read a newspaper on Canadian newsprint because the supplies of softwood pulp are so great here. Our iron ore deposits are extensive, as are their deposits of coking coal; thus, we trade, and both nations are major world steel producers.

The U.S.A. needs Canadian nickel and asbestos, just as this country needs American tools and machinery to extract them.

Even before the signing of the Free Trade Agreement and the North American Free Trade Agreement, 80 percent of the goods crossing the U.S.-Canada border were free of duty. The 1965 Auto Pact enabled vehicles and parts to cross the border with few tariff restrictions. The auto industry accounts for 28 percent of the total trade between the two countries, over $116 billion in 1997.

1. (a) How important, statistically, is U.S. trade with Canada?
   (b) What crosses the U.S.-Canada border other than trade goods?
   (c) Name at least ten products — imports or exports — which cross the Canada-U.S. border, other than those named in the text.
2. How are tariffs and the Auto Pact related?
3. Why can it be said that "the Canadian and U.S. economies are complementary"?

New Words to Know
Bilateral
Merchandise

7-13a Canada's top 10 imports.

		TOTAL VALUE ($ MILLION)
1.	MACHINERY & TRANSPORT EQUIPMENT	102 768
2.	BASIC MANUFACTURES	25 590
3.	MISCELLANEOUS MANUFACTURED ARTICLES	24 098
4.	CHEMICALS & RELATED PRODUCTS	15 242
5.	FOOD & LIVE ANIMALS	10 649
6.	MINERAL FUELS & RELATED MATERIALS	7 029
7.	CRUDE MATERIALS EXCLUDING FUELS	6 313
8.	GOODS NOT CLASSIFIED BY KIND	6 303
9.	BEVERAGES & TOBACCO	925
10.	ANIMAL & VEGETABLE OILS, FATS, WAXES	682
	CANADIAN TOTAL	**199 599**

This graph shows the importance of the automotive industry. Obviously item one and six are directly linked to that activity. Which other items could well be linked to cars and trucks? The U.S.A. as our most important source of imports, is also involved in all of the imports listed, often being the main source. Why are some products so high in value and others so low?

7-13b Canada's top 10 exports.

		TOTAL VALUE ($ MILLION)
1.	MACHINERY & TRANSPORT EQUIPMENT	89 593
2.	BASIC MANUFACTURES	34 318
3.	CRUDE MATERIALS EXCLUDING FUELS	26 356
4.	MINERAL FUELS & RELATED MATERIALS	21 693
5.	GOODS NOT CLASSIFIED BY KIND	13 754
6.	FOOD & LIVE ANIMALS	13 701
7.	CHEMICALS & RELATED PRODUCTS	11 852
8.	MISCELLANEOUS MANUFACTURED ARTICLES	10 851
9.	BEVERAGES & TOBACCO	1 207
10.	ANIMAL & VEGETABLE OILS, FATS, WAXES	554
	CANADIAN TOTAL	**223 879**

Again the importance of the automotive industry is reflected in this list of exports. Determine the percentage of the first item of the total value in each chart; it is surprisingly high. Canada's exports include many crude materials — food minerals, metals, wood — such as in items three to seven above. Why is it unfortunate that Canada exports so many crude materials?

7-13c *A unique Canada-centred world map. Drawn with the aid of a computer, this map is used by the Department of External Affairs as a logo. Where do you think the lines of latitude are on such a projection?*

Canadian Companies Well Connected to the World

Although the Canadian government does play some role in the import and export of goods to and from this country (the Department of Foreign Affairs and International Trade has 130 offices and 828 trade officers around the world), any study of Canadian exports and imports is a study of Canadian businesses in the private sector buying or selling products from or to private sector businesses outside Canada. Charts 7-13 a and b give some indication of Canada's leading exports and imports. The profiles which follow give some indication of the businesses involved in these transactions.

Alcan Aluminium Limited

Alcan Aluminium, a Canadian Corporation, was founded in 1902. Today, it has operations and sales offices in more than thirty countries, and directly employs over 33,000 people. The word ALCAN and the Alcan symbol are trademarked in over 100 countries, and Alcan is involved in all aspects of the aluminium industry. Alcan Group activities include bauxite mining, alumina refining, power generation, aluminium smelting, manufacturing and recycling, and research and technology.

Canadian Aboriginal Products International

Canadian Aboriginal Products International (CAPI) takes traditional Aboriginal crafts and markets them in a high-tech way. Founded in 1996, the company markets high-quality traditional crafts like baskets, quill boxes, pottery, carvings, leather goods and dreamcatchers to target markets across Canada and internationally. CAPI is using the Internet and trade show circuit to discover new outlets, as the company forms a crucial link between producers – often small communities of aging Aboriginal women whose handicrafts are in danger of disappearing – and a sophisticated consumer market that includes Europe and

7-14 *Canadian Aboriginal Products International founders Helen Bobiwash and Michael Jacobs.*

Australasia. CAPI plans to provide gift items for the Sydney 2000 Olympics in Australia, and is vying for a contract to provide items for the World Lacrosse Championships in Baltimore next year. In 1997, the firm was given a warm reception by Maori business people in New Zealand.

ATI Technologies Inc., or ATI

Founded in 1985, ATI Technologies Inc, based in Thornhill, Ontario, is a world leader in supplying high-quality graphics acceleration technology for both the IBM-compatible and Macintosh markets. ATI products include components and add-in boards that accelerate 2D, 3D and video images, enabling video games, computer-assisted design and full-screen video replay on the computer. ATI's products, based on ATI's RAGE(TM) graphics chip, can be found in the computers of all major computer manufacturers around the world, and in retail stores under such names as All-In-Wonder(TM) Pro, XPERT(TM) 98, and Xclaim 3D(TM). In fiscal 1998, ATI's worldwide revenues will exceed $1 billion Canadian. In addition to its Canadian headquarters, ATI has offices in the U.S., Germany, Ireland, France, the United Kingdom, Japan and Malaysia.

Bombardier

Bombardier, originator of the famed snowmobile, is based in Montreal. Active in the fields of aerospace, recreational products, transportation, technical support and maintenance, and financial and real estate services, the company employs over 47,000 employees, and is active on five continents, with

more than 88% of its revenues generated outside Canada. Bombardier's mass transit vehicles are found in North America, Europe, and Asia. Bombardier's areospace group offers Learjet(TM) and Challenger(TM) business aircraft, and the Canadair Regional Jet(TM), among others.

7-15a *One of 70 Rapid Transit (ART) MK11 vehicles developed by Bombardier for a light-rail transit system in Kuala Lumpur, Malaysia.*

7-15b *A Bombardier Ski-Doo MX Z600 snowmobile with the innovative ZX platform for improved cornering and overall balance.*

7-16 *Fording Coal operates the world's largest dragline – the Marion 8750 – (taller than a 20-storey building, with a boom 27 metres longer than a Canadian football field) at its Genesee strip mine 70 kilometres west of Edmonton. The Genessee mine produces some 3 500 000 million tonnes of coal each year. This picture places the Marion inside Edmonton's Commonweath Stadium.*

7-17 *Angela Neill, a software engineer at ATI, installs an ATI graphics accelerator add-in board in a PC. The board enables video games, 3D designing and even TV reception on the computer.*

Fording Coal Limited

Fording, one of Canada's leading producers of high quality coals and industrial materials, with a payroll of over 1600 persons, is a member of the Canadian Pacific Group of Companies. The company operates coal mines in British Columbia and Alberta, and holds potash interests in Saskatchewan. Fording, based in Calgary, services the steel industry, electric power utilities, ceramics and specialized plastics producers, and has an annual metallurgical and coal product capacity in excess of 20 000 000 tonnes. Fording's principal markets are the Pacific Rim countries.

Lavalin

The SNC-Lavalin Group is the world's largest Canadian engineering construction firm, and one of the biggest in the world. It provides engineering, procurement, project management, and project financing services in the industrial, transport, power, infrastructure and buildings, telecommunications, environment and defense sectors. Founded in 1911, and headquartered in Montreal, SNC-Lavalin has more than 6 000 employees in offices across Canada, and in some 30 other countries. In 1998 it was working on projects in nearly 100 countries. The company is listed on the Montreal and Toronto stock exchanges.

Yogen Früz World-Wide Inc.

Yogen-Früz is the world's largest franchisor of frozen yogurt outlets. Introduced as a public company in 1994, Yogen Früz, distributor of the Yogen Früz, I Can't Believe Its Yogurt, and Bresler's yogurts, Java Coast coffees and Golden Swirl frozen yogurt smoothies,currently has 3 611 outlets in 81 countries worldwide, including South Korea, Venezuela, Hungary, Poland, Peru, Chile, Holland, and the United States. Founders Michael, Aaron and Simon Serruya were named Canada's Young Entrepreneurs of the Year in 1997.

7-18 *La Grande 3, Quebec. SNC-Lavalin achievements in power range from major phases of the James Bay hydroelectric development in Quebec to power projects in India, Indonesia, South Korea, and many other parts of the world.*

1. (a) **Name those provinces which have, as at least one of their top three exports, products from the forest industry.**
   (b) **Name the provinces in which forest products are the number-one export.**
2. **Why can Canadians now feel that they are no longer just "hewers of wood and drawers of water"?**
3. **"Canada doesn't export or import." Explain.**
4. (a) **Name the provinces in which each of the top five companies noted in the text has its headquarters.**
   (b) **Choose any one of the five companies noted, and briefly describe its international connections.**
   (c) **How did it make you feel, as a Canadian, when you read about these companies?**

7-20 Exports, imports and trade balance, 1994-1997.

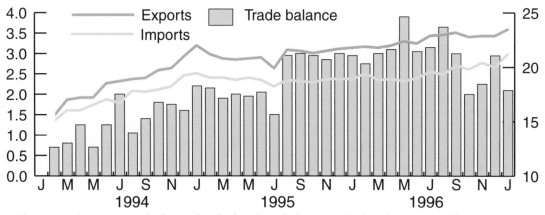

The exports/imports graph shows clearly that Canada has over the last few years sold more goods to the world than it has purchased. It also shows that both figures keep increasing. The trade balance graph shows how trade varies throughout each year.

Visible and Invisible Trade

So far this chapter has only discussed **visible trade**, which is the trade in merchandise or goods. There is also **invisible trade**, which involves services. Here are some examples:

Canada *receives* money for invisible exports when foreign tourists spend Canadian dollars here. Those dollars were purchased with their own currency — **yen, pounds sterling** or **Deutschemarks**. Canadians, however, generally spend more tourist dollars abroad than foreign tourists spend here.

If a Canadian company ships goods to the U.K. or Japan, there is a good chance they will travel in a ship of the **merchant navy** of one of those countries. Canada has few ocean-going merchant ships, so foreign ships carry most of this country's goods. The insurance and banking costs needed to insure and pay for the Canadian shipment may also

be paid to foreign companies. In other words, Canada pays foreigners for services, rather than for physical goods.

Invisible trade also includes **dividend** and **interest** payments between countries. If the shares of a company in Canada are actually owned by a citizen of the U.S.A. or Britain, then any dividends paid on those shares must be sent to those shareholders. Similarly, if a Canadian company or government has borrowed money from foreign sources, the interest on that money must be sent to the foreign owners.

For almost all of its history, the **net trade flow** of both dividend and interest payments has not been in Canada's favour. This country has had a **trade deficit** rather than a **trade surplus** in these payments, because more foreigners own Canadian stocks and bonds than the other way around.

The Balance of Trade

It is easy to understand that when Canada buys goods from, and sells goods to foreign countries, there will be a rough balance between imports and exports. This enables nations to pay for the products they buy from each other. However, since there is also an invisible trade in services, which involves the massive flow of money as well, determining this country's real **current trade balance** is not so simple. To determine whether Canada is running a surplus or a deficit in its current trade account, economists have to total *all* of Canada's trade — the visible flow of goods *and* the invisible flow of money.

Usually Canada runs a surplus in merchandise trade, although we do buy more from certain selected countries than we sell to them. We sold Japan $111 158 worth of goods in 1997, and we bought from Japan $10 443 000. Canada ran a

367

trade-goods surplus of $715 000 with Japan for that year. In reality, however, our trade balance with Japan was much lower because Japanese banks and firms invested in Canadian stock and bonds, causing millions of dividend and interest dollars to flow out of Canada to Japan. Charts 7-10 and 7-11 show several other countries with which Canada traded in 1997. Which countries ran a trade deficit with Canada? Which a trade surplus?

One of Canada's most serious problems in international trade is the fact that we owe a huge amount of money to other nations – $583 billion in 1998 – up from $220 billion in 1988. Canada's per capita debt is $19 440. In other words, to eliminate the national debt, every Canadian would have to pay is $19 440. The Federal Government pays interest charges of $42 billion dollars per year to service this debt.

1. **Shipping costs generally don't help Canada's trade figures. Why is this the case?**
2. **(a) Which are the three major types of invisible trade?**
   (b) Why is Canada's invisible trade generally not in our favour?
3. **Find out what kinds of goods Canada is importing from Mexico, Taiwan, South Korea, and the People's Republic of China. What kinds of goods are we exporting to these countries?**
4. **(a) Using charts 7-10 and 7-11, name four countries with which Canada ran a surplus in merchandise trade.**
   (b) Name four countries with which Canada ran a deficit.
5. **Explain how interest payments can cause a trade deficit.**

Asia-Pacific Economic Community – APEC

Founded in 1994, as an association of sovereign states bordering the Pacific Ocean, the Asia-Pacific Economic Community is dedicated to achieving free trade, and economic development in the Pacific region through assistance to small and medium-sized business; sustainable growth; infrastructure development; future technologies; creation of stable and efficient capital markets, and the development of human resources. APEC has 18 member states (Australia; Brunei Darussalam; Canada; Chile; People's Republic of China; Hong Kong, China; Indonesia; Japan; Republic of Korea; Malaysia; Mexico; New Zealand; Papua New Guinea; Philippines; Singapore; Chinese Taipei; Thailand, and the United States). The APEC region is home to over two billion people. It creates over half of the world's gross domestic product, and accounts for over half of the world's , energy consumption, food consumption, and pollution. APEC heads of state meet annually (Canada hosted the 1997 summit in Vancouver), and APEC issues are addressed throughout the year through the work of international APEC committees and task forces.

Canadian Prime Minister, Jean Chretien, walks through Birmingham Museum with other G8 leaders, from left to right: Helmut Kohl (Germany), Boris Yeltsin (Russia), Tony Blair (United Kingdom), Jacques Chirac (France), Romano Prodi (Italy), Ryutaro Hashimoto (Japan), Bill Clinton (U.S.A.) and Jacques Santer (European Union). Leaders of the Group of Eight gathered in this central England city, in May 1998, for their annual summit which included talks about India's recent underground nuclear tests and third world debt. The G8 countries are among the world's leading and democratic economies and with shared values and objectives. Which of these leaders is still in power today?

7-21

Ups and Downs of the Dollar

One other flow of money should be noted here, though it is not considered part of invisible trade. When Canadians own and manage factories and real estate in any other country, it is known as **direct foreign investment**.

For a long time, Canadians were concerned that too much of Canada's industry and real estate was being sold to foreigners. The situation reversed during the 1980's as Canadians invested more abroad than foreigners invested in Canada (see Fig 7-23). This fact is still true today, although total foreign investment in Canada (over $100 billion) exceeds what Canadian companies and individuals own abroad.

Trade balance coupled with direct investments have a major influence on the value of the Canadian dollar. If people in other countries want to buy Canadian goods, stocks or real estate, the resultant demand sends the value of the Canadian dollar up. Canadians also buy goods abroad, creating a demand for foreign currency. When the demand in Canada for foreign currency matches the demand overseas for Canadian currency, the dollar is more or less stable.

Many other factors can effect the value of our dollar. A favourable trade balance (when Canada sells more exports than it buys imports) increases its value. Tourism earnings make a difference. When Canadians spend more money abroad than tourists spend in Canada, the demand for foreign currency exceeds the demand for our dollar, and its value drops. Economic uncertainty on the international level, war, or political conflict can also cause devaluation (the Canadian dollar dropped markedly before the last Quebec

7-23a   Foreign direct investment in Canada by source, 1976-1996. ($ millions)

	U.S.A.	U.K.	Other EU	Japan	Other OECD	All Other	TOTAL
1976	1708	350	290	37	20	120	2418
1980	5433	227	733	105	106	186	6790
1986	-643	2889	689	473	354	202	3964
1990	3451	2002	1982	861	155	395	8847
1996	5759	396	2006	605	-151	112	8726

7-23b   Foreign direct investment by Canadians, 1976-1996. ($ millions)

	U.S.A.	U.K.	Other EU	Japan	Other OECD	All Other	TOTAL
1976	-560	-16	-79	7	-98	-244	-990
1980	-2954	-684	-114	-27	-277	-736	-4792
1986	-3362	813	-330	-148	-152	-1686	-4864
1990	-3010	-1400	-470	-317	-241	-672	-6110
1996	-6319	-204	-1277	56	-1017	-2832	-11593

Though in general the totals increase (note the two exceptions), it is harder to find patterns in this maze of statistics. Some do exist, however. The general direction shows an increse in both charts, and the importance of the U.S. investment in Canada is obvious in every year except one. The rise of Japan as a source of capital is spectacular though Canadians have few investments in Japan.

referendum). Businesses are not eager to invest in countries facing an uncertain future because of low profits or even the risk of complete investment loss. Political issues influence currency values as do interest rates. Rising interest rates attract capital from abroad, causing the dollar to rise as well.

In 1997/8, political and economic uncertainty in a number of Asian countries led to a decrease in their purchases of raw materials – coal, oil, grains and metals. Canada as a major supplier (raw materials comprise almost 42 percent of Canadian exports), was particularly hard hit. Our exports dropped between 20 – 30 percent.

Economic uncertainty drives investors to seek refuge in a strong currency – in this case, the U.S. dollar. Canada's dollar began to drop sharply in summer 1998, reaching an all time low of 65 cents against the U.S. dollar.

Canadians were shocked when they discovered it cost $1.50 CDN to buy $1.00 U.S. Although the Canadian government has a variety of weapons to prevent our dollar from fluctuating too widely, it can only do so much.

Money and trade flow freely in a free world and Canada's ability to compete in the world marketplace is crucial to its currency and its prosperity, and although the Canadian dollar fell against the U.S. dollar, it rose against the Japanese yen, the German mark, the Italian lire and the French franc.

It is hard to believe that once one Canadian dollar would buy $1.04 U.S., but that was true in 1974!

When the Canadian dollar rises in value, important changes occur:

1. Canadian goods become more expensive for people in other countries. Exports drop and unemployment rises.

7-24 The Canadian Dollar in U.S. Funds

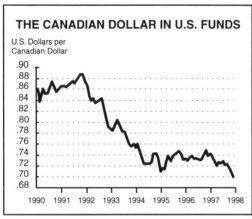

THE CANADIAN DOLLAR IN U.S. FUNDS

U.S. Dollars per
Canadian Dollar

Source: The Alliance of Manufacturers & Exporters Canada

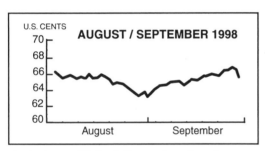

U.S. CENTS

AUGUST / SEPTEMBER 1998

And where will the dollar go from here?

2. Foreign goods bought in Canada become less expensive, because Canadian funds can buy more.

3. Canadian companies with long term agreements to provide goods and services at a fixed price in U.S. dollars lose heavily when the Canadian dollar increases in value against the U.S. dollar. It is estimated that for every one-cent rise in our dollar, our exporters lose $1.3 billion!

4. Canadians buying funds for a vacation in the U.S. can buy more U.S. dollars.

5. Anyone holding a Canadian dollar can buy more on the international market as our money appreciates in value.

1. Explain how a deficit of $2 billion in Canada's tourist trade would influence the value of the Canadian dollar.
2. Explain, in your own words, how the $4.23 of Canadian money in your wallet could increase in value sitting on your dresser overnight.
3. Which of the following strengthens the Canadian dollar:
   (a) interest paid to foreign investors?
   (b) a major sale of aircraft abroad?
   (c) a Canadian developer building a gigantic shopping mall in the US.

Free Trade or Protectionism?

Probably the greatest issue in international trade is, and always has been, whether a country should trade freely with other nations, or protect its national economy from foreign-made goods. **Free trade** obviously implies that there are no restrictions placed on international trade by national governments. **Protectionism**, on the other hand, implies that the economy of a nation is protected from the goods or products of other countries. Most countries follow neither path exclusively, but have some goods trading freely and some goods protected.

Those who prefer free trade seek the advantages of cheaper goods from other countries, and the benefits of selling their own goods back. Free-traders argue that each country should concentrate on what it produces better and more cheaply than other countries. If all countries did so, and then traded their goods, all countries would enjoy the lowest possible prices for the products they purchased. These ideas are old ones

— they were propounded by Adam Smith, the great Scottish economist, over 200 years ago. It was the belief in these ideas that led to the founding of the European Economic Community and the European Free Trade Association.

Those who believe in protectionism quickly point out snags in this argument. They claim, for example, that unrestricted free trade would mean that some countries produced only food and raw materials. Other countries would do the manufacturing, and therefore have more jobs for their people. It is also true that manufactured goods have high value, and that only nations producing such goods would be wealthy if all trade were totally free. Some would say that if a country had no steel or manufacturing plants, it couldn't produce weapons to defend itself. Some would argue, as well, that many nations can't be trusted, and that it can be very dangerous to depend on trading arrangements that are likely to be broken later. It seems more sensible, protectionists believe, for a nation to develop its own economy, within its own borders, protected from foreign influence.

Protection from foreign trade can be put in place in a variety of ways:

1. An outright **ban** on trade from selected, or even all, countries is seldom used, for almost every country needs goods it can't produce within its own borders. However, some items from certain countries are banned by some other countries. For example, for moral and political reasons many countries have banned products from the Republic of South Africa.

2. **A quota** system permits only so

many tonnes of food, or so many thousands of cars, for example, to enter the country each year. This is a fairly common device, and Canada has often used it. This country limited the number of Japanese cars that could be sold here every year during the mid-1980s. As a result, Japanese and South Korean automakers now manufacture some cars here.

3. **Tariffs** can be placed on goods as they enter the country. A tariff, sometimes called a **duty**, is simply a tax on imported goods. The tariff has to be added to the cost of the product by its retailers. It protects domestic manufacturers by raising the cost of imported goods beyond that of goods produced domestically. This is a very popular method of reducing imports, and is used by almost all countries, including Canada. In order to get around tariffs, some countries will give subsidies to their farmers or manufacturers. This enables them to sell their exports at a competitive or lower price even with the tariff added on.

In the 1930s, world trade actually decreased year after year. The Second World War came at the end of the 1930s, and many people claim that the high tariffs and other trade restrictions during that decade were a major cause of that war. However, during the 1950s, 1960s and 1970s, world trade grew at the phenomenal rate of about seven percent a year. The growth rate of exports world-wide actually grew faster than the growth rate of the production of goods in those years.

World Trade has grown tremendously in the last two decades, its value being in the thousands of billions of dollars.

1. (a) **Explain two of the major advantages of free trade.**
   (b) **Explain two of the major advantages of protectionism.**
2. (a) **How does a trade ban differ from a trade quota?**
   (b) **Why is the quota generally more popular?**
3. **Explain how tariffs operate and why they are imposed on certain goods.**

New Words to Know

Free trade	Trade quota
Protectionism	Tariff
Trade ban	Duty

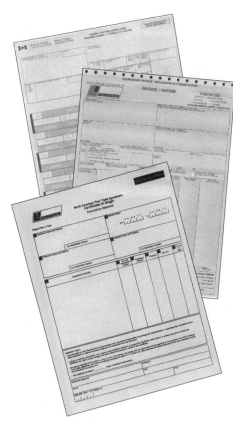

The International Monetary Fund

The International Monetary Fund (IMF) is a financially cooperative association of 182 sovereign nations who consult regularly with one another to maintain international order in the buying and selling of national currencies, so that "payments in foreign moneys can take place between countries smoothly and without delay." Established in 1944 (operational in 1946), the IMF was forged from the economic devastation of the 1930s when international trade dropped by 63 percent, and prices fell 48 percent. The IMF facilitates international trade growth and balance. It works to lessen the gaps between the world's rich and poor nations through financing and technical assistance.

Each IMF member contributes money to the fund, based on a quota system. The quotas are adjusted every five years (most recently in 1997) according to the IMF's needs, and its members' ability to pay. The U.S. quota is the largest, at 18 percent of the total or $35 billion USD. Canada's quota is almost 3% or $5.8 billion USD. These quotas become the IMF's resource pool from which it draws as necessary to lend to members in financial difficulties. By 1998. IMF quotas were $193 billion USD, with a proposal on the table to raise them to $280 billion USD.

Each member of the IMF has the right to borrow three times more than it has paid in quota, but must undertake a series of reforms that will…prepare the ground for high quality economic growth. In addition, the IMF has special short-term financing without limit for members faced with a "Sudden and disruptive loss of market confidence." At the end of August 1998, the IMF had credits and outstanding loans with 60 countries totalling $63 billion USD. Its accounting unit is the SDR (special drawing right). One SDR was worth $1.36 USD in 1998.

Based in Washington, D.C, the IMF is governed by a Board of Governors (one from each member), which meets annually. The 24 member Executive Board enacts IMF policies, while the Managing Director, currently Michell Camdessus (France), is responsible for IMF staff —2 600 persons from 122 countries.

The North American Free Trade Agreement

On January 1, 1994 the largest trade agreement ever signed was initiated between Canada, the United States, and Mexico. The North American Free Trade Agreement (NAFTA) is a comprehensive trade agreement intended to improve all aspects of doing business within North America. NAFTA not only provides for the virtual elimination of tariffs and other trade barriers, it contains provisions to reduce government regulation of investment between the countries, make the provision of services easier, and resolve trade disputes that might arise in the future.

Free trade is not a new idea in world history, nor in Canada-American relations. Even before Confederation, the British North American colonies had free trade with the U.S.A. under a Reciprocity Treaty between 1854 and 1866. In 1911, another free trade agreement was negotiated by the Liberal government before both government and treaty were defeated in a federal election.

World events provided a major reason for the North American free trade deal. The European Union (EU), formerly referred to as the European Economic Community were first formed as long ago as the 1950s. By the late 1990s, the EU intends to become a fully tariff-free area. Israel has signed a free trade agreement with the United States, and Japan and the United States have continually worked towards lessening their trade barriers. Even on their own the U.S.A. and Japan form huge markets of 250 and 125 million people respectively. With its tiny domestic

Report uses facts to explode the myths manufactured by Canadian protectionists
Canada has come out a winner thanks to free trade agreements

A new report being published Monday by the Fraser Institute demolishes the doomsayers who once predicted free trade would hurt Canadian jobs, incomes, and capital investment.

The Vancouver-based institute's policy analysts Fazil Mihlar and Marc Law have examined the impact of the 10-year-old Canada-U.S. Free Trade Agreement and the later North American Free Trade Agreement. Their conclusion: "The protectionist rhetoric is simply mythology unsupported by fact."

For instance, free trade doesn't kill jobs. Job creation is up, and unemployment is down. Canada's jobless rate remains relatively high - we have an expanding labor force and rigidities in our labor markets. But at the end of 1997 employment totaled 14.1 million compared with 12.8 million in 1988. Last year, 324,000 full-time jobs were created. Job losses in the early 1990s were not related to free trade, but to a general recession that afflicted other economies too.

Average weekly earnings and average hourly wages have been rising in Canada in recent years, not falling.

As a share of gross domestic product, manufacturing has declined only marginally since 1988 to 17.3% from 19.2%. This would be expected considering a fast-growing services sector. Manufacturing's relatively constant share is hardly a sign of deterioration. While not mentioned in the study, manufacturing employment has remained fairly steady since 1988 except for declines during the recession. Since then, capital investment in manufacturing has increased strongly once more.

Meanwhile, merchandise exports to the U.S. have soared - to a record $250 billion last year, more than double the level of 1988, and still increasing, latest figures show.

The study contends that food safety and health standards have not been undermined by NAFTA, and that environmental quality in Canada is actually improving, not worsening. Some had feared the reverse.

The agricultural sector hasn't been hurt either. It's still highly protected. But as tariffs fall the study predicts our farmers will become more competitive. Any producer losses will be offset by consumer gains.

Concerns about an erosion of national sovereignty have been debunked too. As a proportion of GDP, our public sector is actually larger now - 45% of GDP versus 39% in the early 1980s. The implications of this may be worrisome, but certainly the trend doesn't signal diminished influence of governments in Canada.

There have been about 30 trade disputes adjudicated under both agreements involving U.S. measures against Canada. The major cases have been resolved in Canada's favor. We haven't been bullied around as some like to suggest.

Free trade hasn't resulted in an exodus of firms from Canada to Mexico to take advantage of low wages. While our wage rates are higher, our workers are proportionately more productive. Unit labor costs are what matter to business. Ours have been below Mexico's since 1992.

The study makes the point that free trade raises the level of national income and increases economic well-being because it allows countries to specialize in their comparative advantages. Free trade, say the authors, creates more wealth than if Canada had to produce everything on its own.

Some theories hold that it's possible under certain conditions to increase a country's overall wealth through export subsidies and import restrictions. However, in an increasingly global economy, implementing successful interventionist policies is difficult. The gains are limited. As U.S. economist Paul Krugman has argued, retaliatory trade wars, for instance, can do great harm, making it foolish to opt for any policy other than free trade.

"The practical case in the real world for free trade remains strong," the Fraser Institute's Mihlar says. The study also raises an interesting point about the non-economic benefits of free trade pacts for natural trading areas such as the U.S. and Canada. Agreeing to avoid trade wars helps governments resist the pressures of special-interest groups, it argues.

A major rationale from Canada's point of view in doing a deal with the U.S. was to counter a growing array of non-tariff barriers. These hadn't been dealt with adequately under the worldwide Uruguay Round trade talks.

The free-trade arrangements have also helped inject more competition into the Canadian economy. This benefits consumers and sharpens the ability of our firms to compete at home and abroad.

The study looks at a lot of issues. But its bottom line is that free trade is not a zero-sum proposition under which there are winners and losers. "It is a positive sum game - no country loses," its authors conclude. Their report makes a compelling case for pursuing other free-trade arrangements. Fortunately, this is the government's policy.

Neville Nankivell is The Financial Post's editor-at-large, based in Ottawa.

NAFTA woes: Firms fire workers in Canada, hire thousands in Mexico

By LAURA EGGERTSON, OTTAWA BUREAU

OTTAWA LEADING MULTINATIONAL corporations that slashed jobs in Canada created more than twice as many positions in Mexico in the first decade of free trade, according to a draft report obtained by The Star.

General Motors, Ford, Allied Signal, United Technologies, General Electric, Rockwell, Dupont and 3M cut 18,462 jobs from their Canadian operations in the years after Canada signed the trade deal with the United States on Jan. 1, 1988.

But they hired 47,045 people from 1900-96 in Mexico, says the report for the International Labour Organization in Geneva.

Canada, the United States and Mexico passed the North American Free Trade Agreement – which built on the Canada-U.S. deal – in 1993 and it took effect Jan. 1, 1994.

The study supports the contention of those who opposed both Canada-U.S.free trade and NAFTA: Companies would divert investment from Canada to lower-wage destinations.

"The general thesis is that workers from all three countries have been adversely affected" in the free-trade environment, said Bruce Campbell, co-author of the study.

9,854 GM workers lost their jobs in Canada when the firm cut its workforce here by 23 per cent between 1990 and 1996 (the years covered by the latest statistics).

At the same time, says the study, GM hired 20,836 people at its low-wage Mexican maquiladora plants. (Maquiladoras are assembly plants in a free-trade zone set up along the Mexico-U.S. border to encourage foreign investment.)

Unlike U.S. workers, Canadians who lost their jobs as a result of free trade did not get government assistance to equip them for a new career – something ex-prime minister Brian Mulroney has said was a mistake.

While it was downsizing in Canada, GM cut its U.S. workforce by 10 per cent, according to employment figures Campbell obtained for 10 companies.

Only two of those 10 maintained or increased their level of employment in Canada from 1990 to 1996.

"There are some winners, but the winners are a restricted minority," Campbell said.

IBM is one of the winners. The U.S. based firm added 582 jobs in Canada and hired 1,200 people at its electronic assembly facilities in Mexico.

Chrysler Canada's employment levels have remained steady at 14,000 since 1990. But Canada also lost investment from Chrysler, as the company increased its Mexican workforce by 8,331 jobs, up from 5,669.

Free trade supporters – including the federal Liberals, who opposed the deal while in opposition – contend that export-related jobs have offset manufacturing job losses.

Campbell, executive director of the Canadian Centre for Policy Alternatives, argues the effects for Canadian workers go beyond job losses.

Many workers can find only part-time work.

But firms and stockholders have done well in the 10 years since the Canada-U.S. trade deal was signed.

The increase in productivity that accompanied job cuts in Canada has benefited senior executives and shareholders more than the workers, Campbell says.

After the cutbacks, for example, productivity at GM Canada increased 88 per cent. Compensation for the chief executive officer grew 247 per cent, while wages in the auto sector increased by 33 per cent.

Other examples:
• At Ford Canada, the company cut 2,182 jobs, or 8 per cent of the workforce. But employment jumped by 22 per cent in the United States and more than doubled in the Mexican maquiladoras, where the company hired 11,688 more workers.

Productivity at Ford Canada also grew by 88 per cent. But CEO compensation rose 645 per cent, compared with the 33 per cent wage gains that Canadian auto workers made.

• Allied Signal cut its Canadian workforce by 21 per cent, or 688 jobs. At the same time, Allied hired 4,800 employees in the maquiladoras, a jump of more than 500 per cent in its Mexico operation.

Allied productivity also rose by 35 per cent in the company's Canadian affiliate. CEO compensation grew by 550 per cent, while workers in the sector gained about 22 per cent in wages.

• United Technologies cut its employment force at Canadian subsidiary Pratt and Whitney by 15 per cent, or 1,442 jobs. The company slashed its U.S. employment by 29 per cent. But in Mexico, United added 5,607 jobs over the same period.

• General Electric reduced its Canadian workforce by 32 per cent, or 3,019 jobs. The company cut its U.S. workforce by 15 per cent. In Mexico GE expanded its labour force by 27 per cent, adding 1,766 employees in the same time period.

GE workers increased their wages by about 16 per cent as productivity increased by 112 per cent. The company's CEO compensation grew by more than 500 per cent.

reprinted with permisssion from the Toronto Star

---

market of 30 million, Canada would find it hard to compete with these giants, its major trading partners.

In the preamble to NAFTA, the following objectives were stated:
• to eliminate barriers to trade in, and facilitate the cross-border movement of goods and services between Canada, the United States, and Mexico;
• to promote conditions of fair competition in the free trade area;
• to increase substantially investment opportunities
• to provide adequate and effective protection and enforcement of intellectual property rights.
• to create effective procedures for the implementation and application of the agreement, and for its joint administration and the resolution of disputes;

• to establish a framework for further trilateral, regional, and multilateral cooperation to expand and enhance the benefits of the agreement.

The agreement set a schedule for tariff reduction: some tariffs, generally those that were already low, would be eliminated immediately; others, with high tariff protection, such as industrial products, would lose their protection gradually over ten years. This gave businesses time to adjust to the new conditions.

NAFTA will create losers as well as winners in Canada. The arguments for and against the deal make it clear that there are excellent points on both sides of the issue. It is clear, however, that economic change is something Canada and other nations have to learn to live with. Both governments and unions must address the

dislocations and opportunities brought about by international trade deals. On the horizon for Canada is a potential expansion of the North American Free Trade Agreement to the more advanced economies of South America to create a hemispheric trade agreement.

1. **How does Canada's population compare to that of its major trading partners and blocs?**
2. **Explain in your own words the first objective mentioned as part of the preamble to the NAFTA?**
3. **(a) Explain what is meant by "winners and losers" in regard to NAFTA?**
   (b) Who are some of the persons who might benefit, and who might be disadvantaged by NAFTA? Why?

New Words To Know
Intellectual Property Rights
Multilateral

GATT: For Fairness in International Trade

After the Second World War, the United States, Britain, and Canada were the principal proponents of a new trading order. The end result in 1947 was the General Agreement on Tariffs and Trade (GATT) which established a new trading order based on reciprocity, non-discrimination, and multilateralism. The agreement was designed to lower the protectionist barriers that had limited international trade in the decade before the Second World War and had prolonged the Great Depression. GATT was to have become an agency of the U.N., but the world's greatest trading nation, the U.S.A. refused to join. As a result, GATT never became part of the U.N. even though the U.S.A. eventually did become a member. GATT, as an institution, was succeeded by the formation of the World Trade Organization in 1996. Over 100 nations are expected to eventually be members of the WTO.

The GATT served both as a treaty and as an institution, with its institutional functions being later absorbed by the WTO. As a treaty, GATT sets out a code of rules for the conduct of trade. As an institution, it oversaw the application of the trade rules and provided a forum in which the countries could discuss trade problems and negotiate reductions in trade barriers. These administrative functions have been incorporated into the mandate of the WTO.

The GATT has had a tremendous influence on international trade. Its greatest success has been in reducing tariffs and thereby increasing world trade — though its purpose has been fair trade, not free

trade. Since 1947 the average tariffs in industrialized countries have dropped 40 percent to five percent. Each member-nation is expected to deal with others according to a **most-favoured nation** policy. That means that each member has the right to trade with all others on the most favourable terms each country provides to any other. The group has grown since it was founded: 124 nations participated in the 1994 Uruguay Round of the GATT Agreement. Many of the member nations of the WTO are developing countries, and GATT signatories include both former Communist states like Russia and current Communist countries such as China. As it grows, the WTO's ability to affect world trade increases. Nations feel obligated to obey WTO rulings on trade disputes. If they don't they could lose huge markets in the WTO group, or even be asked to leave it.

7-28   Harvesting herring on Canada's west coast. *A GATT ruling denied Canada's policy that prohibited the export of unprocessed herring.*

Canada has had to make certain changes in some of its policies in order to conform to GATT. In one such case, GATT ruled the Canada's refusal to allow the export of unprocessed herring and salmon from the West Coast was discriminatory. In another instance, the group found that this country's provinces discriminated against foreign wines, beers and liquors. The Canadian policies had to be amended. The reason for Canada's agreement to change them was given clearly by Pat Carney, Canada's former Minister of International Trade. She said in the House of Commons in March 1988 that "Canada expects other countries to live by the GATT rules, but we cannot do so unless we are prepared to accept those rules ourselves." Canada's government, like that of other countries, must go along with the international pressure. If Canada did not comply with GATT rulings, then other nations could retaliate. If they raised tariffs against Canadian products, for example, then this country's trade and prosperity would suffer.

Canada, too, has taken cases to GATT regarding unfair trade. In one case Canada asked for a ruling on Japan's policy of charging an eight-percent tariff on Canadian lumber, while charging no tariff on lumber from the U.S.A. International trade will benefit if all nations will agree to allow a truly impartial body to rule on trade conflicts. This is a much better way to resolve such conflicts than trade wars, or any other kinds of war. Both of the latter have been used too often in the past to settle trade disputes.

New Words to Know
Most-favoured nation

Political, Strategic and Other Links

Every month well over 35,000 e-mails, telegrams, and telephone calls are received at the Lester B. Pearson Building in Ottawa. This building houses the Department of Foreign Affairs and International Trade, the division of the government that is responsible for Canada's relations with other countries. These communications are sent by Canada's **embassies, consulates** and other posts abroad, and deal with a huge variety of problems. Some examples are below:

• OPEC announces new oil price hike.
• More aid needed for earthquake victims in Turkey.
•West coast fisherman restless over U.S. delay in ratifying fisheries agreement.
• Children abandoned by mother in Germany need to locate father.

The Pearson Building receives an average of 50 such communications every hour. Many require a quick answer. In order to deal with this huge flow of telecommunications travel, a large staff is busy in Ottawa and in Canada's embassies in foreign countries. The Department of Foreign Affairs and International Trade has a staff of between four and five thousand employees divided equally between those working in Canada and those serving abroad. The Canadians abroad are posted in almost 200 offices dealing with 164 countries and international organizations, including U.N. agencies. These figures say a great deal about Canada's involvement in international affairs. In 1919 this country's entire external affairs department had eight people in Ottawa, and no representatives abroad. It is a huge department today, and deals with many thorny issues every day, just a few examples of which are mentioned here: immigration, international trade, refugees, aid to developing countries, Arctic sovereignty, international terrorism, west coast fishing rights.

The list could be almost endless and there isn't enough space to deal with even the above items in full. However, a brief look at some of these items is instructive.

7-29   Lester B. Pearson Building, Department of Foreign Affairs and International Trade, Ottawa. *This building houses the huge staff needed to administer the thousands of connections between the government of Canada and the rest of the world. It is the headquarters of the Department of External Affairs, and is named after former Prime Minister Pearson who had also served as a most-respected Minister of External Affairs.*

Arctic Sovereignty: Unlike many nations, Canada has very few conflicts regarding its international boundaries. Canada is fortunate, of course, in having only two land boundaries, both with the U.S.A. Canada's relations with its southern neighbour have been a model to the world for decades. However, in the seas and Arctic islands north of Canada's mainland, there remains some question of **sovereignty**.

The possible existence of oil under the northern seas is raising the region's profile, among both the oil companies and the foreign governments that back them. The border between Canada and the U.S.A. in the Beaufort Sea, and thus ownership of Arctic waters, is being contested. Making this problem worse is the fact that U.S. and Russian submarines can move freely under the Arctic ice. Canada in the late 1980s had no way of knowing which country's ships were in what it claimed to be its waters. Canada lacks the military equipment and personnel to monitor, let alone stop, this traffic, uphold its sovereignty, enforce its pollution laws, or protect its exclusive Arctic fishing zone.

Another area of concern in the **Northwest Passage**, which runs through the Arctic islands. (See map 1-3.) Canada claims that these waters which lie between Canadian islands, are part of Canada's **internal waters**. The United States and other maritime powers claim that the Northwest Passage is an **international strait**, and that Canada cannot prevent other nations from using it for **innocent passage**. There are straits in a number of parts of the world that are as narrow as the Northwest Passage. They are considered to be international straits rather than

7-30 *Pangnirtung, Nunavut, population 1 000, is located on Baffin Island's south east coast.*

Nunavut

Some 10 000 Inuit voted on a deal in November of 1997 to create Canada's third territory. In the 1970's a group of Inuit leaders started working to create Nunavut, which means "our land" in Inuktitut, the Inuit language. Nunavut covers a huge section of the eastern Arctic from the tree line to the north pole and its population is largely Inuit – unlike the western Arctic. The agreement took 15 years to negotiate and gives title to about 350 000 km2 of land and $1.5 billion over 14 years to an estimated 17 500 Inuit in exchange for the surrender of other land claims. The agreement also gives the Inuit the right to hunt, trap and fish over an area twice as big as B.C.

The new territory based in Iqaluit is expected to have a similar organization, administration and government to the existing Northwest Territory. The new administration will start in April 1999 with a staff that is 50 percent Inuit which should quickly rise to 85 percent to match the distribution of population. Nunavut will also have three "official languages" English, French and Inuktitut.

However, by mid-1998 the plan was running behind schedule. The problem was not the planning, it was the funding. Costs had doubled for things like decentralization, severance pay and recruiting new employees. Funding cuts by the Federal government still have a huge impact because 98 percent of all costs are paid by Ottawa. Despite these issues, the Inuit of the eastern Arctic are looking forward to the spring of 1999. John Amagloalik, one of those responsible for Nunavut, says, "People will have a government they can relate to – a government that speaks and understands their language and understands their culture and priorities."

internal waters because they have been used for many years and continue to be used, by ships of all nations. The Northwest Passage has not been used much because of the problems of ice and bad weather. However, as naval technology improves and oil exploration spreads, Arctic waters become increasingly important to navigation.

A further problem concerns environmental challenges in the Arctic. Canada has a particular role in defending and developing the Arctic environment, an area where international cooperation and national commitment is vital and is just beginning. Through enhanced international cooperation and national commitment, Canada seeks to confront environmental issues such as global climate change, the dumping of nuclear waste into the Arctic Ocean from aging Russian power plants, and the improvement of the Arctic environment and the health and livelihood of the region's inhabitants.

Because of these questions about Canada's sovereignty in the Arctic, the government is trying to increase Canada's presence there. The focus in the Canadian Arctic is increasingly on non-traditional security threats. Canada's recent appointment of an Ambassador for Circumpolar Affairs will increase the focus on such threats. The goal is to create an Arctic Council to meet the challenge of sustainable development in the North and to deal with the critical issues faced by all Arctic countries. Also, the government proposed building a powerful icebreaker, known as a "polar 8" class ship, that would be able to operate year-round in the high Arctic. The government has also proposed a permanent military base for Nanisivik on northern Baffin Island. There is already a settlement there, around a lead-and zinc mine, with a deep harbour and a 2000-m airstrip. The base is to serve as a "recognition of the strategic importance of the North, and as an assertion of Arctic sovereignty."

Fishing rights Another threat to Canadian sovereignty over the years has been the matter of fishing rights. Canada claims fishing rights or territorial waters for a 340 km zone from its shores. This claim has been disputed recently by Spanish and American trawlers on both the east and the west coasts of this country. One American dispute centres on west coast salmon. British Columbia and Alaska have been wrangling over salmon quotas, and who has what rights to fish where. British Columbia contends that Alaska fisherman are destroying the salmon population by overfishing. Alaska fishermen contend that British Columbia does not have a secure knowledge of the actual size of the salmon populations and the possible factors affecting its growth, and that the province has not been protecting salmon habitat. Canadian fishermen blockaded an Alaskan ferry in port, refusing to depart until the dispute had been resolved. Government intervention and public pressure called off the blockade. The Spanish, fishing on Canada's east coast, moved within the Canadian 340 km territorial limit because they did not recognize it. Warned, one trawler continued to fish within Canadian territorial waters, until finally it was boarded, the ship confiscated, and the catch and gear seized. Canadians cut the nets of subsequent Spanish trespassers. Fishing disputes off the east coast of Canada are not uncommon. In 1993, shots were fired between Canadian and American fishermen off George's Banks, historically known as one of the richer fishing areas.

1. **(a) Name five major international topics from today's newspaper or in tonight's radio or television newscast which could involve or concern the Department of Foreign Affairs and International Trade (DFAIT)?**
   (b) How might Canada's DFAIT officials become actively involved in the international news of today?
   (c) Are any items mentioned above as concerning DFAIT still in the news?
2. **Explain the term "Arctic sovereignty," and give several reasons why it is such a complex issue. Use map 1-3.**
3. **(a) Why would Canadian military personnel have to be particularly careful in their actions if a Russian or American submarine was discovered beneath the Arctic ice in Canadian waters?**
   (b) Should Canada monitor economic or military activities in the Arctic archipelago or not? Give your reasons.
4. **When does the new territory of Nunavut officially come into being? How did it come about? What is its importance to the rest of the country? to the Inuit?**

New Words to Know
Embassy
Consulate
Sovereignty
Northwest Passage
Internal waters
International strait
Innocent passage
Inuktitut
Iqaluit
Nunavut

International terrorism Canada has been involved with many other countries over **terrorism**.

Every year throughout the world, terrorists **hijack** airplanes, detonate bombs and take hostages. Sometimes they succeed, sometimes they are stopped. Usually their victims are innocent bystanders. The terrorists are trying to publicize their cause or use blackmail to gain what they want, such as to force the release of other terrorists being held in prison. The strict security precautions at Canada's airports — especially its international ones — are one of this country's responses to terrorism.

Map 7-32 shows how complicated and "international" some terrorist activities have become. In 1985, someone in Canada — probably in Toronto — planted a bomb in an Air India Boeing 747. The bomb exploded over the Atlantic Ocean just off the coast of Ireland, killing all 329 people aboard. The same day, two baggage handlers at Tokyo's Narita International Airport were killed when another bomb, also traced to Canada — probably Vancouver — exploded after the aircraft landed. It is believed that Sikh terrorists, who want to establish a separate Sikh state in India, planted both bombs. The U.K. became involved with these incidents in 1988 when it agreed to **extradite** one of the suspects in the bombing to Canada to face criminal charges.

In another case, the Canadian government attempted to deport a Palestinian living in Brantford, Ontario. He had been convicted of terrorist activities in Greece over a decade earlier. According to the government, he had lied to officials in order to secure entry into Canada.

7-32 International terrorism affects Canada.

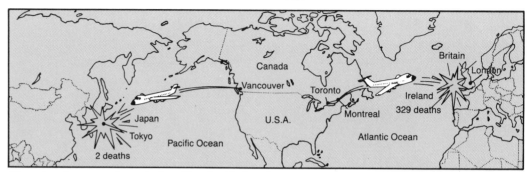

Terrorism knows no boundaries. The double bombings in June 1985 directly involved India, Canada, Japan, Ireland and the U.K. Several other countries were also involved through their nationals flying in the two aircraft: one from Vancouver; one from Toronto and Montreal.

The Iranian Embassy in Ottawa was stormed and briefly occupied by members of the anti-regime organization Mujahedin-e-Khalq (MEK) on April 5, 1992. The Iranian Ambassador was slightly injured in a struggle with one of the assailants during the attempted take-over. On July 11, 1995 and July 13, 1995 letter bombs were sent to a scientist at the Alberta Genetics laboratory and to the Mackenzie Institute in Toronto.

The threat to Canadians and Canadian interests abroad from international terrorism continues to be largely a matter of being "in the wrong place at the wrong time." In terms of Canada, the threat is principally associated with homeland issues. Persons residing in Canada with links to international or former homeland domestic groups engage in a wide range of support activities, including providing safe haven, logistical support for operations, fund raising, and the procurement of weapons and material. Terrorist events such as those mentioned above demonstrate that Canada is by no means safe from world terrorism. Canadians do not have to be directly involved in world conflicts to be affected by them.

A new development in the clandestine worlds of terrorism and espionage has been the growth of economic terrorism and espionage. Accelerated economic interdependence and international competition have emerged as major sources of tension and conflict among world powers. In this uncertain environment, developed countries eager to maintain their standards of living and developing countries equally determined to improve their own, are under pressure to use whatever means they have to improve their productivity and ensure their economic security. One such means is economic espionage, which can be described as illegal or clandestine activity to gain access to economic information. As Canada is a world leader in many technology-intensive fields, a number of Canadian companies operating in these sectors have been targeted by foreign governments to obtain economic or commercial advantages. The damage to Canadian interests takes the form of lost contracts, jobs and markets, and overall, a diminished competitive advantage.

7-33 *Eric Wohlberg, from Burlington, Ontario won a Gold Medal in the 42 km time-trial bicycle race at the 1998 Commonwealth Games in Kuala Lumpur, Malaysia. Wohlberg completed the circuit in 53 minutes, 15 seconds, averaging over 47 kilometres an hour. Prior to the race, he cycled for over an hour to reach the racetrack because he was leery of the ability of outside sources to get him to the race on time. Canada sent 277 athletes to the games.*

The Commonwealth and La Francophonie

These groups constitute two of Canada's links with other nations. This country has belonged to both since their inception. Canada is the only major industrial country that is a member of both.

The Commonwealth is a loose association of the many nations which were once part of the British Empire. The term "external affairs" was chosen in 1909 to define the department of government associated with international policy because the Canadian government did not regard the rest of the British Empire as "foreign" to Canadians. For the same reason, Canada sends high commissioners rather than ambassadors to Commonwealth countries.

There are 50 sovereign member states of the Commonwealth (although two of them, Sierra Leone and Nigeria, have had their memberships suspended as a result of political unrest and violence). Britain, Canada, Australia, New Zealand, South Africa and Newfoundland (which resigned its membership when it joined Canada in 1949) were founding members of the Commonwealth in 1931. Altogether, the Commonwealth covers one-quarter of the earth's surface, and contains well over one billion people.

Today, membership in the Commonwealth includes Asian, African, Caribbean and Middle Eastern states, all of whom are dedicated to five basic principles:
- international peace
- personal freedom under the law
- racial equality and the combat of racial discrimination
- elimination of colonial domination
- decrease in the gap between the world's rich and poor nations.

Commonwealth Heads of Government meetings (or summits) are held every two years. Decisions are made by consensus rather than formal vote.

The Commonwealth encourages friendship and cooperation among countries. Some members are wealthier than others; the richer members channel a great deal of aid to the poorer members, in a non-threatening way. The Commonwealth has developed youth, scholarship and fellowship programs, and the Commonwealth Games. Canada has taken a leading role in the Commonwealth: it hosted its annual meetings in 1973 and 1987.

La Francophonie is rather different from the Commonwealth. Its member-nations share, in varying degrees, the French language and culture. It consists of 51 states and governments from Europe, Africa, the Middle East, the Caribbean, the Americas, and Asia. Two Canadian provinces, New Brunswick and Quebec. Both these provinces are officially bilingual, and share in the deliberations of La Francophonie when it meets. Like the Commonwealth, this group includes both developed countries such as France, Belgium and

Canada, and less-developed ones such as Zaire, Mali, Guinea, Haiti and Vietnam. Like the Commonwealth, La Francophonie has inspired a great deal of cooperation among its members in matters of trade, economic growth, technology, sports and education.

1. Why is Canada unique in regard to the Commonwealth and La Francophonie?
2. What is special about the terms "external affairs" and "high commissioner" in relation to the Commonwealth?
3. What are some advantages of Canada's membership in the Commonwealth?
4. Why are the New Brunswick and Quebec provincial governments included in La Francophonie?

New Words to Know
High commissioner

Other International Organizations

Canada's involvement in world organizations goes far beyond The Commonwealth and La Francophonie. Many of the international groups in which Canada is involved are listed in chart 7-36. They show that Canada is heavily involved with scores of other nations on a variety of fronts. The **International Civil Aviation Organization (ICAO)** (Montreal) has 161 member nations. It is the only U.N. agency headquartered in Canada. Many others, such as the **World Trade Organization (WTO)** (Geneva, Switzerland), and the **Food and Agricultural Organization (FAO)** (Rome, Italy), are managed by the U.N. Others, such as the **European Union (EU)**, the **Organization for Economic Cooperation and Development (OECD)**, are separate bodies. Two military alliances of which Canada is a part are dealt with later.

The chart does not list the hundreds of organizations which operate separately from the government. These include churches, which are heavily involved in aid and missionary work, and groups such as **Amnesty International** and the **International Olympic Committee (IOC)**. The former agency, based in London, England, works to secure freedom from imprisonment and torture for political prisoners throughout the world. As for the Olympics, Canadians have played their full part in recent games. Canada sends a large contingent of athletes to both the Summer and Winter Games. Calgary hosted the XVth Winter Games in February 1988, and Montreal the Summer Games in July 1975.

Canada Hosts First Arctic Council Ministerial Meeting in Iqaluit

Foreign Affairs Minister Lloyd Axworthy and Minister of Indian Affairs and Northern Development Jane Stewart will host the first ministerial meeting of the Arctic Council in Iqaluit, in Canada's eastern Arctic, from September 17 – 18, 1998.

"The Arctic region is a fundamental part of Canada's society, economy, and geography. As we make progress with northern Aboriginal and non-Aboriginal peoples on the many issues before us in the North, we must continue to strengthen partnerships with our circumpolar neighbours. The two agendas are inextricably linked," said Minister Stewart.

Founded in 1996, the Arctic Council brings together eight circumpolar countries: Canada; Denmark/Greenland; Finland; Iceland; Norway; Russia; Sweden; and the United States, plus Permanent Participation

delegations from northern Indigenous peoples: the Inuit Circumpolar Conference; the Saami Council; and the Russian Association of Indigenous Peoples of the North. The Aleut International Association is expected to be admitted to the Council shortly.

The Council has a clear mandate to improve the economic, social and cultural well-being of northern peoples. It coordinates four working committees:
- Arctic monitoring/Assessment programs
- Conservation of Arctic Flora and Fauna
- Emergency Prevention, Preparedness and Response
- Protection of Arctic Marine Environment

Canada chaired the Council from 1996-1998 under the leadership of Mary Simon, Canada's Ambassador for Circumpolar Affairs.

7-35 Headquarters of the United Nations, New York City. *The low building in front houses the General Assembly where international decisions are made. The skyscraper houses the Secretariat, which administers U.N. affairs and carries out General Assembly policies. Canada is currently serving its fifth elected term on the United Nations Security Council.*

7-36 Canada's many official international links.

The following is a summary of Canada's diplomatic links with other countries, largely formed through the Department of External Affairs.

Resident Ambassadors — in 58 countries	WHO — Geneva UNESCO — Paris
Non-resident Ambassadors* — in 53 countries	WMO — Geneva IAEA — Vienna
Resident High Commissioners — in 18 Commonwealth countries	GATT — Geneva ICAO — Montreal
	FAO — Rome
Non-resident High Commissioners — in 28 Commonwealth countries	Permanent Delegations — North Atlantic Treaty Organization (NATO) — Brussels
Commissioners — in Hong Kong only	Organization for Economic Co-operation and Development — Paris
Non-resident Commissioners — in 7 Commonwealth countries	Mutual and Balanced Force Reduction Talks — Vienna
Consuls-General** — in 28 cities	
Consuls** — in 22 countries and cities	Organization of American States (OAS) — Washington
Military Missions — in Berlin only	
Permanent U.N. Missions — in 12 agencies in addition to the Headquarters of the U.N. in New York City, in agencies such as:	Other Missions — to 4 European Community organizations, including the European Economic Community (EEC) in Brussels.

*Non-resident means that Canada's representative to some countries lives in another country, providing service from there.

**Consuls-General and Consuls are representatives of Canadian government, but of a lower rank than Ambassadors.

1. **Determine the meaning and function of the U.N. agencies listed above.**
2. **Map the various locations of the various missions and delegations on a world map.**

Worldwide Pollution It has been noted by world leaders that threats to the environment recognize no boundaries. Their urgent nature requires strengthened international cooperation among all countries. Canada has been a leader in attempts to strengthen international controls on pollution. Canadians are concerned about worldwide environmental problems, such as the destruction of the **ozone layer**, the decimation of rain forests, and the reduction of greenhouse gas emissions. Saving the environment depends on international cooperation.

Canada contributes to the special fund of the United Nations Environment Program. This body deals with a great range of environmental concerns, both global and regional. They include those already mentioned, as well as toxic-chemical pollution, the encroachment of deserts on farmland, environmental law, sea pollution, and so on. Under the **World Meteorological Organization (WMO)**, Canada has established nine stations – 11 are planned – that will monitor air pollution in rural areas. Canada has also signed several international agreements involving issues such as sulphur emissions and long-range transboundary air pollution. The Centre for Inland Waters in Hamilton, Ontario has close ties with another U.N. agency, the **World Health Organization (WHO)**.

Many scientists believe that the burning of fossil fuels cause heat-trapping gas emissions that will radically alter the earth's climate over the coming decades. In December 1997, a Canadian delegation travelled to Kyoto, Japan to meet with negotiators from 159 other countries to reduce the world's use of fossil fuels. As part of the convention, Canada will need to cut its emissions of greenhouse gases by six percent to below 1990 levels – the benchmark year for measuring – to be reached between 2008 and 2012. As the potential ravages of climate change are no longer dismissed as unknowable, Canada and the rest of the world must adopt strict environmental policies to face these challenges.

Not only governments are involved in environmental protection. There are private groups as well, such as the **World Wildlife Fund (WWF),** the world's largest private conservation group. It raises funds in 26 countries, runs projects in 130 and has five million members worldwide.

1. **Choose at least three organizations to which Canada maintains permanent missions, and briefly describe their work.**

2. **Some people believe the Olympics are a waste of money. Explain why you either agree or disagree with their view. Briefly explain the part played by Montreal, Toronto and Calgary in recent Olympic history.**

3. **Describe Canada's involvement with the U.N. in regard to environmental agencies and action.**

New Words to Know
Amnesty International
International Olympic Committee (IOC)
Ozone layer

7-37b *The Great Lakes Water Quality Agreement, which calls for the restoration and maintenance of "the chemical, physical, and biological integrity of the Great Lakes Basin Ecosystem," is administered by the Canadian/American International Joint Commission's Water Quality and Science Advisory Boards. Water in the lakes is monitored regularly through sampling, and through the use of biological indicators.*

7-37 Research vessel operated by Canada's Centre for Inland Waters, Hamilton, Ontario. *This ship is only one of many means by which the Centre monitors conditions in the Great Lakes.*

In 1988, Canada spent just over two percent of the G.N.P. on defence, whereas the U.S.A. spent six percent, Britain five percent and West Germany over three percent. However, in the 1980s, Canada's defense spending increased. During the 1990s, however, there have been drastic cuts to the defense budget.

Between 1945 and 1991 the country spent over $100 billion on defence: 85 percent of that total was related to two of Canada's major alliances – NATO and NORAD.

NATO: The defence of Europe

In 1949 the **North Atlantic Treaty Organization (NATO)** was formed, because the nations of Western Europe, along with Canada and the U.S.A., feared an attack from the U.S.S.R., which controlled Eastern Europe. NATO is a mutual-defence pact: if any member-nation is attacked, the others will come to its aid. The treaty has remained in effect for almost 50 years because its 16 members once believed that it would deter a possible attack by the U.S.S.R. and its East European allies. The combined armies of the NATO countries face the combined forces of the **Warsaw Pact**, the communist equivalent of NATO.

In the 1980s the Soviet leader, Mikhail Gorbachev, did much to improve East-West relations and reduce tensions along the border. Much of the world was hopeful that the massive arms build-up between the two power blocs was beginning to reverse itself. Events in the early nineties eliminated the threat to Western Europe.

7-38 *The Canadian Embassy in Washington, D.C., designed by Canadian architect Arthur Erikson, was built between 1986-1989. It is an unusual building, occupying an irregular triangular site which was the last remaining property on prestigious Pennsylvania Avenue.*

Revolutionary changes within Eastern Europe and the U.S.S.R. suggested that the communist countries were moving toward more independence, and toward more social and economic freedom. These events lead to the fall of communism in both the U.S.S.R. and its satellites.

With the fall of communism, the Iron Curtain between the western and communist nations totally collapsed. A breath of freedom swept through Eastern Europe and the former U.S.S.R.

The latter has now broken into several independent republics.

Several Eastern & Central European nations, former satellites such as Poland and Hungary, are now interested in possibly joining NATO and even the new Russia and a separate Ukraine want a "special partnership" with NATO. Though the threat from the former U.S.S.R. has disappeared, the NATO nations still see a value in remaining linked together in this basically military alliance.

7-39 A Canadian CF-18 fighter plane over Germany. *This aircraft is part of Canada's contribution to the NATO forces based in Europe. It flies out of Canadian Forces Base Baden-Sollingen in Germany, and is seen here over the Hohenzollern Castle.*

NORAD: The Defence of North America

For over four decades, Canada and the United States have shared in the protection of the North American land mass, and its offshore waters and air approaches. In 1959, the two countries signed several defence agreements, the most important of which established the **North American Aerospace Defence Command (NORAD)**. The first NORAD agreement provided for a joint Canada-U.S. defence against long-range Soviet bombers and established a warning system to detect the launch of nuclear intercontinental ballistic missiles (ICBMs). The 1996 renewal of the NORAD agreement reflects the changing global situation. The new mission provides air sovereignty and air defence for North America. This includes the monitoring of man-made objects in space, and detection, validation, and warning of attack against North America by aircraft missiles, or man-made space vehicles.

The NORAD commander-in-chief is responsible to both the U.S. and Canadian military chiefs of staff. Canada, though an equal political partner, is a junior partner in terms of the number of aircraft it makes available. NORAD forces – including those of Canada – are on a five-minute alert basis every hour of the day. The system is controlled from NORAD headquarters in a tunnel complex deep in Cheyenne Mountain near Colorado Springs, Colorado. Canadian service personnel are regularly on duty at the headquarters.

Canada's geographic position, with its east-west expanse from both coasts, to its northern arctic territories, makes it vital to North America's defence. Canada works closely with the U.S. in monitoring northern airspace and, when necessary, warning off aircraft that intrude there. In order to provide warning of an attack over the North Pole, a line of long-range warning stations, known as the Distant Early Warning Line, or DEW Line, was built and is maintained by both Canada and the U.S. across Northern Canada and Alaska. The original line was put in place in 1957. In the late 1980s it was replaced with more advanced equipment, including satellite monitoring systems. If a station detects missiles or aircraft of unknown origin, it sends a message to NORAD headquarters in Colorado, and to the operations centre in North Bay, Ontario.

1. Canada is a contributor to the arms race. Show why some consider its contribution to be minor, however.
2. (a) What is NATO? When and why was it formed?
   (b) Take a vote in your class for or against Canada's membership in NATO. How does the class poll compare with the national average? If there is a major difference, how might it be explained?
   (c) Name five countries in NATO and five from the former Warsaw Pact.
3. (a) What is NORAD? What does it do?
   (b) Why do you think Canadians would support NATO or NORAD?

Canada: A major peacekeeper There are few, if any, aspects of Canada's international involvement in which Canadians take more pride than this country's role in **peacekeeping**. Canada does not pose a threat to other nations. It is only a minor military power, with a small defence force relative to its population. It also has a history of non-aggression: it has never tried to control, let alone invade, other countries. All of this makes Canada an ideal peacekeeper in the world.

Since 1947, over 100 000 members of the Canadian armed forces have served in 16 peacekeeping missions under the auspices of the U.N., as well as four other peacekeeping assignments in which the U.N. was not involved. In most instances the peacekeeping has involved military personnel, acting as observers, supervising a **truce** between two hostile groups. The observers don't become involved in any fighting; they *do* try to ensure that the truce isn't broken. When there is a dispute in the area, they can determine which side was the **aggressor nation**. Canada's peacekeepers are instructed never to attack, though they are permitted to defend themselves. In all, 104 Canadians have been killed in various parts of the world while keeping the peace. Since the Korean War in the early 1950s, Canada has been involved in more peacekeeping missions than any other country.

Much of the credit for the development of Canada's peacekeeping role can be given to Lester B. Pearson. He served as Minister of External Affairs for much of the 1950s, and as Prime Minister from 1963 to 1968. He also won the **Nobel Prize for Peace** in 1956, for his efforts to defuse the Suez Crisis in the Middle East; largely through his efforts, the United Nations created an emergency peacekeeping force that year, to monitor the truce in that area. That mission remains a model for the missions the U.N. still sends to various other war regions. In 1988 the U.N. peacekeeping forces as a group were awarded the Nobel Prize for Peace.

Canadian forces have been involved in a variety of countries, such as:

Korea Canada was a member of the U.N. armistice team that sought a solution to the thorny problem of the two Koreas. Korea was the site of what, for the U.N., was a very unusual peacekeeping role: the Korean War of the early 1950s was actually fought between the "peacekeeping" U.N. contingent and North Korean forces supported by Chinese troops. Communist North Korea attacked South Korea; when the latter appealed for help, many nations sent their armed forces to its defence, under the auspices of the U.N. The U.S.A., the U.K., Canada, Turkey and several other countries fought in this unusual war.

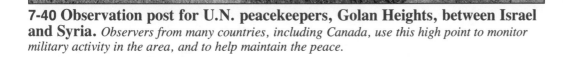

7-40 Observation post for U.N. peacekeepers, Golan Heights, between Israel and Syria. *Observers from many countries, including Canada, use this high point to monitor military activity in the area, and to help maintain the peace.*

India and Pakistan Disputes between these two countries on the Asian sub-continent have been monitored by U.N. observer groups, which have always included Canadians. Canada has always provided air support for these missions.

Iran and Iraq In August 1988, Canada agreed to help the U.N. to maintain the truce that followed the bloody Gulf War between these two countries. Canada sent over 500 armed-forces personnel, most of whom provided communications for the U.N. observer team, which was drawn from 22 other countries besides Canada. Canadians provided well over half of the total U.N. force, which was commanded by a Yugoslavian general.

Though Canada in the early nineties removed its small military presence in Europe, most Canadians still see value in remaining in NATO. NATO has taken on a new peacekeeping (and peacemaking) role. When ethnic conflict between Serbs, Croats and Bosnians reached a serious level in the former Yugoslavia, NATO sent a large force into the region. The force, called IFOR (Implementation Force) included over 1 000 Canadian troops, sent over in 1996 to remain indefinitely until order was restored, and the killings stopped.

Other Peacekeeping Actions By Canada

Canada has also provided, under U.N. auspices, peacekeepers for Somalia, Haiti and Central Africa. In Haiti over 750 peacekeepers from Canada (many of them policemen) were sent in to restore order in 1996. Earlier in the decade, in 1992–93 Canada's airborne regiment was flown to Somalia to assist a nation suffering from

7-41 *The International Peace Garden (spanning the Canada – U.S. border between Boissevain, Manitoba and Dunseith, North Dakota, a few kilometers from the geographic centre of North America) was opened in 1932 to celebrate the peaceful harmony between Canada and the United States. The 931 ha. site contains a wildlife refuge, picnic areas, and an everchanging selection of floral gardens featuring over 150 000 flowers. It attracts over 200 000 visitors a year. Canada and the U.S.A. also share a peace park, the Waterton-Glacier International Peace Park, on the borders of Alberta and Montana.*

civil war and starvation. Unfortunately, this mission resulted in the misconduct of a small number of soldiers, and the deaths of two young Somalis. The result in Canada in 1996 was a board of inquiry, trials of soldiers and a tarnished image for Canada's peacekeepers. Canada also sent troops in late '96 to rescue Rwandans Hutu refugees from Zaire where they forced starvation.

1. Why does Canada make an ideal "peacekeeper to the world"?
2. Why do nations not want to be labelled "aggressors"?
3. On a world map, mark on and name the countries in this section where Canadians have served as peacekeepers.

New Words to Know

Peacekeeping	Aggressor nation
Truce	
Nobel Prize for Peace	

7-42 Canadian troop carrier. *Vehicles such as this are useful in various peacekeeping roles. Its wide tracks also make it suited to Canadian winter conditions.*

7-43 Helicopter preparing to land on the destroyer HMCS *Iroquois* off Canada's east coast. *Land-based aircraft, as well as helicopters based on ships, monitor activity off our east and west coasts. This ensures that our fishing and environmental policies are respected, as well as our sovereignty in Canadian waters.*

Canada's Armed Forces

In 1994, the Liberal government introduced a Department of National Defence White Paper which identified the fundamental mission of Canada's Armed Forces "to defend Canada and Canadian interests and values while contributing to international peace and security." Canada's principal defence roles are:

• Defending Canada: protecting Canada's national territory; ensuring an appropriate level of emergency preparedness; and assisting in national emergencies.

• Defending North America in cooperation with the United States: protecting the Canadian approaches to the continent in partnership with the U.S., particularly through NORAD; promoting Arctic security; and pursuing opportunities for defence cooperation with the U.S. in other areas.

• Contributing to International Peace and Security: participating in a full range of multilateral operations through the UN, NATO, other regional organizations and coalitions of like-minded countries; supporting humanitarian relief efforts and restoration of conflict-devastated areas; and participating in arms control. Despite the wide mandate given to the Canadian forces, the Department of National Defence has been a significant target for government budget cuts. Although Canada spent $9.92 billion on defence in 1998, this represents only 1.2 percent of Canada's Gross Domestic Product. By comparison, the United States spent 3.4 percent of its GDP on defence.

Canada's military personnel is set at 60 000 men and women which are divided into three main branches: maritime command, land forces, and air forces. Although this is a comparatively small number, the Canadian Forces is a professional force, totally supported by volunteers. Few Canadians dispute the need for this country to patrol its coasts and northern regions, and to maintain its sovereignty; but the costs of maintaining even a small force is obviously very high.

Canadians for peace Many Canadians do more than question the build-up of Canada's armed forces; they also support, with money and volunteer work, Canada's peace organizations. There are over 200 of these. They vary in their methods and in their emphasis on particular issues, but as a whole they work for world peace by educating Canadians and pressuring the government about the dangers of the arms race and of nuclear war.

The development of nuclear weapons helped end the Second World War. However, "the Bomb" also created serious problems for the world's peoples and nations. The destructive force of nuclear weapons is now so great that the world population could be wiped out in even a limited nuclear war. The problem is not just that many cities would be incinerated, but that clouds of radiation would be carried to every corner of the globe.

The atomic bomb that was dropped on Hiroshima in Japan on August 6, 1945, destroyed 90 percent of the city's buildings, killed 75 000 people instantly and left 100 000 injured. People who survived the bomb continue, even 40 years later, to suffer pain from its effects; many have died years after the event from the radiation they received on that fateful day. A one-megaton bomb has over 70 times the destructive power of the Hiroshima bomb. Both the U.S.A. and the U.S.S.R. had thousands of megatons in their nuclear arsenals in the late 1980s! The breakup of the Soviet Union meant that the U.S.S.R. arsenal was also broken down and divided among members of the Russian Republics.

This is a cause of great concern. In 1998, both India and Pakistan tested nuclear bombs. Experts believe that Iraq will have nuclear capability very shortly. These are additional causes of concern.

During the 1990s, the United States and Russia signed the Start I and II Treaties formalizing agreements to reduce their nuclear warhead supply by 14 000 units. These destructions are being done. However, as one expert noted during a 1998 Conference on Nuclear Disarmament, there are still 36 000 known nuclear warheads – on launchers or in storage around the world.

It is not only nuclear weapons which create dangers. The world's supply of chemical and biological weapons is another area on which Canadians for peace have focussed their attention. In 1997,

7-44 Canadians also show their support for the peace movement with marches and demonstrations.

the Convention on the Prohibition of the Development, Production, Stockpiling and Use of Chemical Weapons and on their Destruction (CWC) came into force. This is the most comprehensive disarmament treaty ever negotiated. 162 states have signed the Convention and 74 have ratified it. Canada signed in 1995.

There are many peace groups in Canada. Many of them provide educational materials on the nuclear danger and the arms race. Some of these groups are listed below:

- Greenpeace (Vancouver, British Columbia)
- Project Ploughshares (Waterloo, Ontario)
- Canadian Peace Alliance (Toronto, Ontario)
- Pembina Institute (Drayton Valley, Alberta)
- Canada Disarmament Information (Toronto)
- Centre for Peace Studies (Hamilton)
- Marquis Project (Brandon, Manitoba)

1. **Explain briefly how sovereignty and defence spending are related.**
2. **Use as many examples as possible to prove that Canada's armed forces are very international.**
3. **Equipment for our armed forces comes from places such as Vancouver, St. Jean, Kingston and Calgary. Why is it essential that defence contracts be awarded to all regions of the country?**
4. **"Nuclear weapons are so dangerous to the world that they should never be used again." This important topic could be used for debate, small-group discussion, or the preparation of a report of 200 words by each student.**

The Aid Link

Every year Canadians give billions of aid dollars to people in other countries. When this aid comes from the government, it is designated as **Official Development Aid (ODA)** and is administered by the **Canadian International Development Agency (CIDA)**, which was formed in 1968. Government aid is paid out of Canadian taxes. It reflects the concern that the majority of Canadians have about conditions in poorer nations.

Canada's federal aid to other countries has a long history. Following the Second World War, much of Canada's aid went to poorer parts of the British Commonwealth. The **Colombo Plan**, developed in 1951 to assist nations on the Indian subcontinent, was the first large-scale international development project, but others soon followed.

Canadian aid, which is delivered as goods and services, the transfer of knowledge and skills, and money, now assists countries in the Caribbean, Africa (particularly those countries in the Commonwealth and La Francophonie), Latin America, and Asia.

CIDA's objective is "to work with developing countries and countries in transition to develop the tools to eventually meet their own needs."

CIDA has six areas of priority:
- basic human needs
- women in development
- infrastructure services
- human rights, democracy
- private-sector development
- the environment

CIDA also supports foreign aid projects in over 100 of the world's poorest countries.

7-45 Drawing well water, Kenya, Africa. *The provision of clean water has been the aim in scores of CIDA projects.*

CIDA operates four geographic branches in support of its aid programs: Africa and the Middle East; Asia, the Americas, and Central and Eastern Europe.

CIDA uses the Official Development Aid Charter when determining how it should spend the money set aside for aid. The charter has four basic principles:
1. Put poverty first – the poorest countries should receive the greatest amount of aid.
2. Help people help themselves – aid should not solve the problem, but help the people solve it for themselves,
3. Development – the main objective.
4. Partnership – to make Canada's aid more effective, much of it is channeled through other relief agencies such as the United Nations, Canada's **non-government organizations** or **NGOs**, and other international organizations.

This objective means that aid packages are planned with full knowledge of the geography of the recipient area, and that all projects work towards **sustainable development**, or development in tune with a both the present and the future view of the environment and its available resources.

7-48 Standard of living: Canada and the Third World.

	CANADA	THIRD WORLD
% ENROLLED IN HIGHER EDUCATION	44%	3%
GNP (U.S.$) PER CAPITA	13 680	200
CALORIE SUPPLY	3 432	2 073
LIFE EXPECTANCY (YEARS)		
MALE	72	49
FEMALE	79	51
CHILD DEATH RATE PER THOUSAND		
(AGES 1 TO 4)	LESS THAN 1	19
ACCESS TO DRINKING WATER	99%	61% (URBAN)
		31% (RURAL)
% OF POPULATION HAVING ACCESS	100%	54.4%
TO LOCAL HEALTH CARE		
% OF CHILDREN UNDER 5 SUFFERING	—	61.2%
MALNUTRITION		
NUMBER OF PEOPLE PER DOCTOR	550	17 350
TELEPHONES PER 100 PEOPLE	66.4	.3
AUTOMOBILES PER 100 PEOPLE	43	.3

These figures illustrate better than words the gulf between the have-nots of the world and those haves called Canadians.

7-46   Watering crops in Botswana, Africa. *Drinkable water is essential for human life; water also is essential to raise crops. Famine is an ever-present danger, and too often a reality, in the semi-arid areas of Africa.*

7-47   Desertification

Desertification is the term scientists use to describe the spread of deserts into non-desert areas as a result of climate change or human activity. Desertification is a serious problem in the world today, particularly in Africa, where thousands of hectares of land formally used for farming and grazing have become barren desert. Desertification occurs all over the world. Canada's three western provinces have large areas where desertification has occurred.

The loss of arable land can be halted with proper care, scientific help, and careful resource management. The United Nations Convention to Combat Desertification was established in 1994. It has been ratified by 124 countries including Canada, and became effective in 1996. Its member nations have pledged to work together, sharing scientific and technical research, to develop national action programmes to fight the desertification process around the world. The United Nations' *World Day to Combat Desertification* is June 17.

1. **What is ODA, and what is the source of its funds?**
2. **Briefly discuss whether you agree or disagree with the ODA objective.**

New Words to Know
Official Developemt Aid
Non-governmental organizations (NGOs)
Canadian International Development
  Agency (CIDA)
Colombo Plan
Sustainable development

Chart 7-49 illustrates the top 15 recipients of Canadian **bilateral aid** from 1995 – 1997. Bilateral aid is direct aid; recipients are not required to purchase Canadian goods and services in order to receive it. **Multilateral aid** or **tied aid** requires that the recipient country purchase goods and services from the aid-giving country (Canada's food aid is often multilateral aid).

7-49 Top 15 Recepients of Canadian Bilateral Aid 1994 – 1997 (in millions of dollars)

	1994-95		1995-96		1996-97	
1	Egypt	130.03	Egypt	89.24	Egypt	155.71
2	China	92.37	Bangladesh	74.22	Bangladesh	67.94
3	Bangladesh	56.06	China	70.88	China	52.69
4	Cote dIvoire	55.69	India	51.74	Haiti	42.27
5	Tanzania	40.19	Cote dIvoire	31.72	Rwanda	31.31
6	Cameroon	34.18	Ghana	30.91	Ghana	25.78
7	Haiti	33.86	Haiti	30.80	Indonesia	25.16
8	Philippines	31.04	Peru	25.82	Philippines	23.83
9	Rwanda	30.88	Philippines	22.84	Senegal	23.73
10	Peru	30.69	Indonesia	22.31	Peru	23.06
11	Ghana	30.50	Cameroon	20.45	Cote dIvoire	22.84
12	Indonesia	30.31	Mali	19.89	Ex-Yugoslavia	20.8
13	India	29.52	Bolivia	19.85	Vietnam	17.68
14	Ethiopia	28.44	Senegal	19.62	Cameroon	16.78
15	Zambia	28.08	Ex-Yugoslavia	19.46	India	15.97

CIDA's established priorities for its aid allow over 50 percent to go to Commonwealth and Francophone countries, with concentrated amounts (greater funds to fewer recipients).

Chart 7-50 depicts those organizations which oversee Canadian ODA assistance – $2.1 billion in 1997-8. 77 percent of the funds are managed by CIDA. The two slices of the pie *not* given over to CIDA went to financial institutions and to special projects by smaller, separate agencies. The chart shows the vast scope and complexity of CIDA's contribution to international aid.

Aid Provided by NGOs

Although the federal government supplies most of Canada's aid monies, it has only been funding aid programs since the 1940s. Canadian NGOs have been funding aid programs since the 19th century, when Canada's churches began to solicit funds in aid of the world's poor.

There are over 220 NGOs in Canada, employing over 2 400 paid staff to administer aid monies, and manage the fund-raising. There are over 500 Canadians working overseas directly for NGOs on Canadian funded projects and programs – assisting in agricultural projects, performing surgery, advising and training local peoples, and so on. In addition, there are well over 32 000 volunteers assisting full-time staff in Canada, and over 1 500 volunteers overseas.

Some of the largest and best known NGOs are the Red Cross, UNICEF, Care, OXFAM, CUSO, the Foster Parents' Plan, and various churches. Some of these used to be mere fund-raising branches of international organizations. Most, now, are independent and manage their own funds and projects.

Some NGOs are huge, with budgets of over $24 million. The World University Service of Canada and World Vision Canada both have budgets this size. Most NGOs are much smaller – many raise less than $100 000 a year. The government provides about 40 percent of the NGO's budgets. Federal funds are usually granted on a matching basis: if the NGO raises $50 000, the government gives it another $50 000. Canada pioneered this approach and many countries now use it.

Canada's NGOs distribute their aid in three basic ways:

Overseas development projects For decades, most NGO funds have gone to development projects. Recently, NGOs have begun sending fewer of their own workers; instead, they send advisors, engineers, managers and technicians to train local people to do the work themselves and develop more self-reliance in the communities their projects serve. As one person put it, they are trying to work themselves out of a job; once local people are trained to do the work, foreign-aid workers shouldn't be needed.

7-50 Canada's official aid budget, 1997-1998.

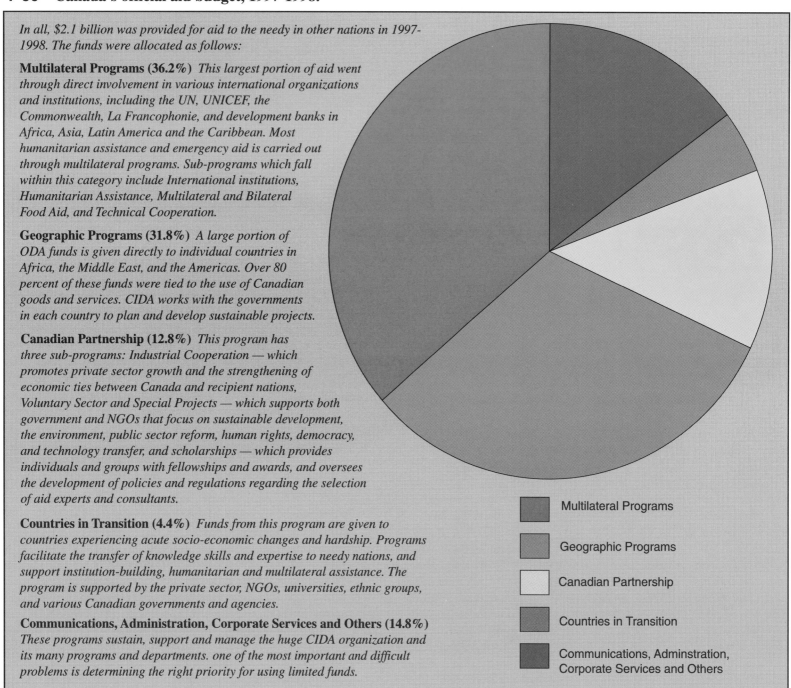

In all, $2.1 billion was provided for aid to the needy in other nations in 1997-1998. The funds were allocated as follows:

Multilateral Programs (36.2%) *This largest portion of aid went through direct involvement in various international organizations and institutions, including the UN, UNICEF, the Commonwealth, La Francophonie, and development banks in Africa, Asia, Latin America and the Caribbean. Most humanitarian assistance and emergency aid is carried out through multilateral programs. Sub-programs which fall within this category include International institutions, Humanitarian Assistance, Multilateral and Bilateral Food Aid, and Technical Cooperation.*

Geographic Programs (31.8%) *A large portion of ODA funds is given directly to individual countries in Africa, the Middle East, and the Americas. Over 80 percent of these funds were tied to the use of Canadian goods and services. CIDA works with the governments in each country to plan and develop sustainable projects.*

Canadian Partnership (12.8%) *This program has three sub-programs: Industrial Cooperation — which promotes private sector growth and the strengthening of economic ties between Canada and recipient nations, Voluntary Sector and Special Projects — which supports both government and NGOs that focus on sustainable development, the environment, public sector reform, human rights, democracy, and technology transfer, and scholarships — which provides individuals and groups with fellowships and awards, and oversees the development of policies and regulations regarding the selection of aid experts and consultants.*

Countries in Transition (4.4%) *Funds from this program are given to countries experiencing acute socio-economic changes and hardship. Programs facilitate the transfer of knowledge skills and expertise to needy nations, and support institution-building, humanitarian and multilateral assistance. The program is supported by the private sector, NGOs, universities, ethnic groups, and various Canadian governments and agencies.*

Communications, Administration, Corporate Services and Others (14.8%) *These programs sustain, support and manage the huge CIDA organization and its many programs and departments. one of the most important and difficult problems is determining the right priority for using limited funds.*

Multilateral Programs

Geographic Programs

Canadian Partnership

Countries in Transition

Communications, Adminstration, Corporate Services and Others

7-51 School supported by Canadian funds, Sri Lanka.
Education is an important part of the answer to problems in the Third World. This school illustrates the importance of "self-help."

7-52 Cropland in Niger. *Improvements in agriculture are a major thrust in CIDA aid. These may range from improved seeds and fertilizers, to better methods of irrigation and harvesting.*

Education is the second major focus of NGO aid. Much of this is **developmental education**, that is, the type that trains the local people in the skills needed to improve their industries, and thus their economies and living standards. Of course, such programs do not neglect formal education, particularly for young people. Canadian aid has established hundreds of schools in the past few decades.

Public policy advocacy This type of aid is relatively new for the NGOs. Years ago, NGO workers began to realize that no matter how hard they tried to help a Senegalese or Indian farmer, the prices those farmers received for their produce were often determined by the policies of foreign governments. These often reduced or cancelled the usefulness of the aid. The governments of wealthier

7-53 Typical yearly aid contributions (in tonnes) from provinces of Canada

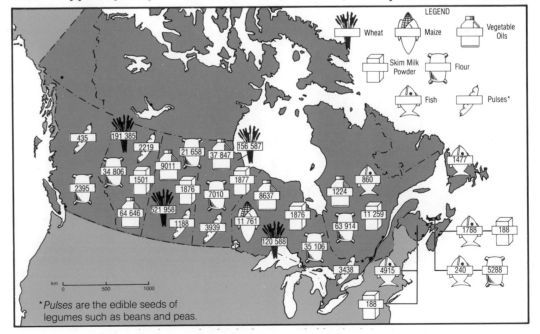

LEGEND

Wheat · Maize · Vegetable Oils · Skim Milk Powder · Flour · Fish · Pulses*

435 · 191 385 · 2219 · 21 658 · 37 847 · 156 587 · 34 806 · 9011 · 1477 · 1501 · 1877 · 860 · 2395 · 1876 · 7010 · 8637 · 1224 · 64 646 · 521 958 · 11 259 · 1188 · 3939 · 11 761 · 1876 · 63 914 · 1788 · 188 · 120 588 · 35 106 · 3438 · 4915 · 240 · 5288 · 188

km 0   500   1000

Pulses are the edible seeds of legumes such as beans and peas.

countries had to be shown that their policies — particularly trade policies — were often hurting millions of people in the underdeveloped world. Often poorer nations are dependent on one cash crop like cacao or peanuts; when a wealthier nation places restrictions on the importing of such produce, poorer nations can suffer serious economic hardship.

Canadian NGOs now have an umbrella organization, the **Canadian Council for International Co-operation (CCIC)**. It works with government committees and politicians on behalf of the NGOs, and lobbies for policies that will help the developing world.

Not all programs of this country's NGOs can be slotted into these three categories. Sometimes emergencies such as a famine, flood, or hurricane, lead to a sudden outpouring of aid. In 1988, Hurricane Gilbert struck Jamaica leaving devastation in its wake — 500 000 homeless, 80 percent of the island's roofs blown away, communications and electrical lines reduced to a shambles. The federal government almost immediately announced a $7.6 million grant for Jamaica, and another $1 million to match the relief money raised by humanitarian groups in Canada. This country's large Jamaican community,

assisted by the NGOs, churches and other Canadians, raised funds to send huge quantities of food, clothing, bedding and other aid. Such aid is above and beyond the long-term projects, whenever and wherever disaster strikes. Canada's NGOs have been involved in scores of countries. India and Haiti have traditionally been large recipients of aid. The Caribbean also receives a large share of NGO aid, a reflection of Canada's strong ties with that region. Asian countries are receiving less now than they have historically, Latin America more, and Africa substantially more.

Africa has received worldwide attention. Poverty, disease, drought, political instability, mass exodus and overcrowded refugee camps, genocide, climate changes, and desertification have made that continent's needs more acute than ever, and more difficult to meet.

Africa receives a large share of Canadian federal aid. NGOs are helping as well. The five year $150 million CIDA Africa 2000 plan is drawing to a close, with over half of the allotted funds having been dispensed to NGOs.

The Canadian government and our NGOs are more effective when they are working together.

7-54   Poor housing in the Philippine Islands. *Housing is of an incredibly low standard in much of the Third World. Canadian aid from CIDA and NGO projects attempts to improve conditions such as those seen here, and some which are even worse.*

1. (a) **Study Fig 7-49. Why do you think Egypt is at the top of the bilateral aid list? For the last three years? What percentage of the top 15 aid recipients is found in Africa? Asia? The Caribbean?**

7-55 Bilateral Aid, 30 major participants in 1995 (Million of $US).

Recipient Countries	Total Bilateral Aid	Canada's Rank Aid Donors	Canada's % of Total Bilateral Aid
Egypt	1 689.4	5	3.7
Bangladesh	712.9	7	5.3
China	2 531.2	7	2.1
Haiti	509.6	4	3.6
Rwanda	338.1	6	5.0
Ghana	358.5	7	6.3
Indonesia	1 303.3	9	1.2
Philippines	748.2	6	2..6
Senegal	397.0	6	3.8
Peru	319.0	6	5.3
Cote d'Ivoire	726.5	5	3.1
Ex-Yugoslavia	1 047.4	10	1.0
Vietnam	549.1	11	1.1
Cameroon	345.3	3	3.8
India	1 051.2	10	1.2
Mali	284.5	6	5.7
South Africa	318.4	9	2.8
Thailand	826.7	7	1.8
Guinea	219.8	7	2.0
Mozambique	701.5	16	0.8
Pakistan	360.1	12-13	0.9
Malawi	220.5	6	3.8
Tanzania	586.6	7	4.3
Congo	105.0	6-7	0.2
Zambia	439.4	8	4.8
Zimbabwe	347.7	10	3.6
West Bank and Gaza	183.0	17	0.3
Bolivia	475.3	9	2.4
Burkina Faso	251.2	7	3.2
Nepal	265.9	10-11	1.4

Source: OECD, Geographical Distribution of Financial Flows to Aid Recipients 1991-1995, Paris, 1997.

(b) What percentage of aid recipients in Fig 7-55 are found in Africa, Asia, the Caribbean? Compare your answers with those in 1a.

2. What is desertification. Cite examples of its impact on the world economy.

3. Divide into groups and research one specific NGO.
Find out the history, goals, and geographical area of operation for each NGO. Each group should present its findings to the class.

New Words to Know
Developmental education
Public policy advocacy
Canadian Council for International Co-operation (CCIC)

Does Canada Do Enough?

This is not an easy question to answer. The opinions of aid donors and aid recipients probably differ. We can, however, compare Canada's aid donations with those of other countries as in Fig. 7-55. There are interesting discrepancies. Canada contributes very little, relatively speaking, to the West Bank and Gaza, however, our aid to Cameroon ranks as the third highest of all major participants. Canada's policy of providing major aid support to members of the Commonwealth and Francophonie affects these figures as well. It is interesting to note that Canada's total aid budget has been dropping in recent years.

7-56 "Dust devil" erodes topsoil from the dry lands of Niger, Africa.
The even more destructive dust storms also affect this area. Removal of topsoil makes recovery from drought very difficult.

Christian Blind Mission International

Since its beginning in 1908, CBMI has grown to become a world leader in practical aid to people with disbilities in developing countries.

A major NGO in Canada, CBMI is an international aid organization with offices in eight other countries, including the USA, Germany, Australia and the UK. Internationally, the organization receives annual donations of over $80 million, from hundreds of thousands of donors. The money is used to help disabled people, particularly blind people, in over 1 000 projects in more than 100 needy countries throughout the world.

CBMI is able to provide assistance to more than 8 million people in need each year. Among these are over 7 million eye patients, including:

- 2 million children whose eyes are examined for blinding vitamin A deficiency and other childhood eye diseases.
- almost half a million visually impaired people who receive spectacles.
- 400 000 trachoma patients who are treated with tetracycline eye ointment or given eyelid surgery.
- 268 000 blind people who have their sight restored through cataract surgery.
- 107 000 people in danger of losing their sight, who are saved from blind ness by an operation.
- tens of thousands of people threatened by river blindness receive protection through the distribution of Mectizan tablets.

In addition, CBMI cares for:

- 70 000 ear, nose and throat patients
- 60 000 orthopaedic and polio patients
- 40 000 leprosy and TB patients
- 800 000 other patients

- 160 000 blind and otherwise disabled men, women and children who receive education and training through CBMI's partners.

CBMI-Canada has its head office in Stouffville, Ontario, where it raises and manages funds for its overseas work.

In 1997, Canadian supporters of CBMI gave over $7.5 million to support 62 projects in 24 countries.

In addition to this primary mandate, CBMI offers Christian literature on cassette tape to blind and print-disabled Canadians from coast to coast, free of charge, through its Talking Book Library.

Also at this location is CBMI's Conference Centre and its "A World of Goods" International Gift Store. The Gift Store sells handcrafted items made by disabled artisans at CBMI projects throughout the world. By providing an extended market for these craftsmen and -women, CBMI is helping them to earn a living with dignity. More than 107 000 sight-restoring cataract surgeries are performed at CBMI-Canada-sponsored projects each year, and the number is growing. Largely through the practical compassion of Canadian donors, thousands of destitute men, women and children are given the precious gift of sight and the hope of a new life.

At CBMI-Canada's 25 partner schools, education and community-based programs, over 12,000 people with disabilities, primarily children, are given special education and training – these are individuals who would otherwise face a lifetime of indignity, isolation and dependence.

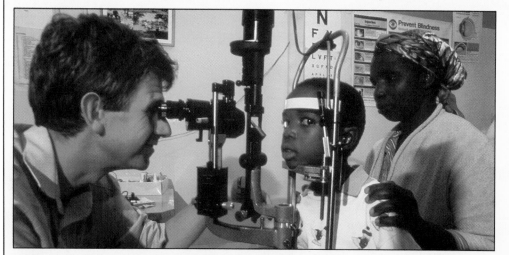

7-57 Canadian aid helps blind people in Zimbabwe. *Based out of the nearby Morgenster Mission Hospital, the Masvingo Eye Program is supported by the Canadian branch of Christian Blind Mission International. Each year the Eye Program is able to treat some 20,000 eye patients, in many cases restoring sight through cataract surgery.*

Environmental Preservation: Debt-for-Nature Swaps

Recently, environmentalists and policy-makers have found that a nation's debt problem and environmental problems could be addressed through an innovative public-financing scheme, a debt-for-nature swaps (DNS). The idea of debt-for-nature swaps arose in response to two crises: the crushing debt of lesser-developed countries, and their environmental problems. It was suggested that the debt crisis could be used to ease environmental problems. Nations wishing to protect their natural resources could reduce their debt liabilities by exchanging their foreign debt for investment projects. The idea of "swapping" debt for some kind of domestic asset that provides a guarantee that the donor or investor directly be able to protect the country's natural resource base for the greater benefit of humanity has an immediate appeal. The basic premise of a DNS is that a certain amount of foreign debt will be forgiven in exchange for an investment in a program designed to improve natural resources management, conservation, and sustainable development in the debtor country. The debt link serves primarily as an added incentive to encourage countries to take the steps that they would ultimately want to pursue anyway in the interest of environmentally sustainable development.

Debt-for-nature swaps involve the acquisition by a conservation organization or government of part of the external debt of a developing country. The debt is then exchanged for the local currency from the developing country for use in environmentally-related projects. Alternatively, the debt may be cancelled in exchange for a commitment by the debtor government to invest in conservation and resource management programs.

Debt-for-nature swaps are not intended to provide debt relief on any significant scale. Their objective is first and foremost to provide funding for conservation and better uses of natural resources and it is incorrect to consider these swaps in terms of a solution to the debt crisis. The net effect of the DNS program being implemented in certain countries is that dollars previously sent to international banks are now being used to protect conservation areas, to reforest, to safeguard biodiversity, and to develop awareness of the need for environmental protection, so that future generations in the lesser developed countries may enjoy greater development potential and a somewhat lighter external debt.

Comparative Debt Structure of The World's Developing Countries 1980 - 1996 (in US dollars)

	1980	1996
Total Debt	603 321	2 095 428
Long Term (LTD)	445 300	1 650 097
Public & guaranteed	377 032	1 397 096
Private & nonguaranteed	68 268	253 000
Use of Internat Monetary Fd	11 564	60 107
Short Term	146 458	385 224
Interest arrears on LTD	1 022	39 401
Principal arrears on LTD	1 963	10 586
Export Credit	0	387 426
Net Resource Flows	84 037	281 603
To LTD	66 419	87 618
Foreign direct investment	4 420	118 960
Grants	19 494	48 467
Interest on LTD	32 355	77 323
GND	2 839 930	5 815 129
Exports	710 721	1 525 100
Imports	686 859	1 608 280
Total Reschduled	0	29 324
Debt stock	0	5 092
Principal	0	16 108
(official)	0	6 315)
(private)	0	9 793)
Interest	0	7 395
Principal forgiven	0	5 935
Interest forgiven	0	1 856
Debt reduction	0	10 540
(through debt buyback)	0	4 313)
Currency percentage		
German Mark	6.5	7.5
French Franc	5.4	4.6
Japanese Yen	5.9	12.1
Pound Sterling	3.3	1.4
Swiss Franc	1.5	1.0
U.S. Dollar	48.1	46.1
Multiple Currency	10.5	14.2
All other	8.2	8.4

Source: World Bank

1. **What would be a fair way to determine which of the nations in chart 7-55 are richer (or poorer) than Canada?**
2. **Should Canada's ODA be increased? Why or why not? Should the federal government give more to Canada's NGOs?**

The Tourist Link

"The real benefits to be gained from the industry are the human benefits — the benefits of social understanding, and cultural awareness that tourism promotes to millions of people every day." John E. Cleghorn, president, the Royal Bank of Canada.

Tourism in Canada is a big business. Total domestic and international tourism spending in this country reached $44 billion in 1997.

Receipts from international visitors totalled close to $12.4 billion. The Canadian tourist industry employed over 600 000 people. Foreigners made 17.6 million overnight trips to this country in 1997; Americans accounted for 13.4 million of these trips. Japanese tourists made 565 700 trips to this country in 1997; United Kingdom visitors recorded 733 600 visits of at least one day, while Asians made 1 375 800 visits. Canada is the ninth most popular tourist destination in the world according to estimates from the World Tourism Organization.

Canadians made 19.1 million overnight trips out of the country, including 15.1 million trips to the U.S.

Canada's 4 000 travel agencies, which are concentrated in Ontario and Quebec, had sales in excess of $10 billion in 1995. Canada had 871 tour operators and wholesale travel businesses in 1995 with total sales of $3.7 billion. Over 75 percent of tour operators sales are tour packages to destinations outside the country.

Social understanding and cultural awareness are two of the main benefits of tourism. Tourists understand a foreign country more clearly after a visit there.

Travelling allows people to observe and appreciate the difference between people and places. Pope John Paul II has praised tourism because it "can help overcome many real prejudices and foster new bonds of fraternity." This concept is well known in the tourist business as industry workers often refer to one class of visitors as "VFRs," meaning "visiting friends and relatives."

Fig 7-61 shows the top 12 source countries for tourists coming to Canada. Note the dominant number of American visitors. The drop in value of the Canadian dollar has spurred American tourists to travel here. Americans made 775 000 overnight car trips to Canada in June 1998, and 306 000 airline flights to this country in the same month.

Many tourists visiting Canada head for urban areas. Canadian cities have a wide selection of attractions, historic sites, festivals, cultural events, and activities. Toronto's theatre district is world-famous, as is Montreal's Jazz Festival, Calgary's Stampede, Vancouver's

7-59 Ontario Place, Toronto, Ontario. *Built on, and between artificial islands along Toronto's lakeshore, Ontario Place is only one of the many recreational attractions which draw millions of foreign visitors to Canada annually. The spherical building is the Cinesphere where films are shown on a huge screen.*

Chinatown (North America's second largest after San Francisco), Halifax's Nova Scotia International Tattoo, and Winnipeg's Royal Canadian Ballet to name a few major attractions.

But Canada also attracts tourists because of its superbly scenic outdoors. This country has 39 National Parks totalling over 224 000 km2, three national marine conservation areas, and well over 1 000 provincial parks and/or recreation areas.

Parks Canada' mandate is *to protect for all time representative natural areas of Canadian significance . . . to encourage public understanding, appreciation and enjoyment of this natural heritage so as to leave it unimpaired for future generations.*

Canada also has 12 World Heritage Sites, international testimony to the quality of our landscapes. The World Heritage Site designation was established by the United Nations (UNESCO) in the 1970s to ensure that the most significant sites on earth, would be preserved for all time under a special international designation. Today, there are 506 of these locations.

Canadians love to travel abroad as the statistics on map 7-60 indicate. They have the money; they also share the same reasons to travel as many of the people who visit here. Many Canadians feel it is important to visit their country of birth or of their ancestors. A large number of Canadians also visit the U.S.A., mostly for recreation; however, a great many also go to visit relatives and friends.

1. **Plan a two-week vacation in any country except Canada or the U.S.A. Get brochures and airline schedules from a travel agent. Read about your destination in the library, and use all the information to form some general plans for the two weeks. What will you do? Where in the country will you go, and how? Where will you stay and what will you eat? How much will it cost?**

2. **How does travel to foreign countries promote peace?**

3. **(a) Write a 200-word brochure that would encourage people from Europe to visit Canada.**
   (b) Write a similar brochure to encourage students to visit a place you have visited in Canada, or in a different country.

4. **Mark the countries listed in chart 7-61 on a world map. Mark the figures on the map, beside each country's name, as well as a coloured bar to show the statistics proportionally.**

7-60 Canadian International Travel Account, 1986-1995

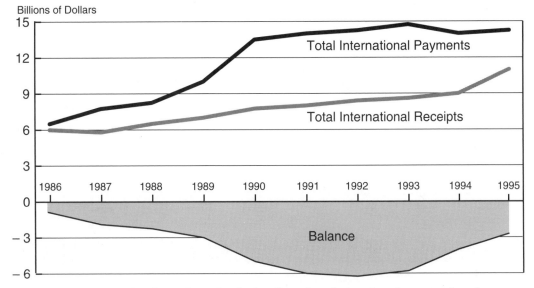

Billions of Dollars

The two line graphs above show clearly that Canadians have spent, for over a decade, more in their travels abroad than visitors have spent here. Below the base line the extent of the balance, a deficit, is indicated.

7-61 Source of the largest number of visitors to Canada of one night or more

COUNTRIES	NUMBER OF VISITORS (THOUSANDS)
U.S.A.	12 543
UNITED KINGDOM	576
JAPAN	481
FRANCE	410
GERMANY	367
HONG KONG	130
AUSTRALIA	117
ITALY	96
NETHERLANDS	90
SWITZERLAND	89
MEXICO	81
WEST INDIES	72
TOTAL - ALL VISITORS	**16 334**

This chart provides only the top 12 source countries. Obviously the U.S.A. is the leading source country; it supplies over 90 percent of our foreign visitors. Japan has twice the population of the U.K., yet the latter provides more visitors. What might explain this fact? Why are Switzerland and the Netherlands, rather small countries, represented in our top ten?

The Family Link

Most Canadians are linked to countries outside of Canada through their families. These family ties are among the most powerful links that exist between Canada and other countries. It is very difficult to separate the lifestyle and culture of Canadians from their family backgrounds. It is in the family, after all, that these things are learned. The church we attend, the foods we especially enjoy, the languages we speak, the sports we follow, even our attitudes and the very names we bear may well be related to our family and its national origin.

Of course, it is only natural that a **new Canadian** (a person born in another country) or a **first generation Canadian** (a member of the first generation of a family to be born in Canada) will have closer ties to the country of origin than later generations. The influence of parents born in the Ukraine, or Scotland, or Jamaica, or Italy will be much greater on their children than on their grandchildren. However, it is surprising how strong the influence can remain. In language, religion, family names, festivals celebrated, attitudes and other aspects of culture, such influences may persist for many generations. Many countries have policies of assimilation, and their governments discourage multiculturalism to ensure the development of a single national culture. In Canada, where multiculturalism is not only accepted but encouraged by governments, the various cultural and national influences will persist even longer than in other countries.

7-62 Cultural festivals from many lands express Canada's international background. *This Chinese New Year parade in Vancouver is another sign of the varied cultural and family backgrounds of Canadians. Other signs are also evident in the photo.*

Canadians and their Geography

Canada underwent many sweeping changes during the twentieth century. We became a highly multicultural society, much of our population migrated from farms to cities, we moved into the electronic age, and exploited our natural resources to forge a strong and prosperous nation. In fact, one could say that since 1900, the face of the nation changed.

Human Heritage

Today, Canada maintains one of the most diverse and multicultural populations in the world. This was not always so. Prior to 1900, our population consisted primarily of Aboriginal Canadians and French and English speaking peoples of Western European descent. Throughout this century, millions of migrants from such areas as Africa, Asia, Eastern Europe and Central and South America left their countries of origin to seek opportunity and security elsewhere. Canada welcomed these peoples, and offered them a level of freedom and prosperity that was not readily available in their homelands. The opening of the west in the first half of the century meant inexpensive plots for land hungry farmers, and throughout much of this century, there was a constant need for skilled and wage labour. For these reasons, many immigrants flocked to our shores and made Canada their home. As a result, we have all been rewarded with a rich, multicultural society, a country that maintains a worldwide reputation for tolerance and peaceful co-existence.

This is not to say that problems do not exist.

In the twenty-first century Canadians must continue to address social inequality. The interests of Canada's Aboriginal people will continue to be a priority. New agreements and models of **Aboriginal Self Government** will need to be negotiated. The desire of many Quebecois to separate (**separatism**) will continue to threaten national unity. Policies and programs designed to eliminate discrimination and racism must continue to be implemented if we are to preserve our **cultural diversity** and peaceful co-existence. Only then can different cultures continue to work and prosper in Canada.

From Farm to Factory to Service Economy

Canada changed from an **agricultural society** to an **industrialized society** during the twentieth century. New agricultural techniques and transportation networks made farms more efficient. Fewer workers were required to grow, harvest and truck food to market. Rural people began moving to Canada's urban centres where they sought employment in mills and factories. The industrial sector expanded and immigrants flocked to cities. As a result, our urban centres grew at a rapid pace, a process which is known as **urbanization**.

Urbanization has come with costs, however. Cities are oftentimes overcrowded, smog-ridden, and individuals have not always been able to locate permanent work. Inadequate transportation networks also plague the daily lives of urban dwellers. How to properly house, employ, and transport urban people remains one of Canada's greatest challenges.

Today we live in a highly urbanized society that is quickly moving away from industrial production. International trade agreements such as NAFTA allow Canadian companies to sell commodities across international borders with little or no restrictions. In turn, many Canadian companies moved plants to places where manufacturing and labour costs are less expensive. We are currently shifting away from selling commodities such as clothes and furniture, to selling services, such as food, entertainment, and vacations. Less industry means less industrial pollution. But many employees have suffered. Often, service industry jobs require little skill; they can be tedious, low paying, and are often performed on part-time basis. Employees do not achieve much economic security, and working part-time or on contract most often means working without benefits or hopes of upward mobility.

The Electronic Age

Electronics touch almost every part of our lives. Where once information was processed and delivered by hand, computers today do much of this work. It assists in managing our railways and traffic light systems; sending satellites to space; checking our spelling, building our cars and delivering our e-mail. Electronics have made the process and delivery of information both extremely efficient and cost effective, and have given us access to a whole world of new and exiting information through the Internet. But the use of modern electronics does have its drawbacks. The personal computer has made the corporate secretary a thing of the past. Computer assisted robots have contributed

7-64 *Bridges* by Raymond Moriyama.

BRIDGES

There are many kinds of bridges.
There are bridges of the mind —
conceptual and philosophic bridges.
There are bridges of the heart —
bridges of love, bridges of friendship
and of the spirit.
There are bridges of vision —
bridges to the future, bridges of hope
and of promise.

The building of a bridge begins —
not from one side, but from each side.
Flexibility is a must —
a rigid bridge will not long endure.
Patience is a necessity —
for to design a lasting structure
takes time.
What makes the best foundation?
Where shall it be placed?
Why?

Nations build bridges —
forge spans to
ideas
goods
people.
People build bridges —
use bridges
are bridges
to the future.

Moriyama is a world-famous Canadian architect of Japanese origin. The Ontario Science Centre building in Toronto is only one of his striking designs.

to unemployment in many other areas such as car and truck manufacturing. Computer use has also altered traditional face-to-face working relations. While working from home, away from our colleagues and teachers, we are in jeopardy of becoming a nation disconnected, where human relations are directed, not through talk and a handshake, but through e-mail and electronic data transfer.

Land and Prosperity

During the last one hundred years, Canadians have transformed the land to facilitate population and industrial growth. Railway, hydro, and telephone lines weave from town-to-town, city-to-city, coast-to-coast. The 8 000 km Trans-Canada highway links the nation from Newfoundland to British Columbia; forests have been razed to make room for farming and oil, gas and mineral exploration. Dams harness water energy and transform it into hydro-electricity; locks have made rivers navigational, and factories and power stations line rivers, lakes and dot the cityscape and countryside.

Canadians point to these achievements with pride. Indeed, because of our progress, we enjoy one of the highest **standards of living** in the world.

Yet the benefits Canadians derive from this prosperity come with a price. Geographically speaking, what we really have done is manipulated our **environment**, and sometimes abused it. We have changed our landscape by exploiting the natural resources of our land and oceans. We have altered the water flow of our rivers by building dams and hydro stations. We have polluted our atmosphere with our cars and industrial plants. These projects

resulted in a variety of problems and consequences: air and water pollution, overfishing and over-lumbering, urban development on what was once good farm land, conventional hydro-electricity mismanagement, resource development on disputed Aboriginal lands, residential, industrial, chemical and nuclear waste, over-use of chemical fertilizers, and over-use of medicines and hormones on dairy and animal farms.

The price of progress has been high. We now look at these consequences and refer to them as "challenges" we must solve. The Canadian Government is doing its part. In 1995, it established an act that required all Ministries to table *Sustainable Development Strategies* by December, 1997. Each Ministry is required to spell out exactly how it aims to balance economic growth, social development, and environmental protection. The Department of Foreign Affairs, for instance, is currently looking for ways to reach sustainable

development both here and abroad. Its aims are "to work with all levels of government... to ensure an integrated approach to the implementation of sustainable development practices through consensus building, training, and drafting of legislation." The Department will also work with other countries toward common goals, "through negotiation, mediation, and promotion of dialogue." These goals are important because the exploitation of our resources cannot go on forever. Sustainable development will become very important in the next one hundred years, and will require the efforts of individuals, governments, and nations worldwide.

The face of Canada has changed since 1900. We have experienced great changes to our geography, economy, and society. We have become a diverse and prosperous nation and the envy of much of the world. But as we have seen, we have also created some very serious issues. These are the challenges of the twenty-first century.

Mind Benders and Extenders

1. "Canada benefits from being involved in the world and affected by the world." Discuss either the positive or the negative side of this statement, as if preparing an argument for a debate.

2. Canada's involvement with other countries raises a number of issues on which Canadians hold opposing opinions. As a student, you should become involved in such issues: you should learn to separate fact from opinion, and to respect the rights of others to hold different opinions. Try holding class debates on one or more of the following topics:
   • Is free trade with the U.S.A. good for Canada?
   • Should there be more or less immigration to Canada?
   • Should Canada increase or decrease foreign aid?
   • Should Canada stay in NATO?
   • Should Canada stay in NORAD?
   • Should Canada use government action to reduce the amount of foreign investment in this country?

3. Select one of the topics in question 2 and prepare an essay, report or oral presentation. Include aids such as maps, graphs, charts, photos, slides and so on in your project.

4. Adam Smith in his famous book *The Wealth of Nations* (1763) claimed that "It is as foolish for a nation as for a person to make that which can be bought for less." Explain this statement in your own words. Also explain, using Canadian examples, how well his principle works in international trade.

5. Prepare a report of 500 to 600 words on "Canada's Trade with the U.S.A."

6. Prepare a report of 500 to 700 words on the Suez Crisis of 1956. Show how "international" this crisis became, and how Lester Pearson helped defuse it.

7. Prepare an essay of 500 to 1000 words comparing the Commonwealth and La Francophonie, and emphasizing the role Canada plays in both organizations.

8. One of the biggest concerns of many peace groups in Canada and around the world is the phenomenon called "nuclear winter." Do research to find out what this is, and what its cause and effects would be. Then write a report of 300 to 500 words about it.

9. Taiwan has just begun to produce VCRs that are being sold in Canada at prices half those of Japanese or South Korean VCRs. The government of Canada is wondering whether tariffs should be placed on these VCRs, since the workers who made them received such low wages.

   In small groups, simulate a committee meeting called by the federal government's trade department. The following roles should be played by different students: federal government trade official (who acts as chairman), Canadian union official, Taiwanese union official, representative of the Taiwanese manufacturer, representative of a South Korean manufacturer, representative of a Canadian consumer group. At the end of the discussion, the chairman or a small group of students should summarize the views of the different parties, and announce the decision reached and the reasons for it.

10. Summarize, in a report of 200 to 300 words, the various kinds of foreign aid made available by Canada's federal government.

11. Write an article of 200 to 300 words for your school newspaper, telling the student readers why a school project, linked to one of Canada's NGOs, should be started to aid the needy peoples of the world.

12. Which countries have the other students in your class visited? How did they react to those countries? Share these experiences in small groups.

Topographic Map Symbols

Scale 1:50,000 Échelle

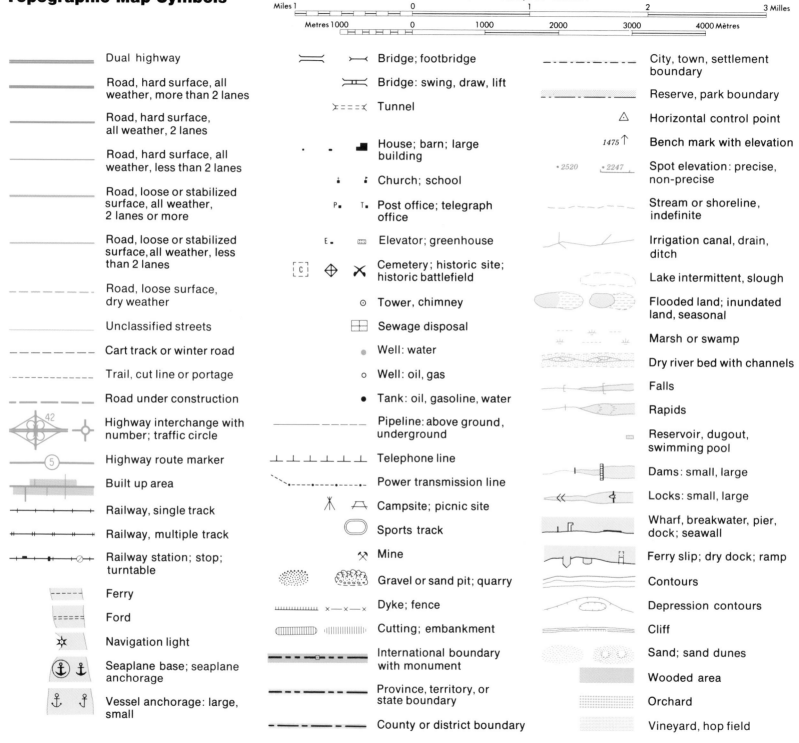

Miles 1   0   1   2   3 Milles

Metres 1000   0   1000   2000   3000   4000 Mètres

	Dual highway
	Road, hard surface, all weather, more than 2 lanes
	Road, hard surface, all weather, 2 lanes
	Road, hard surface, all weather, less than 2 lanes
	Road, loose or stabilized surface, all weather, 2 lanes or more
	Road, loose or stabilized surface, all weather, less than 2 lanes
	Road, loose surface, dry weather
	Unclassified streets
	Cart track or winter road
	Trail, cut line or portage
	Road under construction
42	Highway interchange with number; traffic circle
5	Highway route marker
	Built up area
	Railway, single track
	Railway, multiple track
	Railway station; stop; turntable
	Ferry
	Ford
☆	Navigation light
	Seaplane base; seaplane anchorage
	Vessel anchorage: large, small

	Bridge; footbridge
	Bridge: swing, draw, lift
	Tunnel
	House; barn; large building
	Church; school
P. T.	Post office; telegraph office
E.	Elevator; greenhouse
C	Cemetery; historic site; historic battlefield
☉	Tower, chimney
	Sewage disposal
	Well: water
○	Well: oil, gas
●	Tank: oil, gasoline, water
	Pipeline: above ground, underground
	Telephone line
	Power transmission line
	Campsite; picnic site
○	Sports track
	Mine
	Gravel or sand pit; quarry
	Dyke; fence
	Cutting; embankment
	International boundary with monument
	Province, territory, or state boundary
	County or district boundary

	City, town, settlement boundary
	Reserve, park boundary
△	Horizontal control point
1475 ↑	Bench mark with elevation
•2520 •2247	Spot elevation: precise, non-precise
	Stream or shoreline, indefinite
	Irrigation canal, drain, ditch
	Lake intermittent, slough
	Flooded land; inundated land, seasonal
	Marsh or swamp
	Dry river bed with channels
	Falls
	Rapids
	Reservoir, dugout, swimming pool
	Dams: small, large
	Locks: small, large
	Wharf, breakwater, pier, dock; seawall
	Ferry slip; dry dock; ramp
	Contours
	Depression contours
	Cliff
	Sand; sand dunes
	Wooded area
	Orchard
	Vineyard, hop field

Glossary

Aboriginal person one descended from the original inhabitants of an area

Absolute location location given in altitude and longitude

Acid rain rain with a higher than natural acidity

Agents of erosion forces, like water, wind and ice, which wear away the land

Aggregate in statistics, a total; in construction, sand and gravel, particularly when mixed with tar to produce asphalt

Aggressor nation one that starts a war

Agribusiness any business involved in any agriculture

Air drainage process whereby colder air flows down the slopes of a valley to accumulate at the lowest point

Air mass large body of air with similar characteristics throughout

Alloy combination of two or more metals, producing a new one with better qualities than the originals

Alpine glacier a slow-moving flow of ice produced by the buildup of large masses of snow in mountain areas

Alpine orogeny the process of mountain building

Alumina partly processed bauxite used to make aluminum

Amalgamate to combine into one unit, as with municipalities or companies

Amnesty International international organization based in London, England working to help free political prisoners around the world

Anneal to make a substance, such as steel, tougher and less brittle, by heating and cooling

Annexation process by which one community gains political control over another

Antarctic Circle 66 1/2° South line of latitude

Arctic Council economic, social, and cultural of eight nations bordering the polar regions and representatives of the Aboriginal peoples who live in those regions

Anthropologist scientist who studies early human development

Anticline dome-shaped folds of sedimentary rock

Asian Pacific Economic Community (APEC) an association of 18 nations bordering on the Pacific Ocean

Assimilation policy of blending different cultural entities into a homogenuous entity

Apartheid historical policy of "separate development" of black and white cultures in South Africa

Aquaculture raising of aquatic plants and/or animals commercially; sometimes called "fish farming"

Artic archipelago northern group of Canadian islands

Arctic Circle 66 1/2° North line of latitude

Arithmetic graph graph in which a value along an axis increases by simple addition

Arterial road major urban area road

Asbestos fibrous mineral that will not burn

Assay to analyze and estimate the aluable portion of an ore

Assembly plant factory where final products are put together

Assisted relative immigrant, not of the immediate family, given assurance of assistance from a Canadian relative

Associate term for managers and workers in Honda's plants

Atmosphere envelope of gases surrounding the earth

Automated guided vehicles vehicles to move cars in auto plants, using a Swedish team concept

Automation computerized machines operating by themselves in the workplace

Auto Pact agreement between Canada and U.S.A. allowing almost free flow of cars and parts between them

Autoplex large GM auto plant in Oshawa, Ontario

Baby boom era from 1945 to the early 1960s when the North American birth rate was unusually high

Bacteria microscopic, one-celled organisms

Badlands area of hills and deep gulleys formed by erosion, with sparse vegetation

Basalt most common volcanic rock, which cools on the earth's surface as lava

Base metal non-ferrous commercial metal, such as copper, considered inferior to the precious metals, such as gold

Basic growth number of workers added to a community who export value in goods greater than the value of imports needed to produce them

Basic oxygen furnace one in which oxygen is used to produce steel

Batholith large intrusion of igneous rock into other rocks

Bench spiraling ledge on the sides of an open-pit mine used as a road to transport ore to the surface

Bench mark brass plate used to mark the elevation of a place above sea level

Big Three three large U.S. automakers: GM, Ford, Chrysler

Bilateral having two sides

Biodegradable describes a substance that nature can decay or break down

Bio mass total volume of living matter

Bitumen tar-like oil found in oil sands

Blast furnace large container in which blasts or air assist in smelting iron

Blue-collar trades skilled or semi-skilled jobs involving manual labour

Bluff steep cliff

Boom logs tied into a large raft

Boreal forest northern coniferous forest

Boutique small retail store usually specializing in a limited line of goods

Brain trust group of highly educated, skilled experts organized to give advice

Branch-plant economy one with many industries controlled by foreign companies

Breach infraction, violation of a right

Break-of-bulk a place where goods are repackaged, then shipped on

Buck to cut trees into transportable lengths

Bulk commodities goods generally shipped in large quantities, such as sand, coal, wheat, iron ore, etc.

Business cycle alternating periods of high and low economic activity

Cage type of elevator used in underground mine to move miners and materials up and down a shaft

Canadian Council for International Co-operation (CCIC) body which lobbies governments on behalf of many non-governmental aid organizations

Canadian International Development Agency (CIDA) agency which administers foreign aid funds provided by the Canadian government

Capillary action movement of water upward through small spaces, as in soil

Capital large quantities of investment money

Capitalistic economic system system in which private citizens can own businesses, and in which supply and demand determine prices

Capital goods equipment used to produce consumer goods, industrial machinery, et cetera

Capital -intensive requiring a great investment of money

Capital wealth in any form available for the production of more wealth

Cap rock a resistant layer of rock which protects underlying layers from erosion

Car float special truck transport to carry cars

Causeway solid raised road or path through water or a wet area

Census count of all persons present in a country at a specific time

Census Metropolitan Area (CMA) large continuous urban area with a population of over 100,000 which is rated as a single unit by Statistics Canada, even though it may be made up of several political divisions

Census tract smallest urban area used for census data collection

Central business district (CBD) that area of the city, usually in the centre, with the highest concentration of offices, stores entertainment facilities, hotels and high-rise buildings

Chemical weathering breaking down of material by chemical action

Chief executive officer top manager in a major company

Chinook warm, dry, fast-moving wind that occasionally sweeps down the eastward slope of the Rocky Mountains causing a rapid rise in temperature

Circumpolar map a map centered on one of the poles

City centre location of a large concentration of commercial activities; the largest city centre is in the CBD

Civil servant provincial or federal government employee

Clear-cutting removal of all trees in a given forested area

Climate average conditions of the atmosphere over a long period of time

Climatologist scientist who studies climate

Climograph graph showing the average monthly temperature and precipitation conditions of a place

Cloud a mass of condensed water vapor in the atmosphere

Coke substance formed by baking coal to remove impurities

Cold War unfriendly relationship between the communist and democratic countries following World War II

Colombo Plan early international air program, developed by the Commonwealth in the 1950s

Commerce buying and selling of goods

The Commonwealth group of nations, formerly part of the British Empire, who consult each other regularly, and work together in numerous co-operative projects

Communist economic system one in which the state owns and controls the means of production, decides what is produced, and sets prices

Commuter person who travels daily between home and the place of work

Commutershed area from which a community draws its daily commuters

Company bond agreement whereby a company borrows money and agrees to pay it back with interest, at a specified time

Company stock certificates of ownership of a company

Concentrate ore treated to leave a higher amount of valuable content

Condense to change from a gas to a liquid or a solid

Condominium housing unit, usually an apartment that is owned individually, while the grounds and common areas of the building are owned collectively

Confluence point where a tributary joins a river or stream

Coniferous trees trees with cones and often needlelike leaves: evergreens

Consortium countries or companies united for a specific purpose, particularly a venture requiring large amounts of money

Consulate office of a foreign country, set up to deal with commercial matters and the problems of individual citizens

Consumed water water withdrawn from the natural sources and not returned to them

Consumer goods goods bought and used by the general public

Container large standard-sized metal box used to transport a variety of goods, by ship, truck, train or plane

Container terminal area of a port equipped to handle containers

Continental effect drying, extreme-temperature effect of large continental areas on climate

Continental ice cap massive, long-term build up of ice covering a huge area

Continental plate one of the huge, slowly moving blocks of material which form the earth's crust

Continental shelf shallow ocean floor bordering a continent

Contour interval vertical distance between two adjacent contour lines

Contour line brown line on a topographic map along which all points have the same elevation

Contour profile diagram drawn from a topographic map to show the shape of the land as seen from the side

Contraceptive device to prevent conception of offspring

Convection current circular movement in a gas or liquid created by uneven heating

Conventional oil and gas areas areas not requiring special extraction or production techniques

Convectional precipitation precipitation caused when air that has been heated by the ground rises and cools

Convoy group of merchant ships sailing together under protection of warships

Co-operative organization, usually of farmers or fishermen, in which the workers own an enterprise and share the work, and the profits or losses of the company

Co-op students those who work part-time in an industry as part of their school program

Cordillera continuous chain of mountain ranges in a land mass

Core sample solid cylinder of rock made by a hollow drill

Coriolis force effect of the earth's rotation, which tends to make winds veer to the right in the northern hemisphere and to the left in the southern hemisphere

Correlation chart graph depicting the relationship between two or more factors

Cosmopolitan having a great variety of racial and cultural groups

Country of origin last country that a refugee came from (not including countries where a short stop-over may have been made)

Crop diversification reliance on a variety of crops rather than just one

Crown corporation company owned and operated by a government in Canada

Crude birth rate the total number of live births in a 12 month period, divided by the total population and multiplied by 1000 to give the annual number of births per 1000 population

Crude death rate annual number of deaths per 1000 population, calculated in a manner similar to that used for the crude birth rate

Crude mineral mineral which has not been processed

Crust the cool, rigid surface of the earth

Cul-de-sac dead-end street

Cultural attributes shared traditions and customs of people from a particular country or region

Cultural group people with observable differences from other groups in lifestyle, tradition and culture

Culture shock effect on a person of suddenly moving into a new society with different customs, traditions, language and laws

Cumulo-nimbus cloud heavy, massive cloud that rises vertically, often causing heavy rain and thunder

Cumulus cloud dense, usually white cloud with dome-shaped top and flat bottom

Current trade balance the amount of exports from a country compared to the amount it imports

Cycle of erosion the three processes which break down the surface of the earth, move the material, and deposit it in a new location

Decay rotting or decomposition of a material

Decennial census population count at ten-year intervals

Deciduous trees trees which shed their leaves annually

Delta triangular area of land deposited by a river at its mouth

Demography study of a human population, mainly though statistics, to find patterns and trends over time

Density of population average number of people in a given area

Dependency load that part of a population that is not in the work force, calculated by adding the total number of people age 14 and under and those over 65

Deposition the laying down of eroded material in a new location

Depression (economic) period when the economy is in a severe slowdown with high unemployment and closing of industries

Depression (meteorological) a low-pressure area of rising air

Desertification spread of deserts into non-desert areas as a result of climate change or human activity

Deutschemark basic unit in currency of West Germany

Developmental education Canada's aid program aimed at upgrading industrial skills

DEW line chain of radar stations across Canada's north to provide early warning of serial attack

Diamond extremely hard mineral, the crystal of carbon

Direct foreign investment money invested in the ownership and management of industries abroad

Dispersed settlement settlement pattern in which areas of population are separated from one another

Displaced person someone who has lost his/her home and citizenship and may be unable or unwilling to return home

Displacement movement along a fault line

Distribution way in which things are arranged or located

Diversified producing many different kinds of goods

Dividend share of the profits of a company paid from time to time to owners of stock

Domestic market trade and commerce within a country

Dormant volcano volcano which is not active, but may become so

Drag line dredge

Drainage basin area drained by a river and all of its tributaries

Drought long period of unusually dry weather

Dry dock place used to build or repair ships out of the water

Dry hole well that does not produce oil or gas in commercial quantities

Durum wheat type of spring wheat used mainly to make pasta products

Duty a trade tariff; taxes on imported goods

Earthquake shaking of the earth's crust caused by an erupting volcano or movement along a fault

Earth's axis line between the poles around which the earth rotates

Easting first three figures in a map reference giving the east-west location

Economic sanction penalty imposed by one country on another, such as banning trade or investment

Economic sense a situation in which a profit can be made

Economies of scale savings gained through buying, selling or processing commodities in large quantities

Economist person who studies the production, distribution and consumption of wealth

Ecosphere global system of living things and their environment

Ecozone an area where organisms and their physical environment exist as a system

Electrolysis electric process by which steel is tin-plated

Element one of the basic substances that make up all matter

Elevation altitude above (or below) sea level

Embassy mission sent by one country to another to handle official relations between the two governments

Emigrate to leave a country to live permanently in another

Employment rate number of unemployed divided into the total labour force

Entrepreneur person who starts and develops businesses; in Canada, a business immigrant with special privileges under the "independent applicant" classification

Enumeration day specific date when census or electoral data is recorded

Equalization payments federal payments made to some provinces to equalize the standard of living across Canada

Equator great circle of the earth perpendicular to, and midway between the poles

Equinox two days, on or about

September 21 and March 21, when all places have equal periods of day and night

Era (see Geologic era)

Erosion the process of wearing down the earth's landscape

Ethnic pertaining to a person's race or national origin

European Economic Community (EEC) group of countries in western Europe that have agreed to eliminate tariffs among them, to follow similar economic policies, and to allow workers to move freely among all member countries

European Union as economic association of 15 European nations

Evergreens coniferous trees that do not lose all their leaves at once

Expropriate means by which governments may acquire land for public use by forcibly buying it from the owners, at a fair assessed value

Extensive farming farming that produces low yields, requires large farms, considerable equipment and minimal labour

Extractive industry an industry that digs, drills, nets, traps or removes a natural resource

Extradite to return an accused person or escaped criminal to the jurisdiction of another country

Extrusion magma which has cooled on the earth's surface

Facsimile (FAX) machine one that transmits copies of material electronically from one place to another

Factory freezer trawler large fishing vessel that catches, cuts, grades, packs and freezes its catch while at sea

Family reunification bringing together of a family that has been separated

Fault fracture in rocks along which there has been movement of the earth's crust

Faulting movement of the earth's crust along a fracture in rocks

Fault scarp steep cliff caused by a fault

Fauna animal life in a particular region

Federal political system system in which a central government and a

number of local governments each exercise different powers within the same country

Feldspar common hard red mineral, the most common in the earth's crust

Feller-buncher forest harvesting machine that cuts and stack logs

Fertility rate number of children born annually for every woman of childbearing age

Fibre optics thin strands of glass though which information travels in the form of light

Fire suppression control and elimination of a forest fire

First generation Canadian person whose parents are New Canadians

Fishing bank shallow area of continental shelf that attracts fish

Fixed link permanent connection across a body of water provided by a bridge, tunnel or causeway rather than a ferry

Flat car flat railway car without sides

Flat rolled steel steel rolled flat in a steel mill, particularly for use in auto bodies

Flora plants found in a particular region

Fold mountains long mountain chains pushed into folds by horizontal

Foliage leaves on a tree or plant

Food and Agricultural Organization (FAO) U.N. agency located in Rome

Food land high-quality land where food can be grown.

Foothills stretch of hills lying between a mountain chain and a plain

Forage crop crop grown as food for livestock

Foreign investment money invested in one country by people of another country

Foreign market market outside of a country

Fossil remains of a plant or animal preserved in the earth's crust

Fossil fuel substance such as coal or oil derived from the remains of plants and animals which produces energy when burned

The Four Tigers nickname given to

four newly industrialized countries in Asia: Hong Kong, Singapore, South Korea, Taiwan

Franchised dealer person who has a contract with a company to use its name and sell its products

La Francophonie group of French-speaking countries which consult and co-operate in various activities

Free trade trade carried on between countries without any artificial barriers, such as tariffs

Frigate warship smaller than a destroyer

Front the leading edge of an air mass

Frontal precipitation precipitation caused when air masses of different temperatures meet

Frontier oil and gas areas potential oil - and -gas-producing areas in Canada's geographic frontiers, e.g., Mackenzie Delta, East Coast

Fuel bundle container of concentrated uranium used as fuel in a nuclear plant

Galvanize to coat steel with zinc

General Agreement on Tariffs and Trade (GATT) agreement by most trading nations to enforce fair-trading policies

General cargo ordinary manufactured and packaged goods transported by ship

Generator device which produces electricity when rotated

Geologic eras huge periods of time, into which the earth's history is divided

Geologic periods subdivisions of time which make up the longer eras

Geology science that studies the history, composition and structure of the earth's crust

Geophysics branch of science that combines geology and physics

Ghost town one that has been abandoned by people

Glacial lake lake formed by glacial melt-water

Global sourcing securing parts for products from many countries

Golden Horseshoe wealthy highly populated, urban and industrial zone from Niagara Falls top Oshawa at western end of Lake

Ontario

Grade percentage of a mineral in an ore body

Grain elevator structure in which grain is stored

Granite common igneous rock consisting of three minerals: quartz, feldspar and one other, often mica

Granite gneiss metamorphic rock formed from granite

Great circle route the shortest, most direct route connecting two points on the surface of the globe

Greenwich Meridian north-south line on the earth's surface passing through Greenwich, England, from which all east-west distances are measured

Grid set of lines that intersect forming a pattern of squares

Gross National Product (GNP) value of all the goods and services produced in a country in one year

Groundfish fish that live on or near the ocean floor

Ground moraine glacial deposits spread out fairly evenly across the landscape

Ground water water retained beneath the earth's surface

Group of Seven (G-7) seven democratic nations - U.S.A., U.K., France, West Germany, Italy, Japan, Canada - with advanced industrial economies, whose leaders meet annually to discuss polices

Gulf air warm, moist, subtropical air formed over the Gulf of Mexico

Habitant name for an early French Farmer who settled southern Quebec.

Hail precipitation in the form of small balls of ice

Hardwoods deciduous trees

Headframe structure at the top of a mine shaft to support equipment used to raise ore and lower and raise miners and equipment

Head office the central administration of a company that has more than one location

Head of navigation farthest point up a river that can be reached by ships

Heartland region with a large, concentrated population, considerable

manufacturing, and influence over surrounding regions

Heat equator location of the sun's direct rays at any time

Heat wave popular term for a period of exceptionally hot weather

Heavy industry industries producing bulky, heavy goods, usually using large quantities of raw materials and large machines and facilities

Heavy oil crude petroleum of very thick consistency

Hemisphere half a globe

Herbicide chemical used to kill unwanted plants

Hercynian orogeny mountain-building period which folded the Appalachians

High an area of high pressure in the atmosphere

High Commissioner ambassador from one Commonwealth country to another

High pressure area body of air denser and heavier than the surrounding air

High-rise office or residential building over 10 stories in height

Hijack to take over control of a ship or plane by force

Hinterland area with low population with mostly primary industries connected to a large centre

Homestead land granted to a pioneer settler under the Homestead Act in return for a small fee and a promise to build a house and clear and plow a certain amount of land each year

Horticultural crop vegetables, fruit and/or flowers

Household any person or group of people occupying a dwelling or apartment

Hub central point where many different things, especially transportation systems, come together

Humidity amount of water vapor in the air

Hump mound over which railway cars are pushed so as to let them run by gravity to the desired place in a railway yard

Humus dead and decaying plant and animal material

Hydrocarbons substances found in the organic remains of plants and animals, which form the basic component in fossil fuels

Hydro-electric generating station plant that produces electricity from the force of moving water

Hydrogen the lightest chemical element, one of two elements (the other is oxygen) that combine to form water

Hydrologic cycle circulation of water from the earth's surface through the atmosphere and back

Hydrologist scientist who studies surface and ground water

Icebreaker powerful ship with reinforced hull that can clear a path through thick ice

Igneous rock rock formed from the cooling of magma

Illegal immigrant person from another country who enters Canada without going through any immigration process

Immigrant person from another country who has been granted the legal right to live in Canada

Immigrant reception area area in a town or city which provides a first home and assistance for new immigrants

Immigrate to enter another country to live there permanently

Impact study research report on the results of making a change in a community or environment

Incentive grants money from a government to help industries get started

Industrial crop crop grown for manufacturing and not for food.

Industrial mineral non-metalic minerals, such as salt or asbestos, used by industry and manufacturing

Industrial park pre-planned, environmentally attractive industrial district usually found in the suburbs and well-separated from conflicting land uses

Industrial revolution time, particularly the late 1700s in England, when the introduction of machines and factories began to greatly increase the size and the output of industries

Industrial sector section of the economy

Industry organized activities which comprise the economy

Infant mortality rate annual number of deaths of babies under one year of age per 1000 live births

Information age recent years when information has become so widely and easily dispersed

Information industries industries involved in preparing, storing and retrieving information

Infrastructure foundation or framework or backbone of a country or organization - e.g. transportation or communications network

Ingot large slab of metal produced in a smelter

Innocent passage use of a waterway for peaceful commercial purposes

Insecticide chemical used to kill insects

Integrated industrial complex area of great industrial concentration which processes raw materials thought to a finished product

Integrated sawmill sawmill with other operations included, e.g., processing of pulp and paper

Intensive farming farming that produces high yields of high-value products on small farms utilizing large amounts of labour, machinery, irrigation and fertilizer

Inter-city between cities

Interest a fixed fee paid for the use of money

Intermodal relating to two or more different types of transportation which meet to exchange cargo

Intermodal transportation system which uses more than one type of transportation to carry the same goods

Internal combustion engine engine, as in a car or truck, in which gas and air are burned to move cylinders, which provide power

Internal waters waterway completely within the borders of a country

International Civil Aviation Organization (ICAO) U.N. agency located in Montreal and dealing with commercial aviation

International Olympic Committee

(IOC) body which decides when, where and how the Olympic Games take place

International strait strait freely open to commercial use by any nation's ships

Interstate highway part of the U.S. expressway network

Intrusion magma which has forced its way into, but not out on top of, the earth's crust

Inventory goods available for sale

Invertebrate an animal without a backbone

Invisible trade money earned by a country on services such as tourism, and in interest and dividend payments from other countries

Iron Curtain term coined by Winston Churchill for the border between Western Europe and the communist East after 1945

Iron ore pellets concentrations of iron like small marbles, the first step in refining ore into metal

Isohyet line on a map along which the precipitation is equal

Just-In-Time (JIT) System arrangement by which parts arrive in a factory just before they are needed for assembly

Joint crack or fissure in a mass of rock

Kame conical hill or short ridge of stratified sand and gravel deposited by a retreating glacier

Labour force participation rate total number of people who could be employed as a percentage of total population over the age of 15

Labour-intensive requiring a great deal of human labour

Lakes ship designed to haul freight, usually bulk cargo, on the Great Lakes

Lakescape landscape with many lakes

Landbridge narrow route across a water body; that becomes dry land when the water level drops

Landfill site garbage dump

Landmark obvious feature used to determine location

Land of the Midnight Sun polar regions where there is at least one

day when the sun doesn't set during the summer

Land speculator person who purchases property to resell for a profit

Land-use pattern way in which people occupy and use the land

Latitude lines imaginary lines running parallel east and west around the globe, which measure angular distance north and south of the equator

Lava flow magma which has flowed onto the earth's surface Leaching dissolving and carrying of material by water downwards through the soil

Leapfrogging growth in urban areas which leaves open spaces between developments

Leeward side protected from the wind

Less Developed Country (LDC) country with relatively little industry

Lichens simple plants which grow on rocks

Life expectancy at birth predicted length of a person's life as of the time of birth

Light industry industries producing light, simple goods, usually using processed or semi processed materials, and small, environmentally clean facilities

Light Rapid Transit (LRT) mass-transit system with capacity midway between that of streetcars and subway system

Limestone sedimentary rock, chiefly of carbonate of lime

Local market market close to the place of production

Lockport dolomite tough limestone that caps the Niagara Escarpment

Logan's Line fault line separating the Appalachians and the St. Lawrence Lowlands

Logistics movements of goods and equipment

Long lot deep narrow farm in Quebec

Longitude lines imaginary lines from pole to pole on the earth's surface, which measure angular distances east and wet, starting from Greenwich, England

Low an area of low pressure in the atmosphere

Low pressure area area of the atmosphere which is lighter and less dense than the surrounding air

Magma molten rock within the earth

Magnetometer instrument used by geologists to measure the magnetic property of rocks

Mantle portion of the earth extending from just below the crust halfway to the centre

Map reference six figures representing a location on a topographic map

Maritime (marine) effect moderating, humid effect of large bodies of water on climate

Market any place where goods and services can be bought and sold

Marshalling yard place where railway cars are assembled into trains

Mass communication communication to large number of people through media like books, newspapers, radio and television

Mass production making of things efficiently in large quantities

Mean sea level the average level of the oceans, from which all elevations are measured

Mean mid-point between extreme measurements; an average

Mean temperature mid-point between high and low temperatures

Meander wide bend in a winding river

Mechanical weathering physical breaking up of rock

Megadealer dealer with agreements to sell cars for several firms

Mercator map type of map in which the latitude and longitude lines all meet at right angles

Merchandise goods or commodities for sale

Merchantable volume the amount of wood that is worth harvesting from a tree

Merchant navy ships used for commerce, i.e., not warships

Meridians of longitude longitude lines, so-called from the Latin term for mid-day, because the sun is always directly over a longitude line at noon

Metallic mineral mineral containing a metal

Metamorphic rock rock changed by heat and/or pressure from one form into another form

Meteorologist scientist who studies weather

Metis people of mixed Indian and non-Indian background

Metropolis any large, influential city

Microwaves short radio waves

Mid-ocean ridge long, narrow mountain range in the middle of the ocean, formed by upwelling magma

Migrant person who keeps moving from one place to another

Migration movement of people from one place to another

Mill plant where ore is processed into concentrates

Minamata disease nerve damage caused by the build-up of mercury in the body

Mineral naturally occurring, inorganic material made up of one or several elements in a chemical union

Mixed farming farming which raises both crops and livestock

Mobile spar tall pole and a series of pulleys mounted on a truck used to pull logs to a central loading point in mountainous terrain

Model T early mass-produced car developed by Henry Ford

Moraine deposits of rubble left behind by a glacier or ice cap

Mortgage property pledged as security for loan repayment

Most-favored nation a country granted a favourable trading privileges by another country

Mother factory ship vessel designed to completely process fish caught by trawlers

Mother tongue language first learned and still understood

Mouth end of a river where it empties into a lake or ocean

Multicultural society society that accepts and promotes diversity in language, culture and heritage

Multi-function harvesting machine machine that fells, delimbs, bucks and stacks logs

Multilateral having many sides

Multiplier effect effect in which

some workers are added to a community, then others are needed to provide services for them, further increasing the growth

Municipality political area smaller than a province

Muskey swampy ground in northern regions

National market trade and commerce within a country

Natural increase population growth resulting from an excess of births over deaths, rather than from immigration

Natural land use establishment of land ownership based on individual decisions and the natural features of the landscape rather than on an organized plan

Natural vegetation vegetation that has developed without human intervention

Net migration difference between the number of emigrants from and the number of immigrants to a country

Net trade flow difference between money paid for imports and money earned on exports by a country

New Canadian person born in another country who has been granted Canadian citizenship

Newly industrialized countries (NICs) countries which have recently built up their industries to a high degree

Nobel Prize for Peace prestigious award for achievements in promoting peace among the people of the earth

Non-basic growth increased economic activity resulting from basic growth

Non-ferrous minerals minerals other than iron

Non-governmental organizations (NGOs) private organizations giving aid to other countries

Non-metallic mineral mineral without any metal content

North American Aerospace Defense Command (NORAD) joint military organization of U.S.A. and Canada, to defend North America's airspace

North Atlantic Treaty Organization (NATO) mutual defense treaty involving U.S.A., Canada and 14 European countries

Northing last three figures in a map reference, giving the north-south location

Northwest Passage route by which ships can sail through the Arctic archipelago

Nuclear power station plant equipped to produce electricity by using uranium

Nuclear reactor structure in a nuclear power station where uranium atoms are split

Ocean currents "rivers" of warm and cold water moving through oceans

Oenology science or study of wines and winemaking

Official development is (ODA) foreign aid provided by the Canadian government

Open-pit mining digging a big hole in the ground to get at ore which is near the surface

Orbit path traced by each planet revolving around the sun or by a satellite revolving around a planet

Ore body mineral deposit from which one or more minerals can be extracted at a profit

Organization for Economic Co-operation and Development (OECD) organization, supported by most industrial countries, that works towards economic benefit for all members

Outcrop small part of a large body of rock sticking out above the surrounding ground

Overburden soil, plant life or water that covers an ore body

Overcapacity ability to produce more goods than can be sold

Oxygen an important chemical element that combines with many others, for instance hydrogen, with which it forms water

Ozone layer high layer of the atmosphere that protects earth from some of sun's harmful rays

Pangaea master continent which included all the earth's land masses

200 million years ago

Paradox statement which appears to be self-contradictory

Parallels of latitude latitude lines, so called because each is a fixed distance north or south of the equator and never meets any other line of latitude

Parent company company which controls branch plants and offices in other locations

Parent material rock from which soil is derived

Parklands areas of grass with scattered trees

PCB chemical compounds of polychlorinated biphenyl's which are harmful to the environment

Peacekeeping arrangement, usually made by the U.N. whereby forces from neutral countries, such as Canada, monitor peace agreements and keep enemies from fighting

Pelagic living in the open ocean, far from land

Peneplain an area worn down by erosion until it is almost a flat plain

Penstock pipe that carries water to the turbine in a hydro-electricity plant

Per capita for each person

Period (see Geologic period)

Peripheral located at the edges of something rather than at its centre or spread across it

Permafrost permanently frozen soil

Permeable rock rock that allows liquids and gases to pass through it

Pesticide chemical or other substance used to destroy animals or plants that people do not like

Petrochemicals chemicals obtained from refining crude oil

Petroleum crude unrefined oil derived from rocks

Phytoplankton microscopic plant life suspended in the ocean

Pickle process by which steel is bathed in acid, removing surface scale

Piggy-back to transport truck trailers on railway cars

Pig iron iron produced by a blast furnace

Plane of the orbit flat circular sur-

face whose edge would be the earth's orbit

Planning principles general rules that guide decisions affecting new developments in an urban area

Plateau elevated area of fairly level land

Plate tectonics study of the earth's crust, in particular the large, solid blocks which are believed to float on the liquid mantle, together making up that crust

Pleistocene Ice Age fairly recent geological period when huge ice sheets covered much of North America and other continents

Plywood building material composed of thin sheets of wood glued together under pressure

Podzols ash-grey soils associated with coniferous forests

Points test screening process to evaluate an applicant's suitability for immigration to Canada

Polar Zone area between the Arctic (or Antarctic) Circle and the pole

Population graph diagram showing the structure of a population by age and sex

Portaging carrying a boat between two navigable bodies of water

Port of entry place on the border where there are customs and immigration officials to control people wishing to enter the country

Postal code six-character symbol used by Canada Post to identify a specific area for mail delivery

Potash crude form of potassium carbonate used in fertilizers

Pound sterling basic unit of currency in U.K.

Prairie natural region of generally treeless plains covered in grasslands

Prairie levels three sub-regions at different elevations on the Canadian prairies

Pre-Cambrian shield large area of very old, very hard rocks underlying much of Canada

Precious metal metals, such as gold and silver, valued for their rarity and beauty, rather than for their usefulness

Precipitation any form of water falling from the atmosphere

Prescribed burn planned fire set

to rid the forest floor of slash or unwanted trees after clear-cut logging

Primary energy source natural source of energy, such as fossil fuels, uranium, falling water

Primary industries those dealing directly with the extraction of natural resources

Primary manufacturing processing of raw material

Prime farmland cropland of the highest quality

Prime Meridian longitude line passing through Greenwich, England, called the prime, or first, meridian because it is 0° east or west

Principal meridians lines of longitude designated as starting lines for the prairie township survey system

Productive forests forests capable of growing trees that can be harvested at a profit

Protectionism belief that national economics should be protected from competition of foreign goods

Public policy advocacy efforts by aid groups to influence governments to assist poor of the world

Push and pull factors forces that attract people to migrate to another area

Quad saw saw with four blades

Quartz hard mineral, the second most common in the earth's crust

Quaternary sector fourth group of industries in the economy, including the information services

Quota maximum limit on production or trade

Radiation energy emitted by a body, such as the sun or radioactive material

Radio transmitting sound without wire by using radio waves

Rang French word for a row of long, narrow farm lots surveyed perpendicular to a river, as in Quebec

Range distribution of east-west prairie townships

Rapid transit urban transportation system with its own right-of-way using subway trains, elevated trains or both

Recession period when there is a mild economic slow-down

Reciprocity Treaty Canada-U.S. treaty, signed in 1854, enabling most natural products to cross the border freely

Refinery plant that processes raw or semi-processed material into a pure substance

Reforestation replanting of a forest by humans

Refugee person who has fled his or her homeland to escape from life-threatening danger or persecution

Regeneration renewal of the forest by natural or artificial means

Region area on the earth's surface which has something in common

Registered Indian another term for a Status Indian

Relative humidity amount of water vapor in the air's capacity for holding water vapor

Relative location location given by referring to another place or places

Relief feature physical feature that stands above the general landscape

Relief precipitation precipitation caused when air is forced to flow up and over a high place

Renewable resource resource that has the ability to replace or restore itself, e.g., plants, fish

Renovate to repair and modernize an older building

Representative fraction scale given as distance on the map compared to the actual distance it represents, e.g., 1:50 000

Research and Development (R & D) seeking out and refining of new technology

Reserve a track of land set aside by treaty for the use and occupancy of an Indian band

Retailer merchant who sells small quantities of goods to consumers

Rift valley long valley caused by fault action along each side

Ring of Fire term applied to the region of earthquake and volcanic activity around the Pacific Ocean

Ripple effect various secondary changes which may result from one original change

River basin area drained by a river

Rock naturally occurring inorganic combination of mineral matter

Rock in solution rock dissolved in a stream

Rock in suspension pieces of rock carried in a stream

Roe fish eggs

Ro-Ro ship ship designed so that tractor trailers can drive on and off without special equipment

Roture a long, narrow farm based on the French-Canadian long-lot survey system

Runoff rain water that flows on the earth's surface rather than being absorbed by the ground

Rural-urban fringe area adjacent to an urban area where there is a mixture of urban and rural land uses

Rust belt ironic term referring to U.S. steel-manufacturing areas, which declined during the 1980s

Sand unconsolidated rock made up of tiny pieces of weathered granite

Sand dune mound of wind-blown sand

Sandstone sedimentary rock made of sand particles cemented together

Scale ratio of the distance on the ground

Scoop tram vehicle used to load and haul ore from the stope to the skip in underground mines

Scrubber pollution-control device used to remove sulphur dioxide from smoke-stack emissions

Secondary energy source energy produced from a primary energy source, e.g., electricity

Secondary industries those dealing with manufacturing or construction

Secondary manufacturing industries which turn semi-processed materials into final products

Section a square mile (259 ha) land area that is one 36th of a prairie township

Sector of industry all companies that make similar types of products

Securities firm firm buying and selling stocks, bonds and other investments

Sedimentary rock rock formed from material that has settled to the bottom of a body of water

Seedling young tree

Seed-tree system variation of clear-cutting in which individual trees or clumps of trees are left standing to provide seeds for regeneration

Seigneur major landowner in New France

Seigneury large area of land granted to an import person in New France

Seismic of or related to earthquakes or similar disturbances, and the effects they produce in the earth's crust

Seismic reflection study method of locating oil and gas deposits by analyzing shock waves from artificial explosions as they are reflected off underground rock formations

Selective cutting removal of small mature stands or individual trees, leaving the remaining trees to fill in the logged area

Self-sufficient economy national economy which has all the resources needed to survive with few or no imports

Semidesert area of sparse precipitation, but with more moisture than a full desert

Semi-fabricated (or semi-processed) material a natural product that has been partly refined or treated, but is not ready to be used by industry

Semi-logarithmic graph graph in which one scale increases by multiplication

Semi-submersible a structure capable of being partially submerged and partially above water

Service industries industries that provide help to people, rather than producing manufactured goods

Settlement pattern characteristic way land is used by people living in an area

Shaft mining mining that uses vertical and horizontal tunnels to extract ore from deep underground

Shale sedimentary rock made from layers of clay

Shellfish marine life with a shell or hard outer covering, e.g., lobster, clam, oyster

Shield massive area of Pre-Cambrian rock around which a continent has formed, e.g., the Canadian Shield

Shopping district an area with a large concentration of stores

Silicon second most-common element in the earth's crust

Silviculture art and science of growing trees

Single front early township survey system used in Ontario

Skidder tractor-like machine used to drag logs from the cutting area to the loading area

Skip bucket used to hoist ore to the surface in an underground mine

Slag impurities drawn from a blast furnace

Slash debris such as branches, tops and unwanted trees left after an area has been logged

Sleet precipitation in the form of snow or hail, with rain

Slough small shallow pond

Smelter plant which removes a metallic mineral from ore

Snow precipitation in the form of very small ice crystals

Softwoods coniferous trees

Soil loose material supporting and nourishing plants

Soil horizon section in a soil profile

Soil profile vertical cross-section of a soil down to the base rock

Source place where a stream or river begins

Sovereignty possessing supreme authority, political entity

Spawn to produce and lay (fish) eggs

Specialty shopping retail stores that offer unique services or goods

Sponsor to take legal responsibility for a close family member immigrating to Canada

Spring wheat wheat planted in the spring and harvested in the fall

Spruce budworm destructive caterpillar that feeds on the buds and new foliage of spruce trees

Statistics Canada federal government agency responsible for the collection of data about Canada, including the census

Status Indian person defined by the Indian Act of 1952 as one who "is registered as an Indian,"{on an official list, which is held in Ottawa by the Department of Indian Affairs}

Steel alloy of iron with carbon and/or other metals that is harder and more workable than iron

Steel belt zone in U.S.A. with many steel mills

Stern trawler large fishing vessel that pulls a dragnet along the ocean floor

Stock exchange place where stocks are bought and sold

Stope rock face underground from which ore is removed

Strain specific variety of a crop or livestock

Strata layers of rock, generally sedimentary

Stratus cloud a flat, low-lying cloud layer

Stratigraphic concerning the sequence and position of strata in the earth's crust

Strip-commercial grouping of commercial activities, side by side, in a ribbon along a major roadway in an urban area

Strip-cutting trees cut in strips rather than patches or blocks

Structural material one used in building and construction

Stubble cut stalks of grain left in a field after harvesting

Subdivision one part of a larger area, in particular, an area in a suburb; also, the term for one sixteenth of a prairie section of land

Subduction zone area where the edge of one tectonic plate is pushed downward under an other, and melts into the mantle to become magma

Subsidy money provided by a government, usually for a farm or a company, to help it stay in business

Summer fallow field left uncultivated for a year in order to build up its subsoil moisture

Sun's direct rays those rays which strike the earth at or close to 90°

Sun's indirect rays those rays which strike the earth at an angle under 90°

Sustainable development development which considers present and future use of the environment and its resources

Survey system planned system of measuring and dividing an area into lots

Symbol representation on a map of a feature on the ground

Synthetic crude oil oil extracted from the processing of the oil sands

Tailings waste materials left over from processing ore

Tailrace pipe or channel carrying water away from a turbine or water wheel

Tall grass vegetation type providing excellent black soils when it decomposes

Tariff tax placed on imported goods

Tar sands area in north Alberta where crude petroleum is found mixed with sands

Technology the knowledge and tools used by humans to produce the physical things they need

Tectonic forces forces that affect the structure and movement of the earth's crust

Telecommunication communication sent electronically by wire, optical fibre of through the air

Telegraph system capable of sending sound patterns by wire

Telephone system capable of sending voice by wire

Temper to harden steel by heat

Temperate rainforest forest with a high precipitation and moderate temperatures

Temperate Zone area between the Tropics and the Arctic and Antarctic circles

Tender tree fruit fruit with soft, delicate skin, which grows on trees

Territorial waters waters lying off a country's coast which are considered to be part of the country

Terrorism violent actions by small extremist groups to try to influence governments by threatening and hurting innocent people

Tertiary industries industries that provide services to people, businesses or governments

Thermal-electric generating station plant which produces electricity by burning fuels

Tied aid Canadian government assistance requiring the purchase of Canadian goods and services

Time zone area about 15° of longitude in width within which every place uses the same time

Tinplate sheet steel covered with a thin coat of tin

Township rectangular or square section of land subdivided into smaller lots for sale to settlers

Toxic chemical chemical which is dangerous or poisonous to humans or the environment

Toxin any poisonous matters generated in living or dead organisms that cause many diseases

Trade ban government prohibition of trade with some countries or in certain goods

Trade deficit situation in which a country has bought more in goods or services than it has sold in exports

Trade quota regulation limiting the amount of a product that may be imported (or exported)

Trade surplus positive trade balance resulting when a country has earned more selling exports than it has spent on imports

Trans-shipment point place where goods are transferred from one type of transportation to another

Transitional zone area along the borders between regions where features of both regions are found

Transpiration process by which the leaves of plants and trees add water vapor to the air

Transportation corridor path where different transportation systems exist side by side, such as railway tracks running beside an expressway

Transportation node place where several transportation routes meet

Tree line line that marks the furthest extent of conditions that allow large trees to grow

Tributary river or stream that flows into another river or stream

Trolley bus bus powered by overhead electrified wires

Tropical zone area between the topics

Tropic of Cancer 23 1/2° North latitude line

Tropic of Capricorn 23 1/2° South latitude line

Truce agreement by nations to cease hostilities, at least temporarily

Trunk line main line of a pipeline, railway or telephone system

Tundra place of limited vegetation and cold, dry climate found in the Arctic regions

Turbine wheel with blades, spun by an energy course

Underground mining mining below the earth's surface using vertical and horizontal shafts

Unit train train that carries large mounts of one product only; the cars remain linked as a unit

Urban sprawl physical spread of an urban area, generally in an unplanned, disorganized manner

Vacancy rate amount of unoccupied office or retail or industrial space

Value added by manufacturing that part of a product's value created during the process of manufacturing, excluding the cost of raw materials

Vertically integrated company company which owns and operates every process required to produce its product

Virgin forest forest which has never been logged

Vintner wine producer

Visible trade trade involving goods or commodities

Visiting friends and relatives (VFRs) people who go to other countries to visit friends and relations, rather than for business or sightseeing

Viticulture art and science of grape growing

Volcano vent in the earth's surface through which magma has erupted, sometimes forming mountains.

Voyageurs men who transported the goods of the fur trade in canoes from the interior to Montreal

Vulcanism those processes by which molten rock affects the earth's crust

Watferboard sheet of wood composed of large, then chips bonded together under intense heat and pressure

Warsaw Pact military defense pact signed in Warsaw, Poland, between the U.S.S.R., Poland, Romania, Bulgaria, Albania, East Germany, Hungary and Czechoslovakia

Watershed land that separates one drainage basin from another

Water table upper limit of ground water

Weather conditions of the atmosphere over a short time period

Weathering the breaking up of rocks by chemical or mechanical means

Wetland swamp marshe and bog, place where the water table is at ground level

Wholesaler merchant who buys large quantities of goods from manufacturers and then resells these goods to retailers

Widget humorous term used to stand for anything manufactured

Wind air moving horizontally across the earth's surface

Windward that side facing the wind

Wirephoto method for transmitting and receiving photography by wire

Withdrawal water water that has been taken from its natural source for human activities

Work bee community gathering, particularly in pioneer times, to accomplish some special task

Work ethic attitude felt by a worker towards his or her job

World Health Organization (WHO) U.N. agency based in Geneva, Switzerland

World Meteorological Organization (WMO) U.N. agency based in Geneva involved in weather study and reporting

World Wildlife Fund (WWF) world's largest organization involved in protecting animals and the environment

Yen basic unit in Japan's currency

Zooplankton microscopic animal life which feeds on phytoplankton

Index

Acknowledgements

Fitzhenry & Whiteside thanks the following people and institutions for their help in assembling this volume

John Shamrock, Greyhound Canada Transportation Corp; Tamra Rezanoff, BC Transit; Bonnie Hayes, Saint John Port Corporation; Ginette Larose, Bombardier, Inc.; Norma Belzil, Alcan; Carole Savard, CIDA; Dave Routledge, Oxford Properties; Beth Jost Reimer, Christian Blind Mission International; Bert Frederick, Port of Montreal; Bruce Reid, Greater Toronto Airports Authority; Marilyn Bolton, Toronto Transit Commission; Ramona Noor, Harbourfront Centre; David Quan, Trizec Hahn; Brian Crawley, Hibernia; Carol Moreau, Ford Canada; Lois Booth, Chrysler Canada; Brian Chadderton, ATI; Bill Gear, Dofasco; Ann Landry, Canadian Highways International Corporation; Margaret Williamson, Canadian Geographic; Roger Bailey, International Peace Garden; Mark Ladouceur. Fording Coal; Isabel Baril, Lavalin; W. G. McMurray, Canadian Pacific; John Cooper, The Ontario Report; John Francis, Confederation Bridge; Mike Murphy, Toronto Real Estate Board; Donna Ryan, Stelco; Pierre Poulin, Royal Mat Inc.; Sandra Darling, Canadian Energy Pipeline Assn.; Joan Speares, Greg Hancock and Steven Ramstead, Dept. of Indian and Northern Affairs; Heinz Wiele, Bedford Institute of Oceanography; Shelly Brown, Edmonton Economic and Development; Tobi Fletcher, Winnipeg 2000; Isabelle Gagne, BCE; Harvey Shear, Environment Canada; Calgary Stampede; Heather Bower, Business Parks Office, Halifax Regional Municipality; Brenda Simmons, PEI Potato Board; Jill Skorochod, Ontario Forest Industries Assoc.; Paul Bevado, Magna; Patricia Currie, Halifax Trade and Convention Centre; John Norris, Ryder Integrated Industries; Real Courcelles, Hydro Quebec; Ed Nasello; Lisa Timpf at Honda Canada.

Graphics

Allen, Robert W. (First Light) 2-44; **Aquin, Benoit** 1-78; **ATI Technologies** 7-17; **Atmospheric Environment Service** 1-55b; **Banister Construction** 4-55; **BCE** 6-64; **BC Ferrys** 6-28; **Bedford Institute of Oceanography** 3-57; **Bell Canada** 6-63; **Black Cultural Centre of Nova Scotia** 2-45 a&b; **Bombardier Regional Aircraft** 7-1,7-15a&b;, **Bonner, Dave** 2-30;, **Buschert, Mel** (Banff Laser Graphics) 3-48; **Calgary Stampede** 3-50; **Callis, Dan** 4-22; **Canadian Aboriginal Products** 7-14; **Canadian Airlines** 6-56; **Canadian Energy Pipelines Assn** 6-53b&c; **Canadian General Electric** 5-11; **Canadian Geographic Magazine (Steve Fick)** 1-66, 3-26,4-75,6-41; **Canadian Highways International Corp** 6-44,6-45; **Canadian National Railway** 6-1,6-37; **Canadian Pacific** 6-19; **Canadian Pacific Railway** 6-31,6-32,6-36; **Canadian Polystyrene Recycling Assn** 5-23b; **Canadian Press (McIntosh)** 1-71; **Canadian Women's Movement Archives** 5-1b; **Centre for Inland Water** 7-37; **Chapman, Fred** (Diorama Stock) 3-44; **Christian Blind Mission International** 7-55; **Chrysler Canada** 5-46, 5-64a&; **CIDA** 7-45,7-46,7-50,7-51,7-52,7-54,7-57; **City of Toronto Archives** 3-30; **Cooke, Kennon** (Valan Photos) 2-37; **Denison Mines** 4-35, 4-36, 4-38, 4-3; **Department External Affairs** 7-13c,7-29,7-38; **Department National Defense** 7-39,7-40,7-42; **Dofasco** 5-45; **Doll, Hollister** 2-45, c&d , 7-65; **D'Orazio, Eugene** 2-46, 2-47 a&b; **Dumas, Jean-Philippe** Pono-press) 1-70; **Dwyer, Janet** (First Light) 4-27, 7-59; **Easterbrook, Ursola** (Diorama Stock) 3-40; **Edmonton Economic and Development Office** 5-35; **Environment Canada and United States Environmental Protection Agency** 4-72,4-73,7-37b; **Fifth House** reprinted with permission from *But Its a Dry Cold: Weathering the Canadian Prairies*,(1998 by Elaine Wheaton 1-80; **Fishery Products International** 4-43,4-44,4-46; **Flewelling, Martin** 4-49b; **Ford Canada** 5-51,5-55,5-57; **Fording Coal** 7-13; **Forest Protection Limited** 4-11; **Fowler, John** (Valan Photos) 7-4; **Gallant, Andre** 1-10; **General Motors Canada** 5-46,5-52,5-53,5-54; **Geographic Air Survey** 2-63; **George, Tony** 2-46; **Gillen, Ian** 3-32; **Glenbow Museum** 4-30;

Go Transit 3-21, 6-33; **Goss, Dawn** 6-42, 6-35; **Greater Toronto Airports Authority** 3-23; **Greyhound Canada Transporation Corp** 6-46; **Halifax Regional Municipality Business Parks Office** 3-60; **Halifax World Trade and Convention Centre** 3-59; **Harbourfront Centre** 3-27; **Hickman, Pam** (Valan Photos) 1-44; **Hildebrandt, B.** 1-22, 5-14, 6-50; **Hibernia** 4-59; **Hilliard, Joan** 2-1; **Honda Canada** 5-65a&b; **Image West** 4-48; **Inco** 3-66, 3-67, 4-34; **International Peace Garden** 7-41; **International Pipeline Co** 6-60; **Irving Oil** 5-25; **Kai Slide Bank** 7-44; **Kerry Design** 1-12, 1-14, 6-59; **Kitchen, Thomas** (Valan Photos) 3-38, 7-62; **Kobelka, Bruce** 3-19, 3-36a; **Kovaliv, Taras** 7-30; **Kuhnigk, Martin** (Valan Photos) 4-2; **Lalit, Sakchai** (Canapress Photo Service) 2-48; **Last, Victor** (Geographical Visual Aids) 1-9, 1-32; **Leighton, Douglas** 3-52; **Lenz, Garth** 1-88; **London Free Press** 3-86; **McCain Foods** 5-21; **McCurdy, Ron** 1-53 ; **McDougall, Duncan** (Diorama Stock) 3-45, 6-18; **McMichael Collection** 1-35; **Macmillan Bloedel** 4-6, 5-1a; **Macleans Magazine** 5-2; **MacPhee, Joan A.** 2-3; **Magna International** 5-59; **Marine Atlantic** 6-27; **Marsh, Alan** (First Light) 3-54; **Marti, Enric** (Canapress Photo Service) 2-49; **Mayama, Kimimasia** (Reuters) 7-33; **Michaelson, Hugh** 1-15, 5-12,5-20,5-62; **Milne, Brian** 1-89; **Milne, Brian** (First Light) 3-63; **Minnesota Historical Society** 6-2; **Morrrow, Pat** (First Light) 7-3; **Murray, Jon** (Vancouver Province) 3-42; **National Archives** 2-36, 2-38, 5-5, 6-30; **National Geographic** 6-60; **Oliver, C.** 2-55, 2-58, 3-24 top, 3-31, 3-36 bottom, 3-37, 3-56, 3-59, 3-64, 3-74, 3-77, 3-81, 3-83, 3-85 b, e, f ; **Oliver, Jill** 3-78, 3-79, 3-85 a, c, d; **Ontario Archives** 2-53; **Ontario Forest Industries Association** 4-7b; **Ontario Hydro** 4-58,4-69; **Provincial Archives of Manitoba** 2-52; **Parker, Jessie** (First Light) 5-13, 6-4; **Parlow, Lorraine** (First Light) 1-23, 4-23, 5-26; **Pearce, Joseph R.** (Valan Photos) 3-6;, **Peterson, Roy** 7-2; **Port of Montreal** 6-11;, **Port of St. John** 7-8a,b&c; **Procaylo** (Canapress Photo Sercices) 2-48; **Rooney, Karen D.** (Valan Photos) 2-14, 5-1c, **Royal Bank of Canada Collection** 6-29, **Rubinstein, Larry** (Canapress Photo Services) 7-21;

Title Photographs

Cover
Fall colours along the Dempster Highway
(Ron Watts – First Light)

Frontspiece
Canola fields outside of Riding Mountain
National Park
(Brian Milne)

Chapter 1
Lake Temagami in Northern Ontario has
1200 islands, the largest of which is Bear
Island - a First Nations' Reserve
(Ben Moise)

Chapter 2
A group of Inuit women and children feast
on frozen Arctic Char at Puvirnituq, Quebec
(Emanuel Lowi)

Chapter 3
Winterlude in Ottawa, the nation's capital
(William P. McElligott)

Chapter 4
Potash, a non-metallic metal used in the
production of fertilizer is mined in southern
Saskatchewan
(Saskatchewan Property Management)

Chapter 5
British Columbia's Whistler Mountain is a
popular tourist area and recreational development hotspot
(Chris Speedie – Tony Stone Image Finders)

Chapter 6
The Canada's Stentor alliance of telephone
companies which manages Canada's coast
to coast fibre optic cable -The Ottawa headquarters of the world's longest.
(William P. McElligott)

Chapter 7
Canada's Women Curlers won Olympic
gold in Nagano, Japan in 1998
Curling has become one of Canada's most
exciting exports
(Ken Faught – Canapress Photo Service)